Standards for Engineering Design and Manufacturing

MECHANICAL ENGINEERING
A Series of Textbooks and Reference Books

Founding Editor

L. L. Faulkner

Columbus Division, Battelle Memorial Institute
and Department of Mechanical Engineering
The Ohio State University
Columbus, Ohio

1. *Spring Designer's Handbook*, Harold Carlson
2. *Computer-Aided Graphics and Design*, Daniel L. Ryan
3. *Lubrication Fundamentals*, J. George Wills
4. *Solar Engineering for Domestic Buildings*, William A. Himmelman
5. *Applied Engineering Mechanics: Statics and Dynamics*, G. Boothroyd and C. Poli
6. *Centrifugal Pump Clinic*, Igor J. Karassik
7. *Computer-Aided Kinetics for Machine Design*, Daniel L. Ryan
8. *Plastics Products Design Handbook, Part A: Materials and Components; Part B: Processes and Design for Processes*, edited by Edward Miller
9. *Turbomachinery: Basic Theory and Applications*, Earl Logan, Jr.
10. *Vibrations of Shells and Plates*, Werner Soedel
11. *Flat and Corrugated Diaphragm Design Handbook*, Mario Di Giovanni
12. *Practical Stress Analysis in Engineering Design*, Alexander Blake
13. *An Introduction to the Design and Behavior of Bolted Joints*, John H. Bickford
14. *Optimal Engineering Design: Principles and Applications*, James N. Siddall
15. *Spring Manufacturing Handbook*, Harold Carlson
16. *Industrial Noise Control: Fundamentals and Applications*, edited by Lewis H. Bell
17. *Gears and Their Vibration: A Basic Approach to Understanding Gear Noise*, J. Derek Smith
18. *Chains for Power Transmission and Material Handling: Design and Applications Handbook*, American Chain Association
19. *Corrosion and Corrosion Protection Handbook*, edited by Philip A. Schweitzer
20. *Gear Drive Systems: Design and Application*, Peter Lynwander
21. *Controlling In-Plant Airborne Contaminants: Systems Design and Calculations*, John D. Constance
22. *CAD/CAM Systems Planning and Implementation*, Charles S. Knox
23. *Probabilistic Engineering Design: Principles and Applications*, James N. Siddall
24. *Traction Drives: Selection and Application*, Frederick W. Heilich III and Eugene E. Shube

25. *Finite Element Methods: An Introduction*, Ronald L. Huston and Chris E. Passerello
26. *Mechanical Fastening of Plastics: An Engineering Handbook*, Brayton Lincoln, Kenneth J. Gomes, and James F. Braden
27. *Lubrication in Practice: Second Edition*, edited by W. S. Robertson
28. *Principles of Automated Drafting*, Daniel L. Ryan
29. *Practical Seal Design*, edited by Leonard J. Martini
30. *Engineering Documentation for CAD/CAM Applications*, Charles S. Knox
31. *Design Dimensioning with Computer Graphics Applications*, Jerome C. Lange
32. *Mechanism Analysis: Simplified Graphical and Analytical Techniques*, Lyndon O. Barton
33. *CAD/CAM Systems: Justification, Implementation, Productivity Measurement*, Edward J. Preston, George W. Crawford, and Mark E. Coticchia
34. *Steam Plant Calculations Manual*, V. Ganapathy
35. *Design Assurance for Engineers and Managers*, John A. Burgess
36. *Heat Transfer Fluids and Systems for Process and Energy Applications*, Jasbir Singh
37. *Potential Flows: Computer Graphic Solutions*, Robert H. Kirchhoff
38. *Computer-Aided Graphics and Design: Second Edition*, Daniel L. Ryan
39. *Electronically Controlled Proportional Valves: Selection and Application*, Michael J. Tonyan, edited by Tobi Goldoftas
40. *Pressure Gauge Handbook*, AMETEK, U.S. Gauge Division, edited by Philip W. Harland
41. *Fabric Filtration for Combustion Sources: Fundamentals and Basic Technology*, R. P. Donovan
42. *Design of Mechanical Joints*, Alexander Blake
43. *CAD/CAM Dictionary*, Edward J. Preston, George W. Crawford, and Mark E. Coticchia
44. *Machinery Adhesives for Locking, Retaining, and Sealing*, Girard S. Haviland
45. *Couplings and Joints: Design, Selection, and Application*, Jon R. Mancuso
46. *Shaft Alignment Handbook*, John Piotrowski
47. *BASIC Programs for Steam Plant Engineers: Boilers, Combustion, Fluid Flow, and Heat Transfer*, V. Ganapathy
48. *Solving Mechanical Design Problems with Computer Graphics*, Jerome C. Lange
49. *Plastics Gearing: Selection and Application*, Clifford E. Adams
50. *Clutches and Brakes: Design and Selection*, William C. Orthwein
51. *Transducers in Mechanical and Electronic Design*, Harry L. Trietley
52. *Metallurgical Applications of Shock-Wave and High-Strain-Rate Phenomena*, edited by Lawrence E. Murr, Karl P. Staudhammer, and Marc A. Meyers
53. *Magnesium Products Design*, Robert S. Busk
54. *How to Integrate CAD/CAM Systems: Management and Technology*, William D. Engelke
55. *Cam Design and Manufacture: Second Edition; with cam design software for the IBM PC and compatibles*, disk included, Preben W. Jensen
56. *Solid-State AC Motor Controls: Selection and Application*, Sylvester Campbell

57. *Fundamentals of Robotics*, David D. Ardayfio
58. *Belt Selection and Application for Engineers*, edited by Wallace D. Erickson
59. *Developing Three-Dimensional CAD Software with the IBM PC*, C. Stan Wei
60. *Organizing Data for CIM Applications*, Charles S. Knox, with contributions by Thomas C. Boos, Ross S. Culverhouse, and Paul F. Muchnicki
61. *Computer-Aided Simulation in Railway Dynamics*, by Rao V. Dukkipati and Joseph R. Amyot
62. *Fiber-Reinforced Composites: Materials, Manufacturing, and Design*, P. K. Mallick
63. *Photoelectric Sensors and Controls: Selection and Application*, Scott M. Juds
64. *Finite Element Analysis with Personal Computers*, Edward R. Champion, Jr. and J. Michael Ensminger
65. *Ultrasonics: Fundamentals, Technology, Applications: Second Edition, Revised and Expanded*, Dale Ensminger
66. *Applied Finite Element Modeling: Practical Problem Solving for Engineers*, Jeffrey M. Steele
67. *Measurement and Instrumentation in Engineering: Principles and Basic Laboratory Experiments*, Francis S. Tse and Ivan E. Morse
68. *Centrifugal Pump Clinic: Second Edition, Revised and Expanded*, Igor J. Karassik
69. *Practical Stress Analysis in Engineering Design: Second Edition, Revised and Expanded*, Alexander Blake
70. *An Introduction to the Design and Behavior of Bolted Joints: Second Edition, Revised and Expanded*, John H. Bickford
71. *High Vacuum Technology: A Practical Guide*, Marsbed H. Hablanian
72. *Pressure Sensors: Selection and Application*, Duane Tandeske
73. *Zinc Handbook: Properties, Processing, and Use in Design*, Frank Porter
74. *Thermal Fatigue of Metals*, Andrzej Weronski and Tadeusz Hejwowski
75. *Classical and Modern Mechanisms for Engineers and Inventors*, Preben W. Jensen
76. *Handbook of Electronic Package Design*, edited by Michael Pecht
77. *Shock-Wave and High-Strain-Rate Phenomena in Materials*, edited by Marc A. Meyers, Lawrence E. Murr, and Karl P. Staudhammer
78. *Industrial Refrigeration: Principles, Design and Applications*, P. C. Koelet
79. *Applied Combustion*, Eugene L. Keating
80. *Engine Oils and Automotive Lubrication*, edited by Wilfried J. Bartz
81. *Mechanism Analysis: Simplified and Graphical Techniques, Second Edition, Revised and Expanded*, Lyndon O. Barton
82. *Fundamental Fluid Mechanics for the Practicing Engineer*, James W. Murdock
83. *Fiber-Reinforced Composites: Materials, Manufacturing, and Design, Second Edition, Revised and Expanded*, P. K. Mallick
84. *Numerical Methods for Engineering Applications*, Edward R. Champion, Jr.
85. *Turbomachinery: Basic Theory and Applications, Second Edition, Revised and Expanded*, Earl Logan, Jr.
86. *Vibrations of Shells and Plates: Second Edition, Revised and Expanded*, Werner Soedel

87. *Steam Plant Calculations Manual: Second Edition, Revised and Expanded*, V. Ganapathy
88. *Industrial Noise Control: Fundamentals and Applications, Second Edition, Revised and Expanded*, Lewis H. Bell and Douglas H. Bell
89. *Finite Elements: Their Design and Performance*, Richard H. MacNeal
90. *Mechanical Properties of Polymers and Composites: Second Edition, Revised and Expanded*, Lawrence E. Nielsen and Robert F. Landel
91. *Mechanical Wear Prediction and Prevention*, Raymond G. Bayer
92. *Mechanical Power Transmission Components*, edited by David W. South and Jon R. Mancuso
93. *Handbook of Turbomachinery*, edited by Earl Logan, Jr.
94. *Engineering Documentation Control Practices and Procedures*, Ray E. Monahan
95. *Refractory Linings Thermomechanical Design and Applications*, Charles A. Schacht
96. *Geometric Dimensioning and Tolerancing: Applications and Techniques for Use in Design, Manufacturing, and Inspection*, James D. Meadows
97. *An Introduction to the Design and Behavior of Bolted Joints: Third Edition, Revised and Expanded*, John H. Bickford
98. *Shaft Alignment Handbook: Second Edition, Revised and Expanded*, John Piotrowski
99. *Computer-Aided Design of Polymer-Matrix Composite Structures*, edited by Suong Van Hoa
100. *Friction Science and Technology*, Peter J. Blau
101. *Introduction to Plastics and Composites: Mechanical Properties and Engineering Applications*, Edward Miller
102. *Practical Fracture Mechanics in Design*, Alexander Blake
103. *Pump Characteristics and Applications*, Michael W. Volk
104. *Optical Principles and Technology for Engineers*, James E. Stewart
105. *Optimizing the Shape of Mechanical Elements and Structures*, A. A. Seireg and Jorge Rodriguez
106. *Kinematics and Dynamics of Machinery*, Vladimír Stejskal and Michael Valásek
107. *Shaft Seals for Dynamic Applications*, Les Horve
108. *Reliability-Based Mechanical Design*, edited by Thomas A. Cruse
109. *Mechanical Fastening, Joining, and Assembly*, James A. Speck
110. *Turbomachinery Fluid Dynamics and Heat Transfer*, edited by Chunill Hah
111. *High-Vacuum Technology: A Practical Guide, Second Edition, Revised and Expanded*, Marsbed H. Hablanian
112. *Geometric Dimensioning and Tolerancing: Workbook and Answerbook*, James D. Meadows
113. *Handbook of Materials Selection for Engineering Applications*, edited by G. T. Murray
114. *Handbook of Thermoplastic Piping System Design*, Thomas Sixsmith and Reinhard Hanselka
115. *Practical Guide to Finite Elements: A Solid Mechanics* Approach, Steven M. Lepi
116. *Applied Computational Fluid Dynamics*, edited by Vijay K. Garg
117. *Fluid Sealing Technology*, Heinz K. Muller and Bernard S. Nau

118. *Friction and Lubrication in Mechanical Design*, A. A. Seireg
119. *Influence Functions and Matrices*, Yuri A. Melnikov
120. *Mechanical Analysis of Electronic Packaging Systems*, Stephen A. McKeown
121. *Couplings and Joints: Design, Selection, and Application, Second Edition, Revised and Expanded*, Jon R. Mancuso
122. *Thermodynamics: Processes and Applications*, Earl Logan, Jr.
123. *Gear Noise and Vibration*, J. Derek Smith
124. *Practical Fluid Mechanics for Engineering Applications*, John J. Bloomer
125. *Handbook of Hydraulic Fluid Technology*, edited by George E. Totten
126. *Heat Exchanger Design Handbook*, T. Kuppan
127. *Designing for Product Sound Quality*, Richard H. Lyon
128. *Probability Applications in Mechanical Design*, Franklin E. Fisher and Joy R. Fisher
129. *Nickel Alloys*, edited by Ulrich Heubner
130. *Rotating Machinery Vibration: Problem Analysis and Troubleshooting*, Maurice L. Adams, Jr.
131. *Formulas for Dynamic Analysis*, Ronald L. Huston and C. Q. Liu
132. *Handbook of Machinery Dynamics*, Lynn L. Faulkner and Earl Logan, Jr.
133. *Rapid Prototyping Technology: Selection and Application*, Kenneth G. Cooper
134. *Reciprocating Machinery Dynamics: Design and Analysis*, Abdulla S. Rangwala
135. *Maintenance Excellence: Optimizing Equipment Life-Cycle Decisions*, edited by John D. Campbell and Andrew K. S. Jardine
136. *Practical Guide to Industrial Boiler Systems*, Ralph L. Vandagriff
137. *Lubrication Fundamentals: Second Edition, Revised and Expanded*, D. M. Pirro and A. A. Wessol
138. *Mechanical Life Cycle Handbook: Good Environmental Design and Manufacturing*, edited by Mahendra S. Hundal
139. *Micromachining of Engineering Materials*, edited by Joseph McGeough
140. *Control Strategies for Dynamic Systems: Design and Implementation*, John H. Lumkes, Jr.
141. *Practical Guide to Pressure Vessel Manufacturing*, Sunil Pullarcot
142. *Nondestructive Evaluation: Theory, Techniques, and Applications*, edited by Peter J. Shull
143. *Diesel Engine Engineering: Thermodynamics, Dynamics, Design, and Control*, Andrei Makartchouk
144. *Handbook of Machine Tool Analysis*, Ioan D. Marinescu, Constantin Ispas, and Dan Boboc
145. *Implementing Concurrent Engineering in Small Companies*, Susan Carlson Skalak
146. *Practical Guide to the Packaging of Electronics: Thermal and Mechanical Design and Analysis*, Ali Jamnia
147. *Bearing Design in Machinery: Engineering Tribology and Lubrication*, Avraham Harnoy
148. *Mechanical Reliability Improvement: Probability and Statistics for Experimental Testing*, R. E. Little
149. *Industrial Boilers and Heat Recovery Steam Generators: Design, Applications, and Calculations*, V. Ganapathy

150. *The CAD Guidebook: A Basic Manual for Understanding and Improving Computer-Aided Design*, Stephen J. Schoonmaker
151. *Industrial Noise Control and Acoustics*, Randall F. Barron
152. *Mechanical Properties of Engineered Materials*, Wolé Soboyejo
153. *Reliability Verification, Testing, and Analysis in Engineering Design*, Gary S. Wasserman
154. *Fundamental Mechanics of Fluids: Third Edition*, I. G. Currie
155. *Intermediate Heat Transfer*, Kau-Fui Vincent Wong
156. *HVAC Water Chillers and Cooling Towers: Fundamentals, Application, and Operation*, Herbert W. Stanford III
157. *Gear Noise and Vibration: Second Edition, Revised and Expanded*, J. Derek Smith
158. *Handbook of Turbomachinery: Second Edition, Revised and Expanded*, edited by Earl Logan, Jr. and Ramendra Roy
159. *Piping and Pipeline Engineering: Design, Construction, Maintenance, Integrity, and Repair*, George A. Antaki
160. *Turbomachinery: Design and Theory*, Rama S. R. Gorla and Aijaz Ahmed Khan
161. *Target Costing: Market-Driven Product Design*, M. Bradford Clifton, Henry M. B. Bird, Robert E. Albano, and Wesley P. Townsend
162. *Fluidized Bed Combustion*, Simeon N. Oka
163. *Theory of Dimensioning: An Introduction to Parameterizing Geometric Models*, Vijay Srinivasan
164. *Handbook of Mechanical Alloy Design*, edited by George E. Totten, Lin Xie, and Kiyoshi Funatani
165. *Structural Analysis of Polymeric Composite Materials*, Mark E. Tuttle
166. *Modeling and Simulation for Material Selection and Mechanical Design*, edited by George E. Totten, Lin Xie, and Kiyoshi Funatani
167. *Handbook of Pneumatic Conveying Engineering*, David Mills, Mark G. Jones, and Vijay K. Agarwal
168. *Clutches and Brakes: Design and Selection, Second Edition*, William C. Orthwein
169. *Fundamentals of Fluid Film Lubrication: Second Edition*, Bernard J. Hamrock, Steven R. Schmid, and Bo O. Jacobson
170. *Handbook of Lead-Free Solder Technology for Microelectronic Assemblies*, edited by Karl J. Puttlitz and Kathleen A. Stalter
171. *Vehicle Stability*, Dean Karnopp
172. *Mechanical Wear Fundamentals and Testing: Second Edition, Revised and Expanded*, Raymond G. Bayer
173. *Liquid Pipeline Hydraulics*, E. Shashi Menon
174. *Solid Fuels Combustion and Gasification*, Marcio L. de Souza-Santos
175. *Mechanical Tolerance Stackup and Analysis*, Bryan R. Fischer
176. *Engineering Design for Wear*, Raymond G. Bayer
177. *Vibrations of Shells and Plates: Third Edition, Revised and Expanded*, Werner Soedel
178. *Refractories Handbook*, edited by Charles A. Schacht
179. *Practical Engineering Failure Analysis*, Hani M. Tawancy, Anwar Ul-Hamid, and Nureddin M. Abbas
180. *Mechanical Alloying and Milling*, C. Suryanarayana
181. *Mechanical Vibration: Analysis, Uncertainties, and Control, Second Edition, Revised and Expanded*, Haym Benaroya

182. *Design of Automatic Machinery*, Stephen J. Derby
183. *Practical Fracture Mechanics in Design: Second Edition, Revised and Expanded*, Arun Shukla
184. *Practical Guide to Designed Experiments*, Paul D. Funkenbusch
185. *Gigacycle Fatigue in Mechanical Practive*, Claude Bathias and Paul C. Paris
186. *Selection of Engineering Materials and Adhesives*, Lawrence W. Fisher
187. *Boundary Methods: Elements, Contours, and Nodes*, Subrata Mukherjee and Yu Xie Mukherjee
188. *Rotordynamics*, Agnieszka (Agnes) Muszńyska
189. *Pump Characteristics and Applications: Second Edition*, Michael W. Volk
190. *Reliability Engineering: Probability Models and Maintenance Methods*, Joel A. Nachlas
191. *Industrial Heating: Principles, Techniques, Materials, Applications, and Design*, Yeshvant V. Deshmukh
192. *Micro Electro Mechanical System Design*, James J. Allen
193. *Probability Models in Engineering and Science*, Haym Benaroya and Seon Han
194. *Damage Mechanics*, George Z. Voyiadjis and Peter I. Kattan
195. *Standard Handbook of Chains: Chains for Power Transmission and Material Handling, Second Edition*, American Chain Association and John L. Wright, Technical Consultant
196. *Standards for Engineering Design and Manufacturing*, Wasim A. Khan and Abdul Raouf
197. *Maintenance, Replacement, and Reliability: Theory and Applications*, Andrew K. S. Jardine and Albert H. C. Tsang
198. *Finite Element Method: Applications in Solids, Structures, and Heat Transfer*, Michael R. Gosz

Standards for Engineering Design and Manufacturing

Wasim A. Khan
Abdul Raouf

CRC Press is an imprint of the
Taylor & Francis Group, an **informa** business

CRC Press
Taylor & Francis Group
6000 Broken Sound Parkway NW, Suite 300
Boca Raton, FL 33487-2742

First issued in paperback 2019

ISBN-13: 978-0-8247-5887-5 (hbk)
ISBN-13: 978-0-367-39154-6 (pbk)

Library of Congress Card Number 2005048580

Library of Congress Cataloging-in-Publication Data

Khan, Wasim A.
Standards for engineering design and manufacturing / Ahmed Wasim Khan, Abdul Raouf.
p. cm.
Includes bibliographical references and index.
ISBN 0-8247-5887-0 (alk. paper)
1. Production management--Standards. I. Raouf. A (Abdul), 1929- II. Title.

TS156.K43 2005
658.5′02′18--dc22 2005048580

Visit the Taylor & Francis Web site at
http://www.taylorandfrancis.com

and the CRC Press Web site at
http://www.crcpress.com

Dedication

To M. Ahrar Khan and Shameha
Wasim A. Khan

To Dr. Razia Raouf
Abdul Raouf

Foreword

Professors Khan and Raouf have recognized the need for a book that meets the requirements of undergraduate students, graduate students, and engineering professionals in the cross-disciplinary aspects of mechanical engineering that relate to computer-based design and manufacture and, in particular, focuses on the need to provide a vehicle for cross-disciplinary integration through standards and codes of practice. These standards relate to the configuration, connectivity, and accessibility of data and information through electronic computer-based methods and protocols. The drive is to have universally recognized standards and methods that permit the use of data that define the selection of materials for discrete component manufacture, the shape, size, and geometry of the product or component, and the connectivity of individual components that form engineering assemblies.

Environmental requirements have led to consideration of material disposal and recyclability, sometimes called total-product life care. Concepts such as modularity, module replaceability, and in-life technical modular updates are being introduced to achieve product technology and performance upgrades and specification extension. The concept of global manufacture, in which discrete components are manufactured at locations around the world to take advantage of labor costs and economies of scale, can only be implemented in the required commercial timescales by use of electronic data transfer systems. However, their integration and their robustness, coupled with their commercial viability, rely entirely on the use of internationally agreed standards and commercial data-security systems.

Technology is fast changing, and the principles on which data and information structures, communication protocols, techniques for numerical analysis in design, and methods for computer integrated manufacture integrate through standards to form a working and commercially secure entity must be emphasized. Thus, this book focuses on the generic principles that underpin the formation and establishment of international standards.

This book addresses the concepts associated with the ways in which new products are conceived and realized and discusses how standards are used, to best advantage, in this process. This book makes good use of examples both to introduce concepts and to demonstrate their use. It is a valuable text for both undergraduate and graduate students and for practicing engineers.

D. R. Hayhurst, Sc.D. (Cantab), FR.Eng.
Professor of Design, Manufacture, and Materials
The University of Manchester

Preface

From the Stone Age to the twenty-first century, engineering design and manufacturing has advanced from the potter's wheel to space probes. Engineering design and manufacturing has become multidisciplinary and encompasses the various divisions of applied sciences, including computer engineering and management science.

In ancient times nature was the sole source of wealth. Adam Smith (1723–1790), David Ricardo (1772–1823), and John Stuart Mill (1806–1873) recognized manufacturing as an element in the creation of wealth and introduced the concept of "vendibility— production for the market." Today's enterprise globalization is a logical step forward into a worldwide market that demands a vast range of discrete products.

The introduction of the concept of interchangeability of parts by Eli Whitney (1765–1825) formalized the use of specified tolerances to machined parts. This may not have been the first event in the use of standards; however, it began a completely new era in the standardization of engineering design and manufacturing tasks.

With the evolution of multidisciplinary engineering design and manufacturing, the need for standardized procedures to develop discrete products has increased manifold. This need encompasses areas such as the methodology associated with design and manufacturing functions, the use of materials, the shape and size of the envisaged product, and the geometry data transfer to other functions in the computer-integrated manufacturing environment. To meet that need, this book has been written to introduce undergraduate students, graduate students, and professionals to multidisciplinary engineering design and manufacturing as a subject that can best be practiced through the application of relevant standards.

This book describes the concepts associated with product-realization processes in modular form by referring to elaborate case studies that cover the multidisciplinary engineering domain. The main focus is on the use of standards. This book is divided into five parts with eleven interrelated chapters that can be consulted according to the requirements of the reader.

Chapter 1 introduces the subject by exploring the history of engineering design and manufacturing. It emphasizes the change in the theoretical approaches to engineering design and manufacturing and their implementation from the Stone Age to the modern world. The modern aspect of design optimization and the relation between design and manufacturing are considered. Use of standards as a global entity is introduced.

Chapter 2 presents various methods for formalizing the information and procedures and introduces pertinent regulatory bodies. It presents a survey of the standards-developing organization related to engineering design and manufacturing of discrete products. The list of standards-developing organizations is supplemented by details on support organizations that develop standards and codes in diverse areas such as occupational safety, fire protection, and illumination. A section of the chapter gives details of the standard-development procedure established by the British Standards Institution (BSI) and other standards-developing organizations. A brief on economic benefits of standards as determined by the German Standards Institute (DIN) is also included.

Chapter 3 presents a case study of the variant design and manufacture of wrenches according to pertinent standards. This case study is a parametric modeling exercise that makes use of CAD/CAM software. Standards used at various stages in this case study for such functions as geometry transfer, cutter-location, part-programming, and digital data-transmission are identified.

Chapter 4 reviews standards related to design methodology, design functions, measurement methods, materials, operation of the product, quality, safety, reliability, and product life cycle. The future of standardization in the design process for discrete products is addressed, and, in particular, the ISO geometry standards series is discussed.

Chapter 5 is a case study that demonstrates the use of standards related to the system approach to multidisciplinary design of discrete products. Various loading conditions, analysis procedures, and expected exposures for the discrete products are mentioned.

Chapter 6 explores standards related to man–machine interaction, operation of the equipment, and measuring equipment precision. Future directions in the standardization of manufacturing equipment are also presented. This aspect specifically addresses the current research carried out at the Manufacturing Engineering Laboratory (MEL) of the National Institute of Science and Technology (NIST) for smart machine tools.

Chapter 7 presents a case study related to the development of process plans for a discrete product. The case study extensively uses company, national, and international standards to perform the task. Procedure sheets for individual components, subassemblies, and the whole product are produced.

Chapter 8 addresses standards for the use of inherently diverse discrete products. This chapter includes standards related to man–machine interaction, human–computer interaction, ergonomics, and standards for various forms of energy input and output from the product.

Chapter 9 provides a procedure for discrete product design and manufacture with consideration to ergonomics. Several standards-developing organizations are surveyed to identify appropriate standards to design and manufacture discrete products with ergonomic constraints.

Chapter 10 presents a review of support standards for core design and manufacturing activities. With the advent of digital computers and Internet technology, modern products normally feature automation functions and are marketed electronically, along with use of other mediums. This chapter encompasses standards related to computer interfacing, communication, software development, and e-business procedures.

Chapter 11 presents a case study that addresses the multidisciplinary mechatronics application. It provides terms and conditions for digital interfacing and control software development for a milling machine by use of appropriate standards.

The subject area addressed in this book covers a very large domain. To keep the book to a manageable size, several common concepts, addressed sufficiently in other books, have been presented in condensed form, so that the core topic, the use of standards for design and manufacturing of discrete products, remains the focus. For other topics, related references are available in the bibliographies.

In most cases, the Web site of a related organization is provided. In some cases, acronyms are used to identify the organization. Such acronyms and search words used in the surveys can easily be resolved at a search engine on the World Wide Web. The search word or the wild card selection affects the output from an online or a CD-based search for standards. The information derived from Web sites of standards organizations is copyrighted by those organizations and was correct at the time of browsing.

Most of the organizations surveyed (as presented in Chapter 2) develop procedures, codes, and standards. The major emphasis of this book is on standards; however, other organizations are listed because of their direct relevance to the design and manufacturing of discrete products. One of the major objectives of this book is to introduce the potential of utilizing the standards in engineering design and manufacturing of discrete products to acquire the benefits of standardization.

The Appendix at the end of the book provides basic information regarding online catalogs related to standard machine elements and standard control elements, digital libraries for standards, and the Web sites related to standards. Comprehensive references are also provided as bibliographies at the end of each chapter to help readers with the wealth of material available in related fields.

Acknowledgments

The authors are deeply indebted to our colleagues Mr. Ghulam Haider Mughal, Mr. Syed Mojiz Raza, and Mr. John Corrigan for their help in designing and producing this book. We also acknowledge the support of the students Mr. Sultan Al-Thanayan and Mr. Haroon Naseer. The patience of our family members is deeply appreciated.

Disclaimer

Every effort has been made to keep the contents of this book accurate in terms of description, examples as given in case studies, intellectual rights of others, and contents of Web sites at the time of browsing. The authors and the publisher are not responsible for any injury, loss of life, or financial loss arising from use of material in this book.

Table of Contents

Part 1
Engineering Design and Manufacturing of Discrete Products—The Scope

Part 1

Engineering Design and Manufacturing of Discrete Products—The Scope

1 Engineering Design and Manufacturing

1.1 INTRODUCTION

The term engineering is a derivation of the Latin word "ingenerate," meaning "to create," or "to contrive." Design means "to plan and execute artistically." Design in engineering perspective requires existence of need—genuine, revised, or generated; an analysis mechanism; and a graphical depiction of the process. To make the object by hand or by use of machinery in large quantities from a design produced earlier is termed "to manufacture." A glance into the past, beyond the Latin roots of the words, to observe various facts that took place in our history, reveals interesting phenomena that are still common in today's engineering design and manufacturing.

The process of guided development began the day primitive humans started using logic to overcome their basic problems. Understanding arose, default by nature or established through experience: that night with or without stars follows day; that clouds may bring rain from which some sort of protection is needed; and that something is eaten, whether vegetation or animal. Perhaps lightning started the first fires, and humans learned to make certain types a spark with stones by rubbing them together. The establishment of that fact, whether planned or accidental, made primitive humans aware that fire makes vegetation and meat tender and makes them taste different—perhaps better. The need existed, and observation and experimentation were the only tools.

Many superstitious factors (such as trust in magic and misunderstanding of cause and effect) have discouraged human beings from overcoming ignorance. The more influential factor, from primitive to modern times, has remained the need to safeguard oneself from predators, to fulfill the basic needs of life, or to dominate a piece of land or group of people. The need has led to progress that has continued from the Stone Age, to the Bronze Age, to the Iron Age, and now to the modern world. Engineering and science have remained exercises in observation and experimentation, over a very long period of time, to satisfy the need. In the process, humans have established laws to further the progress. The need, in all the defined or undefined areas of engineering and sciences, does not restrict itself to any level of development; it forges ahead, through old and new analytical, numerical, and experimental procedures. The need is still the prime factor in the production of art, science, engineering, and technology. Scientists and engineers still venture into the unknown to find long-lasting solutions. Some of the inventions that have made lasting marks on the history of humankind are presented in Table 1.1.

TABLE 1.1
Some Important Inventions

Object	Origin	Evidence of Use Since	Comments
Pottery	All ancient civilizations (China)	7900 B.C.	Engineering material clay
Potter's wheel	China	3100 B.C.	Mechanism
Wheel	Mesopotamia	3500 B.C.	Mechanism
Salamis tablet	Island of Salamis	300 B.C.	Inference engine
Abacus (suan–pan)	China	1200 A.D.	Inference engine
Clock	Italy	1300 A.D.	Time
Printing press	Germany	1450 A.D.	Publishing
Telescope	Holland	1608 A.D.	Observation of planetary movement
Microscope	Holland	1674 A.D.	Observation of microbes
Steam Engine	England	1765 A.D.	Transportation
Telephone	Canada	1876 A.D.	Communication
Automobile	U.S.	1900 A.D.	Transportation
Radio	U.S.	1901 A.D.	Communication
Airplane	U.S.	1903 A.D.	Transportation
Assembly line	U.S.	1908 A.D.	Production
Antibiotics	U.K.	1928 A.D.	Medicine
Nuclear bomb	U.S.	1945 A.D.	Weapons
Digital computer	U.S./U.K.	1940–50 A.D.	Inference engine
Internet	U.S.	1973 A.D.	Communication
Space shuttle	U.S.	1981 A.D.	Space travel

With the evolution of multidisciplinary engineering design and manufacturing from the Industrial Revolution to today, the need to standardize procedures to develop discrete products has increased manifold. This need encompasses areas such as methodology associated with design and manufacturing functions, use of materials, shape and size of the envisaged product, geometry data transfer to all other functions in the computer-integrated manufacturing (CIM) environment, and waste disposal. Although standardization is promoted for most tasks in engineering design and manufacturing, certain tasks cannot be standardized or may not be accessible to the general engineering community. At the same time, standardization may not be seen as a development-restricting factor because of the evaluation procedure associated with the standardization.

Today, formal procedures described as regulations, technical standards, codes of practice, technical specifications, frameworks, and benchmarks in a multidisciplinary engineering environment are practiced in a section of an engineering establishment, the whole engineering establishment, or different facilities run by the same establishment in one country or across the globe and during various interactions between different establishments to achieve the following benefits:

1. Convenience in communication and commerce
2. Convenience in building and operating a viable design system
3. Convenience in building and operating a viable production system
4. Convenience in training the workforce
5. Convenience in maintaining the design tools and production machineries
6. Convenience in establishing reliability of the product based on data from basic design calculation and feedback from after-sales service
7. Readiness in incorporating management strategies
8. Readiness in implementing global practices such as
    - Practices in design
    - Practices in ergonomics
    - Practices in design analysis
    - Practices in production control, including process planning and scheduling
    - Practices in manufacturing
    - Practices in assembly
    - Practices in quality control
    - Practices in maintaining optimal inventories of raw materials, finished goods, spare parts, work-holding devices, tools, and accessories
9. Reduced cost of the product
10. Standard Quality assurance procedures
11. Standard Environment related procedures
12. Ease in vendor development
13. More control over accounting and financial matters
14. Simpler to manage and profitable organization
15. No restriction in developing new products
16. Similarity in product features such as man–machine interface, human–computer interface, input energy forms and their permissible values, any intermediate processes, and output energy forms and their permissible values

The scope of this book is limited to technical standards for products developed by use of discrete components and produced by discrete manufacturing processes.

A discrete component is an object that can be uniquely identified in an assembly or a mechanism (generally called a product). Quoting from Parmley, "The building blocks of mechanical mechanisms consist of many typical individual (discrete) components; just as an electron, proton and neutrons are basic parts of an atom. In each case, these discrete components must be properly selected and precisely arranged in a predetermined pattern to result in a functioning unit (an assembly). As each assembly is fit into a larger and more complex device, the

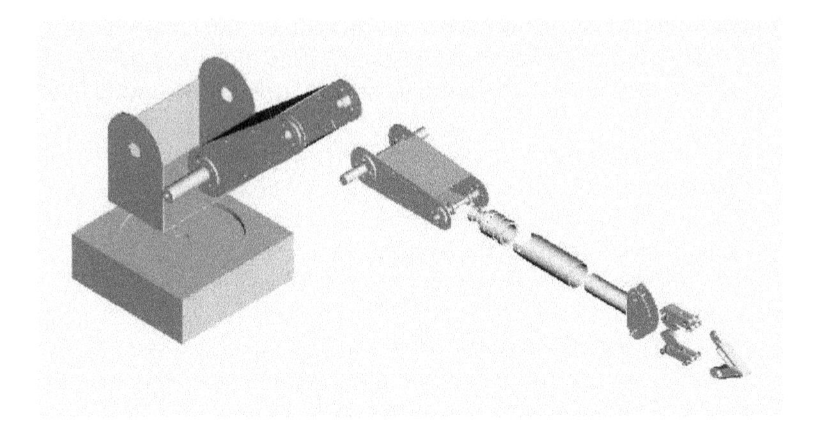

FIGURE 1.1 Exploded view of the revolute robot.

individual component becomes less and less noticeable, until a malfunction occurs." The assembled device is referred to as a mechanism if its components perform movements or is referred to as a structure if the components remain stationary. An exploded view of a simplified revolute robot that shows various components, top view of the mechanism, and the solid model are shown in Figure 1.1 to Figure 1.3. This robot constitutes a basic example of discrete components, assembly, and mechanism. A discrete product that comprises a collection of mechanisms and structures is referred to as a machine.

The basic control diagram of such mechanisms is provided in Figure 1.4. The basic input energy and output energy/work forms from such mechanisms are detailed in Figure 1.5 and Figure 1.6. A general description of the scope of the standards for engineering design and manufacturing is provided in Figure 1.7.

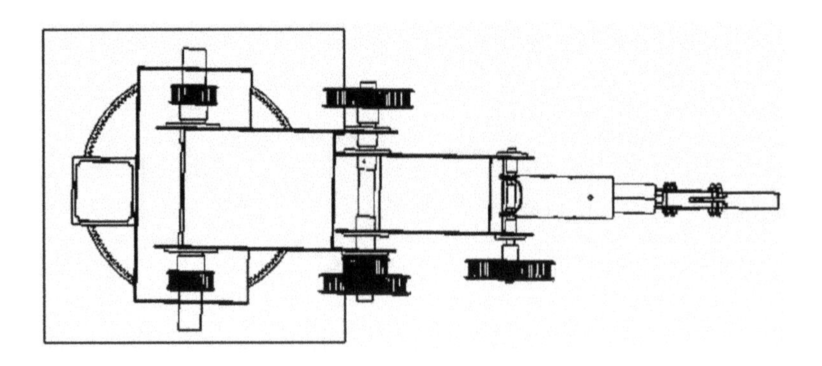

FIGURE 1.2 Top view of the revolute robot.

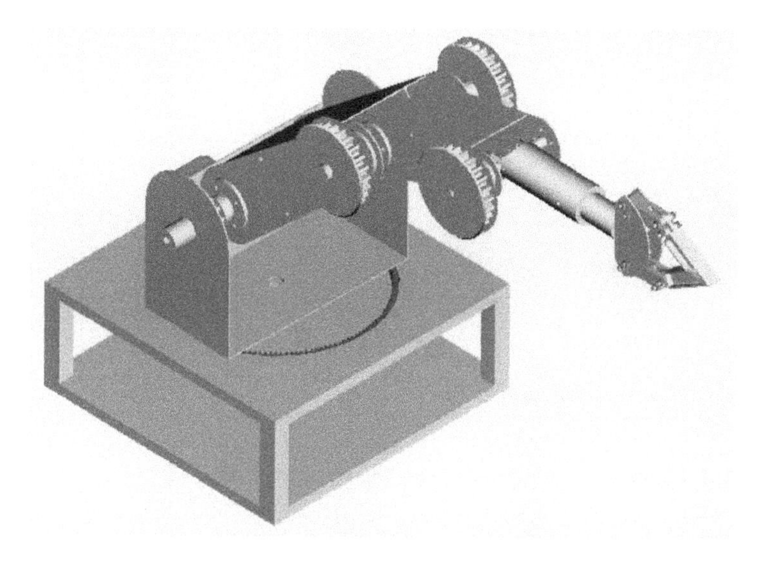

FIGURE 1.3 Solid model of the revolute robot.

The scope of this book is the best described by Figure 1.4 to Figure 1.7. The following sections in this chapter describe the concepts encompassing key elements of design and manufacturing and optimization of design based on multidisciplinary features. The purpose of this effort is to develop a common base that can be used for selection and use of standards.

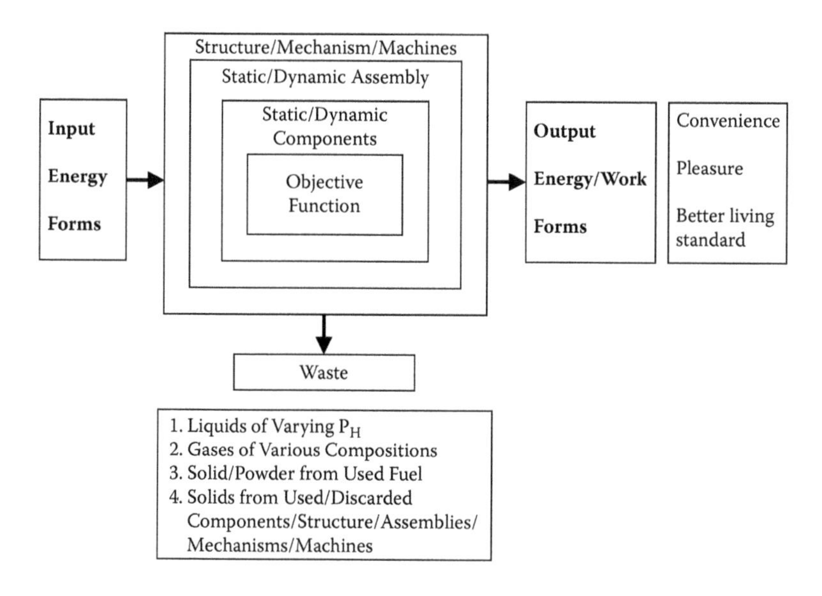

FIGURE 1.4 Complete cycle of the working of discrete product.

1. Hydal
2. Wind
3. Solar
4. Magnetic
5. Thermal (fossil fuel)
6. Thermal (non fossil fuel)
7. Hydraulic
8. Pneumatic
9. Chemical reaction
10. Electrochemical reaction
11. Vibration
12. Gravity
13. Dynamic fluid
14. Biological reaction

FIGURE 1.5 Various forms of input energy to discrete product.

1.2 DESIGN PHILOSOPHY—CONVENTIONAL AND SYSTEM APPROACH

The evolution of humankind into the most primitive social order started when human beings began inventing the methods of verbal communication around 14,000 B.C. This accomplishment was achieved through the basic principle of information transformation by virtue of establishing speech forms. This feat was also the most important step in communication.

Human beings were able to design and manufacture tools (Figure 1.8 and Figure 1.9) during the Stone Age. The design of the tools basically emerged from need followed by observation and experimentation for the required use. Need, contemporary knowledge, and experimentation in various fields at each stage of evolution remained the factor for the presentation and adoption of a particular design philosophy. The need may have been to cross over oceans, to fly through skies, or to explore the universe. Cultures over time learned and applied the current principles of design to the object of their need.

Not until the industrial revolution did individual nations start using defined methodologies for design and manufacture, in particular to maintain the sovereignty

1. Motion
2. Sound
3. Light
4. Radiation of varying intensity
5. Electro weak forces
6. Magnetic fields
7. Heating
8. Cooling

FIGURE 1.6 Various forms of output energy/work forms from discrete product.

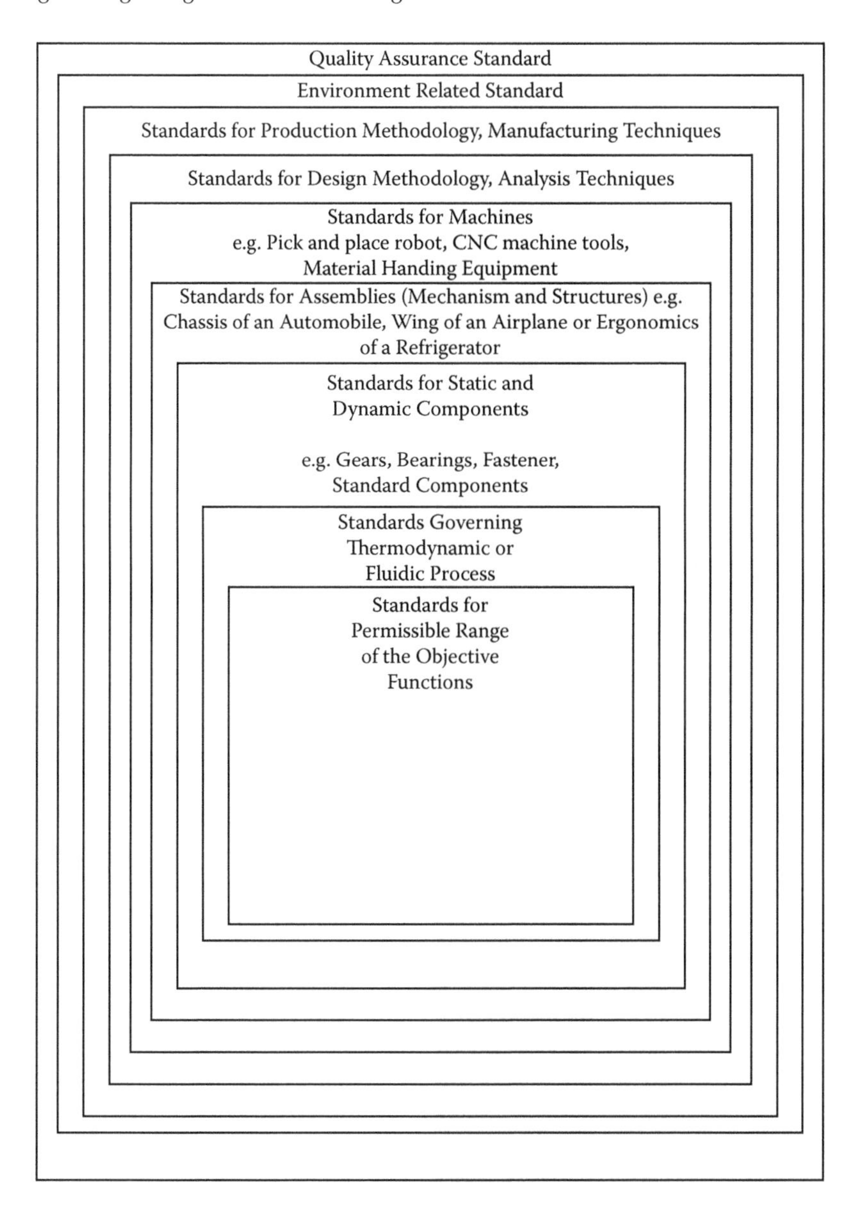

FIGURE 1.7 A general description of scope of standards for engineering design and manufacturing.

of a nation and to reach a certain level of knowledge as compared with other nations. The industrial revolution became the key period for invention and mass production. Automobiles, steam engines, airplanes, and communications media were a few of the important inventions. A number of other inventions made the human life simpler and better when compared with previous centuries of

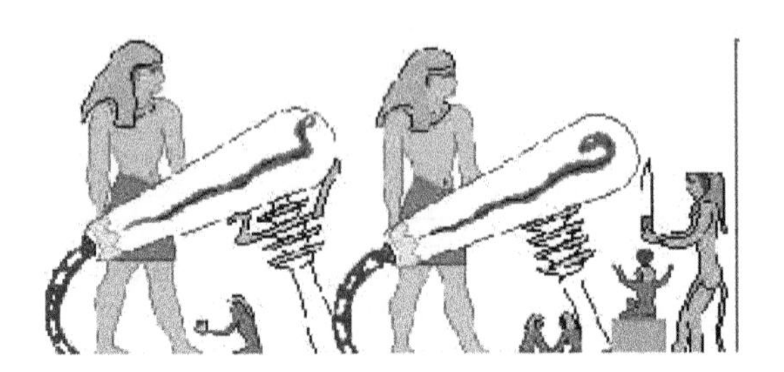

FIGURE 1.8 Ancient wall carving.

evolution. The current age of information, communication, computer, and space travel is founded on the inventions of previous centuries.

The most important inventions were tested during the first and second world wars. These inventions included automobiles and airplanes, along with various kinds of weapons. The theories and technologies that were used at these particular times, and later during the Cold War era and the space race, were adopted by the designers and manufacturers of products used in daily life.

One important development from the Second World War was the lofting principle, a method in computational geometry that allows creation of aerodynamics profiles of airplane wings, an important step in design and design analysis (Figure 1.10). The method relied on computation and was much better than the previously used manual techniques.

The design principle, until recently, remained conventional. It lacked the comprehensive analytical, numerical, and experimental analysis of various steps of the design process. The design cycle, as seen through the centuries of development, mainly relied on need and availability of means to accomplish the targets in hand. Basic principles of observation and experimentation, with both the object

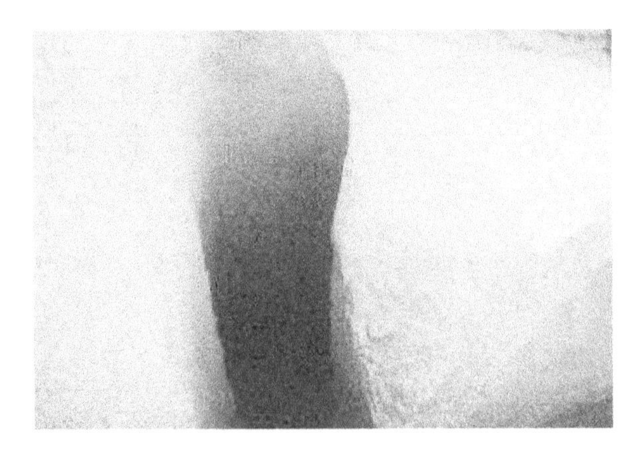

FIGURE 1.9 Drill hole at the Aswan quarries.

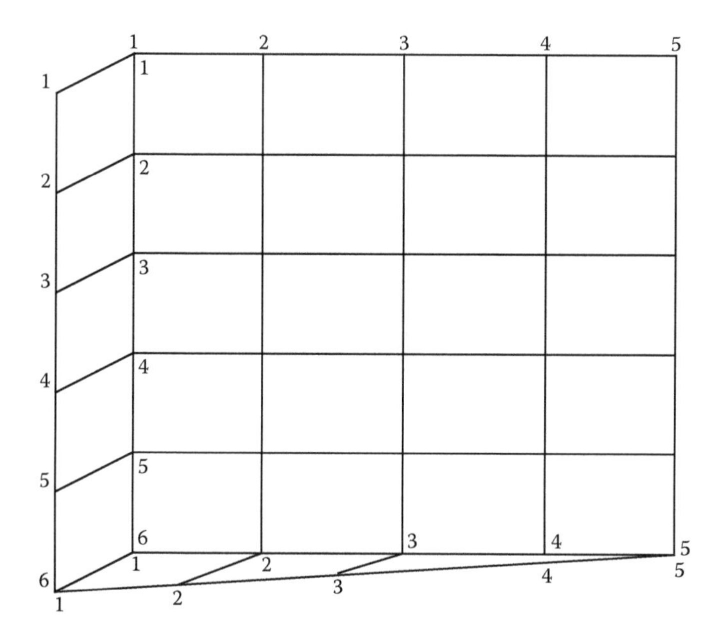

FIGURE 1.10 Lofting by use of longitudinal lines and cross-sectional patches.

to be designed and tools to design it, are still the need of the day. However, the basic principles that were developed by ancient engineers in designing historical artifacts remain in use. Clear demarcation between various functions of design and manufacture has not been made. This demarcation may allow an increase in the efficiency of manpower, the development of focused products, an increase in the production rate of the object, and a decrease in the cost of production.

Planning was first applied to the design of mechanical components during the Industrial Revolution. Although the availability of optimum tools for the design of various objects remained restricted, the design methods have certainly progressed over the decades since that time. Take, for example, the telephone. The technology used for the design and manufacturing of this device varies significantly from that used at its inception.

When we say that a conventional approach to design is being used, we mean that contemporary tools available for the design of objects are not being used. Instead, more primitive tools are being used and demarcation between various design activities is not clear. Even so, the conventional approach has a sequence that gives it the capability to produce a product design at the end. The conventional approach to design is presented in Figure 1.11.

The conventional approach to design does not distinctly divide the exercise into identifiable steps. Over the years it had been applied to highly sensitive products. The procedure attempts to achieve functionality by almost always using excess values for design parameters that result in overdesign. Sometimes, when conservative values

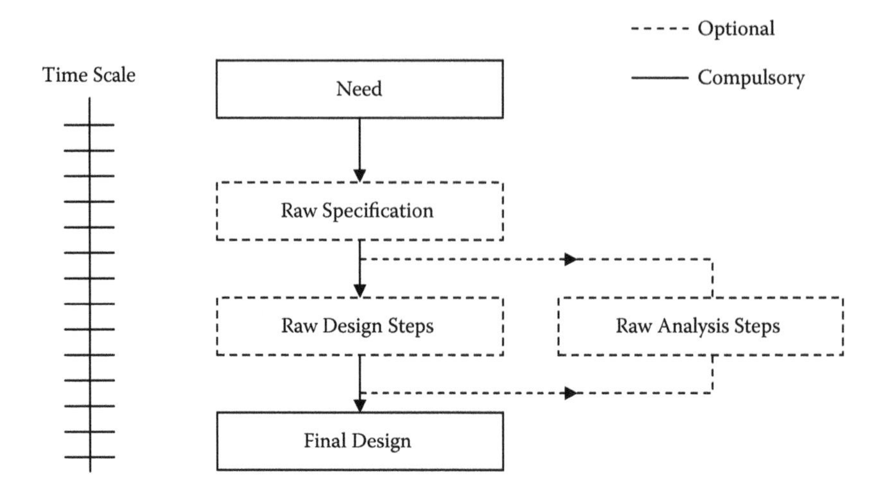

FIGURE 1.11 Conventional approach to design.

are not used for design, the product may result in catastrophe. Cognitive deductions play a more important role in design revisions with large time lapse.

System approach, an organized set of doctrines, ideas, or principles usually intended to explain the arrangement or workings of a systematic whole, inculcated into design practice as the logical division of subsystems, is taking place with a knowledge base that spans centuries. At the same time, the tools available to perform the design have become much more advanced than those used in the conventional approach to design.

The system approach to design, as presented in Figure 1.12, requires a thorough analysis of need, product specification, concept design, embodiment design, and detail design. Each step in the design cycle is subdivided in smaller units. A principle of analysis, synthesis, and evaluation for determination of optimal result for each task or group of tasks is devised. This principle includes task specification, goal setting, and a time frame to achieve a particular target. Methods of transferring information from one step to another in the design cycle and to all other related functions through standard or other interfaces have also been devised. The most important step is the inclusion of the use of standards from the microtask to the macrotask level during the entire design cycle.

Because most of the products used today have a certain level of automation, quality assurance and the environmental aspects associated with it leads to the adoption of multidisciplinary tasks in the design cycle. A team of engineers is now needed to perform design tasks in a specified time frame.

Today, more knowledge has been gathered about the environment than at any other time in history. The concept of product life cycle is becoming increasingly important in the new millennium. Awareness of this fact does not require redefinition of the system approach to design, but the addition of new steps and revision

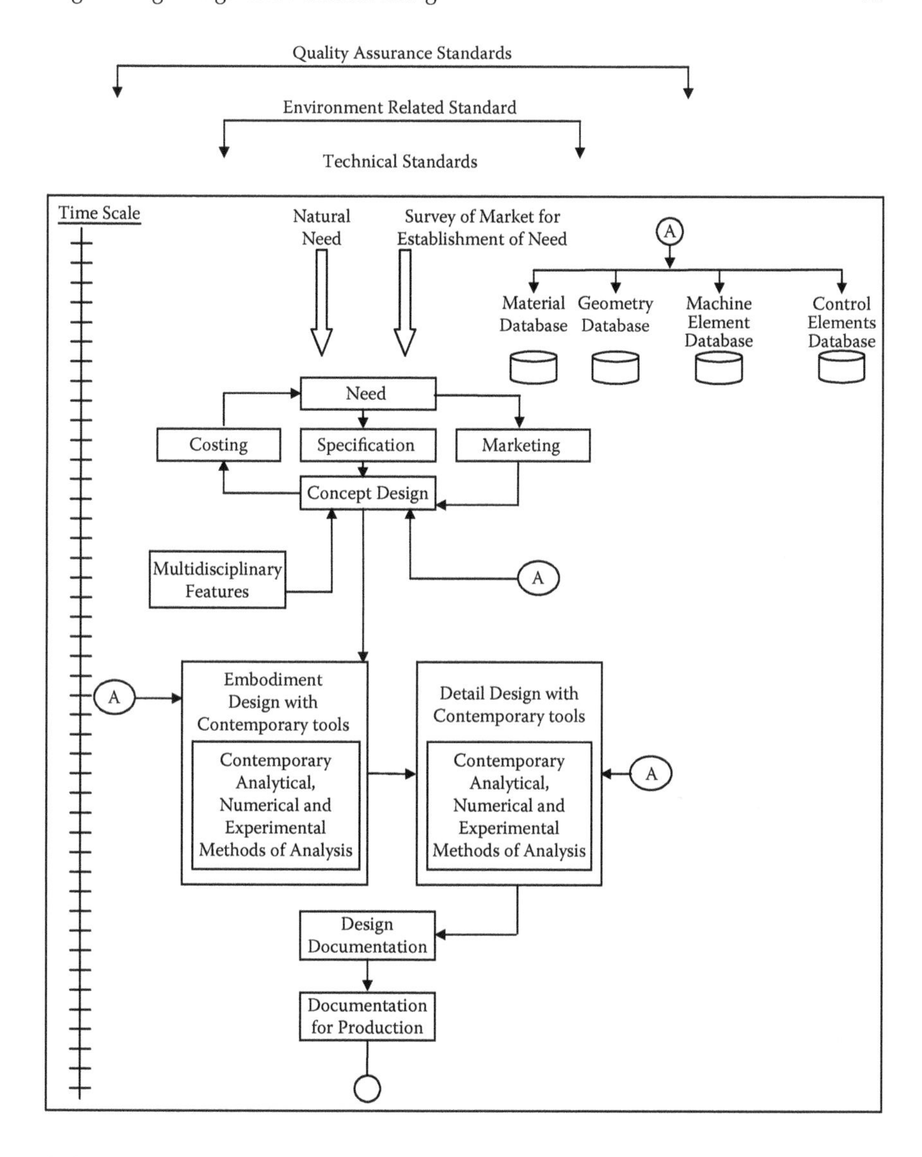

FIGURE 1.12 Contemporary system approach to design.

of certain existing aspects that will make the design cycle correspond to the needs of the environment.

The economies associated with products that have multidisciplinary features and designed on the basis of the products' life cycles are different from economies developed with the conventional design cycle. A cost model for the design has also become mandatory for the design exercise.

The system approach to design is a better form of design, whereby several new functions are introduced. The system approach is better than the conventional approach in many ways. The distinct feature of the system approach is the generation of need in a target market. The product development function of the system approach tries to fill in a vacuum that exists in a particular market. A second important feature of the system approach is the extension of Eli Whitney's (1765–1825) concept of interchangeability. Interchangeability is achieved through the design of modular products by use of pertinent standards. This aspect is very useful in the basic concept of product life cycle and provides ample opportunity for use of standard information.

If the system approach to design, along with aspects such as multidiscipline engineering, quality consciousness, environment consciousness, and cost consciousness is applied to the engineering domain as presented in Figure 1.4 to Figure 1.7, then an enormous opportunity exists for the use of standards. Use of standard information, along with the system approach to design, expedites the final result with more certainty about the projected function, cost, and life of the product.

With the advent of computer-aided design (CAD) and computer-aided manufacturing (CAM), use of databases for design and manufacturing parameters in a particular domain, the availability of standard-component geometry in digital format, and the availability of digital libraries of technical standards from different organization are the trends for the future (Figure 1.13). For more details about standards for the system approach to design of discrete products, see Chapter 4.

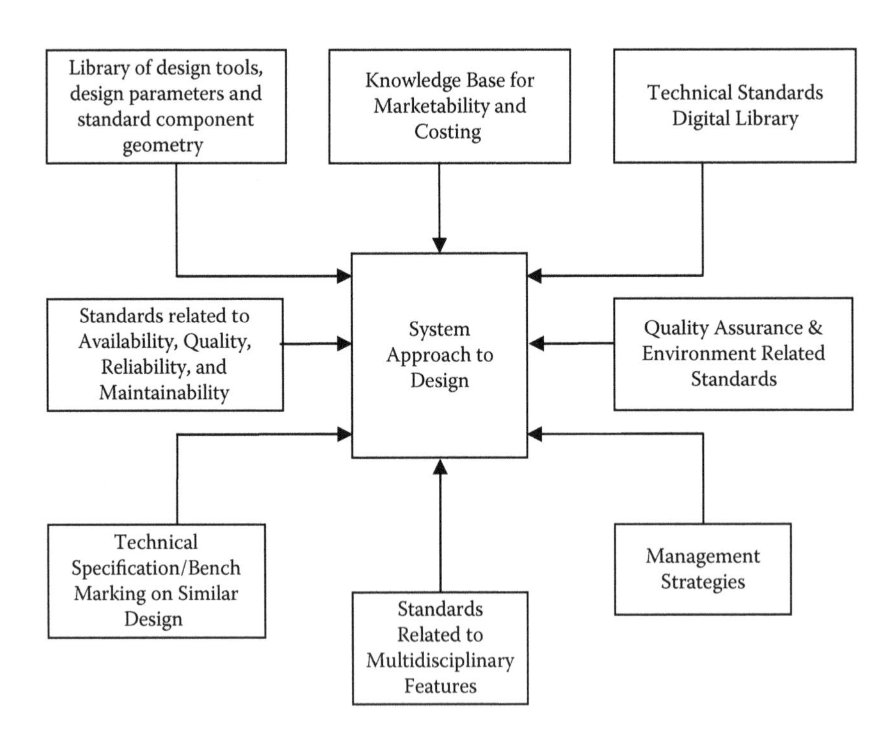

FIGURE 1.13 Possible interactions for contemporary system approach to design.

1.2.1 Design Analysis

Design analysis, a step ahead from hit-and-trial for design and manufacturing, was incorporated into the design and manufacturing cycle to introduce more certainty in all aspects of the life of an object, from drawing board to disposal. The design analysis is the critical procedure that is performed to optimize design and manufacturing parameters such as size, weight, loading, ease of manufacturing, ease of assembly, reliable operation, cost, maintenance, and disposal. Because of the multidisciplinary nature of the mechanisms (discrete products), the design analysis may be focused on one or more of the following areas:

1. Analysis for strength
2. Analysis for bending
3. Analysis for torsion
4. Analysis for buckling
5. Analysis for contact loading
6. Analysis for combined loading
7. Analysis for cyclic loading
8. Analysis for cyclic loading at elevated temperature
9. Analysis for resonance
10. Analysis for thermal effects
11. Analysis for effects of gravity
12. Analysis for the effects of the magnetic field
13. Analysis for the effects of static and dynamic fluid
14. Analysis for the effects of liquids of various pH
15. Analysis for exposure to radiation of various intensity
16. Analysis for exposure to electroweak forces
17. Analysis for exposure to various air or gaseous environments
18. Analysis of electric circuits that power sensors, transducers, and actuators
19. Analysis of electronics circuits used to control sensors, transducers, and actuators
20. Analysis of control software that provides user interface to the product
21. Analysis for human factors (ergonomics)
22. Analysis for ease of manufacturability
23. Analysis for ease of assembly
24. Analysis for packaging and transportation
25. Analysis for market response

The design analysis for mechanical loading is normally performed by use of the following techniques:

1. Analytical procedures
2. Numerical procedures
3. Experimental procedures

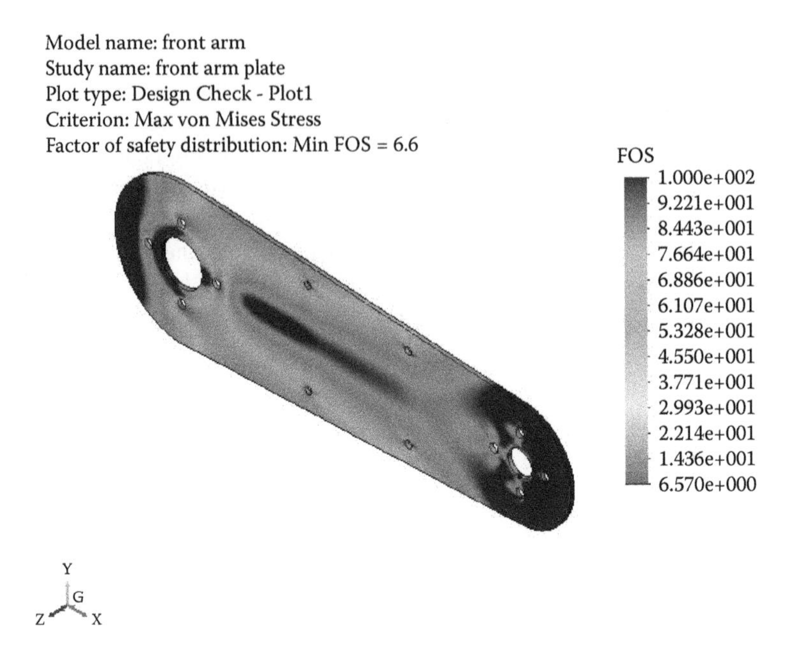

FIGURE 1.14 Finite-element modeling (FEM) of front arm (see Figure 1.1 to Figure 1.3).

The analytical procedure requires synthesis of the problem and formulation of equations that can provide, for example, force balance, mass balance, or heat balance. Simultaneous solution of these equations provides the value of unknown parameters.

Numerical procedures are applied to design when the analytical procedure requires further verification or no solution is possible by use of the analytical procedure. Numerical procedures apply mathematical techniques such as the finite-element method and the boundary element method. These methods mathematically model the approximation of the real problem and provide a solution through rigorous computations that predict displacement of nodes and elements. The solution can be viewed in multicolor models that give intensity of loading of relevant parameters. Figure 1.14 shows a schematic of a finite-element model for a discrete component.

The experimental procedures include use of strain gauges, photo elasticity, and brittle coatings. All of these procedures provide sufficient insight into the loading of an object through use of real objects or miniature models.

Available standards and theories of failure for different design analysis parameters provide certainty that the envisioned design of a product will or will not withstand the loading or exposure for the product's projected lifetime.

Some major multidisciplinary parameters that require analysis for the envisioned design are:

1. Permissible range of electric current
2. Permissible range of EMF
3. Permissible torque for actuators

4. Requirements of man–machine interface
5. Requirements of human–computer interface
6. Suitability of information-transfer protocol
7. Suitability of geometric data-transfer protocol
8. Automation protocol requirements
9. Requirements of networking protocol
10. Use of high-level computer language in transparent interfaces
11. Selection of appropriate mechanical, electrical, and electronics components or products

Different analytical, numerical, and experimental procedures are used to analyze the above multidisciplinary parameters. The list of areas and parameters identified for analysis is not exhaustive. The applicable areas or parameters are dependent on the specific product and the features it supports.

A wealth of standard information is available from digital libraries, online databases, and catalogs for establishing optimal values of mechanical engineering design parameters and multidisciplinary parameters for a discrete product. The reader is referred to details about standards in Chapter 2.

1.2.2 Design for Manufacturing or Assembly

The contemporary system approach to design function covers the design phase from inception of need to design for manufacture (DFM) or design for assembly (DFA), with all the iterations for the sake of design optimization.

DFM takes into account the constraints on the available manufacturing resources and attempts to match them with the requirements of the system approach to design as design. All other subfunctions in design and manufacturing are adjusted on the basis of this one-to-one relationship (Figure 1.15). Similarly, DFA takes into account the constraints on the available assembly resources and attempts to match them with the requirements of the system approach to design. All other subfunctions in design and manufacturing are adjusted on the basis of this one-to-one relationship (Figure 1.16).

The system approach to DFM and DFA defines subfunctions of the design procedure by application of principles of similarity; appropriate standards; and

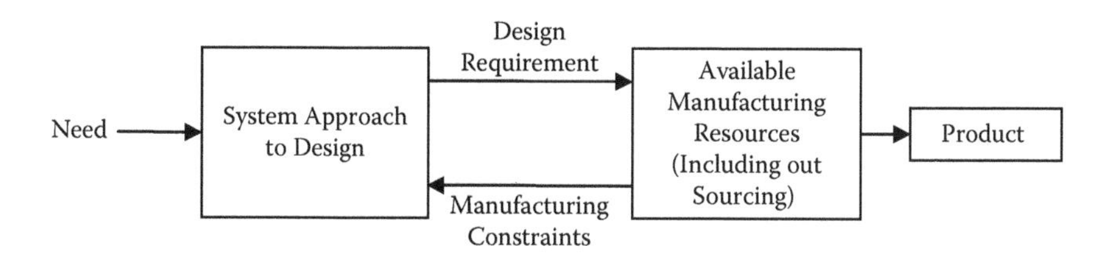

FIGURE 1.15 Design for manufacturing.

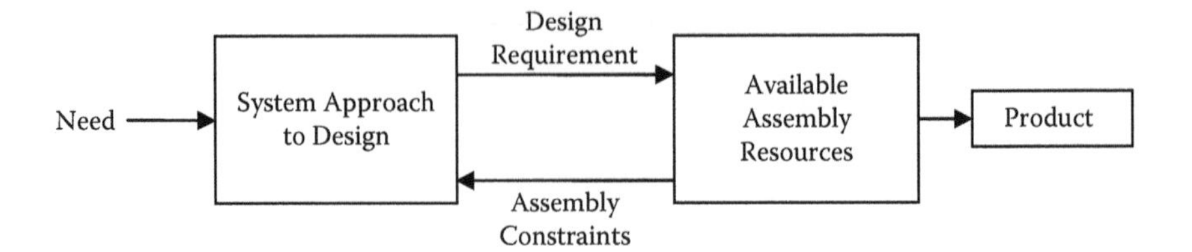

FIGURE 1.16 Design for assembly.

use of analytical, numerical, and experimental methods of analysis function, drawings for the manufacture and assembly functions, and communication between subfunctions and other function that requiring information associated with the design function. The current tools available for DFA and DFM, along with various other interfaces, allow visualization of product, which differs from the conventional approach and provides ease of predicting the cost of the product.

1.2.3 Optimization in Design

Design is repetitive by nature. Repetition may occur in the design analysis step, for DFM, for DFA, or as an exercise in cognitive ergonomics. The repetition may also be performed to optimize the design to meet the requirements of basic physical properties, to conform to certain operational requirements, to meet safety guidelines, to adjust to environmental factors, or to bring the product within certain cost limitations. Use of computers and standard information during the optimization phase expedites the whole process and brings more confidence to the resulting design.

1.2.4 The Cost Model for Design

Engineering design is predominantly concerned with wealth creation. Every design can be a "design to manufacture." This method necessitates a holistic approach, including effective cost. It involves the use of scientific principles, technical information, numeracy, synthesis, analysis, creativity, and decision making. It requires the consideration of human and environmental factors with the maximum practicable economy and efficiency. An overall cost model of design is shown in Figure 1.17.

1.3 MANUFACTURING PHILOSOPHY—CONVENTIONAL AND SYSTEM APPROACH

The subject of manufacturing arose when humans altered the geometry of a workpiece by rotating a tool to cut and remove pieces of material in the production of weapons as early as 700 B.C. In the fifteenth century, the machining of metal began.

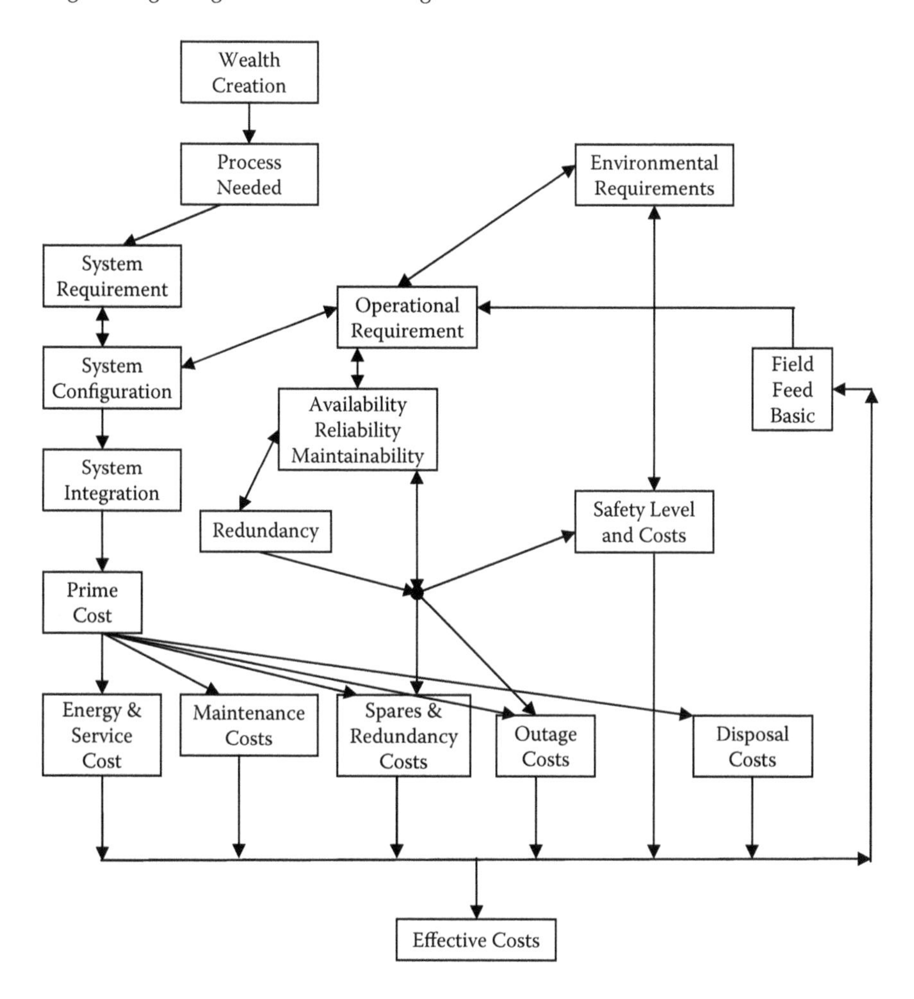

FIGURE 1.17 Cost model for design.

The problems faced by the manufacturer of the first potter wheel are similar to those faced by a manufacturer today: what to produce, how to produce, and when to produce within the available resources of the enterprise. The product made on the manufacturer's machines must be produced during a specified lead time and within certain quantity restrictions. It should be marketable and produce profit for the manufacturing enterprise.

Progression from Stone Age production methods to today's machining centers and flexible manufacturing systems shows that manufacturing is an important element in the creation of wealth. The earlier processes available to convert material into specified objects were few, so little wealth was created. This situation has been significantly changed by the development of new materials, new manufacturing processes, and by the introduction of computerized numerical control (CNC).

As was stated in the case of conventional approach to design, when contemporary tools available for manufacture of an object are not used then the approach to manufacture is regarded as the conventional approach. However, this conventional approach has a logical sequence that gives it the capability to produce the product with required functionality.

The important interface between conventional design and manufacture is implemented through the production planning and control subfunction. Figure 1.18 shows the basic structure for the conventional approach to manufacture.

The system approach to manufacture differs from the conventional approach in terms of the organization of the manufacturing and assembly area. This approach provides a better insight into the dynamics of manufacturing in general and specifically results in an efficient process with higher profit margins (Figure 1.19).

Production subfunctions such as process planning, short-term and long-term scheduling, manufacturing, quality assurance, raw-material and finished-product inventories, inventories for accessories related to a specific manufacturing facility, packaging and storage, and transportation can be analyzed both quantitatively and qualitatively by use of available mathematical techniques.

The role of computers for the analysis of the manufacturing parameters in various subfunctions of the system approach to manufacturing is paramount. The flow of information, the flow of material, and the flow of energy is discussed in detail by Yeomans, Choudary, and Ten Hagen (1986) and Alting (1994).

1.3.1 An Overview of Discrete Manufacturing Processes

Manufacturing processes are physical mechanisms used to alter the shape and form of materials. Manufacturing processes can be classified as shown in Figure 1.20. These categories of manufacturing processes are applied to a variety of engineering materials, which can be classified into nonmetals, metals, ceramics, polymers, and composites. The application of manufacturing process to an individual material is influenced by number of factors, including physical properties of the material and the lot size. A more in-depth discussion on manufacturing processes is provided in Chapter 6. A wide variety of standard information is available about the materials, processes, and the machines that perform manufacturing operation (see Chapter 2).

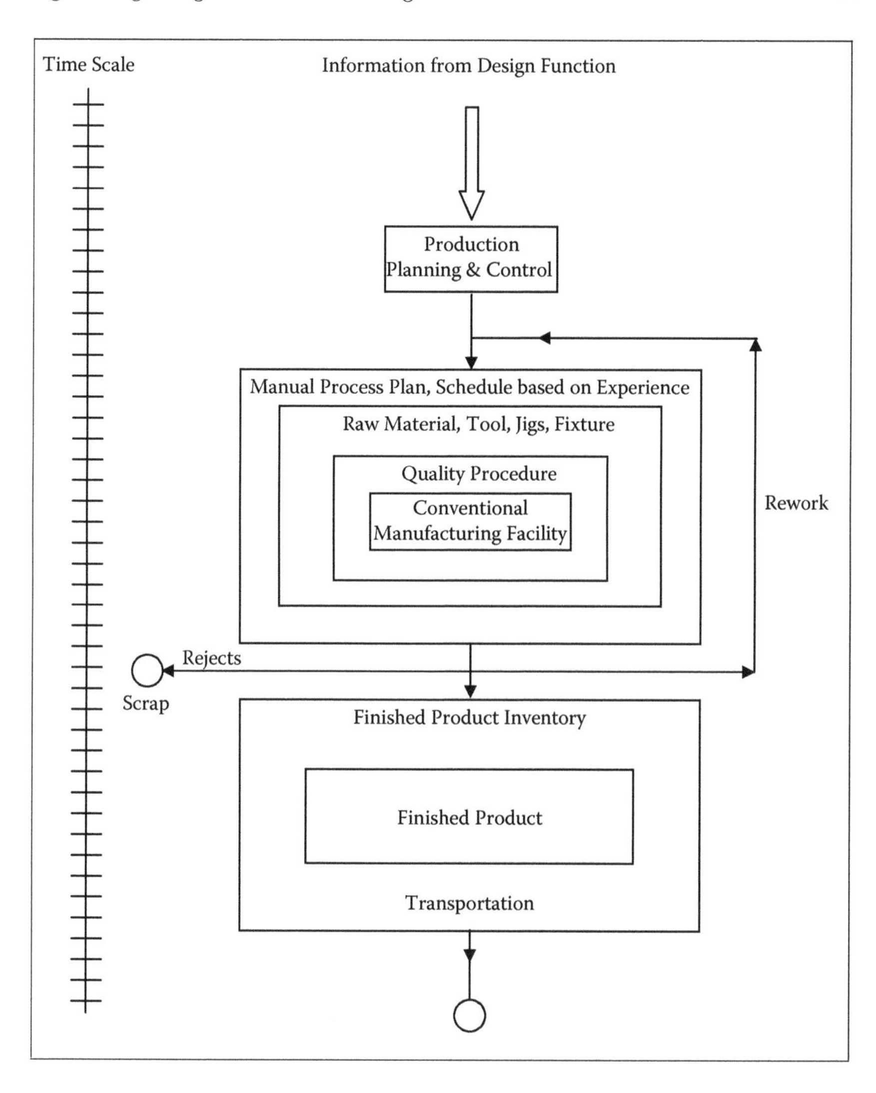

FIGURE 1.18 Conventional approach to manufacture.

1.3.2 Discrete Manufacturing Systems

The discrete manufacturing systems as defined by Chryssolouris (1992) are the following:

Job shop
Project shop
Cellular system
Flow line

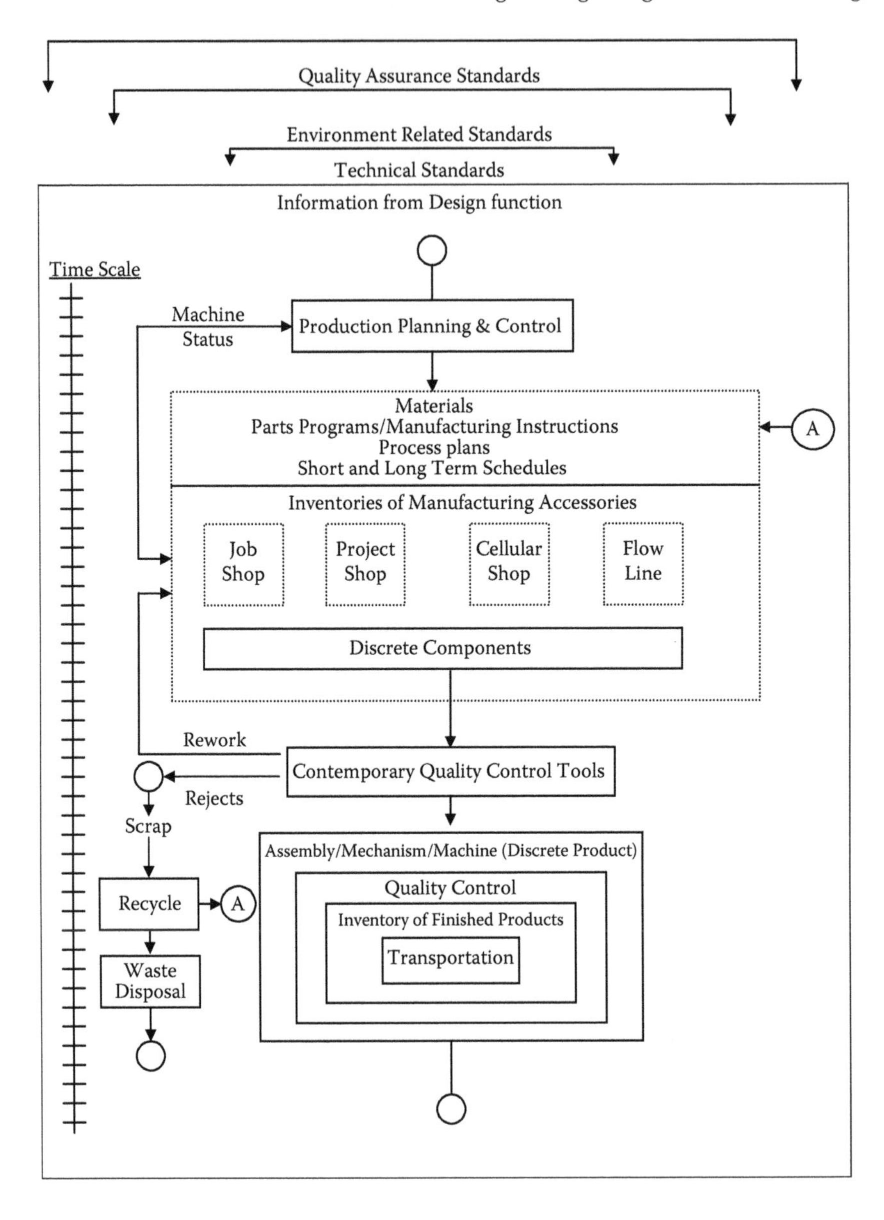

FIGURE 1.19 Contemporary system approach to manufacturing.

The discrete manufacturing system uses five main elements in their construction:

A production philosophy
The physical layout for the manufacturing
A discrete method for flow of raw material and finished product in a manufacturing system

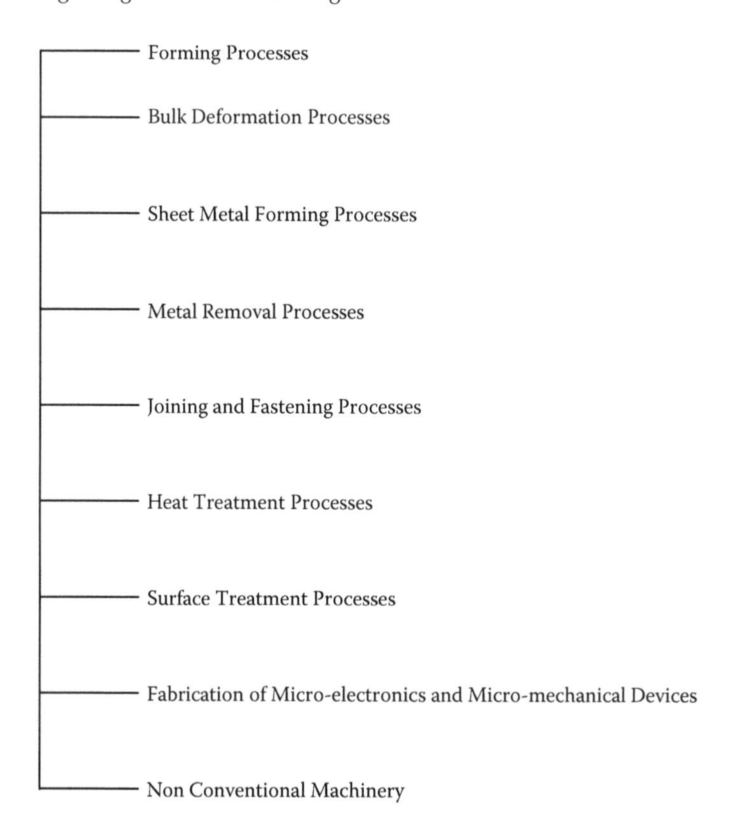

FIGURE 1.20 Process used in discrete manufacturing (adapted with permission from Kalpakjian and Schmid 2003).

A distinct method for flow of information
A method for flow of energy

Discrete manufacturing systems are described in detail in Chapter 6.

1.3.3 Production Control

Hitomi (1975) defines manufacturing as comprising essentially all the tangible operations and production as encompassing both tangible and intangible operations. The tangible operations include metal cutting, metal forming, metal deforming, joining, and other mainstream processes. On the other hand, intangible operations include services such as process planning, scheduling, and inventory management. The production control theories presented over the past decades consider both tangible and intangible factors to improve the efficiency of a system and to increase the profits of an organization. Some of the parameters used in these theories are similarity of products, flow of information, flow of material, manufacturing facilities layouts, elimination of defect, optimization of slowest links, supply of raw material, produced goods inventories, automation, and continued improvement. These theories include the following:

1. Group technology (GT)
2. Just in time (JIT)
3. Material resource planning (MRP)
4. MRPII
5. Enterprise resource planning (ERP)
6. Supply chain management
7. The theory of constraints (TOC)
8. Lean manufacturing
9. Six sigma
10. Computer-integrated manufacturing (CIM)
11. KAIZEN
12. Agile manufacturing

1.4 RELATIONSHIP BETWEEN DESIGN AND MANUFACTURING

The engineering design and manufacturing functions are the most commonly and most elaborately implemented functions in the discrete manufacturing industry. The varying nature of the relationship between engineering design and manufacturing emanates from various sources that span engineering design, manufacturing, and production functions and subfunctions. Geometry of the product is the major factor that augments these relationships.

The design cycle provides a solution to a problem in the form of geometry with or without special features. The geometry represents the product as a whole product with all its components. Design exercises such as DFM and DFA carry more detail for manufacture or assembly with this geometry. Several other functions of the production processes, such as process planning, scheduling, inventory, packaging, and transportation, are dependent on geometry as a direct or indirect parameter.

At a much earlier step, sales and marketing requires estimates of the costs that are also dependent on the geometry. These cost estimates rely on projections of use of certain types of production processes. These projections are based on the geometry of the product being marketed or sold.

With the use of computer-aided techniques in design and manufacturing, the relationship between functions and subfunctions of engineering design and manufacturing can be thoroughly studied and used to provide an efficiently executed design and manufacturing cycle. Figure 1.21 provides a schematic of the basic relation between engineering design, manufacturing, and production functions.

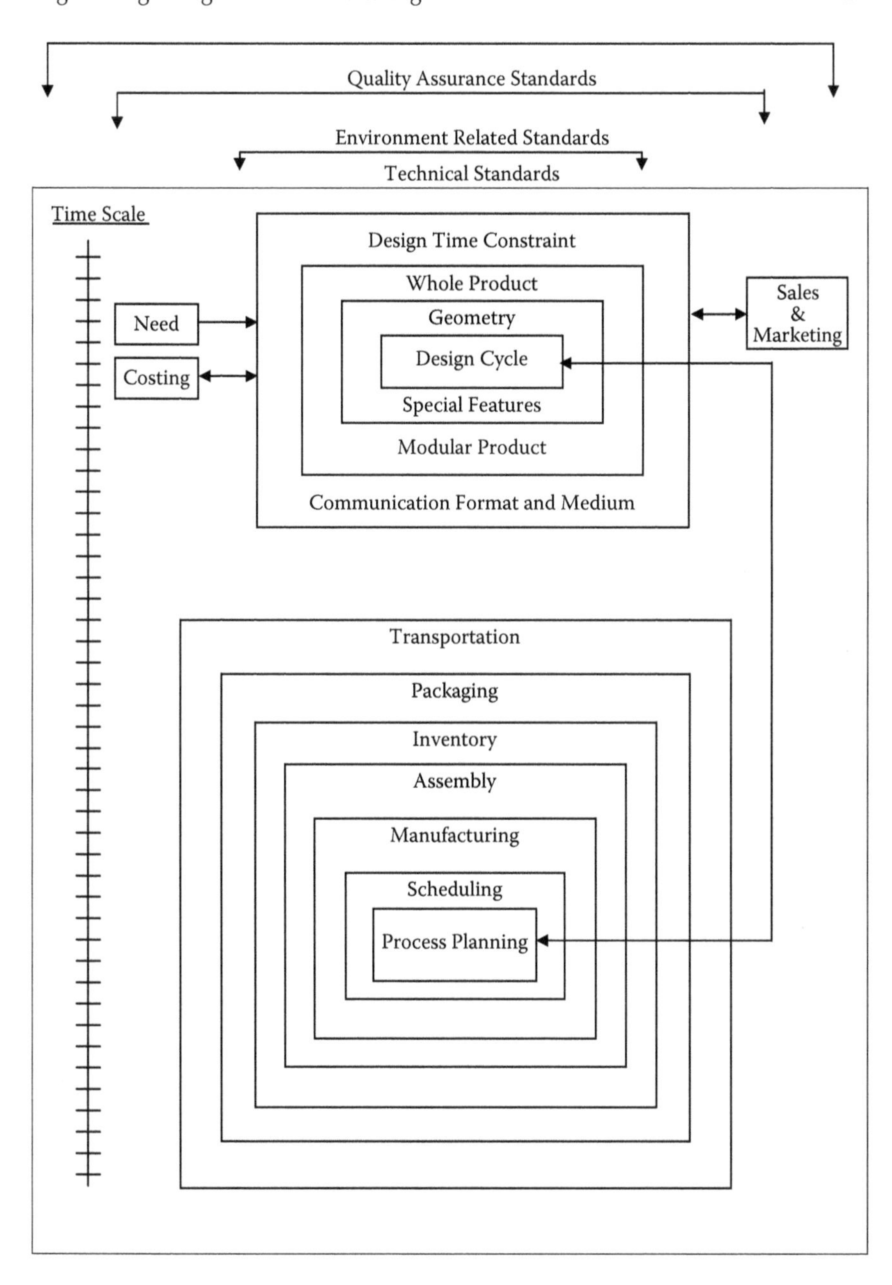

FIGURE 1.21 Relationship between design and manufacturing.

1.5 MULTIDISCIPLINARY ASPECTS OF DESIGN AND MANUFACTURING

The engineering design and manufacturing functions, especially after the Industrial Revolution, have become more complex because of the use of multidisciplinary features. The design cycle that earlier covered none or very few analytical procedures has now reached a stage in which analysis for a product may extend to areas such as those identified in Figure 1.22. Availability of materials and diversified processes in manufacturing has also allowed complex product design. Areas in which interaction between mechanical engineering design and other disciplines of applied science occurs include metallurgical, electrical, electronic, computer, control, civil, chemical and environmental engineering.

Discrete products from toys to complicated mechanisms such as machine tools now have multidisciplinary features. These features originate from the application of the control theory and frequently incorporate microprocessors, memory, various kinds of input/output devices, actuators, sensors, and transducers. Products are equipped with a graphical user interface (GUI) backed by an interpreter or a compiler.

A prime example of the extent to which multidisciplinary aspects are used in design and manufacturing is the microwave oven. Progress in design and manufacturing procedures has allowed a design for microwave thermal characteristics, control of the embedded system, aesthetic features, general mechanical and electrical engineering, and the requisite safety features.

The manufacturing function and subfunctions have remained at par with the progress in contemporary design objectives set in military and consumer products. Figure 1.22 shows the scope to which multidisciplinary design and manufacturing of discrete products are required to accommodate other branches of engineering. The design and manufacturing phase encounter more complications because these features are more advanced. Use of standards in such conditions simplifies the exercise.

1.6 USE OF COMPUTERS IN DESIGN AND MANUFACTURING

With advancement in mathematical methods that allow representation of products on computers, use of computer-aided design and computer-aided manufacturing (CIM) techniques have increased enormously. In this sector of computing, the geometry of the product plays the central role for CAD modeling, for numerical analysis, for production planning, and for manufacturing by numerical control machines.

A CIM system, which is an extension of CAD/CAM strategy, is an expensive but efficient system that extensively utilizes protocols for geometry representation, communication, cutter location data, and part programming.

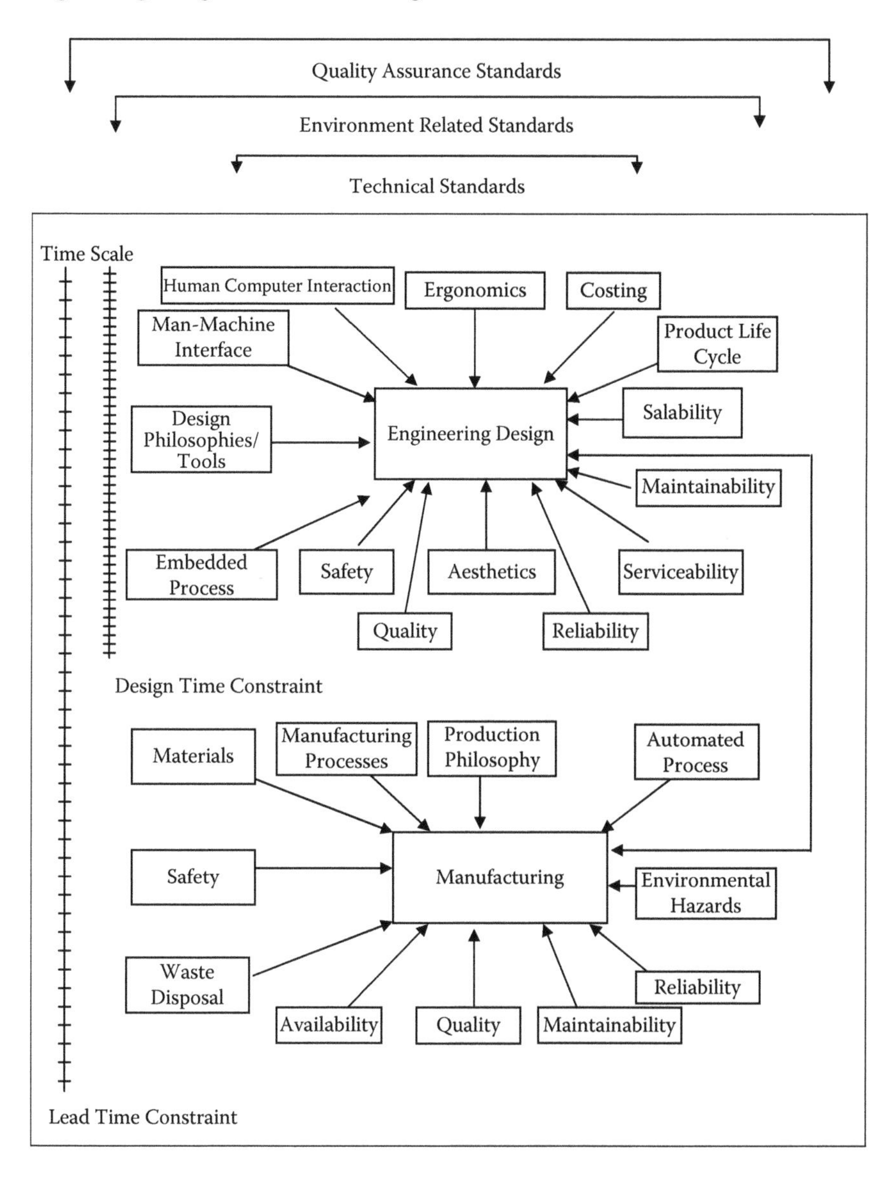

FIGURE 1.22 Multidisciplinary aspects of engineering design and manufacturing.

1.7 QUALITY AND RELIABILITY IN DESIGN AND MANUFACTURING

Quality is a measure of excellence in terms of the selection of a product for design and manufacturing, implementation of the design, manufacture of the product, and performance of the product over its normal life cycle.

Quality control is a management function whereby control is exercised over the quality of the following:

Raw materials, assemblies, produced material, and components
Services related to production
Management, production, and inspection processes

The aim of quality control is to prevent production of defective material or rendering of faulty services. Several techniques are used to implement quality control.

Quality assurance is the function whereby actions are taken to ensure that standards and procedures are adhered to and that delivered products or services meet performance requirements.
Reliability is the ability of an item to perform a required function under given conditions for a given time period.
Reliability engineering is the process of analyzing the expected or actual reliability of a product, process, or service and identifying actions to reduce failures or mitigate their effect. The overall goal of reliability engineering is to reduce repairs, to lower costs, and to maintain a company's reputation. To best meet this goal, reliability engineering should be done at all levels of design and production.
A reliability prediction is the analysis of parts and components in an effort to predict the rate at which an item will fail. A reliability prediction is one of the most common forms of reliability analysis.

The major international and national organizations that develop standards recommend general and specific standards for quality and reliability of products and processes.

1.8 DESIGN AND MANUFACTURING FOR PRODUCT LIFE CYCLE

The concept of product life cycle is becoming increasingly popular because of the recognition that survival of humankind depends upon the quality of the environment. Both machines and products have an impact on the environment. The impact of the machines is initially confined to the factory premises, whereas the impact of the products is widespread. The hazardous wastes from machines can be processed before being allowed to mix with soil, water, and air. On the other hand, thousands of different types of products are accumulated as junk or trash at the end of their useful life. Some of these products are biodegradable, but many are not.

A product designed by use of modular techniques is more appropriate for both manufacture and separation of parts during disassembly after the useful life of the product. This scheme generates economies that are different from both the conventional approach and the ordinary system approach to design and manufacturing. National, international, and professional organizations provide standards related to this comparatively new branch of engineering.

BIBLIOGRAPHY

Almgren, H., Pilot production and manufacturing start-up: The case of Volvo S80, *Int. J. Prod. Res.*, 38(17), 4577–4588, 2000.

Alting, L., *Manufacturing Engineering Processes*, Marcel Dekker, New York, 1994.

Askin, R.G. and Standridge, C.R., *Modeling and Analysis of Manufacturing Systems,* 1st ed., Wiley, New York, 1993.

Bailey, D.E., Modeling work group effectiveness in high-technology manufacturing environments, *IIE Trans.*, 32(4), 361–368, 2000.

Baker, L., Manufacturing Motors in a New World Economy. Equipment Standards, Proceedings of the Electrical/Electronics Insulation Conference, 2001, 63–67.

Bayart, M., Smart devices for manufacturing equipment, *Robotica*, 21(3), 325–333, 2003.

Chang, T. and Wysk, R.A., *An Introduction to the Automated Process Planning System,* Prentice Hall, New Jersey, 1985.

Chattha, J.A., Khan, W.A., and Qureshi B.M., Automatic Coding Facility for Product Life cycle, Proceeding R99–Recovery, Recycling, Re–Integration World Congress, Geneva, Switzerland, 1996.

Chen, D., Developing a Theory of Design through a Multidisciplinary Approach, Proceedings of the IEEE International Conference on Systems, Man and Cybernetics, 3, 618–623, 2002.

Choi, K.K., Youn, B.D., and Tang, J., Structural Durability Design Optimization and Its Reliability Assessment, Proceedings of the ASME Design Engineering Technical Conference, 2, 73–83, 2003.

Chryssolouris, G., *Manufacturing Systems—Theory and Practice*, Springer-Verlag, New York, 1992.

Cross, N., *Engineering Design Methods*: *Strategies for Product Design*, 3rd ed., Wiley, New York, 2000.

Dangayach, G.S. and Deshmukh, S.G., Manufacturing strategy: Experiences from Indian manufacturing companies, *Prod. Plann. Control*, 12(8), 775–786, 2001.

Davis, D., Integrating Engineering, Art, and Business into a Multidisciplinary Architecture Program, ASEE Annual Conference Proceedings, 2003 ASEE Annual Conference and Exposition: Staying in Tune with Engineering Education, 2003, 541–546.

De Berg, M., *Computational Geometry*, Springer-Verlag, Heidelberg, 2000.

DeGarmo, E.P, Black J.T., and Kohser R.A., *Materials and Processes in Manufacturing*, 9th ed., Wiley, New York, 2002.

Dieter, G., *Engineering Design*: *A Materials and Processing Approach*, 3rd ed., McGraw-Hill, New York, 1999.

Feldmann, K. and Rottbauer, H., Electronically networked assembly systems for global manufacturing, *J. Mater. Process. Technol.*, 107(1–3), 319–329, 2000.

Ferreira, J.C.E., Steele, J., Wysk, R.A., and Pasi, D.A., A schema for flexible equipment control in manufacturing systems, *Int. J. Advanced Manuf. Technol.*, 18(6), 410–421, 2001.

Freedenberg, P., Manufacturing technology R and D mandates: A Manhattan project, *Am. Machinist*, 147(10), 28, 2003.

Groover, M.P., *Fundamentals of Modern Manufacturing: Materials, Processes, and Systems*, 2nd ed., Wiley, New York, 2001.

Han, M. and Deng, J., Multidisciplinary Design Optimization Methods for Complex Engineering Systems, Proceedings of the International Conference on Agile Manufacturing, Advances in Agile Manufacturing, ICAM, 2003, 347–354.

Hill, T.J., *Manufacturing Strategy*, 3rd ed., McGraw-Hill/Irwin, New York, 1999.

http:// www.enchantledlearning.com/inventors

Hitomi, K., *Manufacturing System Engineering*, Taylor & Francis, London, 1975.

Kalpakjian, S. and Schmid, S.R., *Manufacturing Engineering and Technology*, 4th ed., Prentice Hall, New York, 2000.

Kalpakjian, S. and Schmid, S.R., *Manufacturing Processes for Engineering Materials,* Prentice Hall, New York, 2003.

Kron, A., Doucet, P., Masson, P., Van Hoenacker, Y., Lapointe, J., and Micheau, P., Unifying Approaches to Mechanical Engineering Design through a Multidisciplinary Effort, ASEE Annual Conference and Exposition: Vive L'ingenieur, 2002, 6415–6427.

Leaver, E.W., Process and manufacturing automation—past, present, and future, *ISA Trans.*, 26(2), 45–50, 1987.

Linde, E., Dolan, D., and Batchelder, M., Mechatronics for Multidisciplinary Teaming, ASEE Annual Conference Proceedings, 2003 ASEE Annual Conference and Exposition: Staying in Tune with Engineering Education, 2003, 6901–6910.

Lindsay, C., Bright, G., and Hippner, M., Advanced material handling system for computer integrated manufacturing, *Robotics Comp. Integrated Manuf.*, 16(6), 437–441, 2000.

Liu, Z., Jiang, S.Q., Tang, B.Y., Zhang, J.H., and Zhong, H., The Study and Realization of SCADA System in Manufacturing Enterprises, Proceedings of the World Congress on Intelligent Control and Automation, 5, 3688–3692, 2000.

Marselli, M., Lean manufacturing/six sigma, *Wire J. Int.*, 37(2), 40–49, 2004.

Mashford, K., Next generation manufacturing, *Manuf. Eng.*, 82(6), 30–34, 2003.

McMasters, J.H. and Cummings, R.M., Airplane design—Past, present, and future, *J. Aircraft*, 39(1), 10–17, 2002.

Meyers, F.E., and Stephens, M.P., *Manufacturing Facilities Design and Material Handling*, 3rd ed., Prentice Hall, New York, 2004.

Oborski, P., Man–machine interactions in advanced manufacturing systems, *Int. J. Advanced Manuf. Technol.*, 23(3–4), 227–232, 2004.

Pahl, G., Beitz, W., Wallace, K., Blessing, L., and Bauert, F., *Engineering Design: A Systematic Approach*, 2nd ed., Springer-Verlag, Heidelberg, 1996.

Parmley, R.O., *Illustrated Source Book of Mechanical Components*, McGraw-Hill, New York, 2000.

Pat, L.M., *The Principles of Materials Selection for Engineering Design*, 1st ed., Prentice Hall, New York, 1998.

Patnaik, S.N., Coroneos, R.M., Hopkins, D.A., and Lavelle, T.M., Lessons learned during solutions of multidisciplinary design optimization problems, *J. Aircraft*, 39(3), 386–393, 2002.

Pidaparti, R.M., The Art of Engineering in Capstone Design, ASEE Annual Conference Proceedings, 2004, 591–598.

Pryputniewicz, R.J., MEMS design education by case studies, *Design Eng.*, 113, 73–79, 2003.

Quesada, G., Gonzalez, M.E., and Diaz-Bernardo, R., Differences in Strategic Manufacturing Priorities among Continents: An Empirical Study, Proceedings of the Annual Meeting of the Decision Sciences Institute, 2002, 1792–1797.

Ramirez–Valdivia, M.T., Christian, P., Govande, V., and Zimmers, E.W. Jr., Design and implementation of a cellular manufacturing process: A simulation modeling approach, *Int. J. Industrial Eng. Theory Appl. Pract.*, 7(4), 281–285, 2000.

Raouf, A., Engineering Design and Natural Resources Management, 6th Saudi Engineering Conference, KFUPM, Dharan, Saudi Arabia, 2002.

Ray, S.R. and Jones, A.T., Manufacturing Interoperability, Proceedings of the 10th ISPE International Conference on Concurrent Engineering Research and Application, Enhanced Interoperable Systems, 2003, 535–540.

Rehg, J.A. and Kraebber, H.W., *Computer Integrated Manufacturing*, 3rd ed., Prentice Hall, New York, 2004.

Riis, J.O., and Johansen, J., Developing a manufacturing vision, *Prod. Plann. Control*, 14(4), 327–337, 2003.

Rolt, L.T.C., *A Short History of Machine Tools*, MIT Press, Cambridge, mass., 1961.

Rudolph, J.E., *Engineering Design*, Prentice Hall, New York, 2004.

Scallan, P. *Process Planning: The Design/Manufacture Interface,* Butterworth-Heinemann, Oxford, 2003.

Schey, J.A., *Introduction to Manufacturing Processes*, 3rd ed., McGraw-Hill, New York, 1999.

Schwartz, K.K., *Design and Wealth Creation*, IEE, U.K. 1990.

Shigley, J., Mischke, C., and Budynas R., *Mechanical Engineering Design*, 7th ed., McGraw-Hill, New York, 2003.

Sielinski, R., Microsoft in manufacturing, *Manufacturing Eng.*, 125(1), 71–82, 2000.

Sullivan, W.G., McDonald, T.N., and Van Aken, E.M., Equipment replacement decisions and lean manufacturing, *Robotics Comput. Integrated Manuf.*, 18(3–4), 255–265, 2002.

Todd, R.H., Allen, D.K., and Alting, L, *Manufacturing Processes Reference Guide*, 1st ed., Industrial Press, New York, 1994.

Tylor, E.B., *Early History of Mankind and the Development of Civilization*, University of Chicago Press, Chicago, 1964.

Vollmann, T.E., Berry, W.L., and Whybark, D.C., *Manufacturing Planning and Control Systems*, 4th ed., McGraw-Hill, New York, 1997.

Yang, H., Research and Progress on Key Technologies of Manufacturing Informationization, Proceedings of the International Conference on Agile Manufacturing, Advances in Agile Manufacturing, ICAM, 2003, 11–16.

Yeomans, R.W., Choudary, A., and Ten Hagen, P.J.W., 1986, *Design Rule for a CIM System*, North Holland Publishers, Amsterdam, 1992.

Zhang, H., Fan, W., and Wu, C., Concurrent design approach and its collaborative environment for integrated complex product development, *J. Integrated Design Process Sci.*, 8(3), 89–97, 2004.

2 Standards and Standards Organizations

2.1 FORMALIZING TECHNICAL INFORMATION AND PROCEDURES

Technical information presented as regulations, acts, standards, codes, technical specifications, frameworks, and benchmarks makes commerce possible. Following are some definitions of formalized technical information:

1. A regulation is an official rule, law, or order that states what may or may not be done or how something must be done.
2. An act is a record or statement of the decision made by a law-making or judicial body, such as Congress.
3. A standard is a level of quality or excellence that is accepted as the norm or by which actual attainments are judged.
4. A code is a system of acceptable laws and regulations that govern procedure or behavior in particular circumstances or within a particular profession.
5. A technical specification is a detailed description of a particular thing that provides the information needed to construct or repair the thing.
6. A framework is set of ideas, principles, agreements, or rules that provide the basis or outline for something that is more fully developed at a later stage.
7. A benchmark is a point of reference from which measurements may be made or something that serves as a standard by which others may be measured.

2.2 DEVELOPMENT OF STANDARDS

The development of standards is addressed by organizations that have recognized authority to establish standards. Within such organizations are standards committees known as technical committees (TCs) that perform the standard-development work. When the TC is not directly attached to the standards-developing organization but is accredited by the organization to develop a standard on its behalf, then the TC is called an accredited standard committee (ASC). When two or more organizations are involved in developing or adopting a common standard, a joint technical committee (JTC) performs the tasks. A TC, a JTC, or an ASC may consult experts in a particular field. These experts are commonly called a technical advisory group (TAG). Subcommittees may assist the main TC. Other committees with different names but similar functions may also exist.

An accredited standard development organization performs all the functions in standard development except that it does not have a charter to promulgate the standard. For this purpose it requires a formal accreditation from a standard developing organization.

Standards development is time-consuming process that may involve several parties and technologies. It is not always about developing a new consensus, but sometimes about rationalizing an existing one.

Establishment of a standard generally proceeds as follows:

1. A need for a new standard is recognized by a local, national, regional, international, or professional standards-developing organization.
2. A determination is made of whether a suitable standard is available from another standards-developing organization.
3. If available, a standard is adopted from another standards-developing organization. (See step 11 for information related to this procedure.)
4. For development of a new standard, a committee[1] is formed that may comprise the following:
    - Experts from commercial organizations who believe the proposed standard will ease related trade in general
    - Experts in the field for which the new standard is required
    - Potential users of the new standard
    - Members of the public who may be affected by the new standard
    - Experts in defining and drafting new standards
5. The composition of the TC, JTC, or ASC is determined in consultation with a TAG or other subcommittee.
6. The draft standard is made available for public reading for a specified period of time.
7. Feasible suggestions from the public are accepted, and reasons for not accepting other suggestions are stated.
8. The standard is adopted locally, nationally, regionally, or internationally for a specified period of time.
9. The standard is maintained during or after the specified adoption period through amendments.
10. The standard is withdrawn or replaced if the information contained in the standard has become obsolete or has been superceded by another standard.
11. If an existing standard is to be adopted from another organization, a JTC is formed to facilitate the introduction of the existing standard.
12. Steps 8 to 10 are followed.

[1] The standards committee may be an independent committee, such as a TC, attached to the standards-developing organization.

2.2.1 A Standard for Standards—BSI Perspective[2]

The British Standards Institute (BSI) has formally defined a procedure for standard development in the form of a standard for standards that comprises three parts:

Part 1: Guide to the Context, Aims, and General Principles (BS 0–1:1997)
Part 2: Recommendation for Committee Procedure (BS 0–2:1997)
Part 3: Specification for Structure, Drafting, and Presentation (BS 0– 3:1997)

Part 1 outlines the principles and procedural safeguards associated with standards. The use of standards in trade descriptions and contracts and how they may be involved in legislation are discussed. Definitions are given for terms such as standardization, standard, consensus, and regulations. Creation of consensus-based documents and voluntary use are compared with adherence through regulation.

Various types of standards are examined, such as vocabularies (glossaries), methods, specifications, codes of practice, guidelines, and recommendations. The contents of the types of standards are analyzed for their normative (i.e., standardizing) and informative elements.

Part 1 identifies the need for standards-developing bodies, whether national, regional, or international organizations, to form collaborative agreements. Part 1 also emphasizes the possible use of standards by a court of law, in which a judge might admit a relevant standard as evidence and consider it while pronouncing sentence.

The aim of standardization, as stated in Part 1, is to provide the following benefits:

Improvement in the quality of goods and services (i.e., their fitness for purpose)
Improvement in the quality of life (i.e., health, safety, and environment)
Efficient use of resources
Enhancement of conditions for trade

The following principles of standard development should be adhered to:

Standard should be needed or wanted
Standard should be used
Standard should be agreed on at the widest level
Standard should be impartial
Standard should be planned

[2] Permission to produce extracts from BSO parts 1, 2, 3 is granted by BSI. British standards can be obtained from BSI customer services, 389 ChisWick High Road, London W4 4AL.

As standards are developed, the following procedural safeguards should be maintained:

Compatibility with international agreements
Consensus
Balanced participation
Transparency
Rules for drafting and presentation
Copyright

The following general information provided by standards is outlined in Part 1:

1. Voluntary status of standard and when use of standard became binding
2. Trade description through manufacturer's or supplier's declaration of conformity
3. Trade description through third-party attestation of conformity
4. Suitability of standards in defining contracts
5. Suitability of standards for public-sector procurement contracts
6. International and regional procurement requirements such as World Trade Organization agreement on government procurement or European Union directives
7. Reference to standards in regulations
8. Exclusive or indicative reference
9. Intergovernmental endorsement
10. Regional legislation

Part 1 also describes the conditions for the development and use of new deliverables in international, regional, and national domains.

Part 2 addresses the organizational aspects of standards development at the national, regional, and international level. The requirements for the organization of BSI technical committees, subcommittees, and panels and the organization of international and European technical committee are described in detail. A typical chain of responsibilities in a BSI committee structure and the sequence of activities in the development of national, international, and European standards are presented. Part 2 also discusses details of the standard-development process and the maintenance of recommendations, including issuing amendments for a standard.

Part 3 outlines the structure specification, drafting, and presentation aspects of developing a standard before it becomes a published document. Consideration is given to factors such as safety of the standard to be proposed, whether an identical standard has already been published, and copyright of the document. A list of normative references to other British, European, and International standards completes Part 3.

All major organizations that develop standards use a variation of procedures described by BSI. Some of these organizations and their Web sites are listed below:

The International Organization for Standardization
http://www.iso.org/iso/en/stdsdevelopment/whowhenhow/how.html
The Institute of Electrical and Electronics Engineers
http://standards.ieee.org/resources/
The National Information Standards Organization (NISO)
http://www.niso.org/creating/index.html
Deutsches Institut für Normung e. V. (DIN)
http://www.dke.de/DKE_en/Electrotechnical+Standardization+in+Germany/Development+of+a+DIN+standard++-.htm

In the industrialized societies, professional organizations have defined standards that have become nationally or internationally accepted in their respective domains. Standard-developing organizations should follow the procedures detailed in BS 0–1:1997, BS 0–2:1997, and BS 0–3:1997. However, the users of the standards have the responsibility of selecting and applying them appropriately.

2.3 STANDARDS-DEVELOPING ORGANIZATIONS

Many organizations are currently engaged in the development of standards related to engineering design and manufacture of discrete products. Other organizations provide standards related to safety, occupational health, illumination, and building conditions. This section is a survey of major standards-developing organizations. A format is used that allows the reader to quickly find information related to the activities, associations, official languages and Web sites of these organizations. More detailed and updated information is available at the Web sites of the individual organizations.

2.3.1 The International Organization for Standardization and Its Affiliates

Organization name: International Organization for Standardization (ISO)
Web address: http://www.iso.ch
Domain: General standards
About ISO: See the ISO Web site.
Related organizations: See the ISO Web site.
Official languages: English, French

Organization name: ISO Member Bodies

Web address: http://www.iso.org/iso/en/aboutiso/isomembers/Member-List. MemberSummary?MEMBERCODE=10

The hyperlink at the ISO Web address as given above automatically connects to the list of member organizations. Each respective standards organization of the member country is connected through a corresponding hyperlink.

Organization name: ISO Correspondent Members

Web address: http://www.iso.org/iso/en/aboutiso/isomembers/Correspon-dent MemberList.MemberSummary?MEMBERCODE=20

The hyperlink at the ISO Web address as given above automatically connects to the list of correspondent members. Each respective standards organization of the correspondent member is connected through its specific hyperlink.

Organization name: ISO Subscriber Members

Web address: http://www.iso.org/iso/en/aboutiso/isomembers/Subscriber MemberList.MemberSummary?MEMBERCODE=30

The hyperlink at the ISO Web address as given above automatically connects to the list of subscriber members. Each respective standards organization of the subscriber member is connected through its respective hyperlink.

2.3.2 Regional Standards-Developing Organizations

Organization name: International Electrotechnical Commission (IEC)[3]

Web address: http://www.iec.ch

Domain: Electrical engineering, electronics, and related technologies standards

About IEC: The IEC is the leading global organization that prepares and publishes international standards for all electrical, electronic, and related technologies. These publications serve as a basis for national standardization and as references for drafting international tenders and contracts. The objectives of IEC are to following:

- To meet the requirements of the global market efficiently
- To ensure primacy and maximum worldwide use of its standards and conformity assessment schemes

[3] Reproduced with the permission of International Electrotechnical Commission (IEC).

- To assess and improve the quality of products and services covered by its standards
- To establish the conditions for the interoperability of complex systems
- To increase the efficiency of industrial processes
- To contribute to the improvement of human health and safety
- To contribute to the protection of the environment

Related organizations: The IEC works closely with international and regional partners such as the International Organization for Standardization (ISO), the International Telecommunication Union (ITU), the World Health Organization (WHO), the International Labor Office (ILO), United Nations Economic Commission for Europe (UNECE), the International Council on Large Electric Systems (CIGRE), the International Maritime Organization (IMO), the International Organization of Legal Metrology (OIML), the Union of the Electricity Industry (EURELECTRIC), the International Federation of Standards Users (IFAN), the European Committee for Electrotechnical Standardization (CENELEC) The Pan American Standards Commission (COPANT), the Euro Asian Interstate Council for Standardization (EASC), and the European Telecommunication Standards Institute (ETSI).
Official languages: English, French

Organization name: International Committee for Information Technology Standards (NCITS)
Web Address: www.ncits.org
Domain: Information technology
About NCITS: See the NCTIS Web site
Related organizations: See the NCTIS Web site
Official language: English

Organization name: International Center for Standards Research (ICSR)
Web address: http://www.standardsresearch.org/
Domain: Standards and standardization
About ICSR: See the ICSR Web site
Related organizations: See the ICSR Web site
Official language: English

Organization name: European Committee for Standardization (CEN)
Web address: http:hr//www.cenorm.be
Domain: Engineering and general
About CEN: See the CEN Web site
Related organizations: See the CEN Web site
Official language: English

Organization name: European Committee for Electrotechnical Standardization (CENELEC)[4]

Web address: http://www.cenelec.be

Domain: Electrical engineering, electronics, and system engineering

About CENELEC: CENELEC was created in 1973 as a nonprofit technical organization under Belgian law and composed of the National Electrotechnical Committees of 23 European countries. In addition, 12 national committees from Central and Eastern Europe are participating in CENELEC with Affiliate status. Their ultimate goal as affiliates is to gain full membership to CENELEC. The mission of CENELEC is to prepare voluntary electrotechnical standards that help develop the Single European Market/European Economic Area for electrical and electronic goods and services, to remove barriers to trade, to create new markets, and cut compliance costs.

Related organizations: The European Association of Electrical Contractors, the European Committee of Electrical Installation Equipment Manufacturers, and the European Association for the Promotion of Cogeneration

Official language: English

Organization name: European Telecommunications Standards Institute (ETSI)

Web address: http://www.etsi.org

Domain: Telecommunication

About ETSI: See the ETSI Web site

Related organizations: See the ETSI Web site

Official language: English

Organization name: Ecma International (formerly the European Computer Manufacturers Association)[5]

Web address: http://www.ecma-international.org

Domain: Information and communication technology and consumer electronics

About Ecma International: Ecma International is an industry association founded in 1961 and dedicated to the standardization of information and communication technology (ICT) systems, and consumer electronics. The aims of Ecma International are the following:

- To develop cooperation among the appropriate national, European, and international organizations to facilitate and standardize the use of ICT systems

[4] Reproduced with the permission of European Committee for Electrotechnical Standardization (CENELEC).

[5] Reproduced with the permission of Ecma International (until 1994 Ecma stood for European Computer Manufacturers Association).

- To encourage the correct use of standards by influencing the environment in which they are applied
- To publish standards and technical reports in electronic and printed form, free of charge, without restrictions for both members and nonmembers

Related organizations: ISO, IEC, and ISO/IEC JTC 1: Information Technology, ITU (International Telecommunication Union)
European Telecommunications Standards Institute (ETSI), CENELEC: European Electrotechnical Committee, and the National Institute of Standards and Technology (NIST) in the U.S.
Official language: English

Organization name: European Association of Aerospace Industry (AECMA)
Web address: http://www.aecma.org
Domain: Aeronautics and astronautics
About AECMA: See the AECMA Web site
Related organizations: See the AECMA Web site
Official language: English

Organization name: Council for Harmonization of Electrotechnical Standardization of the Nations of the Americas (CANENA)[6]
Web address: http://www.canena.org
Domain: System engineering
About CANENA: CANENA was founded in 1992 to foster the harmonization of electrotechnical product standards, conformity-assessment test requirements, and electrical codes. The ultimate objective of CANENA is to have one standard for a product for all of the Western hemisphere, submit the product to a conformance-assessment testing laboratory in one of the countries, and, upon successful completion of the testing, receive listings in all countries.
Related organizations: Not listed
Official language: English

Organization name: Pan American Standards Commission (COPANT)[7]
Web address: http://www.copant.org
Domain: Technical standardization
About COPANT: COPANT is a civil, nonprofit association. It has complete operational autonomy and is of unlimited duration. The basic objectives of COPANT are to promote the development of technical standardization and related activities in its member countries with the aim of promoting

[6] Reproduced with the permission of Council for Harmonization of Electrotechnical Standardization of the Nations of the Americas (CANENA).
[7] Reproduced with the permission of Pan American Standards Commission (COPANT).

their industrial, scientific, and technological development to facilitate the exchange of goods and the provision of services and foster cooperation in intellectual, scientific, and social fields.

Related organizations: The International Organization for Standardization (ISO) and the International Electrotechnical Commission (IEC)

Official languages: Spanish, Portuguese, and English.

Organization name: Pacific Area Standards Congress (PASC)[8]

Web address: http://www.pascnet.org

Domain: System technology

About PASC: The importance of international standardization to trade and commerce is recognized throughout the world. Countries on the Pacific Rim agree on the need for a forum to strengthen international standardization programs of the International Organization for Standardization (ISO) and the International Electrotechnical Commission (IEC) and to improve the ability of Pacific Rim standards organizations to participate in these programs effectively.

Related organizations: the Asia Pacific Laboratory Accreditation Corporation (APLAC), the Asia Pacific Metrology Program (APMP), the Asia Pacific Legal Metrology Forum (APLMF), and the (PAC)

Official language: English

2.3.3 Major National Standards-Developing Organizations

Organization name: American National Standards Institute (ANSI)

Web address: http://www.ansi.org

Domain: General

About ANSI: See the ANSI Web site

Related organizations: See the ANSI Web site

Official language: English

Organization name: British Standards Institute (BSI)

Web address: http://www.bsi-global.com/

Domain: Standards

About BSI: See the BSI Web site

Related organizations: See the BSI Web site

Official language: English

Organization Name: Canadian General Standards Board (CGSB)

Web address: http://w3.pwgsc.gc.ca/cgsb

Domain: Standards

About CGSB: See the CGSB Web site

[8] Reproduced with the permission of Pacific Area Standards Congress (PASC).

Related organizations: See the CGSB Web site
Official language: English

Organization name: China Standardization Administration (SAC)
Web address: http://www.sac.gov.cn
About SAC: See the SAC Web site
Related organizations: See the SAC Web site
Official language: Chinese and English

Organization name: DIN Deutsches Institut für Normung e. V. (DIN)[9]
Web address: www.din.de
Domain: Standards
About DIN: DIN, the German Institute for Standardization, is a registered association founded in 1917. Its head office is in Berlin. Since 1975, it has been recognized by the German government as the national standards body and represents German interests at the international and European level. The mission of DIN is to determine the state of the art in technology, document the results, and make these results available to the public.
Related organizations: The ISO and the CEN
Official languages: German and English (for selected documents)

Organization name: Association Française de Normalisation (AFNOR)[10]
Web address: http://www.afnor.fr/
Domain: General standards
About AFNOR: The AFNOR Group is composed of an association and of two commercially oriented subsidiaries. It is the French National Standardization body. AFNOR was founded in 1926 as a state-approved but private nonprofit organization and is placed under the supervision of the Ministry of Industry. AFNOR works in collaboration with the trade organizations and with a large number of national and regional partners.
Related organizations: The ISO and the CEN
Official languages: French and English

Organization name: Japanese Industrial Standard Committee (JISC)
Web address: http://www.jisc.go.jp/eng/
Domain: General
About JISC: See the JISC Web site
Related organizations: See the JISC Web site
Official language: Japanese and English

Organization name: State Committee of Russian Federation for Standardization, Metrology and Certification (GOST-R)
Web address: http://www.gost.ru/sls/gost.nsf

[9] Reproduced with the permission of DIN Deutsches Institut für Normung e. V. (DIN).
[10] Reproduced with the permission of AFNOR (Association Française de Normalisation).

Domain: Engineering and general
About GOST-R: See the GOST-R Web site
Related organizations: See the GOST-R Web site
Official languages: Russian and English

2.3.4 Professional Standards Organizations Active in Developing Standards

Organization name: Federal Emergency Management Agency (FEMA)
Web address: www.fema.gov
About FEMA: See the FEMA Web site
Related organizations: See the FEMA Web site
Official language: English

Organization name: International Code Council (ICC)
Web address: http://www.iccsafe.org/
Domain: Building safety
About ICC: See the ICC Web site
Related organizations: See the ICC Web Site
Official language: English

Organization name: Department of Energy (DOE), Office of Energy Efficiency and Renewable Energy
Web address: http://www.energycodes.gov/
Domain: Building energy codes
About DOE Office of Energy Efficiency and Renewable Energy: See the DOE Web site
Related organization: See the DOE Web site
Official language: English

Organization name: Occupational Health and Safety Administration (OSHA)
Web address: http://www.osha.gov/
Domain: Occupational Health and Safety
About OSHA: See the OSHA Web site
Related organizations: See the OSHA Web site
Official languages: English and Spanish

Organization name: NSF International (NSF)[11]
Web address: http://www.nsf.org/
Domain: Public health and safety
About NSF International: NSF International is a not-for-profit, nongovernmental organization that provides standards development, product

[11] Reproduced with the permission of NSF International (NSF).

certification, education, and risk-management for public health and safety. NSF International develops national standards, provides learning opportunities through its Center for Public Health Education, and provides third-party conformity assessment services while representing the interests of all stakeholders. The primary stakeholder groups include industry, the regulatory community, and the public at large. NSF International is widely recognized for its scientific and technical expertise in the health and environmental sciences. Its professional staff includes engineers, chemists, toxicologists, and environmental health professionals with broad experience both in public and private organizations.

Related organizations: The American National Standards Institute (ANSI), the Occupational Safety and Health Administration (OSHA), and Standard Council of Canada (SCC)

Official language: English

Organization name: Health and Safety Executive (HSE)[12]

Web address: www.hse.gov.uk/

Domain: Occupational health and safety

About HSE: Britain's Health and Safety Commission (HSC) and the Health and Safety Executive (HSE) are responsible for the management of almost all the risks to health and safety arising from work activity in Britain. The mission of HSE is to protect people's health and safety by ensuring that risks in the changing workplace are properly controlled. HSE enforces health and safety regulations in nuclear installations, mines, factories, farms, hospitals, schools, and offshore gas and oil installations and monitors the safety of the gas grid, the railways, and the movement of dangerous goods and substances. Local authorities are responsible to HSC for enforcement in offices, shops, and other parts of the services sector. Dedicated teams in the policy, technical, and operational divisions of HSE are responsible for negotiating and implementing specific directives, standards and conventions, via domestic legislation. HSE has an international branch with a wide range of responsibilities.

Related organizations: The Directorate General of the Commission, the European Agency for Safety and Health at Work, Eurostat, and European Committee for Standardization

Official language: English

Organization name: Health Level Seven (HL7)[13]

Web address: http://www.hl7.org

Domain: Clinical and administrative health-care data

About HL7: Health Level Seven is one of several ANSI-accredited, standards-developing organizations (SDOs) that operate in the health-care

[12] Reproduced with the permission of Health and Safety Executive (HSE).

[13] Reproduced with the permission of Health Level Seven (HL7).

arena to provide standards for the exchange, management, and integration of data that support clinical patient care and the management, delivery, and evaluation of health-care services.

Related organizations: The American Dental Association (ADA), the American Nursing Informatics Association (ANIA), the American Society for Testing Materials International (ASTM)

Official language: English

Organization name: American Society of Safety Engineers (ASSE)

Web address: www.asse.org

Domain: Industrial engineering

About ASSE: See the ASSE Web site

Related organizations: See the ASSE Web site

Official language: English

Organization name: Safety Equipment Institute (SEI)[14]

Web address: http://www.seinet.org/

Domain: Safety equipment

About SEI: The SEI, headquartered in McLean, Virginia is a private, nonprofit organization established in 1981 to administer the first nongovernmental, third-party certification programs to test and certify a broad range of safety equipment products. The American National Standards Institute (ANSI) in accordance with ISO Guide 65, General Requirements for Bodies Operating Product Certification Systems, accredits SEI certification programs. As a result, SEI complies with 10 ISO guides that pertain to product testing, inspection, and certification. The purpose of the SEI certification programs is to assist government agencies, along with users and manufacturers of safety equipment, in meeting their mutual goals of protecting those who use safety equipment in a work or nonwork environment.

Related organizations: Not listed

Official language: English

Organization name: International Commission on Illumination (CIE)[15]

Web address: http://www.cie.co.at

Domain: Standards related to light and lighting

About CIE: The CIE is devoted to worldwide cooperation and the exchange of information on all matters relating to the science and art of light and lighting, color and vision, and image technology. The CIE is an independent, nonprofit organization with strong technical, scientific, and cultural foundations. The CIE provides a unique bridge between fundamental

[14] Reproduced with the permission of Safety Equipment Institute (SEI).

[15] Reproduced with the permission of International Commission on Illumination (CIE).

research and application. Its technical committees collectively incorporate a broad, international base of scientific knowledge and, together with the lighting application community, forge solutions to the problems encountered by all of our CIE partners. The vision of the CIE is to study visual responses to light and to establish standards of response functions, models, and procedures of specification relevant to photometry, colorimetric, color rendering, visual performance, and visual assessment of light and lighting.

Related organizations: The Indian Society of Lighting Engineers, the Institution of Lighting Engineers, U.K., and the Inter-Society Color Council (U.S.)

Official languages: English, French, and German

Organization name: National Fire Protection Association (NFPA)

Web address: www.nfpa.org

Domain: Fire safety

About NFPA: See the NFPA Web site

Related organizations: See the NFPA Web site

Official language: English

Organization name: Human Factors and Ergonomics Society (HFES)

Web address: http://www.hfes.org

Domain: Industrial engineering

About HFES: See the HFES Web site

Related organizations: See the HFES Web site

Official language: English

Organization name: United States Environment Protection Agency (EPA)

Web address: http://www.epa.gov/

Domain: Environment protection

About EPA: See the EPA Web site

Related organizations: See the EPA Web site

Official languages: English and Spanish

Organization name: International Council on Systems Engineering (INCOSE)[16]

Web address: http://www.incose.org

Domain: System engineering

About INCOSE: INCOSE fosters the definition, understanding, and practice of world-class systems engineering in industry, academia, and government. INCOSE is an international organization formed to develop, nurture, and enhance the interdisciplinary approach and means to enable

[16] Reproduced with the permission of International Council on Systems Engineering (INCOSE).

the realization of successful systems. INCOSE works with industry, academia, and government to accomplish the following:

- Provide a focal point for dissemination of systems engineering knowledge
- Promote collaboration in systems engineering education and research
- Assure the establishment and publication of professional standards, guidelines, and handbooks for the practice of systems engineering
- Encourage governmental and industrial support for research and educational programs that will improve the systems-engineering process and its practice

The goals of INCOSE are the following:

- To provide a focal point for dissemination of systems engineering knowledge
- To promote collaboration in systems engineering education and research
- To assure the establishment of professional standards for integrity in the practice of systems engineering
- To improve the professional status of all persons engaged in the practice of systems engineering
- To encourage governmental and industrial support for research and educational programs that will improve the systems-engineering process and its practice.

The objectives of INCOSE are the following:

- To encourages conferences, workshops, seminars, and courses and sponsor or cosponsor such events as appropriate
- To provide its members with a membership listing and newsletter and initiate bulletins, technical journals, and electronic bulletin boards, when feasible, to improve the dissemination of the systems engineering knowledge base
- To increase the funding of research and educational activities that enhance the practice of systems engineering

Related organizations: Not listed
Official language: English

Organization name: National Institute of Standards and Technology (NIST)[17]
Web address: http://www.nist.gov
Domain: Engineering and technology
About NIST: NIST is a nonregulatory federal agency within the United States government founded in 1901 to develop and promote measurement, standards, and technology to enhance productivity, facilitate trade,

[17] Reproduced with the permission of National Institute of Standards and Technology (NIST).

and improve the quality of life. NIST strives to be the global leader in measurement and enabling technology.
Related organizations: The American Chemical Society (ACS), the American Institute of Physics (AIP), the American Nuclear Society, and the Association of Computing Machinery (ACM)
Official language: English

Organization name: United States Metric Association (USMA)
Web address: http://lamar.colostate.edu/~hillger/
Domain: International system of units
About USMA: See the USMA Web site
Related organizations: See the USMA Web site
Official language: English

Organization name: Western Wood Products Association (WWPA)[18]
Web Address: www.wwpa.org
Domain: Specifications related to wood
About WWPA: WWPA is a trade association that represents softwood lumber manufacturers in the 12 western states, from the Canadian border south to Mexico and from the West Coast to the Black Hills of South Dakota. The Association also provides services in Alaska. WWPA mills produce lumber from western softwood species, including Douglas fir, western larch, western hemlock, true firs, Engelmann spruce, ponderosa pine, lodge pole pine, sugar pine, Idaho white pine, western/inland red cedar and incense cedar. The products manufactured from these species include structural lumber (dimension products used in construction), appearance lumber (selects, finish, and common board graded for their aesthetic qualities), and factory lumber (shop products that are remanufactured into components for doors, windows, molding, and cabinets.)
Related organizations: The American International Forest Products, Inc., the American Wood Council, and the American Forest and Paper Association
Official language: English

Organization name: Aluminum Association, Inc (AA)[19]
Web address: http://www.aluminum.org
Domain: Aluminum and aluminum alloys
About AA: AA is the trade association for producers of primary aluminum, recyclers, and producers of semifabricated aluminum products, as well as suppliers to the industry. AA advances its members' interests by aligning its actions and services with a changing global business environment. AA provides value to its membership through its leadership and services in aggressively promoting the growth of the aluminum industry.

[18] Reproduced with the permission of Western Wood Products Association (WWPA).
[19] Reproduced with the permission of Aluminum Association, Inc (AA).

Related organization: The Economic Strategy Institute (ESI), the European Aluminum Association (EAA), and the Australian Aluminum Council limited (AACL)
Official language: English

Organization name: American Institute of Steel Construction (AISC)[20]
Web address: www.aisc.org
Domain: Mechanical engineering, materials
About AISC: AISC is a not-for-profit technical institute and trade association established in 1921 to serve the structural-steel design community and construction industry in the United States. The mission of the AISC is to make structural steel the material of choice by being the leader in structural-steel–related technical and market-building activities, including: specification and code development, research, education, technical assistance, quality certification, standardization, and market development. The AISC has provided timely and reliable information and service to the steel-construction industry for more than 80 years.
Related organizations: The Metal Service Center Institute (MSCI)
Official language: English

Organization Name: Semiconductor Equipment and Materials International (SEMI)
Web address: http://www.semi.org/
Domain: Semiconductors
About SEMI: See the SEMI Web site
Related organizations: See the SEMI Web site
Official language: English

Organization name: Institute of Nuclear Materials Management (INMM)
Web address: www.inmm.org
About INMM: See the INMM Web site
Related organizations: See the INMM Web site
Official language: English and Spanish

Organization name: American Composites Manufacturing Association (ACMA)
Web address: www.acmanet.org/
Domain: Composites
About ACMA: See the ACMA Web site
Related organizations: See the ACMA Web site
Official language: English

[20] Reproduced with the permission of American Institute of Steel Construction (AISC).

Organization name: ASTM International
Web address: http://www.astm.org
Domain: Materials, mechanical engineering, and manufacturing engineering
About ASTM: See the ASTM Web site
Related organizations: See the ASTM Web site
Official language: English

Organization name: ASM International (ASM)
Web address: www.asm.com
Domain: Materials
About ASM: See the ASM Web site
Official language: English

Organization name: Iron Casting Research Institute (ICRI)
Web address: www.ironcasting.org
Domain: Iron casting
About ICRI: See the ICRI Web site
Related organizations: See the ICRI Web site
Official language: English

Organization name: Metal Powder Industries Federations (MPIF)[21]
Web address: www.mpif.org
Domain: Mechanical engineering and metallurgy
About MPIF: MPIF is a not-for-profit trade association formed by the powder metallurgy industry to promote the advancement of the metal-powder producing and consuming industries.
Related organizations: The APMI and the ISO
Official language: English

Organization name: Metal Treating Institute (MTI)
Web address: http://www.metaltreat.com/
Domain: Metal treatment
About MTI: See the MTI Web site
Related organizations: See the MTI Web site
Official language: English

Organization name: ASME[22]
Web address: www.asme.org
Domain: Mechanical engineering
About ASME: Founded in 1880 as the American Society of Mechanical Engineers, today's ASME is a 120,000-member professional organization focused on technical, educational, and research issues of the engineering

[21] Reproduced with the permission of Metal Powder Industries Federations (MPIF).
[22] Reproduced with the permission of ASME.

and technology community. ASME conducts one of the world's largest technical publishing operations, holds numerous technical conferences worldwide, and offers hundreds of professional development courses each year. ASME sets internationally recognized industrial and manufacturing codes and standards that enhance public safety. The work of the society is performed by its member-elected Board of Governors and through its five Councils, 44 Boards, and hundreds of Committees in 13 regions throughout the world. A combined 600 sections and student sections serve ASME's worldwide membership. ASME promotes the art, science, and practice of mechanical and multidisciplinary engineering and allied sciences to diverse communities throughout the world. The mission the organization is to promote and enhance the technical competency and professional well-being of its members and, through quality programs and activities in mechanical engineering, enable its practitioners to contribute to the well-being of humankind.

Related organizations: Not listed

Official language: English

Organization name: American Bearing Manufacturing Association (ABMA)[23]

Web address: http://www.abma-dc.org

Domain: Mechanical engineering

About ABMA: The American Bearing Manufacturers Association (ABMA) is a nonprofit association that consists of American manufacturers of antifriction bearings, spherical-plain bearings, or major components thereof. The purpose of ABMA is to define national and international standards for bearing products and maintain bearing industry statistics. The ABMA has become the collective voice of the American bearing industry, and it influences government policies and international trade. ABMA member companies manufacture 85% of the bearings produced in the United States.

Related organizations: The American Gear Manufacturing Association (AGMA), the Association for Manufacturing Technology (AMT), the Fluid Sealing Association (FSA), the National Association of Manufacturing (NAM), and the International Trade Administration (ITA)

Official language: English

Organization name: American Gear Manufacturing Association (AGMA)

Web address: www.agma.org

Domain: Mechanical engineering

About AGMA: See the AGMA Web site

[23] Reproduced with the permission of American Bearing Manufacturing Association (ABMA).

Related organizations: See the AGMA Web site
Official language: English

Organization name: Mechanical Power Transmission Association (MPTA)
Web address: www.mpta.org
Domain: Mechanical power transmission
About MPTA: See the MPTA Web site
Related organization: See the MPTA Web site
Official language: English

Organization name: Fluid Controls Institute (FCI)
Web address: http://www.fluidcontrolsinstitute.org/
Domain: Hydraulic and pneumatic components and equipment
About FCI: See the FCI Web site
Related organizations: See the FCI Web site
Official language: English

Organization name: Hydraulic Institute (HI)
Web address: http://www.pumps.org/
Domain: Pumping equipment
About HI: See the HI Web site
Related organizations: See the HI Web site
Official language: English

Organization name: Industrial Fasteners Institute (IFI)
Web address: http://www.industrial-fasteners.org/
Domain: Temporary fasteners
About IFI: See the IFI Web site
Related organizations: See the IFI Web site
Official language: English

Organization name: National Fluid Power Association (NFPA)[24]
Web address: www.nfpa.com
Domain: Mechanical engineering
About NFPA: Since its founding in 1953, NFPA has provided a forum for the fluid power industry. The association promotes fluid power technology and applications, educates the marketplace to its benefits, representing the industry to government and to the broader public both nationally and internationally, develops design and performance standards, and broadens the managerial, manufacturing, engineering, and marketing skills of members' employees.

[24] Reproduced with the permission of National Fluid Power Association (NFPA).

Related organizations: The American Society of Mechanical Engineering (ASME), the American Society for Testing and Materials (ASTM), the Deutsches Institute für Normung e. V. (DIN), the Canadian Fluid Power Association (CFPA), and the British Fluid Power Association (BFPA)
Official language: English

Organization name: American Welding Society (AWS)
Web address: www.aws.org
Domain: Mechanical engineering
About AWS: See the AWS Web site
Related organizations: See the AWS Web site
Official language: English

Organization name: The Instrumentation, Systems, and Automation Society (ISA)[25]
Web address: www.isa.org
Domain: Instrumentation, automation, measurement, and control
About ISA: Founded in 1945 as a nonprofit, educational organization, ISA has expanded its technical and geographical reach to become a resource for 33,000 members and thousands of other professionals and practitioners in more than 110 countries around the world. The society fosters advancement in the theory, design, manufacture, and use of sensors, instruments, computers, and systems for automation in a wide variety of applications. In addition to hosting the largest conferences and exhibitions for automation in the Western Hemisphere, ISA is a leading technical training organization and a respected publisher of books, magazines, and standards. ISA also serves the professional development and certification needs of industry professionals and practitioners with its Certified Automation Professional (CAP), Certified Control Systems Technician (CCST), and Certified Industrial Maintenance Mechanics (CIMM) programs, and the Control Systems Engineers (CSE) license.
Related organizations: The Asia Pacific Federation of Instrumentation and Control Societies (APFICS)
Official language: English

Organization name: American Society of Agricultural Engineers (ASAE)[26]
Web address: www.asae.org
Domain: Agriculture engineering
About ASAE: The American Society of Agricultural Engineers is an educational and scientific organization dedicated to the advancement of

[25] Reproduced with the permission of (ISA) Instrumentation, Systems, and Automation Society.
[26] Reproduced with the permission of American Society of Agricultural Engineers (ASAE).

engineering applicable to agricultural, food, and biological systems. Founded in 1907, ASAE comprises 9,000 members in more than 100 countries.
Related organizations: The International Organization for Standardization (ISO) and the International Electrotechnical Commission (IEC)
Official language: English

Organization name: Association of Home Appliance Manufacturers (AHAM)
Web address: http://www.aham.org/
Domain: Home appliance manufacturers, traders, and users of appliances
About AHAM: See the AHAM Web site
Related organizations: See the AHAM Web site
Official language: English

Organization name: Association for the Advancement of Medical Instrumentation (AAMI)[27]
Web address: http://www.aami.org/
Domain: Medical instrumentation
About AAMI: The Association for the Advancement of Medical Instrumentation (AAMI), founded in 1967, is an alliance of over 6,000 members united by the common goal of increasing the understanding and beneficial use of medical instrumentation. AAMI is the primary source of consensus and timely information on medical instrumentation and technology and is the primary resource for the industry, the professions, and government for national and international standards. AAMI provides multidisciplinary leadership and programs that enhance the ability of the professions, health-care institutions, and industry to understand, develop, manage, and use medical instrumentation and related technologies safely and effectively. AAMI helps its members contain costs, keep informed of new technology and policy developments, add value in health-care organizations, improve professional skills, and enhance patient care. AAMI provides a critical forum for its membership of clinical and biomedical engineers and technicians, physicians, nurses, hospital administrators, educators, researchers, manufacturers, distributors, government representatives, and other health-care professionals. These diverse groups have been instrumental in making AAMI the leading source of essential information on medical devices and equipment for over 30 years.
Related organization: See the AAMI Web site
Official language: English

[27] Reproduced with the permission of Association for the Advancement of Medical Instrumentation (AAMI).

Organization name: American Society of Heating and Refrigeration and Air-Conditioning Engineers (ASHRAE)
Web address: http://www.ashrae.org
Domain: Thermal engineering
About ASHRE: See the ASHRAE Web site
Related organizations: See the ASHRAE Web site
Official language: English

Organization name: Air Conditioning and Refrigeration Institute (ARI)
Web address: www.ari.org
Domain: Thermal engineering
About ARI: See the ARI Web site
Related organizations: See the ARI Web site
Official language: English

Organization name: International Institute of Refrigeration (IIR)[28]
Web address: http://www.iifiir.org
Domain: Refrigeration and Air-conditioning
About IIR: The International Institute of Refrigeration (IIR) is a scientific and technical intergovernmental organization that enables pooling of scientific and industrial know-how in all refrigeration fields on a worldwide scale. The mission of the IIR is to promote knowledge of refrigeration technology and all its applications to address major issues such as food safety, protection of the environment (reduction of global warming and protection of the ozone layer), and the development of the least-developed countries (food and health). The IIR has 61 member countries that are represented by selected commissioners. Other IIR members include corporate or benefactor members (companies, laboratories, and universities) and private (individual) members. The IIR provides its members with tailored services that meet a wide range of needs.
Official languages: French and English

Organization name: National Boards of Boilers and Pressure Vessel Inspectors (National Board)
Web address: www.nationalboard.org
Domain: Boiler and pressure vessels
About National Board: See the National Board Web site
Related organizations: See the National Board Web site
Official language: English

[28] Reproduced with the permission of International Institute of Refrigeration (IIR).

Organization name: Heat Exchange Institute (HEI)
Web address: http://www.heatexchange.org/
Domain: Heat-exchanger and vacuum technology
About HEI: Seen the HEI Web site
Related organizations: See the HEI Web site
Official language: English

Organization name: Society of Automotive Engineers (SAE)[29]
Web address: www.sae.org
Domain: Air, sea, and land transport
About SAE: Since its founding in 1905, SAE has served the professional needs of engineers and the transportation needs of humanity. SAE meets the needs of mobility practitioners who serve in all five phases of the product life cycle: design, manufacturing, operations, maintenance, and disposal/recycling. Members include a worldwide network of technically informed mobility practitioners.
Related organizations: The Aluminum Association, Inc., the American Design Drafting Association (ADDA), the American National Standard Institute (ASNI), and the Association for Manufacturing Technology (AMT)
Official language: English

Organization name: Defense Standardization Program (DSP)
Web address: http://www.dsp.dla.mil
Domain: Engineering and general
About DSP: See the DSP Web site
Related organizations: See the DSP Web site
Official language: English

Organization name: Defense Supply Center Columbus (DSCC) Document Standardization Unit[30]
Web address: http://www.dscc.dla.mil/offices/Doc_control
Domain: Weapons engineering
About DSCC: This Center, now called the Defense Supply Center Columbus, has served in every major military engagement since World War I by reducing process time, reducing cost, and improving quality of land, air, and maritime weapon systems.
Related organizations: The Department of the Air Force, the Department of the Army, and Department of the Navy
Official language: English

---

[29] Reproduced with the permission of Society of Automotive Engineers (SAE).

[30] Reproduced with the permission of Defense Supply Center Columbus (DSCC) Document Standardization Unit.

Organization name: American Institute of Aeronautics and Astronautics (AIAA)[31]
Web address: www.aiaa.org
Domain: Aeronautics, astronautics, and mechanical engineering
About AIAA: Officially formed in 1963 by a merger of the American Rocket Society (ARS) and the Institute of Aerospace Sciences (IAS), AIAA is the principal society of aerospace engineers and scientists. The purpose of the organization is to advance the arts, sciences, and technology of aeronautics and astronautics and to promote the professionalism of those engaged in these pursuits.
Related organizations: The International Organization for Standardization (ISO) and the American National Standards Institute (ANSI)
Official language: English

Organization name: Aerospace Industries Association (AIA)
Web address: http://aia-aerospace.org
Domain: Aerospace Engineering
About AIA: See the AIA Web site
Related organizations: See the AIW Web site
Official language: English

Organization name: Marine e-Business Standards Association (EMSA)
Web address: http://www.emsa.org/
Domain: Marine engineering
About EMSA: See the EMSA Web site
Related organizations: See the EMSA Web site
Official language: English

Organization name: Institute of Electrical and Electronics Engineers (IEEE)[32]
Web address: www.ieee.org
Domain: Electrical engineering, electronics, and computer science and engineering
About IEEE: IEEE promotes the process of creating, developing, integrating, sharing, and applying knowledge about electro- and information technologies and sciences for the benefit of humanity and the profession.
Related organizations: Not listed
Official language: English

---

[31] Reproduced with the permission of American Institute of Aeronautics and Astronautics (AIAA).
[32] Reproduced with the permission of Institute of Electrical and Electronics Engineers (IEEE).

Organization name: National Electrical Manufacturers Association (NEMA)[33]

Web address: http://www.nema.org/

Domain: Electrical equipment

About NEMA: NEMA, created in the fall of 1926 by the merger of the Electric Power Club and the Associated Manufacturers of Electrical Supplies, provides a forum for the standardization of electrical equipment, which enables consumers to select from a range of safe, effective, and compatible electrical products. The organization has also made numerous contributions to the electrical industry by shaping public policy development and operating as a central confidential agency for gathering, compiling, and analyzing market statistics and economics data.

Related organizations: See the NEMA Web site

Official language: English

Organization name: National Electrical Contractors Association (NECA)

Web address: www.neca.org

Domain: Electrical engineering

About NECA: See the NECA Web site

Related organizations: See the NECA Web site

Official language: English

Organization name: Electronic Industries Alliance (EIA)

Web address: www.eia.org

Domain: Electronics

About EIA: See the EIA Web site

Related organizations: See the EIA Web site

Official language: English

Organization name: Consumer Electronics Association (CEA)

Web address: http://www.ce.org

Domain: Electronics

About CEA: See the CEA Web site

Related organizations: See the CEA Web site

Official language: English

Organization name: Association Connecting Electronics Industries (IPC)[34]

Web address: www.ipc.org

Domain: Electronics

[33] Reproduced with the permission of National Electrical Manufacturers Association (NEMA).

[34] Reproduced with the permission of Association Connecting Electronics Industries (IPC).

About IPC: IPC is a trade association dedicated to furthering the competitive excellence and financial success of its members worldwide. IPC is brings together all of the players in the electronics interconnection industry, which includes designers, board manufacturers, assembly companies, suppliers, and original equipment manufacturers. IPC promotes management improvement and technology enhancement programs, the creation of relevant standards, protection of the environment, and pertinent government relations.
Related organizations: The North East Circuits Association (NECA), the American National Standards Institute (ANSI), and the Electronic Industries Alliance (EIA).
Official language: English

Organization name: AIIM International (AIIM)[35]
Web address: http://www.aiim.org/
Domain: Information Technology
About AIIM: AIIM International (the Association for Information and Image Management) was founded in 1943 as the National Microfilm Association. Despite countless revolutions in technologies, the AIIM core focus has remained that of helping users connect with suppliers who can help them apply document and content technologies to improve their internal processes. AIIM produces educational, solution-oriented events and conferences, provides industry information through publications and an online resource center.
Related organizations: The American National Standards Institute (ANSI) and the International Organization for Standardization (ISO)
Official language: English

Organization name: Data Interchange Standards Association (DISA)
Web address: www.disa.org
Domain: Digital data interchange
About DISA: See the DISA Web site
Related organizations: See the DISA Web site
Official language: English

Organization name: Information Technology Industry Council (ITIC)
Web address: www.itic.org
Domain: Information system
About ITIC: See the ITIC Web site
Related organizations: See the ITIC Web site
Official language: English

[35] Reproduced with the permission of AIIM International (AIIM).

Organization name: National Information Standards Organization (NISO)
Web address: www.niso.org
Domain: Standards related to information
About NISO: See the NISO Web site
Related organizations: See the NISO Web site
Official language: English

Organization name: Worldwide Web Consortium
Web address: http://www.w3.org/
Domain: World Wide Web
About W3C: See the W3C Web site
Related organizations: See the W3C Web site
Official language: English

Organization name: American Gas Association (AGA)
Web address: www.aga.org
Domain: Fuel
About AGA: See the AGA Web site
Related organizations: See the AGA Web site
Official language: English

Organization name: Compressed Gas Association (CGA)
Web address: http://www.cganet.com/
Domain: Compressed gases
About CGA: See the CGA Web site
Related organizations: See the CGA Web site
Official language: English

Organization name: American Petroleum Institute (API)[36]
Web address: http://www.api.org
Domain: Petroleum Engineering
About API: As the primary trade association of the petroleum industry in the United States, API represents more than 400 members involved in all aspects of the oil and natural gas industry. API draws on the experience and expertise of its members and staff to support a strong and viable oil and natural gas industry. The API is a member-driven organization that offers large and small companies the opportunity to participate in shaping API programs and policy priorities.
Official language: English

[36] Reproduced with the permission of American Petroleum Institute (API).

Organization name: American Nuclear Society (ANS)
Web address: http://www.ans.org
Domain: Nuclear engineering
About ANS: See the ANS Web site
Related organizations: See the ANS Web site
Official language: English

Organization name: Manufacturing Standardization Society (MSS)
Web address: http://www.mss-hq.com
Domain: Manufacturing engineering
About MSS: See the MSS Web site
Related organizations: See the MSS Web site
Official language: English

Organization name: Consortium for Advanced Manufacturing-International (CAM-I)
Web address: http://www.cam-i.org
Domain: Manufacturing engineering
About CAM-I: See the CAM-I Web site
Related organizations: See the CAM-I Web site
Official language: English

Organization name: American Society for Quality (ASQ)
Web address: www.asq.org
Domain: General engineering
About ASQ: See the ASQ Web site
Related organization: See the ASQ Web site
Official language: English

Organization name: NACE Foundation[37]
Web address: http://www.nace.org
Domain: Corrosion engineering
About NACE: The purpose of the NACE foundation is to create awareness of and interest in corrosion science, education, and career opportunities through dedicated training, education/industry cooperative experiences, summer science camps, scholarships, grants, internships, and teacher workshops.
Related organizations: Not listed
Official language: English

Organization name: Institute of Transportation Engineers (ITE)[38]
Web address: http://www.ite.org/

[37] Reproduced with the permission of NACE Foundation.
[38] Reproduced with the permission of Institute of Transportation Engineers (ITE).

Domain: Transportation engineering
About ITE: The Institute of Transportation Engineers (ITE) is an international individual-member, educational and scientific association. ITE members are traffic engineers, transportation planners, and other professionals who are responsible for meeting society's needs for safe and efficient surface transportation through planning, designing, implementing, operating, and maintaining surface transportation systems worldwide.
Related organizations: Not listed
Official language: English

Organization name: International Cost Engineering Council (ICEC)
Web address: http://www.icoste.org/
Domain: Cost engineering
About ICEC: See the ICEC Web site
Related organizations: See the ICEC Web site
Official language: English

Organization name: Association of Records Managers and Administrations (ARMA)
Web address: http://www.arma.org
Domain: Administration
About ARMA: See the ARMA Web site
Related organizations: See the ARMA Web site
Official language: English

2.3.5 Conformity Assessment

Conformity assessment is a major requirement for most standards. This section briefly describes the scope of activities of organizations engaged in this activity.

Organization name: Inter-American Accreditation Cooperation (IAAC)
Web address: http://iaac-accreditation.org/
Domain: Accreditation
About IAAC: See the IAAC Web site
Related organizations: See the IAAC Web site
Official language: English

Organization name: Industry Cooperation on Standards & Conformity Assessment (ICSCA)
Web address: http://www.icsca.org.au
Domain: Standards and conformity assessment
About ICSCA: See the ICSCA Web site
Related organizations: See the ICSCA
Official language: English

Organization Name: U.S. Product Data Association (US PRO)[39]

Web address: http://www.uspro.org

Domain: Geometric data transfer

About Us Pro: The U.S. Product Data Association (US PRO) is a nonprofit membership organization. Established by industry, US PRO works for industry by providing the management functions for the IGES/PDES Organization (IPO) and its related activities, including the U.S. Technical Advisory Group (TAG) to ISO TC184/SC4. US PRO is accredited by the American National Standards Institute (ANSI) to support the development, publication, and distribution of the IGES and PDES standards in the U.S.

Related Organizations: International Organization for Standardization (ISO), American National Standards Institute (ANSI)

Official Language: English

Organization Name: PDES, Inc.[40]

Web address: http://pdesinc.aticorp.org/

Domain: Geometric data transfer

About PDES Inc: PDES, Inc. is an international industry/government consortium accelerating the development and implementation of ISO 10303, commonly known as STEP (Standard for the Exchange of Product model data). STEP, a key international product data technology, provides an unambiguous, computer sensible description of the physical and functional characteristics of a product throughout its life cycle.

Related Organizations: International Organization for Standardization (ISO)

Official Language: English

2.3.5.1 Conformity Marking

A mark is a shorthand way of providing product information for consumers/users that can give characteristics of a product or evidence of the claim of the manufacturer that the product is in conformity with fixed requirements (marks of conformity/approval).

The practice of marking goes back hundreds of years. Early examples are the assay and hallmarking of precious metals such as gold and silver. Normally, marks are found on the product itself or (mostly for small products) on the packaging. Marks are affixed to products as varied as eggs and butter, razors and high-pressure boilers, computers, and so on. Marks can be affixed on documents (e.g., test reports) as well.

Marks can be voluntary or mandatory. This distinction does not refer to the substance of the information or message and is not always clear-cut. In some

---

[39] Reproduced with the permission of US PRO.

[40] Reproducted with the permission of PDES, Inc.

situations, "voluntary" marks, mostly national marks, are "quasimandatory," for example, because of insurance requirements. In some countries or regions marks are protected by laws.

Marks are based on conformity assessments, testing, and inspection and are placed on products or on documents about products by or on behalf of independent "third parties" as certification bodies, which are mostly owners of the marks. Sometimes a mark is the property of supplier's organization, such as the Woolmark, but is affixed under supervision of a third party.

Suppliers, through self-declarations of conformity, can place marks on products. Marks are used in a variety of different and confusing ways, which sometimes obscures the messages they are trying to deliver. Some examples include CE marking, quality-control stickers, and "house marks" such as UNOX. The ability to distinguish between the types of marks that can appear on products is important.

The ISO/IEC Guide 2 defines a certification mark (a mark of conformity for certification) as a: "protected mark, applied or issued under the rules of a certification system, indicating that confidence is provided that the relevant product, process or service is in conformity with a specific standard or other normative document."

Certification marks can be divided into two categories: total (generic) marks and partial (sector or aspect specific) marks. Total marks apply to all aspects of a product, including its performance. Determining all the performance requirements a product has to meet is a time-consuming and expensive process. Frequently consumer panels, assisted by experts, are used to draft the requirements.

Partial marks apply to some aspects of a product (e.g., safety). These requirements are rather easy to draft because only technical expertise is needed. However, the limited scope of partial marks on products may not be communicated. For example, a product with a "safety mark" may not have been tested for other criteria, including performance.

Logos, brand names, and trademarks are not used as marks of conformance. They provide corporate identity and indicate the origin of a product. They are seen as the most valuable assets a company/supplier has and are used as a marketing tool.

Pictograms are stylized identification symbols. They differ from other types of marks in that the message is represented by an image. Safety-handling information and warnings are usually presented in pictograms.

Labels give to-the-point information about special aspects of products. Manufacturers' claims are also included on labels. Some claims, such as volume, can be seen as mandatory information about the minimum volume of the package.

2.3.6 Users Groups for Standards

With the increased use of standards, many concerned parties desire to influence the standardization process. This section lists the major groups that represent the rights of the standards users.

Organization name: European Association for the Coordination of Consumer Representation in Standardization (ANEC)
Web address: http://www.anec.org
Domain: Consumer participation in standards development process
About ANEC: See the ANEC Web site
Related organizations: See the ANEC Web site
Official language: English

Organization name: International Federation of Standards Users (IFAN).
Web address: http://www.ifan-online.org
Domain: Standards users
About IFAN: See the IFAN Web site
Related organizations: See the IFAN Web site
Official language: English

2.3.7 Commercial Organizations for the Supply of Standards

This section lists the organizations involved in the supply of technical documents and standards.

Organization name: Document Engineering Company, Inc (DECO)
Web address: http://www.doceng.com
Domain: Standards supply
About DECO: See the DECO Web site
Related organizations: See the DECO Web site
Official language: English

Organization name: Global Engineering Documents (IHS)
Web address: http://global.ihs.com/
Domain: Standards supply, publisher
About IHS: See the IHS Web site
Related organizations: See the IHS Web site
Official language: English

Organization name: ILI Infodisk Inc.
Web address: http://www.ili-info.com
Domain: Standards supply
About ILI: See the ILI Web site
Related organizations: See the ILI Web site
Official language: English

Organization name: International Library Service
Web address: http://www.normas.com

Domain: Standards ordinary service
Related organizations: See the International Library Service Web site
Official language: English

Organization name: National Standards System Network (NSSN)
Web address: http://www.nssn.org/
Domain: Standards supply
About NSSN: See the NSSN Web site
Related organizations: See the NSSN Web site
Official language: English

Organization name: Standards and Technical Publishing (STP)
Web address: http://www.stp.com.au/
Domain: Standards and technical literature supply
About STP: See the STP Web site
Related organizations: See the STP Web site
Official language: English

Organization name: Techstreet[41]
Web address: http://www.techstreet.com
Domain: Standards supply service
About Techstreet: Techstreet provides mission-critical information resources and information-management tools for technical professionals. Through Techstreet's Web site, customers can find and purchase from over 500,000 technical-information titles, including one of the world's largest collections of industry standards and specifications.
Related organizations:
- American National Standards Institute (ANSI)
- American Petroleum Institute (API)
- American Society of Mechanical Engineers (ASME)
- ASTM International
- American Welding Society (AWS)
- American Water Works Association (AWWA)
- Canadian Standards Association (CSA)
- Deutsche Institute für Norman (DIN)
- International Electrotechnical Commission (IEC)
- Institute of Electrical and Electronics Engineers (IEEE)
- Instrumentation, Systems and Automation Society (ISA)
- International Committee for Information Technology Standards (INCITS)
- International Organization for Standardization
- National Fire Protection Association (NFPA)

Official language: English

[41] Reproduced with the permission of Techstreet.

Organization name: USA Information System, Inc
Web address: http://www.usainfo.com
Domain: Standard supply service
About USA Information System, Inc.: See the USA Information System, Inc. Web site.
Related organizations: See the USA Information System, Inc. Web site.
Official language: English

2.4 ECONOMIC BENEFITS OF STANDARDIZATION—A BRIEF ON A DIN CASE STUDY[42]

The process of standardization greatly affects the competitive ability and strategies of companies. It provides technological and economic infrastructure for a nation. The international business environment has changed dramatically because of increased globalization. To identify the economic implication of standards and technical information in terms of both form and content in an increasingly globalized environment together with the changing role of standardization at the local, regional, and international levels makes an examination of the standardization process in practical details necessary.

A study designed to establish scientifically the economic benefits of standardization was jointly initiated by DIN, the German Institute of Standardization and the German Federal Ministry of Economics Affairs and Technology (1997–2000). The joint research project "Economic Benefits of Standardization," was carried out simultaneously in Germany, Austria, and Switzerland. The Department of Market-Oriented Business Management and the Department of Political Economics and Research at the Technical University, Dresden, Karlsruhe, and the Frannhofer Institute for Systems and Innovative Research, Karlsruhe were contracted by DIN to conduct this study.

This analysis of economic benefits of standardization takes as its starting point the four main partners in the standardization process: business, private households, the state, and the standards body. The latter partner acts as an intermediary between the other three, which are affected by standardization in different ways.

The study contracted by DIN is one of the most scientific among several studies conducted on technical standardization. This study provides a detailed analysis of the economics benefits of standardization, economic efficiency through standardization, and costs and benefits of standardization at both the microeconomic and the macroeconomic level.

A survey of more than 4,000 companies in Germany, Austria, and Switzerland was conducted for this study, with a response rate of over 17%. Interviews were

[42] The summary of Economic Benefits of Standardization, first edition, 2000 was produced in 2005 by W.A. Khan and A. Raouf with permission of Bewth Verlag GmbH.

held with representatives of public interest groups, private households, and government. The report related to this study is divided into two parts: (1) benefits for business and (2) benefits for the economy as a whole.

2.4.1 Benefits for Business

The study of the benefits to business that result from standardization revealed the following:

1. Companies are generally unaware of the strategic significance of standards.
2. National involvement in standardization is required to influence European and international standards.
3. Where national standards are adopted as European and international standards, participation in standards frequently results in advantages of cost and competitiveness.
4. Involvement in standardization in anticipation of new regulatory legislation leads to avoidance of costs.
5. Competitive advantage is gained more through company standards than through industry-wide or private-industry standards.
6. The advantage of insider knowledge is gained.
7. Insider knowledge is more important than time advantage.
8. Competitive advantage is gained through influencing contents of standards.
9. Conforming to European and international standards should be the export strategy of businesses.
10. Eighty percent of the businesses surveyed do not know the exact cost of adapting to foreign standards.
11. Lower trading costs, simplification of contractual agreements, and lowering trade barriers are the advantages of harmonized European and international standards.
12. Cost and savings that result from the application of European and international standards apply to such factors as workload of the staff through travel, use of foreign languages, adoption of products for European and international markets, opportunities for cooperation, and choice of suppliers.
13. Only 9% of the businesses surveyed were prepared to give actual figures for cost and savings.
14. Transaction costs are reduced by use of standards.
15. Costs of developing company standards and industry-wide standards are not easily quantified.
16. Company standards help lower production costs more than do industry-wide standards.

17. Positive effects on interdepartmental communication result from company and industry-wide standards.
18. Standards have a positive effect on the buying power of companies.
19. Standards offer a wider choice of suppliers with the same degree of quality.
20. Standards are used to exert market pressure on clients.
21. Standards affect relationships with suppliers more than with clients.
22. Industry-wide standards have a positive effect on cooperation between businesses.
23. Strategic alliances require similar use of private-industry standards and industry-wide standards.
24. Cooperation between businesses not only can result in cost reduction but also can result in monopolization.
25. Standards are not a large factor in hindering innovative projects.
26. Participating in standards reduces research and development risks.
27. Participating in standards reduces research and development costs.
28. Relevance of both industry-wide standards and private-industry standards increases with product life spans.
29. Where product life spans are short, little difference exists in the significance of private-industry standards and industry-wide standards.
30. Where product life spans are long, industry-wide standards are more significant than private-industry standards.
31. Lower accidental rates are partly to the result of standards.
32. Participation in the standardization process increases awareness of product safety.
33. Liability risks are reduced through standards.
34. Standards assist government regulators.
35. Companies prefer majority voting.
36. A switch to majority voting is less important than other desired changes to the standardization process.
37. Possible changes to the standardization process include the following:
    - More project management
    - More information for nonparticipants
    - Increased use of electronic media
38. Representatives of minority interest groups, such as consumers, are largely in favor of consensus standard bodies.
39. DIN, ON, and SNV are necessary but are bureaucratic and expensive.
40. All partners in standardization would incur more costs without the help of DIN, ON, and SNV
41. Even in sectors where national regulations are valid, cooperation with DIN is necessary to have an impact on European and international standards development.

2.4.2 Benefits for the Economy as a Whole

The study of the benefits to the economy as a whole that result from standardization revealed the following:

1. Innovation potential alone is not sufficient to maintain competitiveness.
2. An efficient dissemination of innovation via standards is a precondition for economic growth.
3. New standards are more numerous in innovative sectors.
4. German standardization responds adequately to technical change.
5. The lifetime of standards is shorter where technological change is greater.
6. Standards are a positive stimulus for innovation.
7. Standards have a positive influence on innovation.
8. Standards should be withdrawn as soon as they are technically outdated.
9. Leaders in technology should become more involved in standardization.
10. Indicators for technological progress are the following:
    - Patents
    - Expenditure on export licenses
    - Number of standards
11. Standards are as important for economic growth as patents. Diffusion of innovation through standards is a decisive factor.
12. Standards are no longer misused as nontariff trade barriers.
13. The very existence of standards is positive for trade.
14. Cost and quality advantage are conferred to businesses that use standards.
15. Standards have positive effects on foreign trade.
16. Standards are indicators of innovative technological competitiveness.
17. The number of existing standards cannot explain, in all cases, structures in bilateral trade relations.
18. In one third of technological sectors, standards play a positive role in creating export surpluses.
19. Standards have a generally positive effect on exports.
20. International standards encourage intraindustrial trade more than do national standards.
21. Development of the national standards collection has no significant influence on total German export.
22. National German standards are not trade barriers.
23. International standards improve the competitive chances of domestic products.
24. Overall, empirical support exists for the theory that international standards lead to international competitiveness.

Generally, the results of this analysis and those of the company survey tally:

1. Businesses do not regard standards as outdated.
2. Contradictory effects of standards on research and development and even negative effects exist in some sectors.
3. Most businesses benefit from participating in standards development.
4. Standards do not hinder innovation.
5. Standards are internationally respected.
6. Standards make technical specification more transparent.
7. The majority of the businesses use European and international standards because of their positive effect on export.
8. International standards encourage trade.
9. International and European standards are more significant for German export than are national standards.
10. Increased participation in European and international standards development is necessary.
11. Standards encourage technology transfer.
12. Standards make initiation of products and processes easier for foreign competitors.
13. Standards should be concentrated in sectors where greater national innovation potential exists.
14. Results of the macroeconomics analysis show the economic benefits of standardization to be approximately 1% of the gross national product (GNP).
15. Macroeconomic benefits of standardization are greater than the sum of individual advantages.
16. Innovation policies should support standardization.

The major findings of this study are the following:

1. The benefit to the national economy amounts to more than U.S.$15 billion per year.
2. Standards contribute more to economic growth than do patents and licenses.
3. Companies that participate actively in standards development have a head start on their competitors in adapting to market demands and new technologies.
4. Transaction costs are lower when European and international standards are used.
5. Research risks and development costs are reduced for companies that contribute to the standardization process.

The main conclusions of this study are the following:

1. With its broad-based dual approach, new insight into the economics effects of standardization is provided in international context.
2. Examination of specific branches in the necessary detail was not possible.
3. The results of the study can be used for a strategic discussion of the future of standards development.

BIBLIOGRAPHY

Alexander, P.E. Revealing the Inner Sanctum of the Standards Process; Proceedings of the Electrical/Electronics Insulation Conference, Boston, MA, 1985, p. 31–34.

Bhullar, B.A.S., Standardization: The part-time less-understood activity, *Indian Weld. J.*, 17(4), 91–95, 1986.

Blower, R.W., Standards: Are they a benefit? *Power Eng. J.*, 2(3), 137–142, 1988.

Breitenberg, M.A., The ABCs of the U.S. Conformity Assessment System, NISTIR 6014, National Institute of Standards and Technology, http://ts.nist.gov/ts/htdocs/210/ncsci/primer.htm, 1997.

Carpenter, R.R., What value standards? *Stand. News*, 30(5), 30–33, 2002.

Chambord, A.B., European standardization, *ASTM Stand. News*, 14(12), 44–48, 1986.

Christoph, H., ISO-CEN-DIN: Confusion or transparency in standardization? *Schweissen and Schneiden*, 40(2), 18–21, 69–75, 1988.

Conti, T.A., How to find the correct balance between standardization and differentiation, *Qual. Prog.*, 34(4), 119–121, 2001.

Donoghue, E.A., Codes and standards, *Elevator World*, 45(4), 4, 1997.

Endelman, L.L., Effect of standards on new equipment design by new international standards and industry restraints, *Proc. SPIE*, 1346, 90–92, 1990.

Foster, M.E., Japanese industrial standards system: An American's viewpoint, *Annu. Qual. Congr. Trans.* Las Vegas, NV, 6–16, 1994.

Gundlach, H.C.W., Marks of Conformity for Products, EOTC-10-0217/99, Discussion paper, EOTC-Marking Discussion Forum, European Organization for Conformity Assessment, http://www.EOTC.BE/FORUM/MARKING. 1999.

Hafner, E., Specifications and standards, *Precis. Freq. Control*, 2, 297–304, 1985.

Hengst, S., Standardization: a competitive tool, *J. Ship Prod.*, 13(3), 198–206, 1997.

Hitchcock, L., Succeeding as a standards professional, Part 1, *Stand. News*, 30(8), 16–17, 2002.

Hohmann, L., Process control in standards bodies, *IEEE Commn., Mag.*, 27(9), 59–61, 1989.

http://aia-aerospace.org

http://global.ihs.com/

http://iaac-accreditation.org/

http://lamar.colostate.edu/~hillger/

http://w3.pwgsc.gc.ca/cgsb

http://www.aami.org/

http://www.abma-dc.org

http://www.acmanet.org/
http://www.aecma.org
http://www.aerospace-technology.com
http://www.afnor.fr/
http://www.aga.org
http://www.agma.org
http://www.aham.org/
http://www.aiaa.org
http://www.aiim.org/
http://www.aisc.org
http://www.aluminum.org
http://www.anec.org
http://www.ans.org
http://www.ansi.org
http://www.api.org
http://www.ari.org
http://www.arma.org
http://www.asae.org
http://www.ashrae.org
http://www.asm.com
http://www.asme.org
http://www.asq.org
http://www.asse.org
http://www.astm.org
http://www.aws.org
http://www.bsi-global.com/
http://www.cam-i.org
http://www.canena.org
http://www.ce.org
http://www.cenelec.be
http://www.cganet.com/
http://www.cie.co.at
http://www.copant.org
http://www.din.de
http://www.disa.org
http://www.doceng.com
http://www.dscc.dla.mil/offices/Doc_control
http://www.dsp.dla.mil
http://www.ecma-international.org
http://www.eia.org
http://www.energycodes.gov/
http://www.eotc.be/forum/marking
http://www.epa.gov/
http://www.etsi.org
http://www.fema.gov
http://www.fluidcontrolsinstitute.org/
http://www.gost.ru/sls/gost.nsf
http://www.heatexchange.org/

http://www.hfes.org
http://www.hl7.org
http://www.hse.gov.uk/
http://www.iccsafe.org/
http://www.icoste.org/
http://www.icsca.org.au
http://www.iec.ch
http://www.ieee.org
http://www.ifan-online.org
http://www.iifiir.org
http://www.ili-info.com
http://www.incose.org/
http://www.industrial-fasteners.org/
http://www.inmm.org
http://www.ipc.org
http://www.ironcasting.org
http://www.isa.org
http://www.iso.ch
http://www.iso.org/iso/en/aboutiso/isomembers/CorrespondentMemberList MemberSummary?MEMBERCODE=20
http://www.iso.org/iso/en/aboutiso/isomembers/MemberList.MemberSummary? MEMBERCODE=10
http://www.iso.org/iso/en/aboutiso/isomembers/SubscriberMemberList.MemberSummary? MEMBERCODE=30
http://www.ite.org/
http://www.itic.org
http://www.metaltreat.com/
http://www.mpif.org
http://www.mpta.org
http://www.mss-hq.com
http://www.nace.org
http://www.nationalboard.org
http://www.ncits.org
http://www.neca.org
http://www.nema.org/
http://www.nfpa.com
http://www.nfpa.org
http://www.niso.org
http://www.nist.gov
http://www.normas.com
http://www.nsf.org/
http://www.nssn.org/
http://www.osha.gov/
http://www.pascnet.org
http://www.pumps.org/
http://www.sac.gov.cn
http://www.sae.org
http://www.seinet.org/

http://www.semi.org/
http://www.standardsresearch.org/
http://www.stp.com.au/
http://www.techstreet.com
http://www.usainfo.com
http://www.w3.org/
http://www.wwpa.org
http:hr//www.cenorm.be
Iorio, J., IEEE standards: mechanizing the standardization process, *Compt. Stand. Interfaces*, 12(3), 205–207, 1991.
Irving, B., International standardization: A wake-up call for American industry, *Weld. J.* 78(9), 34–39, 1999.
Jakobs, K., Procter, R., and Williams, R., Study of User Participation in Standards Setting, Conference on Human Factors in Computing Systems Proceedings, Vancouver, B.C. 1996, 109–110.
Jakobs, K., Procter, R., and Williams, R., The making of standards: Looking inside the work groups, *IEEE Commn. Mag*. 39(4), 102–107, 2001.
Kearsey, B.N. and Etesse, L., Standardization: The uniting factor, *Electrical Commn.*, 3rd Quarter, 222–230, 1994.
Kirkham, R.J.R., European standardization versus national and world-wide standardization, *Elektrotechnische Zeitschrift*, 107(17), 794–796, 798–799, 1986.
Krechmer, K., The need for openness in standards, *Computer*, 34(6), 100–101, 2001.
Mazda, F., Standardizing on standards, *Telecommunications*, 26(6), 6, 1992.
McAdams, W.A., Standards and the United States, *ASTM Stand. News*, 14(12), 40–43, 1986.
Moran, G. and Spanner, J., Importance of standardization, *Materials Evaluation*, 54(8), 901–904, 1996.
Morrell, M.D., Characteristics of the Standards Professional, Proceedings, Annual Conference, Standards Engineers Society, Minneapolis, MN, 1990, 32–36.
Picariello, P., A global standards strategy by industry for industry, *Stand. News*, 30(10), 26–27, 2002.
Rada, R., Carson, G.S., and Haynes C., Sharing standards: The role of consensus, *Commn. ACM*, 37(4), 15–16, 1994.
Salter, L., User Participation in Standardization New or Merely Recycled? IEEE International Engineering Management Conference, Managing Virtual Enterprises: A Convergence of Communications, Computing, and Energy Technologies, 1996, Vancouver, B.C., 583–590.
Scheidweiler, A., European Standards, the Expectations of the Industry and the Users; IEE Conference Publication, No. 408, 1995, 28–33.
Schmidt, M.W., International standards: A smaller (and easier) world, *Biomed. Instrum. Technol.* 31(3), 282–285, 1997.
Thiard, A., Worldwide standards—the only way, *ASTM Stand. News*, 14(12), 34–37, 1986.
ttp://www.jisc.go.jp/eng/
Upp, E.L., Impact of Standards in Marketing Products or Services in a Global Marketplace, Proceedings, Annual Conference, Standards Engineers Society, Standards Do Make a Difference, 1990, 58–62.
Viganego L.R., Harmonization of Technical Standards, Proceedings, SPE Annual Technical Conference and Exhibition, v Gamma, EOR/General Petroleum Engineering, 1991, 165–177.

Wallace, R.B., Test methods become national standards—eventually, *ASTM Stand. News*, 14(5), 40–44, 1986.

Wende, I. V., German participation in European and international IT-standardization, *Comput. Stand. Interfaces*, 17(1), 7–11, 1995.

Woodgate, J.M., Standards bodies: Processes and access, *IEE Colloquium*, 520, 4/1–4/5, 1998.

Woods, J.R., Consensus perspective on the realities and possibilities of standardization in Europe '92, *ASTM Stand. News* 18(2), 32–33, 1990.

Woollacott, W.J., So You Want to Write Standards, Proceedings, Annual Conference, Standards Engineers Society, Standards Do Make a Difference? 1990, 15–21.

3 Parametric Design and Manufacturing of Hand Tools—A Case Study

3.1 INTRODUCTION

This case study deals with computer-aided parametric design and manufacturing of hand tools. Hand tools are produced to standards, either set at the international or national level or set by a company.

The production of hand tools largely relies on highly skilled labor. The use of computer-aided design (CAD) is minimal. It is usually found only in conjunction with other design and manufacturing functions. The production lead time and throughput time for hand tools are long. One of the major bottlenecks in the production process is the manufacture of forging dies. Approximately 75% of hand tool components or component subassemblies require the use of forging dies. This particular sector of industry produces components that normally fall in the variant-design category of engineering design. Pahl and Beitz have defined variant design as the type of engineering design that involves varying the size and arrangement of certain aspects of the chosen system, while the function and the solution principle remains unchanged.

Two goals identified by the Federation of British Hand Tool Manufacturers are lowering the unit costs and increasing the quality of the product. Close tolerances are another major requirement.

Use of computers to increase the effectiveness of hand tool manufacture is essential. However, this proposition may pose many problems. These problems arise from the lack of availability of the computer-based tool manufacturing products that suit the nature of the industry and the working practices it follows. The computer-based tool manufacturing products are normally designed on a general specification to capture the broadest market. Similarly, the proposed computer-integrated manufacturing (CIM) strategies either have a universal approach or confine themselves to a particular manufacturing environment. The low level of implementation of advanced manufacturing technology is the result of both the lack of awareness of its potential and the high cost of capital involved.

This case study identifies the manufacturing subfunctions that greatly affect production performance in the industry and shows how the use of CAD in an integrated manner improves production and can be applied to other design and manufacture functions. The die-making department is identified as the area of work. For most hand tools, die making is the first step in the manufacturing sequence, and the use of conventional methods in this process contributes to long production lead times. The activities in this department lie within the scope of CAD and computer-aided manufacturing (CAM). The following aims were set for this study:

Customization of general-purpose CAD/CAM software to conduct the variant design of hand tools according to standards
Generation of a parts program to manufacture the electrodischarge machining (EDM) electrodes for forging dies by application of the EIA RS 274 D parts-programming standard
Use of the EIA RS 232 C standard for parts-program communication from design office to CNC milling machine

The following benefits to the industry are expected:

Rapid and inexpensive die design
An expedited die-manufacturing process
Improved product quality and lower unit costs
A model that individual companies can adopt at their own pace and according to their own financial and technical capabilities

3.2 TYPICAL DESIGN AND MANUFACTURE TECHNIQUES

The techniques used to design and manufacture forging dies for hand tools and hand tool subassemblies are largely based on human skills and experience. Here we describe the design and manufacturing technique used by Footprint Tools; however, other hand tool manufacturers generally follow a similar process. The dies are designed and manufactured for hammer forging. The design and manufacture of one die for forging a particular size of the combination wrench is described. However, dies for other hand tools and hand tool subassemblies go through all or most of the same design and manufacturing steps.

The initial specifications of a hand tool or hand tool subassembly are passed to the die-making department. The tool room manager draws sketches of the component from the initial specifications and the appropriate British standards, as shown in Figure 3.1. These sketches outline the exact shape of the component, the necessary dimensions, and the position of the die line.

The next step is the manufacture of the component template and the die-line template. The component template depicts the component in two dimensions, as shown in Figure 3.2; it is manufactured by use of conventional tools and is finished by hand. Similar procedures are used for the manufacture of the die-line template. The component template is used to copy mill the master pattern with two-dimensional geometrical characteristics. The master pattern is then brought

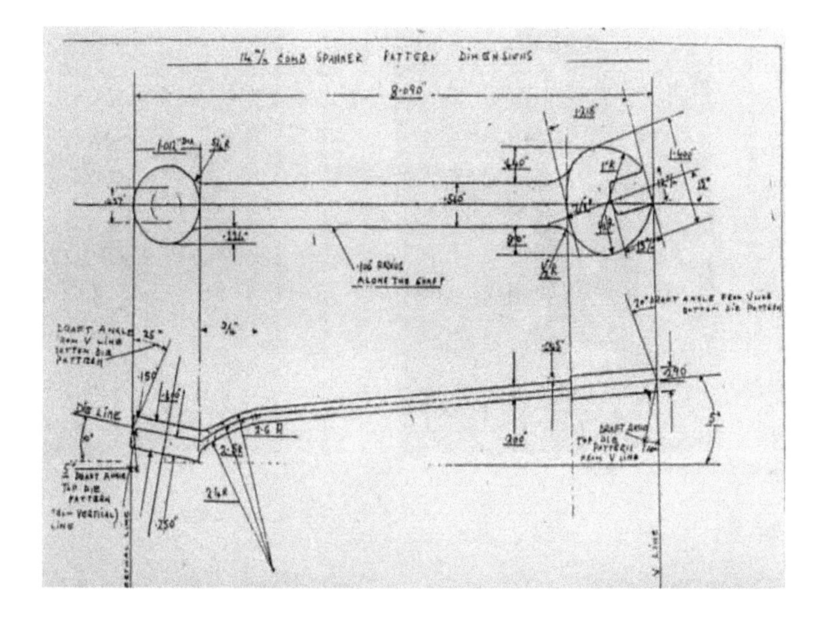

FIGURE 3.1 Preliminary sketch for a combination wrench.

back to the tool room manager so that any three-dimensional features, draft angles, aesthetic features, and the appropriate surface finish can be added, which is done by use of conventional machines and hand finishing. The master pattern (Figure 3.3) is then used to mill the EDM electrode (Figure 3.4). In parallel with the manufacture of the master pattern, the die line is copied at the diestock by use of the die-line template. The diestock material, with the die line machined onto it, then has the forging cavity machined into it by the EDM electrode (Figure 3.5).

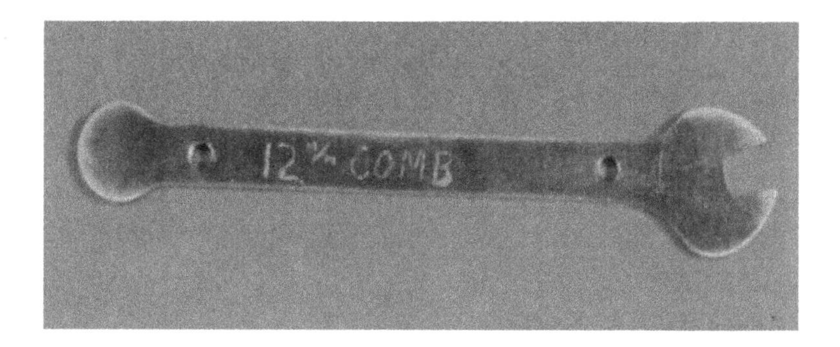

FIGURE 3.2 Two-dimensional templates.

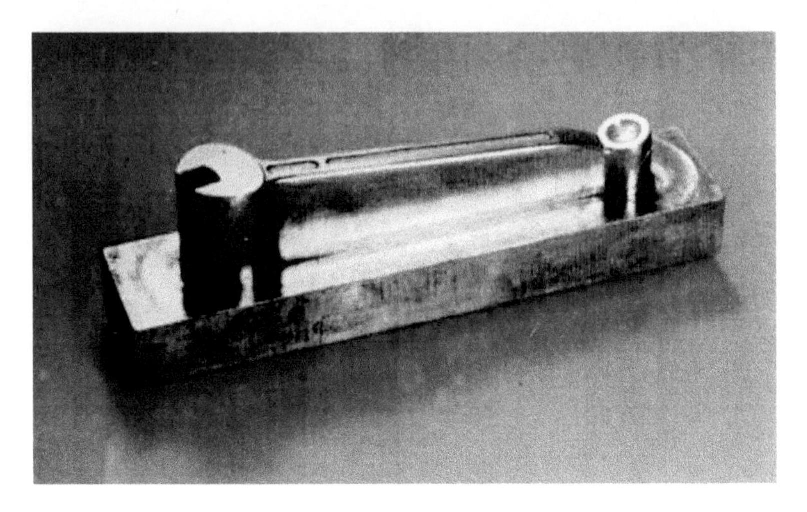

FIGURE 3.3 Master pattern for a combination wrench.

The forging die is then hand finished (Figure 3.6) so that the flash relief angles are incorporated. The top and bottom forging dies are finished to the required standard. At this stage, the decision is made as to whether the dies should be passed to the forging department or whether they require certain alterations. The die-design optimization phase is completed with the production done of one

FIGURE 3.4 Copy milling of EDM electrode.

FIGURE 3.5 EDM of die block.

FIGURE 3.6 Hand finishing of forging dies to incorporate flash-relief angle.

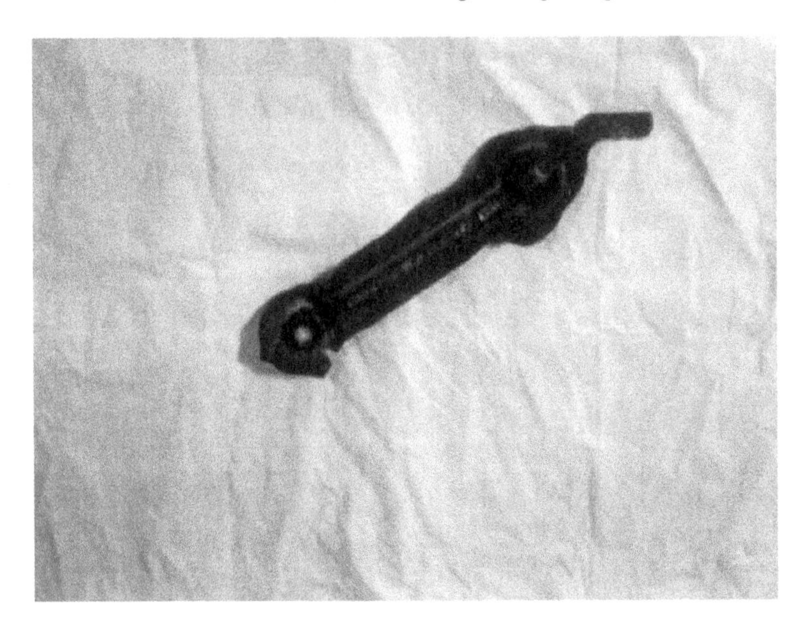

FIGURE 3.7 Forged component.

forged component (Figure 3.7). Any alteration that results from empirical analysis is incorporated in the master pattern. The master pattern is kept for any future copy milling of an EDM electrode for this particular size.

The process of die making in the hand tool industry is highly error prone, because the geometrical shape of the hand tool components is often complex. This fact, coupled with the use of conventional methods to cut the forging cavity, makes the die-making process slow and highly inflexible. It leads to long throughput times that, consequently, affect the production lead time. Delivery dates may be affected and the maintenance of quality is difficult.

3.3 CAD/CAM FOR HAND TOOLS

As stated earlier, hand tool components normally possess variant design characteristics that are strongly constrained by national and company standards. This situation is ideal for the development of a computer algorithm for use in a general-purpose CAD/CAM application to generate three-dimensional models and associated manufacturing equipment for hand tools. The algorithm may include parametric feature definition to achieve the variation in tool size. Modeling would enable the design to be optimized efficiently and economically. Subsequently, the model could be used to generate the part programs for machining either the forging cavity or the EDM electrodes. The full-size component model could also be used in marketing and sales. In a computer-integrated manufacturing environment, the component geometry can be transferred to other design manufacturing functions that require geometrical information. The part program may be directly downloaded to the numerically controlled machine tool.

3.4 SOFTWARE SELECTION

Hand tools are normally complex objects, so the selection of appropriate software to model them is important. The CAD software tools use three geometrical representation schemes: wire frame, constructive solid geometry, and boundary representation. Boundary representation is used in modeling and machining hand tools, and DELCAM software modules perform this task. This software does the following:

1. Defines sculptured surfaces according to BS 7586: 1992, BS 192 Part 1: 1996; BS 192 Part 2: 1998, and BS 3555: 1998 specifications
2. Facilitates application of British standards to the parametric design of hand tools
3. Generates tool paths for milling and spark erosion operations
4. Supports multiaxes machining operations
5. Produces standard cutter-location data (CLDATA) according to BS 5110 specifications
6. Transmits component geometry on standard interfaces such as Initial Graphics Exchange Specification (IGES)
7. Generates finite-element meshes for design analysis
8. Calculates physical information for cost estimation

3.5 PARAMETRIC MODELING OF WRENCHES

Parametric modeling allows automatic modeling of various sizes of a component from a single set of instructions as the input data that define the major dimensions of the component are changed. The parametric modeling of hand tool components involves the following:

Identification of the component subassemblies
Identification of the features in the subassembly
Modeling of the feature
Patching of the feature model to form the subassembly
Modeling of the subassembly
Blending of the subassemblies to form the component

The development and operation of parametric-modeling software is described below.

The wrench has been chosen as the object of this study because it is a component that demonstrates the potential of variant design and because it requires use of all the manufacturing operations in the die-making process. Three variations of the wrench are considered for the modeling study: the combination wrench, the ring wrench, and the open-ended wrench. These wrenches are a combination of three basic functional subassemblies: a C-end, a shank, and a ring. The shank could either be straight or cranked. The combination wrench is an amalgamation of a C-end a cranked shank and a ring, the open-ended wrench is the combination of two C-ends joined by a straight shank, and the ring wrench is the combination of two rings joined by a cranked shank. Single-job command files are created to automatically model the C-end, the straight shank, the cranked shank, and the ring. Separate single-job command files perform automatic orientation and subsequent blending and always take into account the appropriate British standard, and they also forge die characteristics (e.g., draft angles) and certain aesthetic features associated with the company.

The management command file is used to select the input parametric data for the required type of wrench, to access appropriate single-job command files to model individual subassemblies, and to orient and blend the subassemblies together. The steps involved in modeling a combination wrench are described below.

Once the wrench type has been selected, the system asks for the major dimensions of the functional subassemblies; in this case, for the C-end, for the cranked shank, and for the ring. The user enters these dimensions for the required wrench size. These major dimensions are taken from the British standard and are supplemented by the dimensions from the company standards. The system automatically models the C-end and presents the model on the VDU in the X-Y plane. Subsequently, an isometric view can be obtained at the graphic display. The user will then be prompted to check the model validity if the model provides the required geometrical characteristics of the functional subassembly. The user may give a positive response to proceed to the next step. Otherwise, the process may be repeated by re-entering different dimensions for the individual subassembly. The new set of dimensions must still follow the required British standard, and changes are only allowed in those dimensions not otherwise constrained. As soon as the model for the C-end is optimized, the system will perform the modeling for the shank and, subsequently, the modeling for the ring. During the modeling of the shank and the ring, the user will be prompted to check the functional subassembly model validity. When all the functional subassemblies of the combination wrench are modeled, the system will automatically orientate them to give the structural characteristics of the combination wrench. The subsequent operation involves the automatic blending of the C-end with the shank and the blending of ring with the shank. The introduction of intermediate points results in automatic blending. The positions of these points are selected by the system. Bezier curves are introduced by use of these points, and a surface is

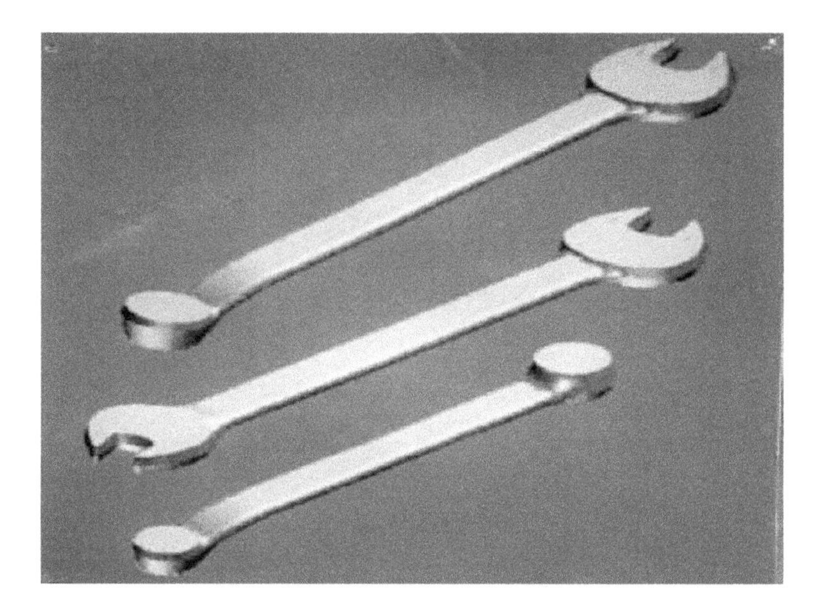

FIGURE 3.8 Variant design of wrenches.

formed that ensures continuity of slopes in different directions at the interface between subassemblies. The complete model, which is the combination of five surfaces, will be displayed at the screen. At this stage, the model can be assessed for final approval. The need for modification will require the modeling procedure to be repeated from the start.

A similar procedure can be used to produce size ranges of the combination wrench. Component variants can also be produced by use of similar methods. A combination wrench, an open-ended wrench, and a ring wrench are shown in Figure 3.8

The approved model of a wrench can be used to create tool paths according to British standard BS 5110, as is shown in Figure 3.9. Part programs for machining either the forging cavities or the graphite EDM electrodes required for the manufacture of the forging cavities can be produced for specific controllers. The master patterns can be generated on a milling machine with draft and flash-relief angles, as are shown in Figure 3.10 and Figure 3.11.

The system can also be used to create full-size images and models for use in marketing and sales. The model geometry can also be passed via the IGES interface to a design analysis software package (e.g., finite-element analysis) or to any other design manufacture function that requires geometry information.

Physical properties of the model can be obtained for cost-estimation purposes. Future development of computer hardware and software technology will lead to the availability of lower-cost software and hardware, which, in turn, will lead to

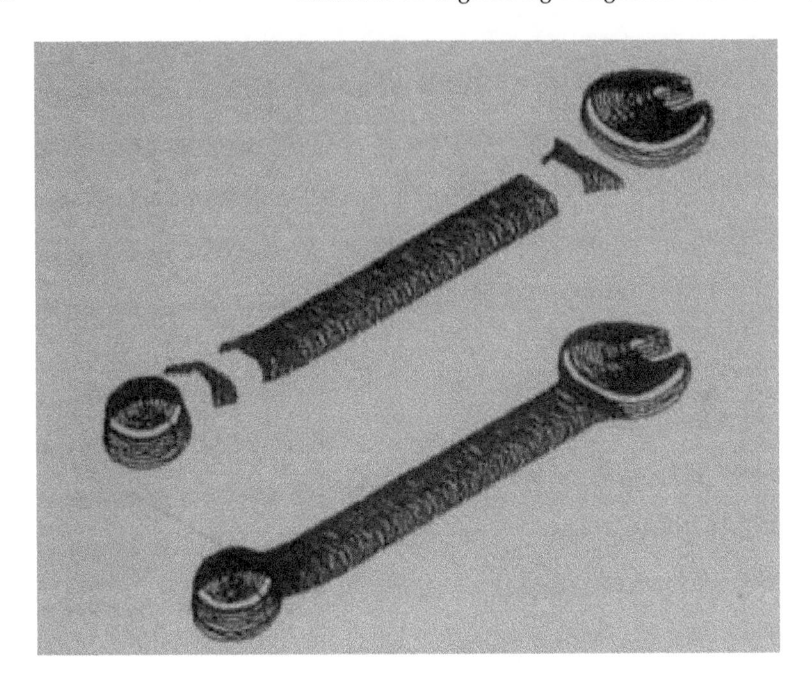

FIGURE 3.9 Tool path for combination wrench.

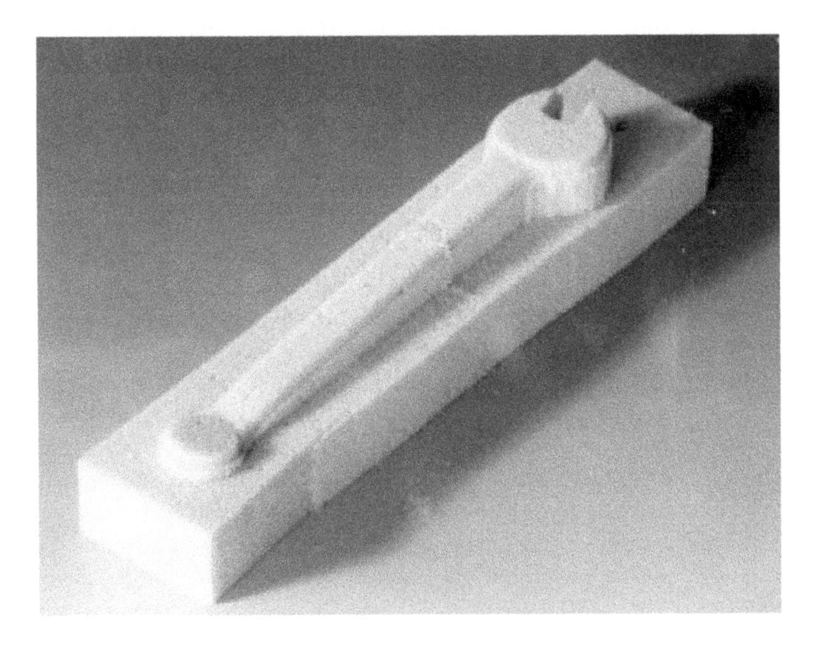

FIGURE 3.10 Master pattern with draft angle machined on high-density foam.

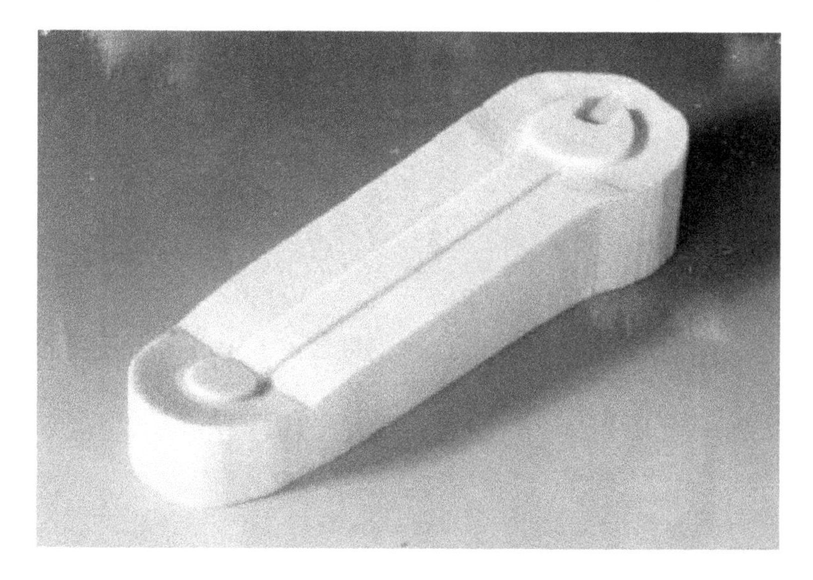

FIGURE 3.11 Master pattern with draft and flash-relief angles machined on high-density foam.

the better awareness of and more widespread industrial use of these techniques, especially in hand tool manufacturing.

3.6 CONCLUSION

The prototype or subsequent production of hand tools can ideally be performed by use of general-purpose CAD/CAM software and appropriate standards. Such software should have the capability of surface definition and should also support manufacturing processes such as milling, turning, and electrodischarge machining.

The CAD module of the CAD/CAM software package should be capable of parametric programming for designing size ranges and shapes. The master pattern should be stored on a computer hard disk rather than kept in inventory as a tool.

The modern techniques of CAD/CAM, in conjunction with appropriate standards, provide flexibility and higher productivity in design and manufacturing of discrete products, which leads to decreased lead time. Some techniques can be used to design and manufacture a variety of hand tools. A small selection of such tools is presented in Figure 3.12.

FIGURE 3.12 A selection of hand tools designed and manufactured by use of procedures described in this case study (courtesy of Footprint Tools, Sheffield, U.K.).

BIBLIOGRAPHY

British Standard B 192-2, Specification for Open-ended Wrenches, Part 2, BS and Whitworth, 1998.

British Standard BS 192 Part 1, Specification for Open-ended Wrenches, Part 1, Metric and Unified, 1996.

British Standard BS 3555, Specification for Ring Wrenches, 1988.

British Standards BS 7586, Specification for Combination Wrenches, 1992.

Cutter Location Data BS5110.

DELCAM User's Manuals.

Electronics Industries Association RS 232 C, Serial Communication Standard.

Electronics Industries Association RS 274 D, Programming Standards.

Excell, M., Toolmaking in the fast lane, *Metalworking Prod.*, 139(8), 45–46, 1995.

Khalil, A. and Powell, A., 5-Axis model making saves time and money, *Prod. Eng. (London)*, 66(2), 15–16, 1987.

Pahl G., Beitz W., Wallace Ken, Blessing L., and Bauert F., *Engineering Design: A Systematic Approach*, 2nd ed., Springer-Verlag Telos, Heidelberg 1996.

Relvas, C. and Simoes, J.A., Optimization of Computer Numerical Control set-up Parameters to Manufacture rapid Prototypes, Proceedings of the Institution of Mechanical Engineers, Part B, *J. Eng. Manuf.*, 218(8), 867–874, 2004.

Schumacher, B., Integration of toolmaking into CAD/CAM systems, *CIRP Ann.* 33(1), 113–116, 1984.

Part 2

Standards in Engineering Design of Discrete Products

4 Review of Standards Related to Techniques for Engineering Design—A Product Perspective

4.1 INTRODUCTION

Discrete products are basically the result of applications of solid mechanics, fluid mechanics, and thermodynamics. The application of solid mechanics includes the creation of components, structures, assemblies, mechanisms, and machines. The application of fluid mechanics includes pneumatics, hydraulics, turbomachinery, and aerodynamics. The application of thermodynamics includes refrigeration, heating, ventilating and air conditioning, steam generation and turbines, internal-combustion engines, and gas turbines. Commonly, products are constructed of components, structures, assemblies, mechanism, and machines based on solid mechanics and utilized in processes based on fluid mechanics or thermodynamics. These products are operated either manually or by use of automatic techniques based on control-system principles.

A characteristic of these discrete products is that their construction normally involves standard mechanical components, standard machine elements, standard control features based on digital electronics, standard electrical accessories, and standard computer software. During their operation, these discrete products may have varying exposures that range from ambient environment to saline water to thermal radiation. Some examples of discrete products are shown in Figure 4.1.

4.2 ENGINEERING DESIGN CLASSIFICATION

The design of discrete products follows established rules for solid mechanics, fluid mechanics, and thermodynamics, along with related extensions of these subject areas. Any other special factors related to the multidisciplinary feature of discrete products, such as control factors, environmental factors, and ergonomic factors, are considered separately or amalgamated into the basic design cycle. The design of discrete products covers the basic input, process, and output

Air-conditioners and Refrigerator

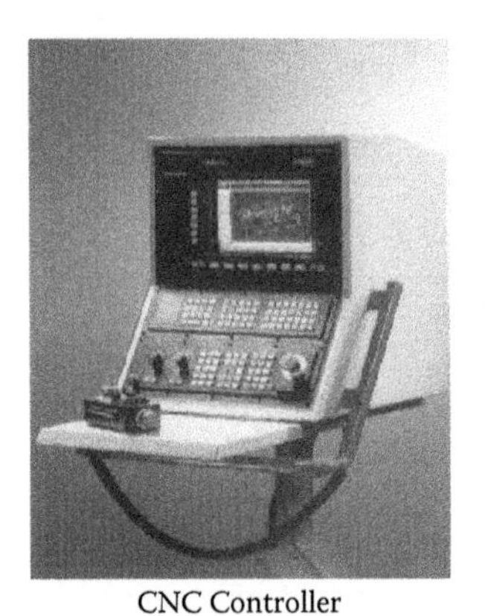

CNC Controller

Blower

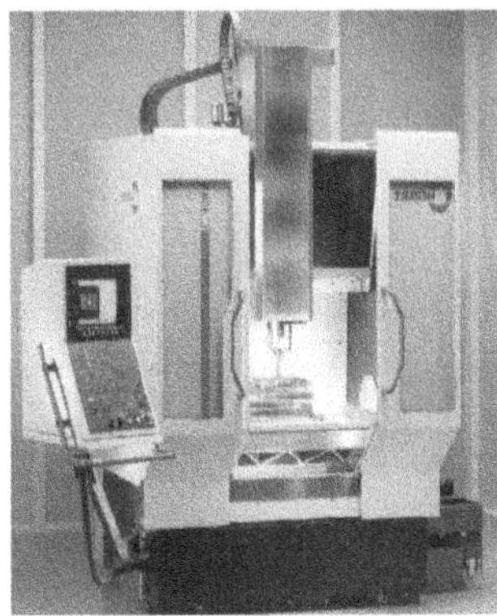

Milling Machine

Helicopter

Space Shuttle

Car

FIGURE 4.1 Examples of discrete products.

sequence. The fluidic and thermodynamic aspects of discrete products normally constitute a process that is physically enclosed by artifacts from solid mechanics. The activation of the process based on fluid mechanics or thermodynamics relies on input energy as is shown in Figure 1.5. The process operates according to the basic objectives of the design. The output from such a system is as shown in Figure 1.6. The complete system constituents are shown in Figure 1.4.

The design principles of discrete products have a general similarity. The design process in this context may be classified as is shown in Figure 4.2.

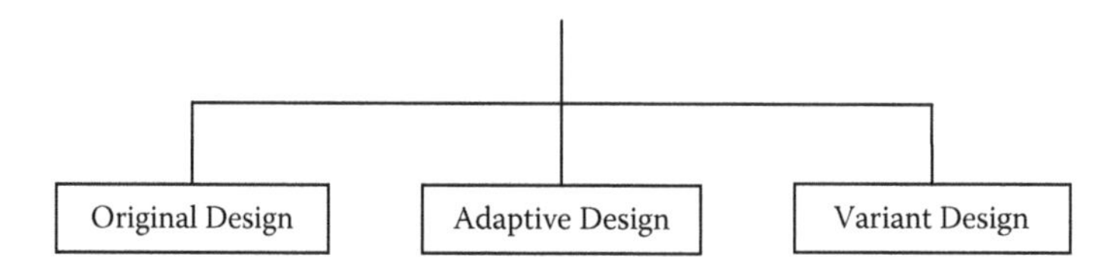

FIGURE 4.2 Engineering design classification.

An original design (typically 25% of all designs) goes through a rigorous analysis of need, specification, and concept design. Costing and marketing are thoroughly studied, and the use of multidisciplinary features (the norm in today's discrete products) is carefully examined (see Figure 4.3).

Adaptive design is normally the application of a known system to a changed task. In adaptive design, emphasis is placed on embodiment and detail design (Figure 4.3). Adaptive design is typically 55% of the total design tasks. An example of products exhibiting adaptive design is shown in Figure 4.4.

Variant design (typically 20% of total design tasks) is based on variation of size or arrangement of certain aspects of a chosen system. Embodiment design and detail design play an important role in this design category (Figure 4.3). Some products that display variant-design characteristics are shown in Figure 4.5. The design of components, structures, assemblies, mechanisms, machines, and processes follow the basic system approach to design in general, with appropriate parameters taken into consideration.

4.3 THREE-PHASE DESIGN PROCESS

The system approach to original design, adaptive design, and variant design follows a three-phase design process that comprises analysis, synthesis, and evaluation both locally (i.e., at each level) and globally (i.e., at the level of implementation of the whole design cycle).

The global analysis explores the underlying domain for viable design. This phase covers identification of need, market, cost, basic specification, and multidisciplinary features for a concept design. The tools used in this exercise may be surveys, CAD models, and real-time simulation. Complete details of the analysis phase of a concept design of discrete products are shown in Figure 4.6.

The synthesis phase is the embodiment design, which relates to the details about the composition of the products. This exercise includes permissible loading/exposure and problems to be encountered during production, during working life, or in the discarded state (see Figure 4.7).

The evaluation phase is performed through the detail design. In detail design, the product specification that originally emerged from the analysis phase (concept design) and was later improved in the synthesis phase (embodiment design) is finalized. This product specification is normally kept frozen for a certain period after its implementation at the production line. Apart from product specification,

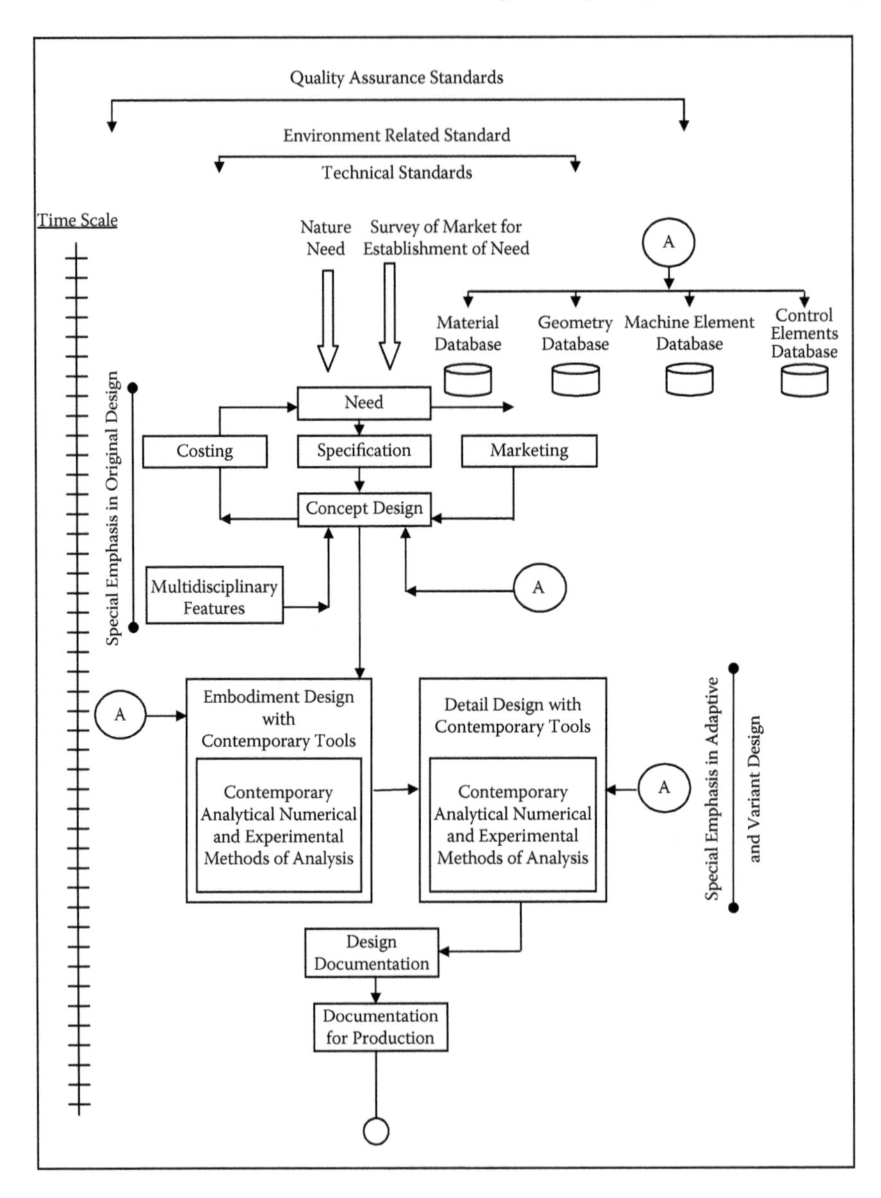

FIGURE 4.3 Contemporary system approach to original design, adaptive design, and variant design.

the other important output from detail design is the production documentation (Figure 4.8).

The design cycle, with all local and global analysis, synthesis, and evaluation stages, is still not complete. The cycle must be perfected through prototyping and pilot production. The product is later supplied to the market and consumer

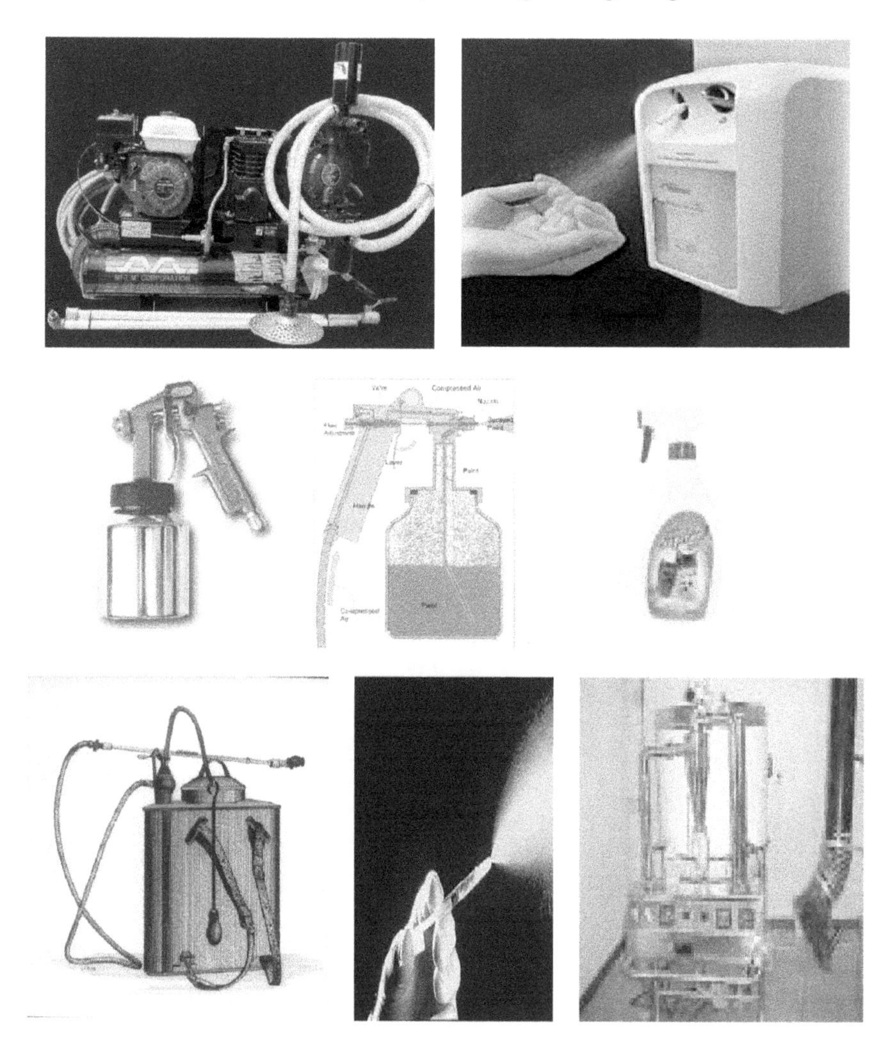

FIGURE 4.4 Products exhibiting adaptive design characteristics.

feedback plays an important role in upgrading the basic design or in development of another product in the same domain/range. This process is explained in Figure 4.9.

4.4 ROLE OF PRODUCT GEOMETRY

A discrete product, whether a structure, mechanism, machine, or process, uses a design scheme based on either original design, adaptive design, or variant design, along with local and global application of analysis, synthesis, and evaluation

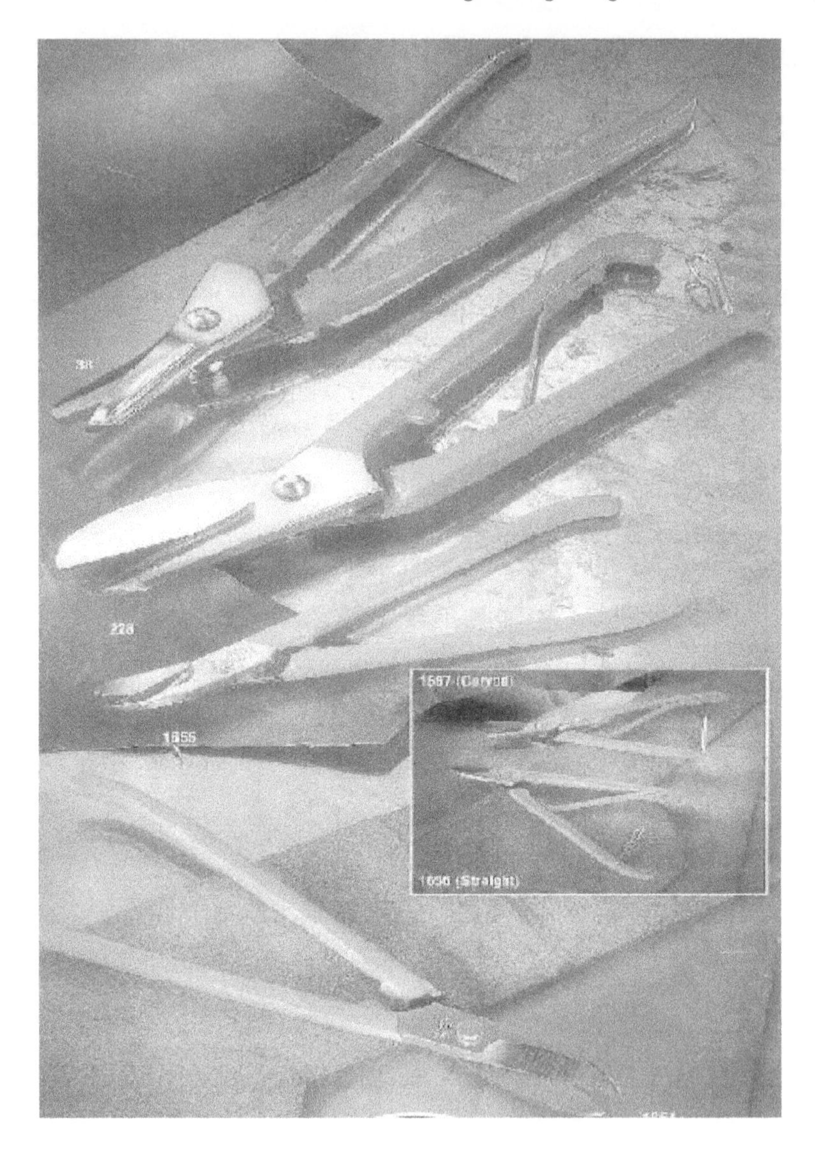

FIGURE 4.5 Products exhibiting variant design characteristics (courtesy of Footprint Tools, Sheffield, U.K.).

principles. The geometry of the product plays a central role in all phases of design because of its immense utility. The utility of the geometry extends to esthetics, man–machine interface, safety, reliability, and maintainability. These parameters are considered from concept to the final product.

The application of standards in the design of discrete products is limited only by the availability of this information. In every domain, at least one standards

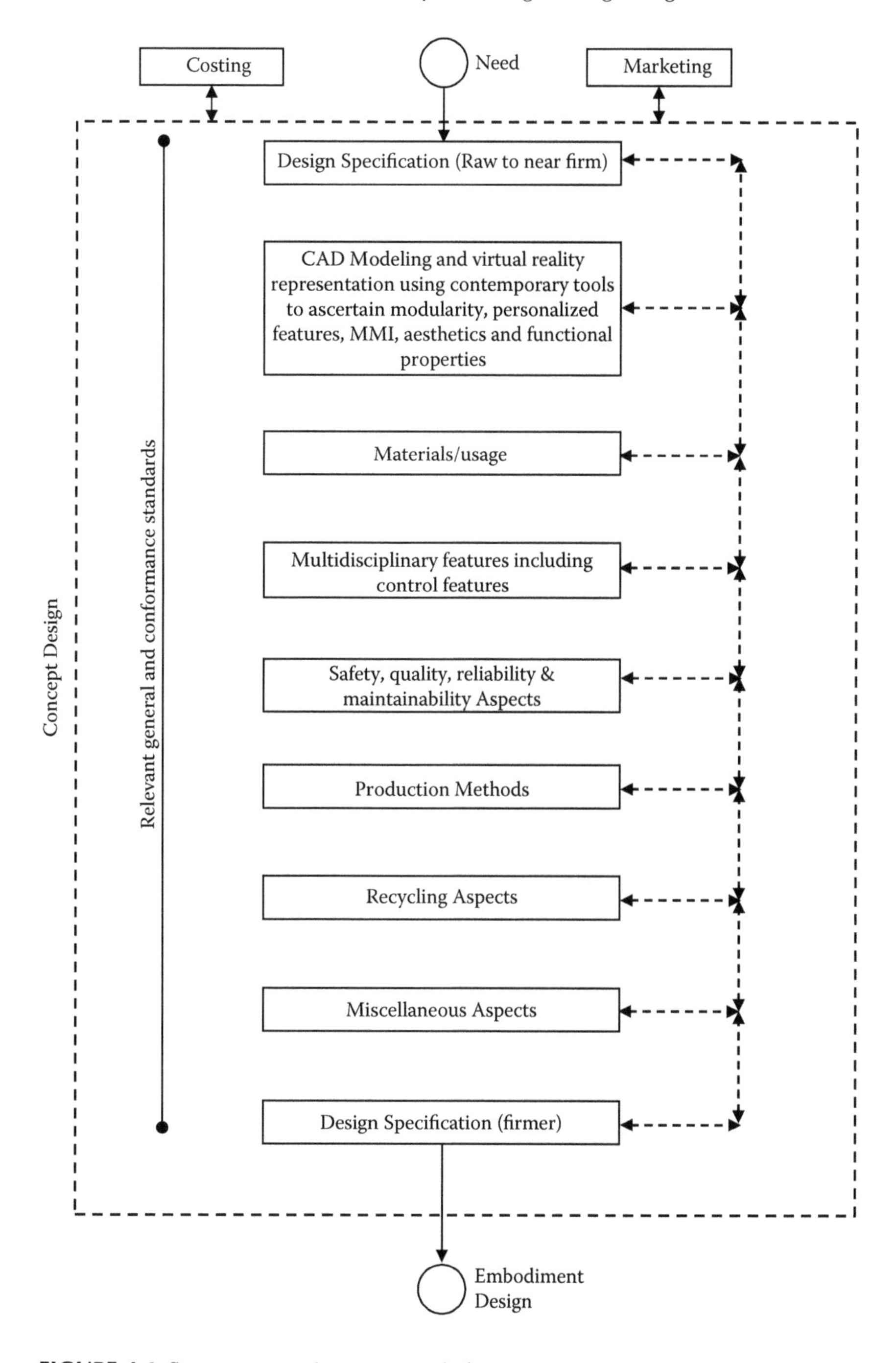

FIGURE 4.6 System approach to concept design.

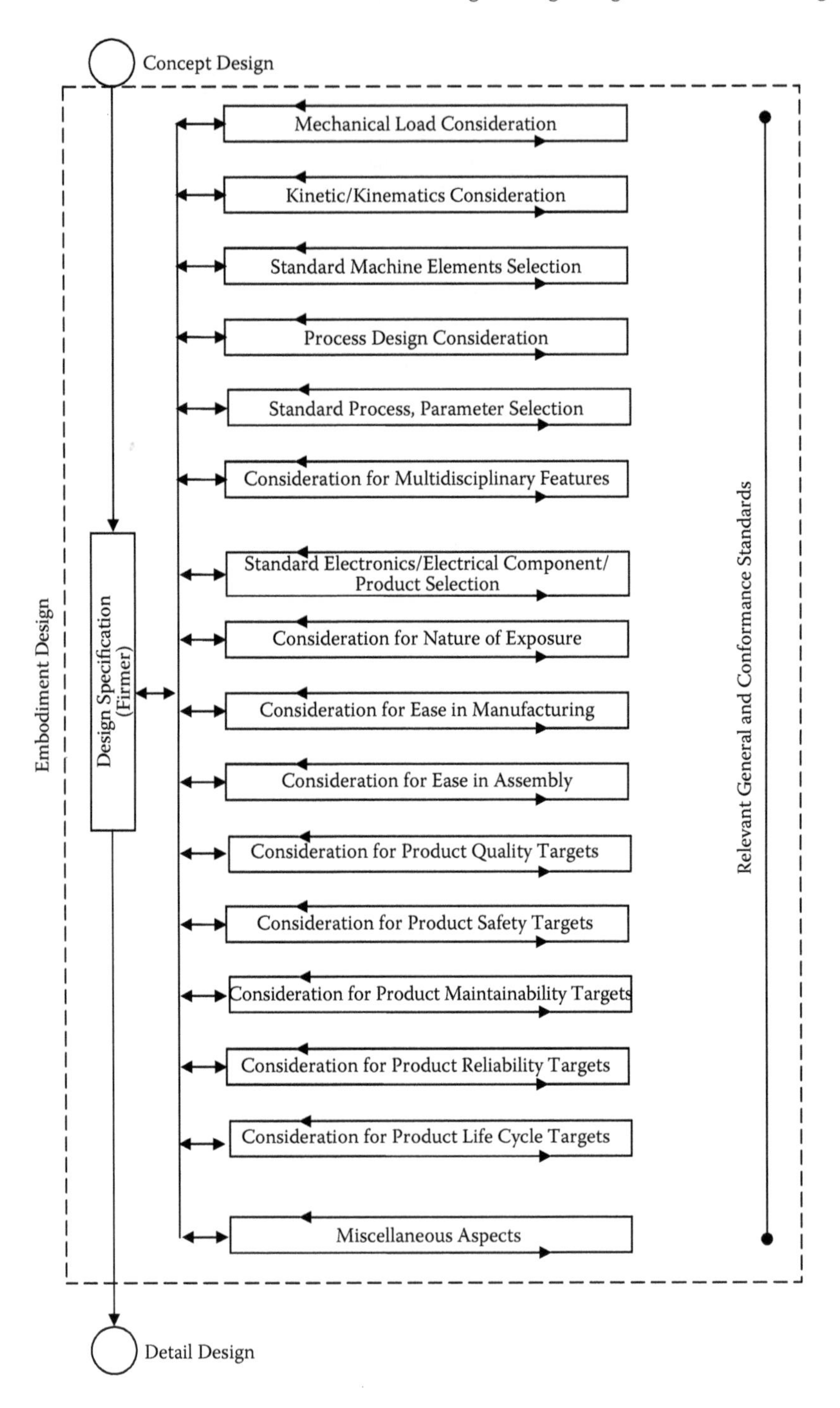

FIGURE 4.7 System approach to embodiment design.

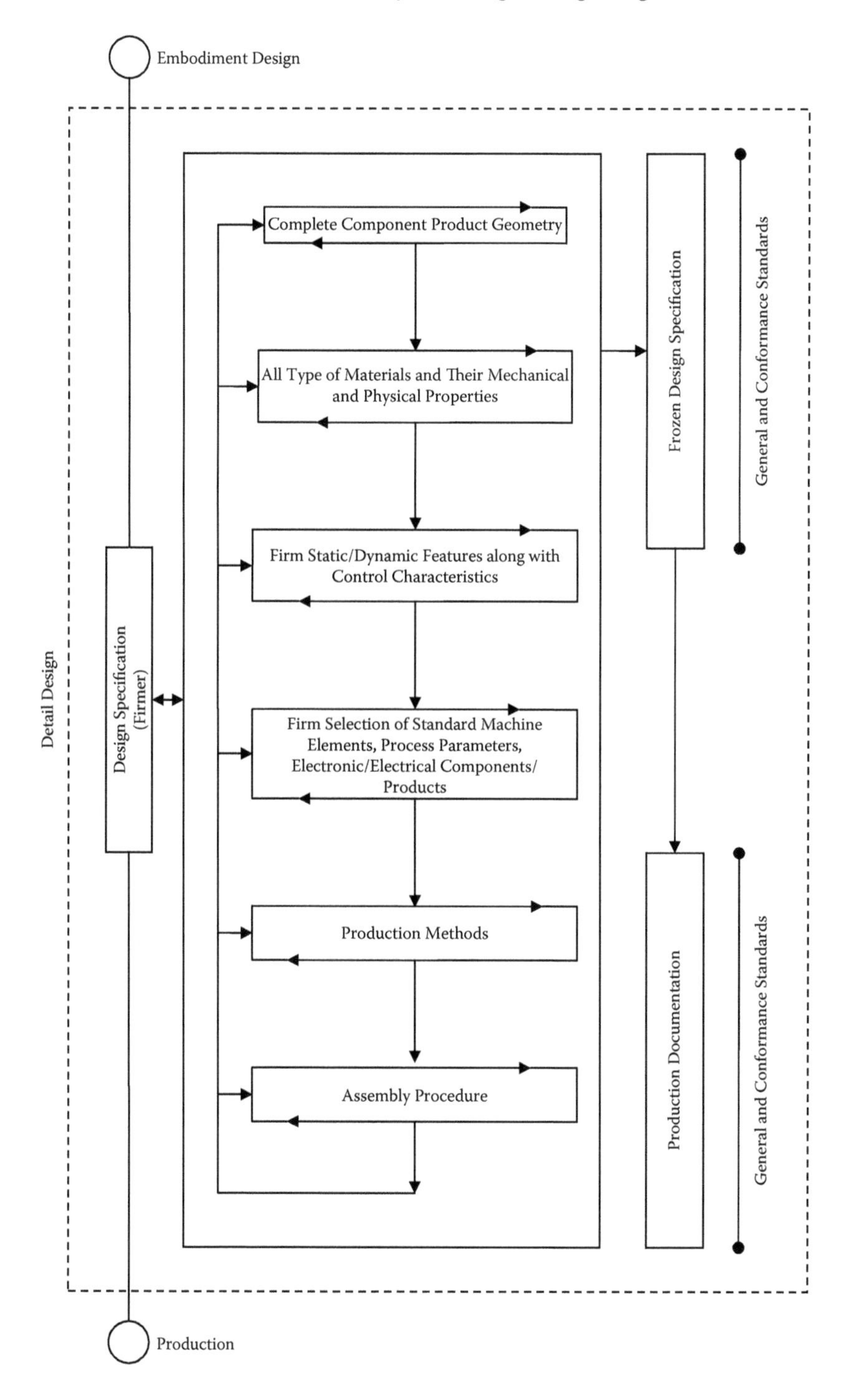

FIGURE 4.8 System approach to detail design.

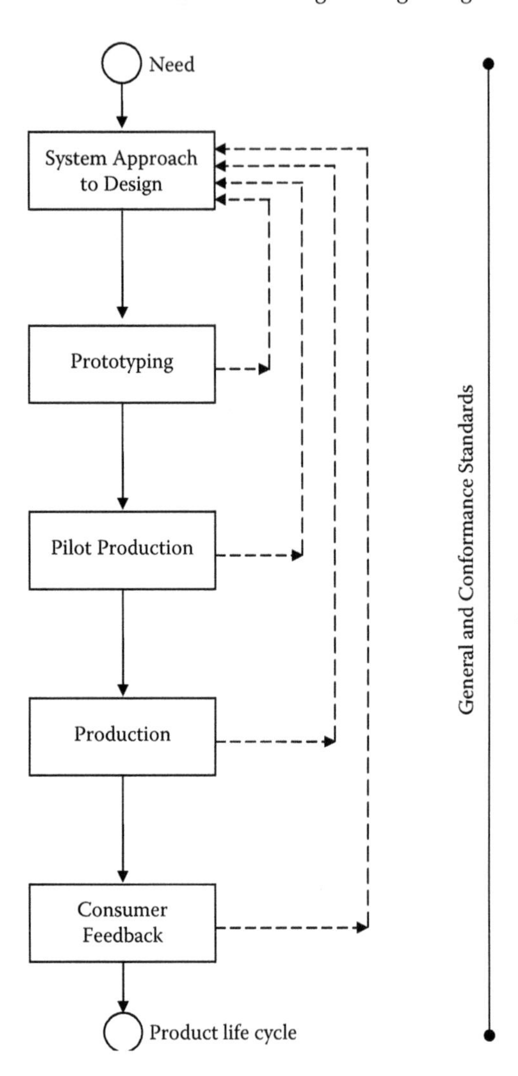

FIGURE 4.9 Universal design cycle.

organization exists to provide information related design parameters and function (see Chapter 2). Standards are also available for the geometry of discrete products. Such standards permit an understanding of products within organizations, between organizations, and by the product users. For example, standards are used to describe the product in first angle, third angle, and isometric projections. Standards are used to show production drawings in standardized formats. Standards are used to transfer the geometry data to different functions of the computer-integrated manufacturing (CIM) environment. These standards

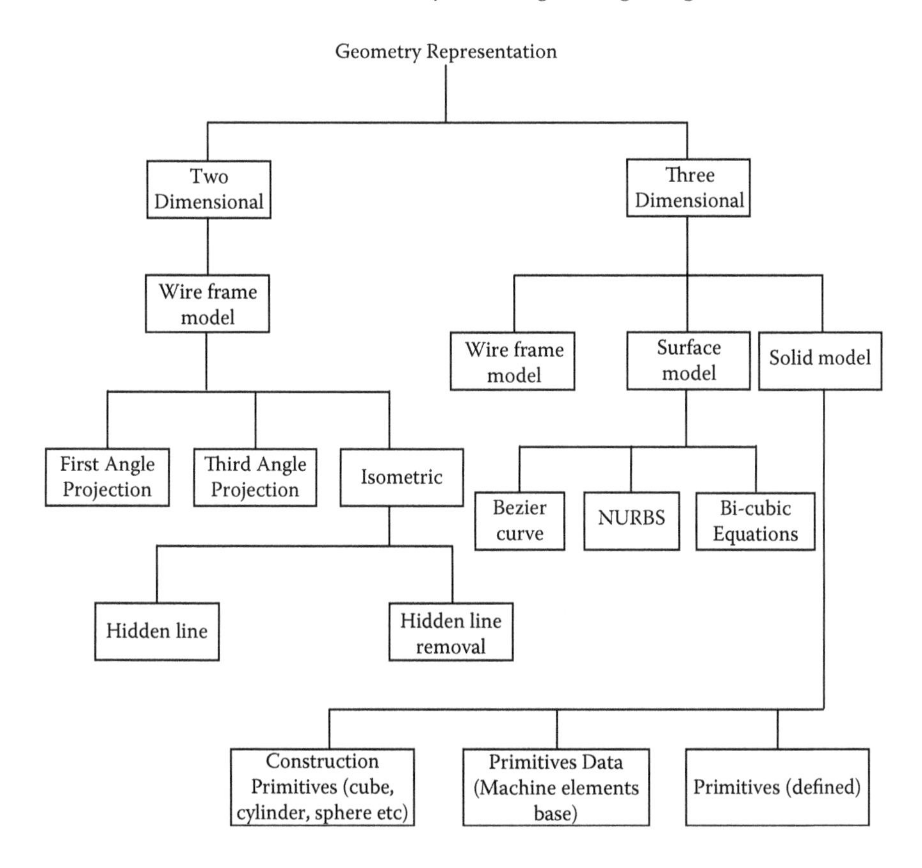

FIGURE 4.10 Geometry representation in two and three dimensions.

originate from international, national, and professional standards-developing organizations. Procedures used to define geometry of the product in a CIM environment are presented in Figure 4.10. Some of the prominent geometry representation and transfer standards are listed in Figure 4.11.

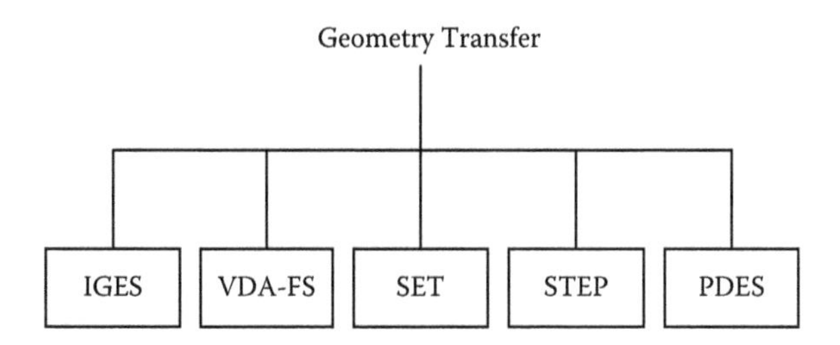

FIGURE 4.11 Common geometry-transfer protocols.

4.5 MULTIDISCIPLINARY FEATURES IN DESIGN

Multidisciplinary features constitute one of the most important aspects of the design cycle for discrete products. These features relate to, in order of priority, mechanical engineering, metallurgy, control engineering, electrical engineering, electronics, computer engineering, and computer science (see Figure 4.12). Relevant information based on company standards, professional standards, and national and international standards is always available to simplify the design cycle. Use of standards in the design cycle brings more certainty and confidence to the design, manufacturing, and operation of the product.

4.6 MECHANICAL LOAD ANALYSIS

Mechanical loading in discrete products is more prominent than other types of loading or exposure. The types of loads and exposures experienced by discrete products are presented in Figure 4.13 and Figure 4.14.

The analysis of mechanical loads requires that these loads be divided as in Figure 4.15. The current available methods of analyzing mechanical loads of various kinds are given in Figure 4.16. The basic principle of mechanical load analysis for safe operation is presented in Figure 4.17.

A wealth of literature is available that presents details of analytical, numerical, and experimental analysis of various types of loads for different ranges in single-load or combined-load applications. Some of the commonly used methods of analysis for general loading conditions and some practical situations are described here.

Figure 4.18 and Figure 4.19 describe the elastic calculation for strength both in elastically determinate and in elastically indeterminate fashion. Figure 4.20 lists the analysis procedure for design for bending. Figure 4.21 outlines some of the cross sections encountered in torsion. Figure 4.22 presents the method used for predicting column buckling under centric and eccentric loading. Figure 4.23 shows a contact-loading problem in different situations. Figure 4.24 provides impact-loading classifications. Figure 4.25 explains the methods for analyzing harmonic and nonharmonic vibration. Figure 4.26 illustrates cyclic loading for fatigue or creep. Standard components made of selected standard materials and made to standard dimensions are available or can be made to order under standard or specified loading conditions. The Appendix at the end of this book provides a list of suppliers engaged in this activity.

Calculations for single-load or multiple-load conditions for components, assembly, mechanism, structure, or machine show that the safe operation or use of discrete products is dependent upon the application of proper (local and global) safety standards to each component and the whole product. The theory of failure, as discussed later, provides important guidelines in the determination of safe operating ranges of mechanical artifacts.

Section 4.9 onward presents examples of standards used in the design of discrete products.

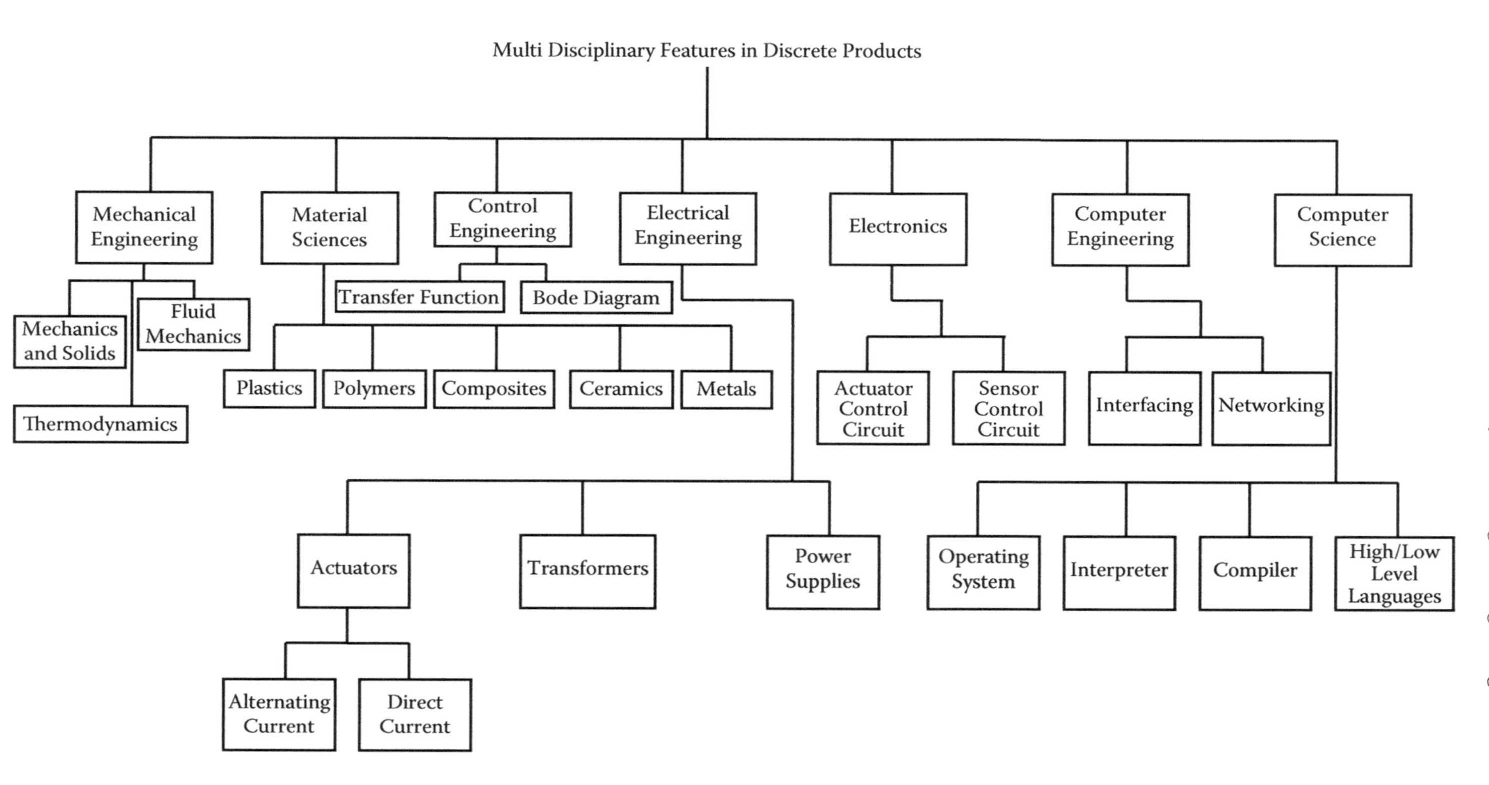

FIGURE 4.12 Multidisciplinary features in discrete products.

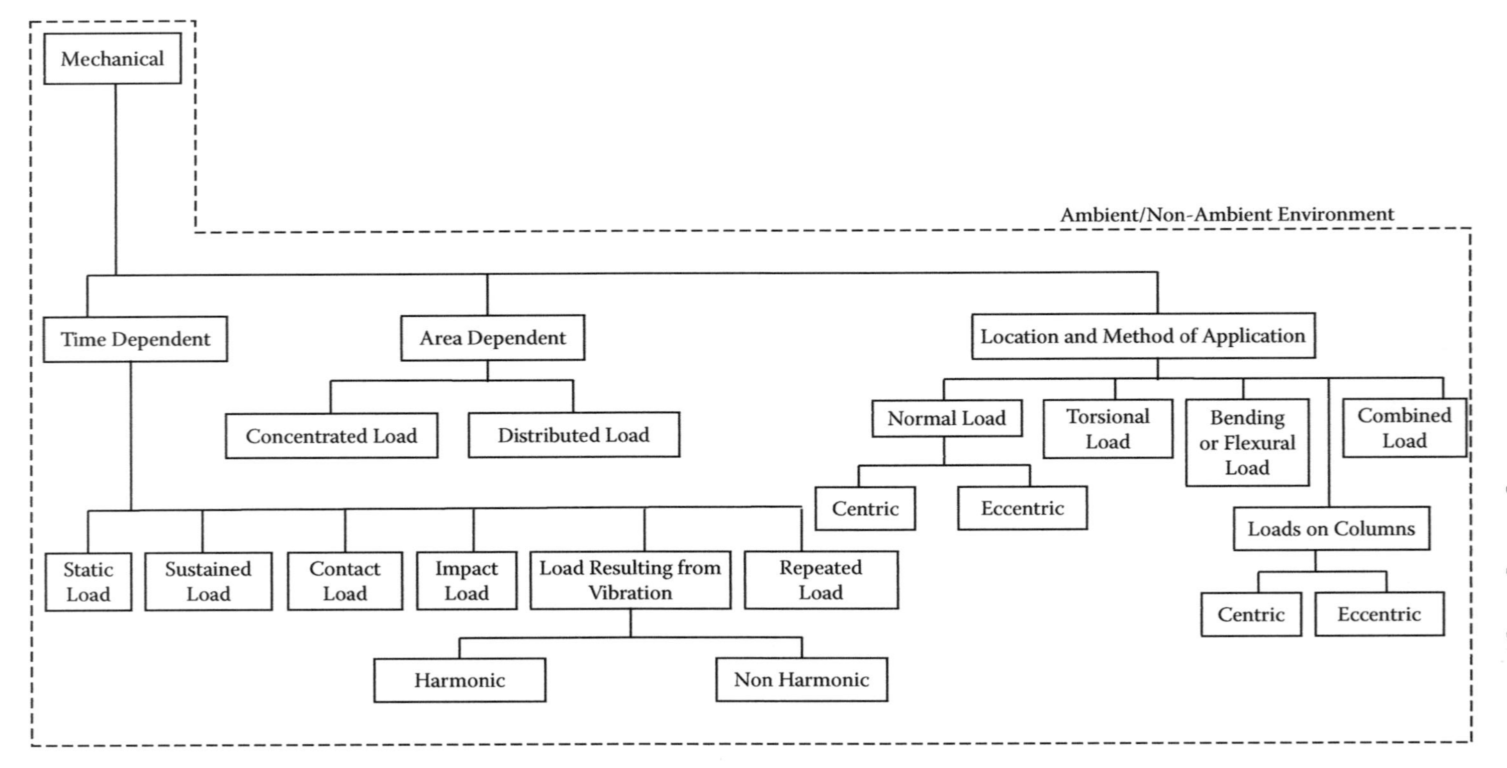

FIGURE 4.13 Mechanical loading for discrete products.

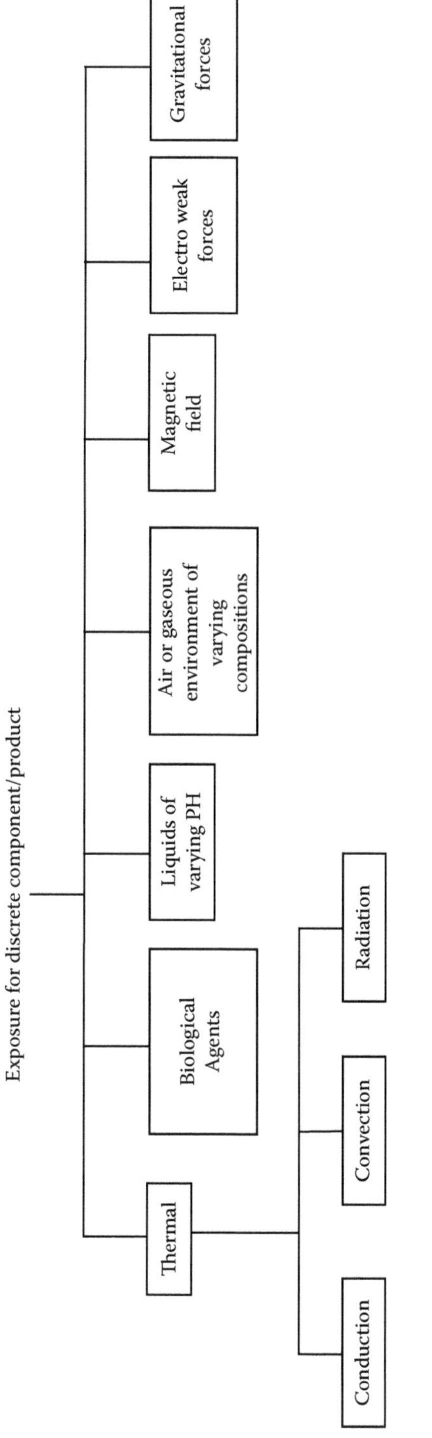

FIGURE 4.14 Exposure for discrete component/product.

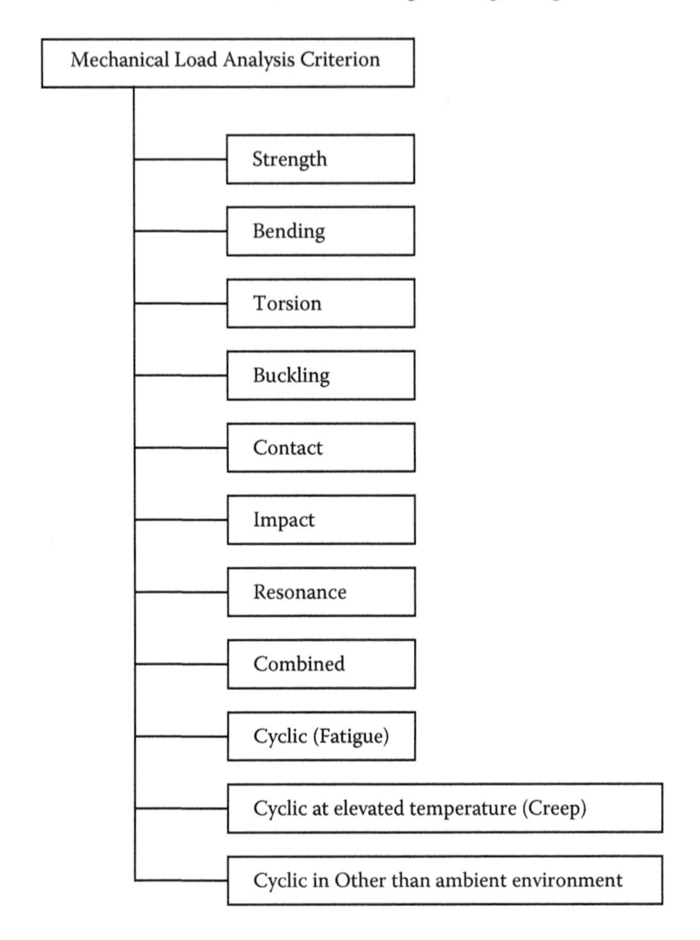

FIGURE 4.15 Classification of mechanical loading analysis.

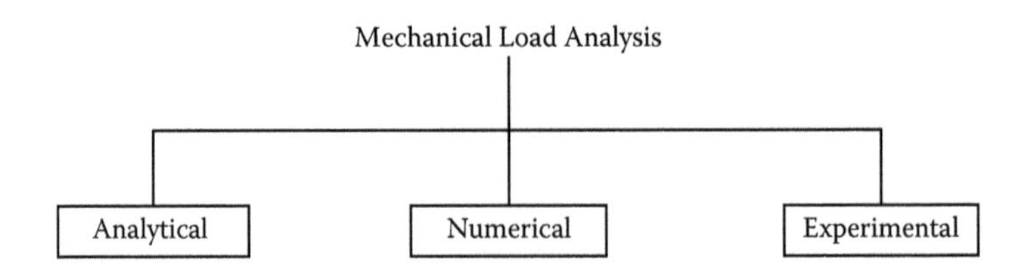

FIGURE 4.16 Methods for the analysis for mechanical load.

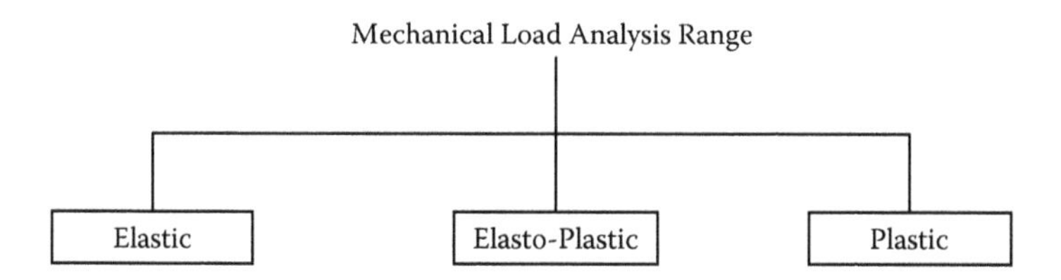

FIGURE 4.17 Mechanical load analysis range.

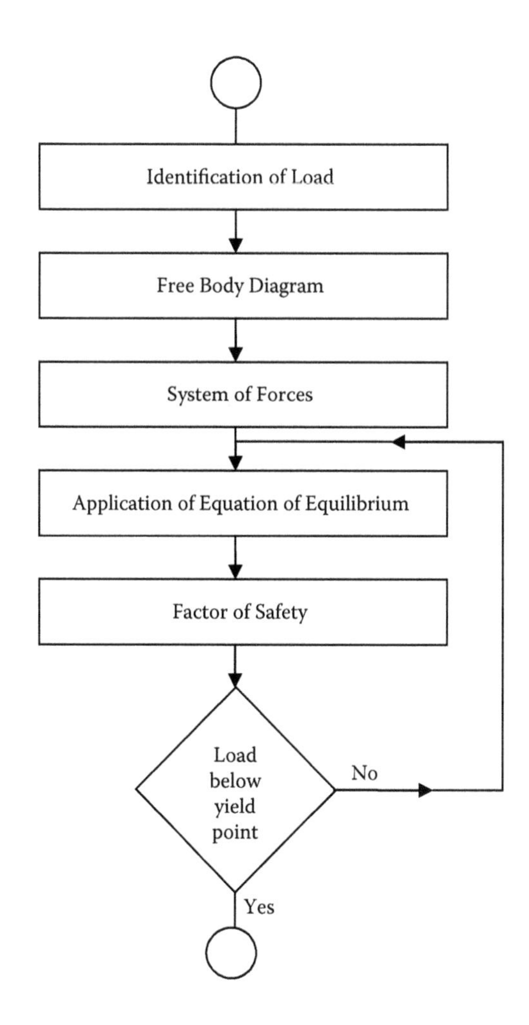

FIGURE 4.18 Statically determinant elastic-calculation procedure for strength.

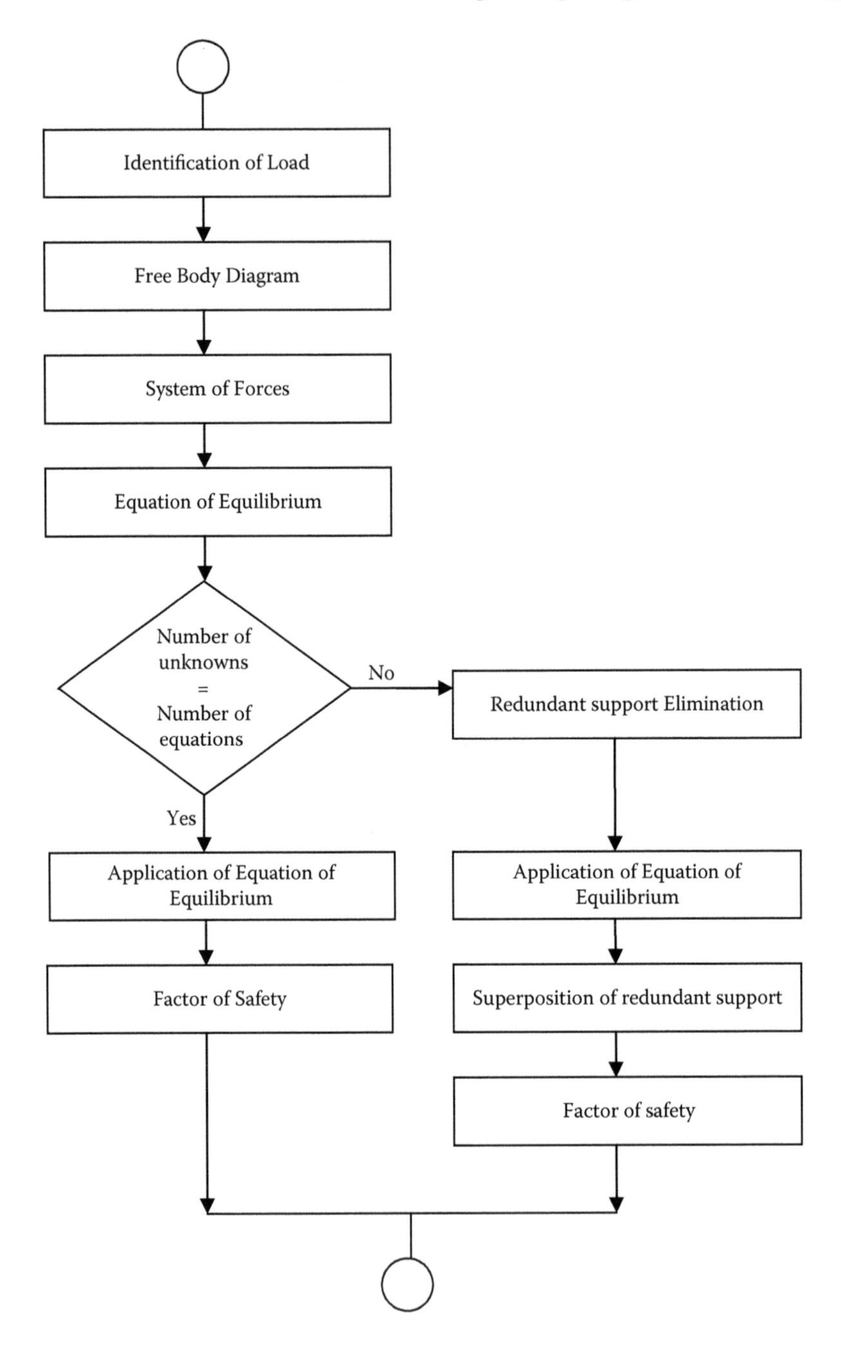

FIGURE 4.19 Statically indeterminate elastic-calculation procedure for strength.

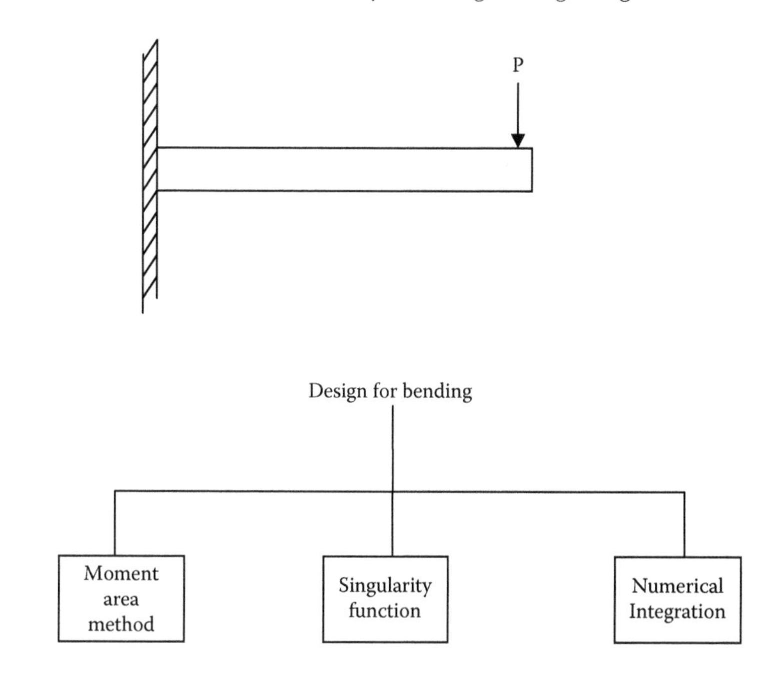

FIGURE 4.20 Analysis methods for design for bending.

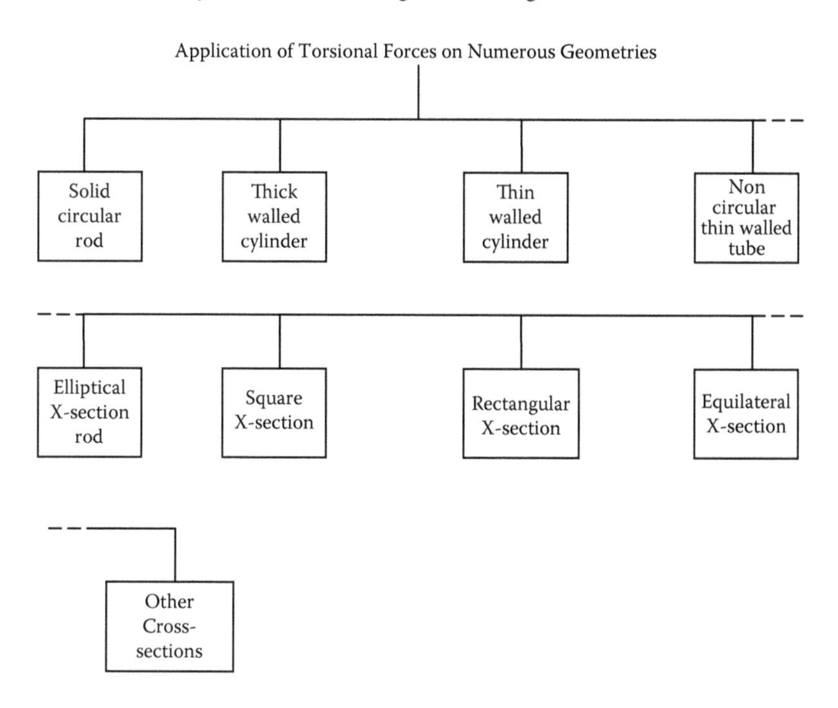

FIGURE 4.21 Cross sections for torsion of mechanical components.

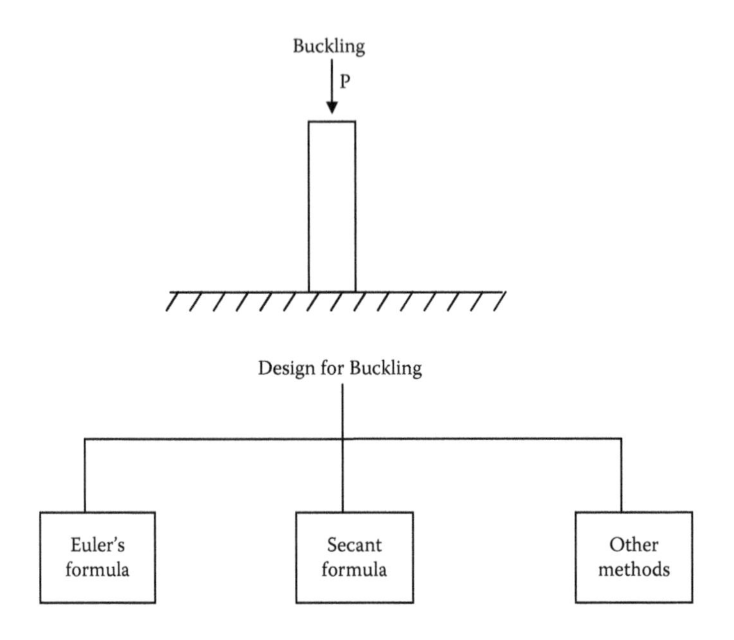

FIGURE 4.22 Analysis methods for buckling of columns.

4.7 THE ENGINEERING MATERIALS

With the availability of a vast variety of materials for the manufacturing of discrete products, use of different materials in a versatile manner has become common. Standards that provide mechanical and physical properties of these materials with their form (cross section) are available from manufacturers and standards organizations. The Appendix provides a list of manufacturers those supply engineering materials.

4.8 THE MACHINE ELEMENTS

Other important factor in the engineering design of discrete products is the machine elements. These elements are used in almost every discrete product. A

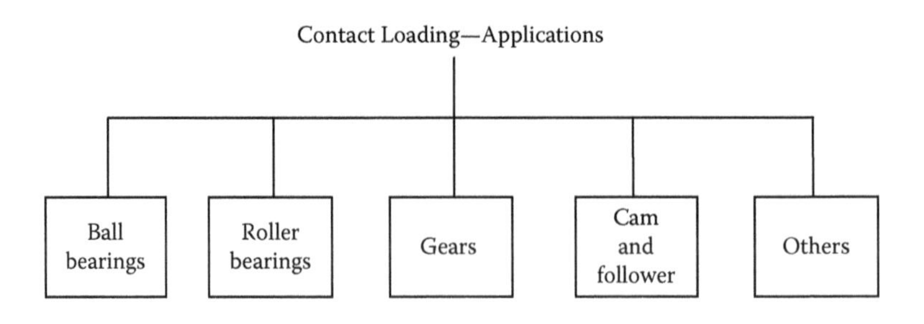

FIGURE 4.23 Application of contact-loading problem.

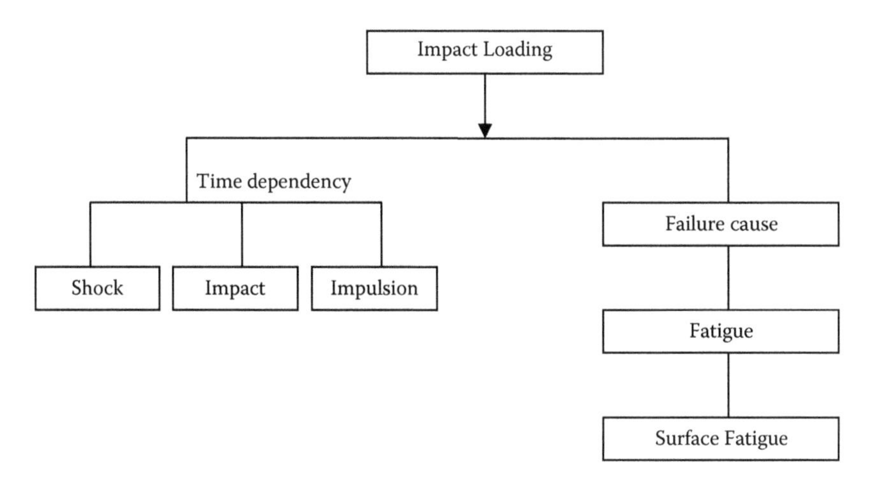

FIGURE 4.24 Time dependency and failure causes in impact loading.

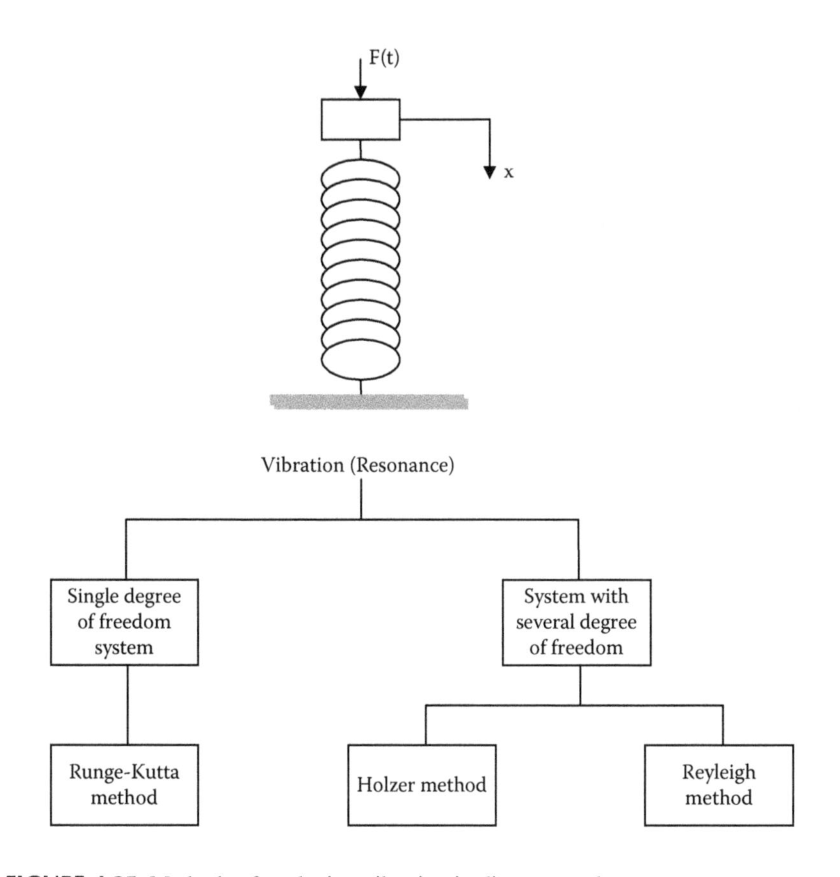

FIGURE 4.25 Methods of analyzing vibration in discrete products.

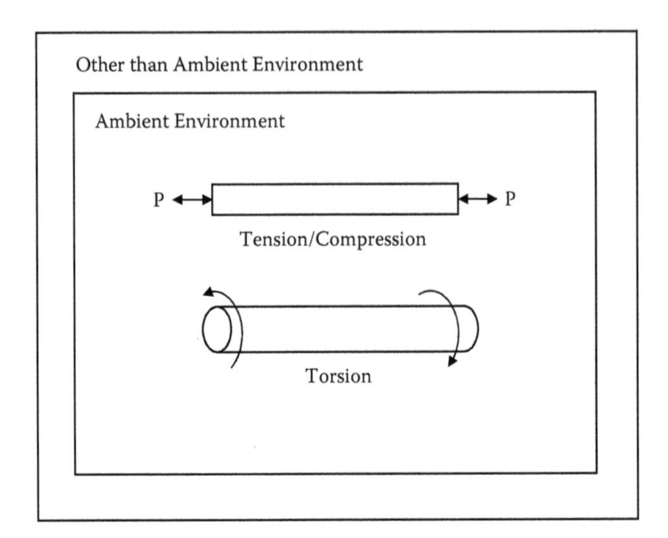

FIGURE 4.26 Fatigue and creep in discrete products.

list of commonly used machine elements is provided in Figure 4.27. Machine elements come in a large variety and are standardized by national standards bodies, by professional standards bodies, and by companies that manufacture these elements. Some common machine elements are shown in Figure 4.28.

The first priority of the designer is to choose a standard machine element for a discrete product under development. This choice may force the designer to change several other parameters, within safety limits, while going through the iterative procedure of design to adjust the chosen standard machine elements. Machine element databases are now commonly provided as part of CAD software packages. The Appendix provides a list of standard machine-element manufacturers.

4.9 THE CONTROL ELEMENT

The control system is yet another of the most important factors in the design of discrete products. The general control-system design is an open-loop or closed-loop configuration. The development of the control system for a discrete product is well governed by contemporary theories. Control-system analysis is performed according to procedures routinely described in the pertinent literature.

The contemporary technology of sensors, transducers, and actuators pose a selection challenge to the designer when the implementation of a control system is necessary. Sensors and transducers that measure various physical parameters are available in different shapes to be accommodated in all types of perceivable spaces, range, measurements, and accuracy to suit differing precision requirements. Standards exist from both the manufacturers and the standards organizations to cover this pivotal element in design. Figure 4.29 provides a list of common types of sensors and transducers, and Figure 4.30 shows some sensing and conditioning equipment.

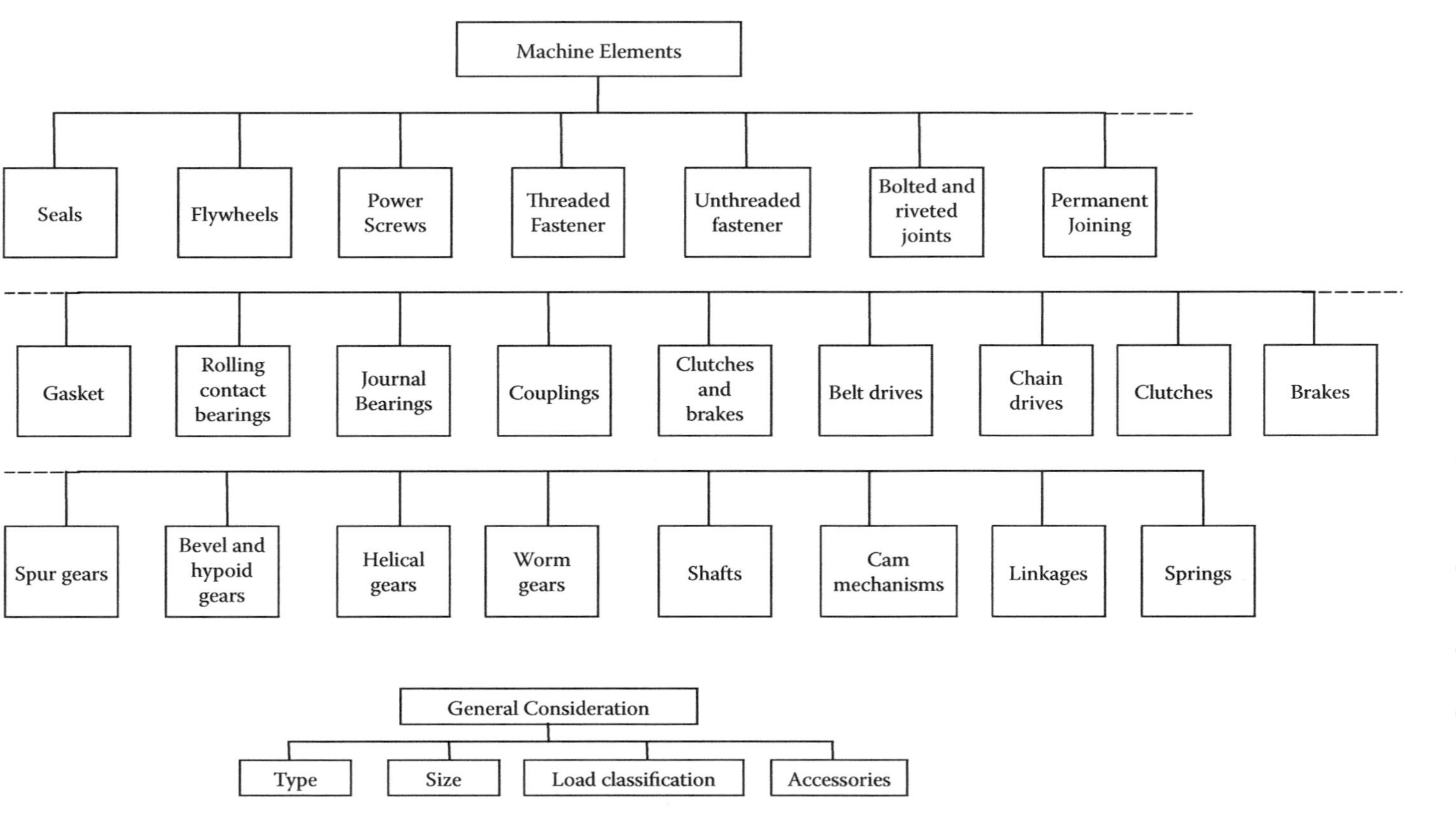

FIGURE 4.27 Machine elements used in discrete products.

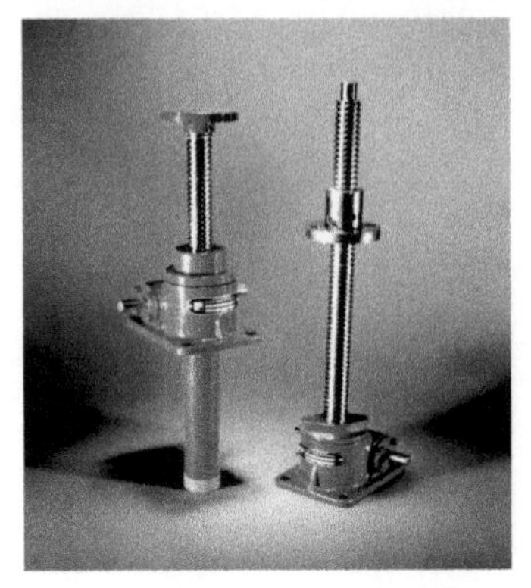

Metric Ball Screw Jacks

Machine Screw Jacks

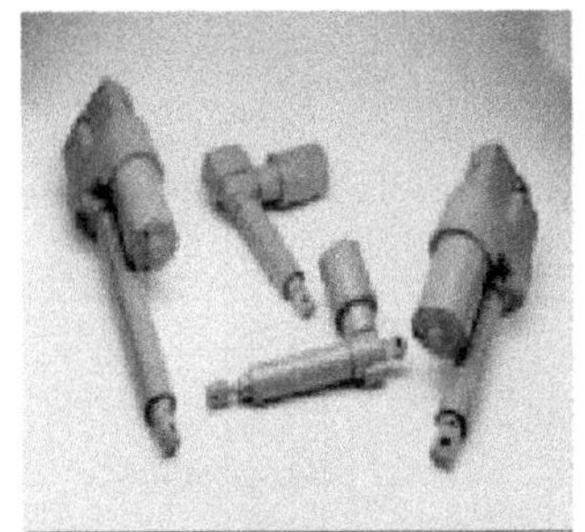

AC Actuator

Mechanical Ratchet Jacks

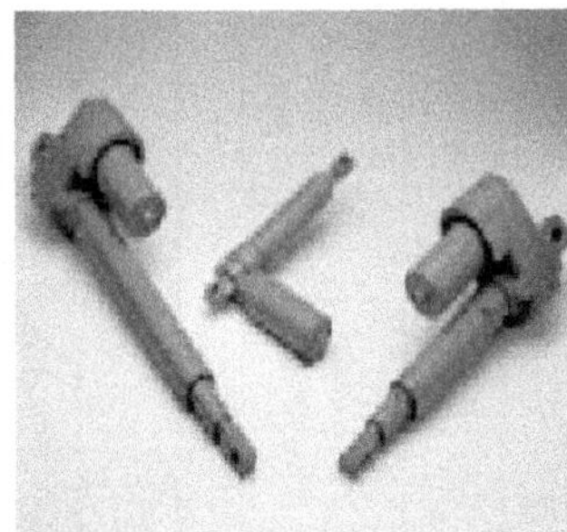

DC Actuator

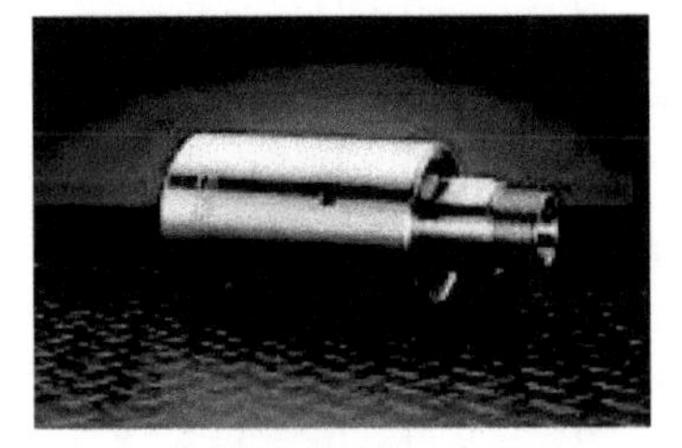

Rotary Union

FIGURE 4.28 Some machine elements (courtesy of Power Jacks Ltd., U.K.).

Actuators provide the required linear or rotary motion in a control system implemented as a discrete product. Actuators are also available in different types, sizes, and accuracy according to standards set by the manufacturers and standards organization. Figure 4.31 shows some common types of actuators. Electrical, electronic, and computer engineering custom-designed and off-the-shelf products are also available in conformance with company, national, and international standards.

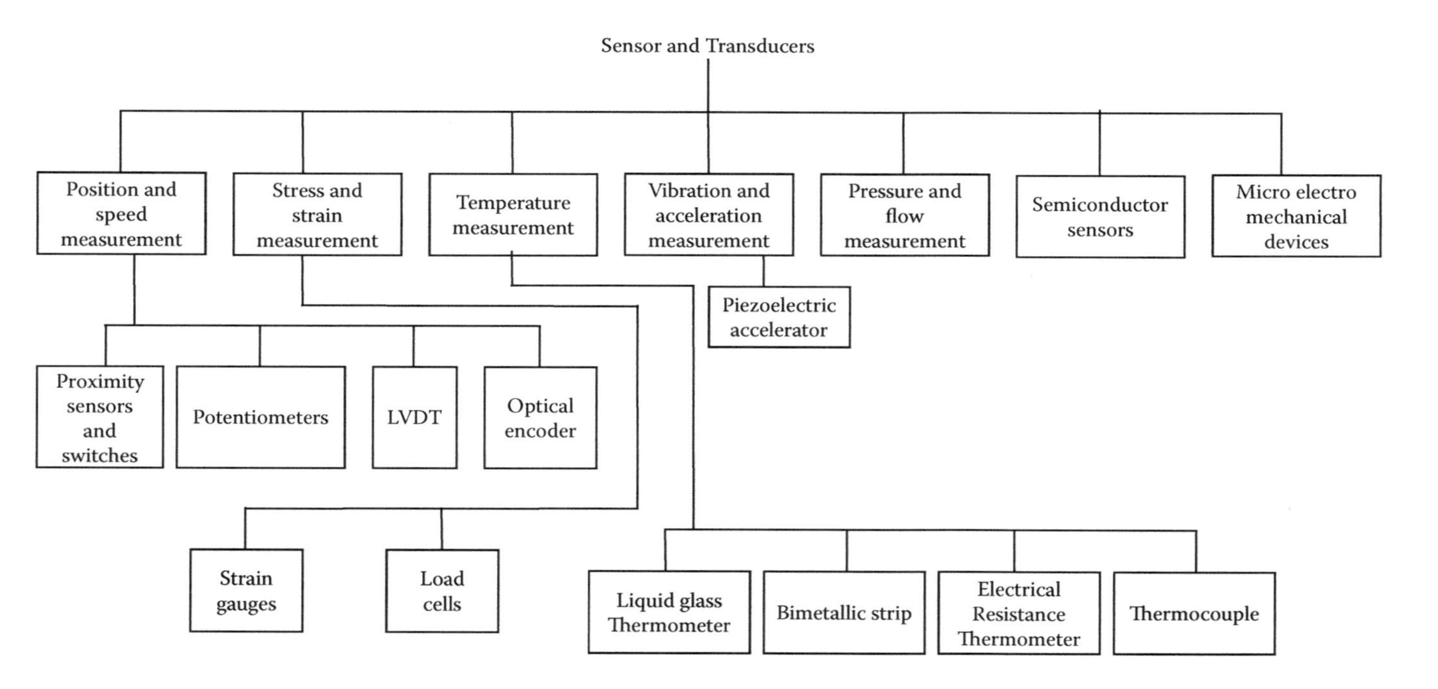

FIGURE 4.29 Sensors and transducers used in discrete products.

Audio power amplifier

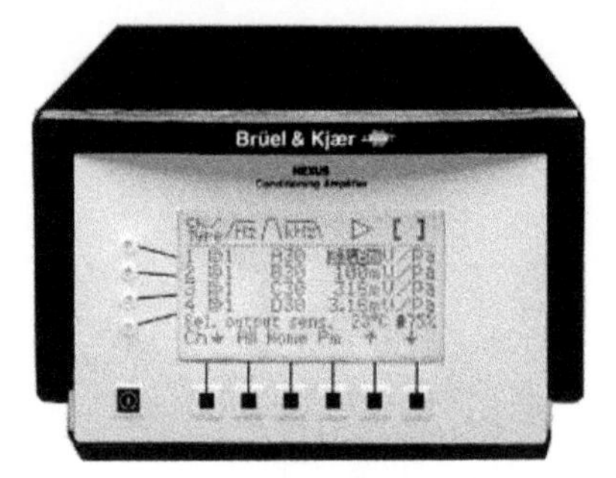

Nexus Conditioning Amplifier

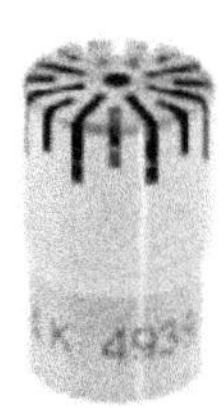

Free-field Microphones

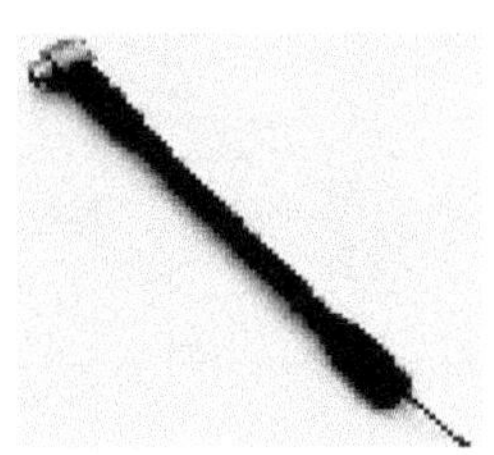

Impact Hammer

Modal exciter

FIGURE 4.30 Some sensing and conditioning equipment (courtesy of Bruel & Kjaer, Denmark).

The use of computer software in discrete products is also governed by well-accepted practices such as object-oriented programming and agile software development. Software products such as operating systems, real-time operating systems (RTOS), interpreters, and compilers are all widely available and conform to company standards.

The Appendix provides a list of manufacturers engaged in the production of control elements.

4.10 COMMON FACTORS RELATED TO OVERALL DESIGN PROCESS

The design of discrete products that comprise, for example, components, structure, mechanism, machine, process, and control system is a complex iterative procedure that requires optimality of objective function from microlevel to macrolevel tasks. Each area, such as mechanism, process, or the control system must be analyzed under the system approach to design with appropriate subsystem abstraction and the use of relevant parameters through analytical, numerical, or experimental procedures. Appropriate standards at each stage should be considered.

The dimensions of the components or the values of the parameters that result from design calculations are compared with a standard value. In some cases, a

FIGURE 4.31 Bosch Servodyn D drive system. (Courtesy of Bosch GmbH.)

direct comparison is available from the relevant standard, whereas in other cases, a deduction must be made. Normally a higher value than the calculated dimension/parameter is chosen. This selection may require altering the value of other dimensions/parameters to adjust the standard value.

The final selection of appropriate materials for a discrete product requires analysis (iterative by nature) for size, weight, and tolerance to mechanical and other types of loads and exposure with or without surface treatment and consideration of recycling and disposal.

The exercise involves performance of all the design steps (i.e., concept design, embodiment design, and detail design) in minute detail and consideration of the availability and adaptability of general mechanical components, machine elements, control elements, and software components. Generally, because of the different scope for design of various features in discrete products, only bridging parameters such as torque required to drive a power screw (a mechanical system) from an electrical motor (an electrical system) are matched, while the rest of the calculations under each design domain remain independent of each other.

The availability of standards for parameters (as in thermodynamics), elements (as in machine elements), and software components (as in software engineering) provides certainty in up-front design. However, each subject area has its own method of declaring a product or part of a product safe or unsafe (e.g., failure theories for mechanical components) (Table 4.1). Methods are also available to perform formal modeling, analysis, and verification of hybrid industrial systems. Similarly, electrical and electronic systems and components failure-analysis

TABLE 4.1
Theories of Failure for Mechanical Loading

Maximum Normal Stress Theory
[W.J.M. Rankine]
"Failure will always occur whenever the greatest tensile stress tends to exceed the uniaxial tensile strength or whenever the largest compressive stress tends to exceed the uniaxial compressive strength."
Maximum normal stress theory correlates well with brittle fracture.

Maximum Shear Stress Theory
[C.A. Coulomb, H. Tresca, J.J. Guest]
"Plastic flow occurs when the maximum shear stress exceeds a critical value."

$$\zeta_{max} = \tfrac{1}{2}(\sigma_1 - \sigma_3) \geq \text{constant } k = \zeta_{cnt}$$

Maximum shear stress theory correlates well with ductile fracture.

Maximum Distortion Energy Theory/Maximum Octahedral Shear Stress Theory
[J.C. Maxwell, M.T. Hueber, R. Von Mises, H. Henky]
"Yielding occurs when the work of deformation per unit volume provided by the system of stresses exceed the critical value for the particular material."
Maximum distortion energy theory correlates well with ductile fracture.

Modified Mohr Theory/Coulomb-Mohr Theory
"An extension of maximum shear stress theory applied to failure of brittle materials."

procedures are also available. Furthermore, the susceptibility to failure in varying environments by mechanical components is much lower than susceptibility to failure by control elements.

4.11 REVIEW OF STANDARDS FOR ENGINEERING DESIGN OF DISCRETE PRODUCTS

The number of related parameters for design of discrete products is so large that listing existing standards related to engineering design of discrete products in this book is not possible. Thus, we define a domain specification, a target standards organization, and the search parameters to constrain the list of available standards. All the headings of various sections in the domain specification for this review are provided in Figure 4.32.

American National Standards Institute (ANSI) is chosen as the target standards organization. A search of the online catalog of the standards organization is performed by use of the search parameter usually from the domain specification. For each domain specification, alternative standards organizations are also listed. The filtered listing of standards, normally diverse, are included here as the achievement of the whole exercise. Sometimes this information may be useful to the reader (designer), whereas at other times, some attempts at the alternative

1. Design of Discrete Products
2. Measurement Standards
3. Standards Related to Materials
4. Standards for the Operation of the Product
5. Safety Standards Related to Product
6. Quality Standards Related to Product
7. Reliability Standards Related to Product
8. Standards Related to Maintainability
9. Product Life Cycle and Emerging Standards

FIGURE 4.32 Domain specification related to standards for design of discrete products.

standards organization may be required. The reader may also refer to the information provided in Chapter 2 about standards, standards organizations, and the search for standards for discrete products under consideration.

The objective of this section is to underline the quantity and diversity of the availability of design parameters as standard values. In a real design exercise, the choice to use the standard values for design parameters is made by the design engineer (team). Subsequently, an exercise as detailed here can be used to search the standard values.

Chapter 2 lists many important standards-developing organizations that provide major parameters that may be considered at a discrete-product design and manufacturing facility.

In most cases, all the standards needed for a discrete product cannot be obtained from a single source. The search then turns to investigation of the committee work of the target organization to find out what directions it is taking, to requesting information on new developments from the target organization, or to contacting other organizations for the alternative standards.

4.12 STANDARDS FOR THE DESIGN OF DISCRETE PRODUCTS

Because of the diversity of discrete products, a large number of standards are available from various organizations for related domains in which the discrete products principally fall.

Specifically, all the major national standards bodies are recommended for search in this category. Generally, the professional standards bodies related to the domain specification are also recommended. These bodies include the following:

1. The American Society of Heating Refrigerating and Air-Conditioning Engineers (ASHRAE)
2. The American Society of Testing and Materials (ASTM)

3. The American Society of Mechanical Engineers (ASME)
4. The Electronics Industries Alliance (EIA)
5. The Human Factor and Ergonomics Society (HFES)
6. The Hydraulic Institute (HI)
7. The Institute of Electrical and Electronics Engineers (IEEE)
8. The Association for the Advancement of Medical Instrumentation (AAMI)
9. The National Fluid Power Association (NFPA)
10. The National Committee for Information Technology Standards (NCITS)
11. The Fluid Control Institute (FCI)
12. The Heat Exchange Institute (HEI)
13. NACE Foundation (NACE)

The following key words can be used to search the online catalog for this area:

1. Design
2. Design <discrete product>
3. Design <component of the discrete product>
4. Design <process>
5. Design <process parameter>

The key words enclosed in angle brackets refer to individual products or processes. A brief listing of the standards from ANSI in this category is presented in Table 4.2.

4.13 MEASUREMENT STANDARDS

The proper function of control features in a discrete product requires measurement of one or more process parameters. The magnitude of the measured parameter values then causes the system to take certain actions. These parameter values can be provided by standards-developing organizations.

All the major national standards bodies are recommended for the search for measurements of the physical parameters and relevant professional standards bodies are recommended for the search for specifications of sensors, transducers, and actuators (and their precision). These bodies include the following:

1. The American Society for Quality (ASQ)
2. The Association Française de Normalization (AFNOR)
3. The Japanese Industrial Standards Committee (JISC)
4. Deutsches Institut für Normung (DIN)
5. The British Standards Institute (BSI)
6. The American National Standards Institute (ANSI)
7. The International Organization for Standardization (ISO)

TABLE 4.2
Selected ANSI Standards Related to Design of Discrete Products

Serial Number	Document Number	Document Title
1	VDI/VDE 3850 Blatt 1	User-Friendly Design of Use Ware for Machines (Foreign Standards)
2	VDI 6030 Blatt 1	Designing Free Heating Surfaces: Fundamentals—Designing of Heating Appliances (Foreign Standards)
3	VDI 2246 Blatt 2	Designing Maintainable Engineered Products—Requirements Catalog (Foreign Standards)
4	VDI 2130	Crank and Rocker Mechanisms: Design and Computation in Four-Link Planar Linkages (Foreign Standards)
5	US PRO ISO 10303-214:2001	Product Data Exchange Using STEP Part 214—Application Protocol: Core Data for Automotive Mechanical Design Processes
6	US PRO ISO 10303-212:2001	Product Data Exchange Using STEP Part 212—Application Protocol: Electrotechnical Design and Installation
7	US PRO ISO 10303-210:2001	Product Data Exchange Using STEP Part 210—Application Protocol: Electronic Assembly Interconnection and Packaging Design
8	US PRO ISO 10303-203:1994	Product Data Exchange Using STEP Part 203—Application Protocol: Configuration Controlled Design
9	SS-EN 954-1	Safety of Machinery: Safety-Related Parts of Control Systems—Part 1: General Principles for Design (Foreign Standard)
10	SS-EN 953	Safety of Machinery—General Requirements for the Design and Construction of Fixed and Movable Guards (Foreign Standard)

A brief listing of ANSI standards in this category is given below. The following key words can be used to search the online catalog for this area:

<process> <parameter>
<sensor>
<transducer>
<actuator>

A short list of the ANSI standards from this category is provided in Table 4.3.

TABLE 4.3
Selected ANSI Standards Related to Input, Process, and Output Parameter Measurement in Discrete Products

Serial Number	Document Number	Document Title
1	VDI/VDE 3511 Blatt 5	Temperature Measurement in Industry—Installation of Thermometers (Foreign Standard)
2	VDI/VDE 2642	Ultrasonic Flow-Rate Measurement of Fluids in Pipes Under Capacity Flow Conditions (Foreign Standard)
3	VDI/VDE 2041	Measurement of Fluid Flow with Primary Devices, Orifice Plates, and Nozzles for Special Applications (Foreign Standard)
4	VDI 3873 Blatt 1	Emission Measurement; Measurement of Polycyclic Aromatic Hydrocarbons (PAH) in Stationary Industrial Plants; Dilution Method (RWTUeV Method); Gas-Chromatographic Determination (Foreign Standard)
5	VDI 3869 Blatt 1	Measurement of Acids and Bases in Ambient Air; Measurement of Sulphuric Acid Aerosols; Precipitation by Thermal Diffusion (Foreign Standard)
6	VDI 3861 Blatt 1	Measurement of Fibrous Particles; Manual Measurement of Asbestos in Flowing Clean Exhaust Gas; Determination of Asbestos Mass Concentration by IR-Spectroscopy (Foreign Standard)

4.14 STANDARDS RELATED TO MATERIALS

Metals, nonmetals, polymers, composites, and ceramics are available in different compositions and forms for use in discrete products. Selection of material for whole or part of the discrete product is an important element of the design cycle. The designer may decide to investigate use of alternative materials to optimize cost, production methods, and recycling after use of the product. Major national standards and professional bodies issue detail standards related to the properties of engineering materials. These bodies include the following:

1. The American Society of Testing and Material (ASTM)
2. The Aluminum Association (AA)
3. The American Institute of Steel Construction (AISC)
4. The American Petroleum Institute (API)
5. The American Nuclear Society (ANS)
6. The American Society of Mechanical Engineers (ASME)
7. The Society for Protective Coatings (SSPC)

8. The Western Wood Products Association (WWPA)
9. The Metal Powder Industries Federation (MPIF)
10. The Materials Information Society (MIS)
11. The Institute of Nuclear Materials Management (INMM)
12. The American Composite Manufacturing Organization (ACMO)

The following key words can be used to search the online catalog for this area:

Metal
Nonmetals
Ferrous metal
Nonferrous metal
Polymers
Ceramics
Composites
Chemical composition of <material>
Physical properties of <material>

Selected standards from ANSI in this category are listed in Table 4.4.

4.15 STANDARDS FOR THE OPERATION OF THE PRODUCT

Discrete products produced by different manufacturers are diverse and so is their operating principle, even if they are performing similar functions. The operation of a product is normally described by the operations manual supplied by the manufacturer. Such manuals are normally sufficient to enable the user to use the product satisfactorily during its working life. However, a conformance certification of the manufacturer and for the product from an organization such as the ISO, CEN, IEC, and other standards bodies guarantees more reliability in the use of the product.

Factors such as input parameters, as given in Figure 1.5, output parameters, as given in Figure 1.6, and waste, as given in Figure 1.4, should always conform, both in magnitude and in intensity, to values that make a product suitable for human use, with or without protective measures. Appropriate safety warnings, as provided by relevant standards organizations, are always useful.

4.16 SAFETY STANDARDS RELATED TO THE PRODUCT

The unsafe use of a discrete product may result in damage to the product or injury to the operator. The safety features, in general, are associated with the input-process-output operations of the products. At the same time, safety features may also relate to handling and safekeeping of the discrete products.

Major international, national, and professional standards organizations generally, and the operations manuals from the manufacturers specifically, provide

TABLE 4.4
Selected ANSI Standards Related to Materials and Their Usage in Discrete Products

Serial Number	Document Number	Document Title
1	VDI/VDE 2616 Blatt 1	Hardness Testing of Metallic Materials (Foreign Standard)
2	VDI 4430	Ecology-Minded Purchasing of Indirect Materials (Foreign Standard)
3	VDI 3958 Blatt 10	Environmental Simulation—Weathering of Materials (Foreign Standard)
4	VDI 3955 Blatt 3	Assessment of Effects on Materials Due to Corrosive Ambient Conditions—Exposure of Natural Stone Samples (Mank's Carrousel) (Foreign Standard)
5	VDI 3955 Blatt 1	Assessment of Effects on Materials Due to Corrosive Ambient Conditions—Exposure of Steel Sheets (Mank's Carrousel) (Foreign Standard)
6	SS-EN 522	Adhesives for Leather and Footwear Materials: Bond Strength—Minimum Requirements and Adhesive Classification (Foreign Standard)
7	NAS 411-1994	Hazardous Materials Management Program
8	NAS 2014-1991	Pin-Rivet, Grooved, and Collar for Composite Materials
9	LN 29898	Aerospace Series: Rigid Cellular Materials, Polyvinyl Chloride and Polymethacrylimide Sheet—Dimensions, Masses (Foreign Standard)
10	JIS H 7501:2002	Method for Evaluation of Tensile Properties of Metallic Super-Plastic Materials (Foreign Standard)
11	JIS H 7002:1989	Glossary of Terms Used in Damping Materials (Foreign Standard)
12	JIS G 4107:1994	Alloy Steel Bolting Materials for High Temperature Service (Foreign Standard)

guidelines related to safety of discrete products. Some of the standards organizations involved in this area are the following:

1. The American Society of Safety Engineers (ASSE)
2. The Occupation Safety and Hazards Administration (OSHA)
3. The Health and Safety Executives (HSE)
4. The Association Française de Normalization (AFNOR)

5. The Japanese Industrial Standards Committee (JISC)
6. Deutsches Institut für Normung (DIN)
7. The British Standards Institute (BSI)
8. The American National Standards Institute (ANSI)
9. The International Organization for Standardization (ISO)
10. The European Committee for Standardization (CEN)
11. The International Electrotechnical Commission (IEC)

The following key words can be used to search the online catalog for this area:

Safety
Safety <product>
Safety <component>
Safety <process>
Safety <process parameter>

Some of the selected ANSI standards related to the safety of discrete product are listed in Table 4.5.

4.17 QUALITY STANDARDS RELATED TO THE PRODUCT

A measure of product quality is the detail that is considered in the design and manufacturing of the product. Obviously, a product designed and manufactured with high quality considerations shall perform better than a product designed without such considerations.

Quality standards are very well covered by international and national standards organizations. The major organizations include the following:

1. The British Standards Institute (BSI)
2. The International Organization for Standardization (ISO)
3. The American Society for Quality (ASQ)
4. The American National Standards Institute (ANSI)
5. The Japanese Industrial Standards Committee (JISC)
6. The European Committee for Standardization (CEN)

The following key words can be used to search the online catalog for this area:

Quality
Quality <product>
Quality <component>
Quality <process>
Quality <process parameter>

A brief listing of ANSI standards in this category is provided in Table 4.6.

TABLE 4.5
Selected ANSI Standards Related to Safety in Discrete Product

Serial Number	Document Number	Document Title
1	VDI/VDE 3542 Blatt 3	Safety Terms for Automation System—Application Information and Examples (Foreign Standards)
2	VDI/VDE 3542 Blatt 2	Safety Terms for Automation Systems—Quality Terms and Definitions (Foreign Standards)
3	VDI/VDE 3542 Blatt 1	Safety Terms for Automation Systems—Quality Terms and Definitions (Foreign Standards)
4	VDI/VDE 2180 Blatt 6	Safeguarding of Industrial Process Plants by Means of Process Control Engineering—Using Safety-Related Programmable Electronic Systems (Foreign Standards)
5	SS-EN 99	Safety of Machinery—The Positioning of Protective Equipment in Respect of Approach Speeds of Parts of the Human Body (Foreign Standards)
6	SS-EN 983	Safety of Machinery: Safety Requirements for Fluid Power Systems and Their Components—Pneumatics (Foreign Standards)
7	SS-EN 982	Safety of Machinery: Safety Requirements for Fluid Power Systems and Their Components—Hydraulic (Foreign Standards)
8	SS-EN 954-1	Safety of Machinery: Safety-Related Parts of Control System—Part-1: General Principles for Design (Foreign Standards)
9	SS-EN 809	Pumps and Pump Units for Liquids—Common Safety Requirements (Foreign Standards)
10	SS-EN 61558-2-7	Safety of Power Transformers, Power Supply Units and Similar—Part 2–7: Particular Requirements for Transformers for Toys (Foreign Standards)
11	SS-EN 61558-2-6	Safety of Power Transformers, Power Supply Units and Similar—Part 2–6: Particular Requirements for Transformers for General Use (Foreign Standards)

TABLE 4.6
Selected ANSI Standards Related to Quality of Discrete Products

Serial Number	Document Number	Document Title
1	VDI 2700 Blatt 5	Safety of Loads on Vehicles—Quality Management Systems (Foreign Standards)
2	SS-EN 1124-1	Pipes and Fittings of Longitudinally Welded Stainless Steel Pipes with Spigot and Socket for Waste Water Systems—Part 1: Requirements, Testing, Quality Control (Foreign Standard)
3	SS-EN 1123-1	Pipes and Fittings of Longitudinally Welded Hot-Dip Galvanized Steel Pipes with Spigot and Socket for Waste Water Systems—Part 1: Requirements, Testing, Quality Control (Foreign Standard)
4	SS-EN 10221	Surface Quality Classes for Hot-Rolled Bars and Rods—Technical Delivery Conditions (Foreign Standard)
5	NASM 1312/25-1997	Fastener Test Methods: Method 25, Driving Recess Torque Quality Conformance Test
6	MSS SP-94-1999	Quality Standard for Merritic and Martensitic Steel Castings for Valves, Flanges, and Fittings and Other Piping Components—Ultrasonic Examination Method
7	MSS SP-93-1999	Quality Standard for Steel Castings and Forgings for Valves, Flanges, and Fittings and Other Piping Components—Ultrasonic Examination Method
8	JIS Z 9901-1998	Quality Systems—Model for Quality Assurance in Design, Development, Production, Installation, and Servicing (Foreign Standard)
9	JIS G 0588-1995	Visual Examination and Classification of Surface Quality for Steel Castings (Foreign Standard)

4.18 RELIABILITY STANDARDS RELATED TO THE PRODUCT

Reliability of the discrete product is a parameter that guarantees trouble-free operation of the product during its life cycle. A wealth of experimental test data is required to project reliability by use of methods such as the Weibull distribution.

Reliability standards are very well covered by international and national standards organizations. Some of these organizations are the following:

1. The British Standards Institute (BSI)
2. The International Organization for Standardization (ISO)
3. The American National Standards Institute (ANSI)
4. The Japanese Industrial Standards Committee (JISC)
5. The European Committee for Standardization (CEN)

A listing of selected standards from ANSI for discrete products is provided in Table 4.7.

The following key words can be used to search the online catalog for this area:

Reliability
Reliability <product>
Reliability <component>
Reliability <process>
Reliability <process parameter>

4.19 STANDARDS RELATED TO MAINTAINABILITY

Standards related to maintainability of discrete products are available from international and major national standards organizations. Some of these organizations are the following:

1. The British Standards Institute (BSI)
2. The International Organization for Standardization (ISO)
3. The American National Standards Institute (ANSI)
4. The Japanese Industrial Standards Committee (JISC)
5. The European Committee for Standardization (CEN)

The following key words can be used to search the online catalog for this area:

Maintainability
Maintainability <product >
Maintainability <component >
Maintainability <process >

A list of selected maintainability standards from ANSI is provided in Table 4.8.

TABLE 4.7
Selected ANSI Standards Related to Reliability of Discrete Products

Serial Number	Document Number	Document Title
1	VG 95215-2	Electronic And Electrical Components; Established Reliability for Nonactive Components (Foreign Standard)
2	VDI/VDE 3542 Blatt 4	Safety Terms for Automation Systems—Reliability and Safety of Complex Systems (Terms) (Foreign Standard)
3	VDI/VDE 2180 Blatt 4	Safeguarding of Industrial Process Plants by Means of Process Control Engineering—Calculating Methods for Reliability Characteristics of Safety Instrumented Systems (Foreign Standard)
4	NFPA/T2.12.11-1-2001	Fluid Power Systems and Components—Reliability Analysis, Reporting Format, and Database Compilation
5	NASM 33602-1999	Bolts, Self-Retaining, Aircraft, Reliability and Maintainability, Design and Usage Requirements For
6	NASM 33588-1999	Nut, Self-Locking, Aircraft, Reliability and Maintainability Usage Requirements For
7	NASM 33522-1999	Rivets, Blind, Structural, Mechanically Locked and Friction Retainer Spindle, (Reliability and Maintainability) Design and Construction Requirement For
8	JIS C 5700:1974	General Rules for Reliability Assured Electronic Components (Foreign Standard)
9	IEC/TR 61586 Ed. 1.0 b:1997	Estimation of the Reliability of Electrical Connectors
10	IEC 62278 Ed. 1.0 b:2002	Railway Applications—Specification and Demonstration of Reliability, Availability, Maintainability, and Safety (RAMS)
11	IEC 61703 Ed. 1.0 b:2001	Mathematical Expressions for Reliability, Availability, Maintainability, and Maintenance Support Terms
12	IEC 61163-2 Ed. 1.0 b:1998	Reliability Stress Screening—Part 2: Electronic Components
13	IEC 61163-1 Ed. 1.0 b:1995	Reliability Stress Screening—Part 1: Repairable Items Manufactured in Lots
14	IEC 61124 Ed. 1.0 b:1997	Reliability Testing—Compliance Tests for Constant Failure Rate and Constant Failure Intensity

(*Continued*)

TABLE 4.7 (Continued)
Selected ANSI Standards Related to Reliability of Discrete Products

Serial Number	Document Number	Document Title
15	ASTM F595-01	Standard Test Methods for Vacuum Cleaner Hose—Durability and Reliability (All-Plastic Hose)
16	ASTM F450-01	Standard Test Methods for Vacuum Cleaner Hose—Durability and Reliability (Plastic Wire Reinforced)
17	ASTM E235-88(1996)e1	Standard Specification for Thermocouples, Sheathed, Type K, for Nuclear or for Other High-Reliability Applications

TABLE 4.8
Selected ANSI Standards Related to Maintainability of Discrete Products

Serial Number	Document Number	Document Title
1	NASM 33602-1999	Bolts, Self-Retaining, Aircraft, Reliability and Maintainability, Design and Usage Requirements For
2	NASM 33588-1999	Nut, Self-Locking, Aircraft, Reliability and Maintainability Usage Requirements For
3	ISO/IEC 2382-14:1997	Information Technology: Vocabulary—Part 14: Reliability, Maintainability and Availability
4	ISO 8107: 1993	Nuclear Power Plants: Maintainability—Terminology
5	ISO 3977-9:1999	Gas Turbines: Procurement—Part 9: Reliability, Availability, Maintainability, and Safety
6	IEC 62278 Ed. 1.0 b:2002	Railway Applications—Specification and Demonstration of Reliability, Availability, Maintainability, and Safety (RAMS)
7	IEC 61703 Ed. 1.0 b: 2001	Mathematical Expression for Reliability, Availability, Maintainability, and Maintenance Support Terms
8	IEC 60863 Ed. 1.0 b: 1986	Presentation of Reliability, Maintainability, and Availability Predictions
9	IEC 60706-6 Ed. 1.0 b: 1994	Guide on Maintainability of Equipment—Part 6: Section 9: Statistical Methods in Maintainability Evaluation
10	IEC 60706-5 Ed. 1.0 b: 1994	Guide on Maintainability of Equipment—Part 5: Section 4: Diagnostic Testing

4.20 PRODUCT LIFE CYCLE AND EMERGING STANDARDS

Consideration of product life cycle includes the use of recyclable materials; modularity in product design, manufacture, and assembly; and less emission of dangerous liquids, solids, and gases to the environment. All major international and national standards organizations support this trend. Some of these organizations are the following:

1. The International Organization for Standardization (ISO)
2. The European Committee for Standardization (CEN)
3. The Association Française de Normalization (AFNOR)
4. The American National Standards Institute (ANSI)
5. The British Standards Institute (BSI)
6. Deutsches Institut für Normung (DIN)
7. The Japanese Industrial Standard Committee (JISC)

Table 4.9 provides a selection of product life-cycle standards from ANSI. The following key words can be used to search the online catalog for this area:

Life cycle
<product> Life cycle

TABLE 4.9
Selected ANSI Standards Related to Product Life Cycle

Serial Number	Document Number	Document Title
1	TR Q 0004:2000	Environment Management: Life Cycle Assessment—Example of Application of ISO 14041 to Goal and Scope Definition and Inventory Analysis (Foreign Standard)
2	JIS Q 14043:2002	Environment Management: Life Cycle Assessment—Life Cycle Interpretation (Foreign Standard)
3	JIS Q 14042:2002	Environment Management: Life Cycle Assessment — Life Cycle Impact Assessment (Foreign Standard)
4	JIS Q 14041:1999	Environment Management: Life Cycle Assessment—Goal and Scope Definition and Inventory Analysis (Foreign Standard)
5	JIS Q 14040:1997	Environment Management: Life Cycle Assessment—Principles and Framework (Foreign Standard)
6	ISO/TS 14048:2002	Environment Management: Life Cycle Assessment—Data Documentation Format
7	ISO/TS 14049:2002	Environment Management: Life Cycle Assessment—Examples of Application Of ISO 14041 to Goal and Scope Definition and Inventory Analysis

4.21 FUTURE DIRECTIONS IN THE STANDARDIZATION OF DESIGN PROCESS FOR DISCRETE PRODUCTS[1]

The design process for discrete products is very diverse. As shown in Chapter 2, several standards-developing organizations are involved in the investigation of the need for new standards for the design process and are working toward developing such standards for implementation at various levels. The current status of these standards can be best monitored by following the committee work of various standards-developing organizations.

The facilities available to designers include digital libraries of machine elements, mechanical components, control elements, software components in CAD/CAM databases, digital libraries of simulations of discrete products, and digital libraries of standards related to the design of discrete products. However, the most important development in the design of discrete products is the continuous upgrading of standards related to the geometry of the products.

The geometry of the discrete products is a common parameter for all functions of the design and manufacturing facility. Several standards, such as IGES, SET, VDA-FS, PDES and STEP, have been used to define the product geometry and to exchange the geometry data across different platforms. Other less-recognized standards are also available for geometric data exchange. Computer-integrated manufacturing philosophy heavily relies on these standards for information transfer to various design and manufacturing functions.

One of the geometry-data transfer standards, Initial Graphics Exchange Specifications (IGES), is used to transfer design illustrations to dissimilar systems. Translators based on IGES standards are used to export design drawings into an IGES file and for importing IGES file into the destination system.

The basic principle utilized by IGES to convert drawings into the neutral IGES format is to describe the product in terms of geometric and nongeometric entities according to the relevant version of IGES standards. Nongeometric information contains annotations, definitions, and organization. The geometric information consists of elements such as points, curves, surfaces, and solids. Both geometric and nongeometric information, when combined, defines the product. IGES is extensively used for geometry-data transfer in existing CIM systems.

However, the work carried out for the development and use of STEP-ISO 10303 is paramount. STEP-ISO 10303 is an international standard for the computer interpretable representation and exchange of product data that has a wider scope than IGES. STEP-ISO 10303 is described by National Institute of Standards and Technology (NIST) as follows:

> The information generated about a product during its design, manufacture, use, maintenance, and disposal is used for many purposes during that life cycle. The use may involve many computer systems, including some that may be located in different organizations. In order to support such uses, organizations need to be able to represent

[1] Extracts reproduced with the permission of National Institute of Science and Technology (NIST), U.S.

their product information in a common computer-interpretable form that is required to remain complete and consistent when exchanged among different computer systems.

The objective is to provide a mechanism that is capable of describing product data throughout the life cycle of a product, independent of any particular system. Thus, STEP-ISO 10303 is suitable not only for neutral file exchange but also for implementing and sharing product databases and archiving.

The geometry information has its digital equivalent in the form of geometry transfer/representation protocols such as IGES or STEP within a suite of CIM programs. It allows geometry transfer between programs such as a CAD modeling package to an aerodynamic modeling package. STEP-ISO is an important future standards series that addresses a wider domain than has been studied before.

BIBLIOGRAPHY

Bedford, A. and Fowler, W., *Engineering Mechanics*: *Statics*, 2nd ed., Addison-Wesley, Reading, mass, 1999.

Bolhouse, V.C., Vision Systems as Control Elements in Production; Proceedings of the Annual Control Engineering Conference, 1986, 122–127.

Carter, J.A., Standards, the future, and designers, *Proc. Hum. Factors Soc.*, 1, 448–452, 1992.

Cliff, M., *Case Studies in Engineering Design*, Butterworth-Heinemann, Oxford, 1998.

Dereli, T. and Filiz, H., A note on the use of STEP for interfacing design to process planning, *CAD Comput. Aided Design*, 34(14), 1075–1085, 2002.

Dimarogonas, A.D., Origins of engineering design, *Vib. Mech. Syst. Hist. Mech. Design* (*ASME*), 63, 1–18, 1993.

Fenves, S.J., Garrett, J.H., Jr., and Hakim, M.M., Representation and processing of design standards: A bifurcation between research and practice, *Proc. Struct. Congr.* 94, 122–126, 1994.

Goodall, R.M. and Austin, S.A., Modelling, placing and controlling active elements in structural engineering design, *Control Eng. Pract.*, 2(5), 743–753, 1994.

Hardwick, M. and Loffredo, D., The STEP international data standard is becoming a manufacturing tool, *Manuf. Eng.*, 126(1), 38–50, 2001.

Hartenberg, R.S. and Denavit, J., *Kenematic Synthesis of Kinkages*, McGraw-Hill, New York, 1964.

Harzheim, L. and Mattheck, C., 3D-shape optimization: Different ways to an optimized design, *Lect. Notes Eng.*, 63, 173–179, 1991.

Hill, D.R., Mechanical engineering in the medieval near east, *Sci. Am.*, 264(5), 100–105, 1991.

Incropera, F.P. and Dewitt, D.P., *Fundamentals of Heat and Mass Transfer*, 5th ed., John Wiley, New York, 2002.

Laura, R.A., David, P., Helfferich, W., and Barone, J., Design considerations for ecosystem restoration, *Int. Water Resour. Eng. Conf. Proc.*, 1, 259–263, 1995.

Levy, R., Science, Technology and Design, *Design Studies*, 6(2), 66–72, 1985.

Linde, E., Dolan, D., and Batchelder, M., Mechatronics for Multidisciplinary Teaming, ASEE Annual Conference Proceedings, Nashville, TN 2003, 6901–6910.

Liu, Z., Jiang, S.Q., Tang, B.Y., Zhang, J.H., and Zhong, H., The study and realization of SCADA system in manufacturing enterprises, *Proc. World Congr. Intelligent Control Automation* (*WCICA*), 5, 3688–3692, 2000.

McMath, I.; Revolutionary sense in machine control, *Eng. Technol.* 7(4), 45, 2004.

McQuiston, F.C., Parker, J.D., and Spitler, J.D., *Heating, Ventilating, and Air-conditioning — Analysis and Design*, 5th ed., John Wiley, New York, 2000.

Ming, X.G., Mak, K.L., and Yan, J.Q., PDES/STEP-based information model for computer-aided process planning, *Robotics Comput. Integrated Manuf.*, 14(5–6), 347–361, 1998.

Mufti, A.A., Morris, M.L., and Spencer, W.B., Data exchange standards for computer-aided engineering and manufacturing, *Int. J. Comput. Appl. Technol.*, 3(2), 70–80, 1990.

Munson, B.R., Young, D.F., and Okishi T.H., *Fundamental of Fluid Mechanics*, 3rd ed., John Wiley, New York, 1998.

Muster, D. and Mistree, F., Engineering design as it moves from an art towards a science: Its impact on the education process. *Int. J. Appl. Eng. Educ.* 5(2), 239–246, 1989.

Qiao, L.-H., Zhang, C., Liu, T.-H., Wang, B.H.-P., and Fischer, G.W., PDES/STEP-based product data preparation procedure for computer-aided process planning, *Comput. Ind.*, 21(1), 11–22, 1993.

Parmley, R.O., *Illustrated Source Book of Mechanical Components*, McGraw-Hill, New York, 2000.

Rawson, K.J., Ethics and fashion in design, *Nav. Architect*, 1–28, February, 1990.

Rivin, E.I., Conceptual developments in design components and machine elements, *J. Mech. Design Trans. ASME*, 117B, 33–41, 1995.

Shahmanesh, N., Design engineering, *Automot. Eng. (London)*, 23(10), 42–54, 1998.

Sharma, R., and Gao, J.X., Implementation of STEP application Protocol 224 in an automated manufacturing planning system, Proceedings of the Institution of Mechanical Engineers, Part B, *J. Eng. Manuf.* 216(9), 1277–1289, 2002.

Smith, R.P., Historical roots of concurrent engineering fundamentals, *IEEE Trans. Eng. Manage.* 44(1), 67–78, 1997.

Sonntag, R.E., Borgnakke, C., and Van Wylen G.J., *Fundamental of Thermodynamics*, 5th ed., John Wiley, New York, 1998.

Tanik, M.M. and Ertas, A., Design as a basis for unification: System interface engineering, *Comput. Appl. Design Abstr. 1992 (ASME)*, 43, 113–120, 1992.

Town, H.C., Control of hydraulic transmission elements, *Power Int.* 32(175), 65–67, 1986.

West, A.A., Harrison, R., Wright, C.D., and Carrott, A.J., Visualization of control logic and physical machine elements within an integrated machine design and control environment, *Mechatronics,* 10(6), 669–698, 2000.

Zha, X.F. and Ji, P., Assembly process sequence planning for STEP based mechanical products: An integrated model and system implementation, *Int. J. Industrial Eng. Theory Appl. Pract.* 10(3), 279–288, 2003.

5 Discrete Product Development with Pertinent Standards—A Case Study

5.1 SYSTEM ENGINEERING FOR DISCRETE PRODUCTS

Whether designed from an original idea or on the basis of an adaptive or variant system, discrete products are first analyzed according to the system-engineering principle. The discrete product is a system that is divided into subsystems, and each subsystem performs a function (process) in response to an initial input and yields an expected output. The input may be provided by another subsystem or by an external source and the output may be supplied to another subsystem or to an external sink. The input or procedure to provide the input to the process, the process details, and the output or the procedure to receive the output from the process is generally based on the relevant principles of mechanical engineering, control engineering, electrical engineering, electronics, and computer engineering or computer science. Expanding further, the relevant principles of mechanical engineering, for example, may derive from solid mechanics, fluid mechanics, and thermodynamics. See Figure 1.4 to Figure 1.6 for the basic layout for the input, process, and output principles that apply to discrete components, structures, assemblies, mechanisms, machines, and various types of processes.

For any discrete product under consideration for engineering design and manufacturing, one of the design classifications applies. In the case of an original design, the system (discrete product) must be developed from scratch. In the case of an adaptive or variant design, some basic information about the proposed system (discrete product) is available to the design engineer.

In each step of the system approach to design (whether envisaged, deduced, or replicated for each component of the discrete product), the three-phase model for design (i.e., analysis, synthesis, and evaluation) applies locally. As the design procedure progresses and machines or various types of processes starts taking shape, a global application of the three-phase model brings more optimality into the design.

Figure 4.6 to Figure 4.8 illustrates the system approach to design clearly in terms of all contemporary aspects of engineering design. It takes into account the concept design, the embodiment design, and the detail design in a macroscopic manner. For a microscopic analysis, every single parameter in each step of the contemporary system approach to design is analyzed, synthesized, evaluated, and compared with any available standard values. Evaluation of the value of parameters involved in the design of a product, whether a machine or process, differs in terms of the use of analytical governing equations, numerical models, or experimental methods in each of the above steps of the system approach to design.

Figure 5.1 illustrates design considerations for dicrete products in a macroscopic manner. The direction of activities moves from outward (local and global analysis, synthesis and evaluation) to inward (objective function).

The visualization of a system (a component, structure, assembly, mechanism, machine, or process) during the design cycle is a basic requirement for the design process. Most systems can be represented graphically by the use of engineering drawing standards such as the DIN technical drawing series or the use of CAD software that applies appropriate standards. Simulations that provide the dynamics of the component, assembly, mechanism, machine, or process may be built by application of general-purpose high-level languages or special software designed to simulate specific classes of discrete products. Computer-generated simulations allow transfer of geometry and other product specifications to any other design and manufacturing functions to establish preliminary models in those areas. Any changes in product specification that result from analysis in one area of design can easily be implemented in all other areas.

Figure 5.2 illustrates the step-by-step inclusion of design aspects as a balancing act between the need and the concept design stage of the system approach to design for a subsystem based on input, process, and output sequence. Similar steps are performed for concept design to embodiment design interface, embodiment design to detail design interface, and detail design to manufacturing interface for each component, subsystem, and the whole discrete product. Values of parameters at each step and at each interface related to transition from component to assembly to mechanism to machine or process are best determined by relevant standards. Figure 5.3 presents all these interfaces graphically.

Because the scope of the use of various engineering concepts in a multidisciplinary product differs, the subsystems must be considered separately. Although, the concepts presented in Figure 5.1 to Figure 5.3 apply to all of the systems shown, concurrency in the design of technically differing subsystems may be necessary in some discrete products. Figure 5.4 illustrates such a process.

5.2 GOVERNING EQUATIONS AND STANDARDS

The analysis to compute the parameters involved in the design of a discrete product is normally governed by analytical equations in the field of application of the parameter, numerical methods to compute the permissible value of the

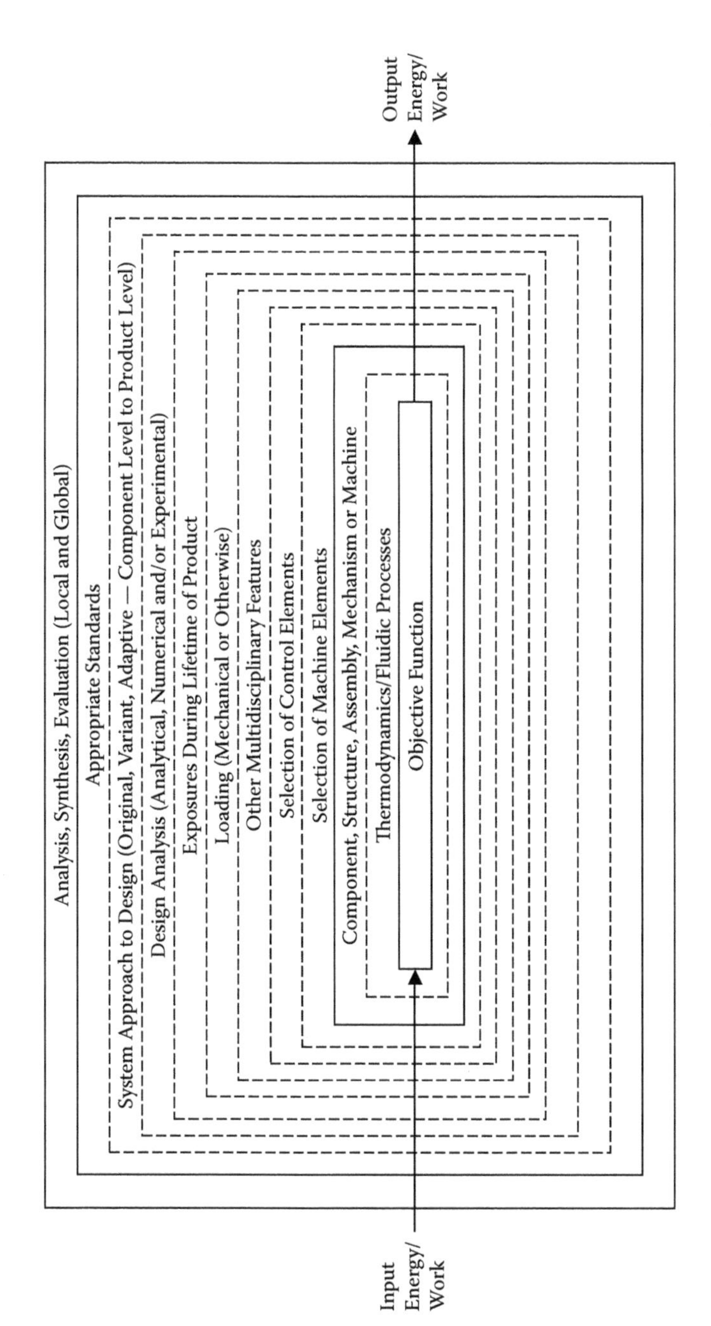

FIGURE 5.1 Design consideration for a discrete product.

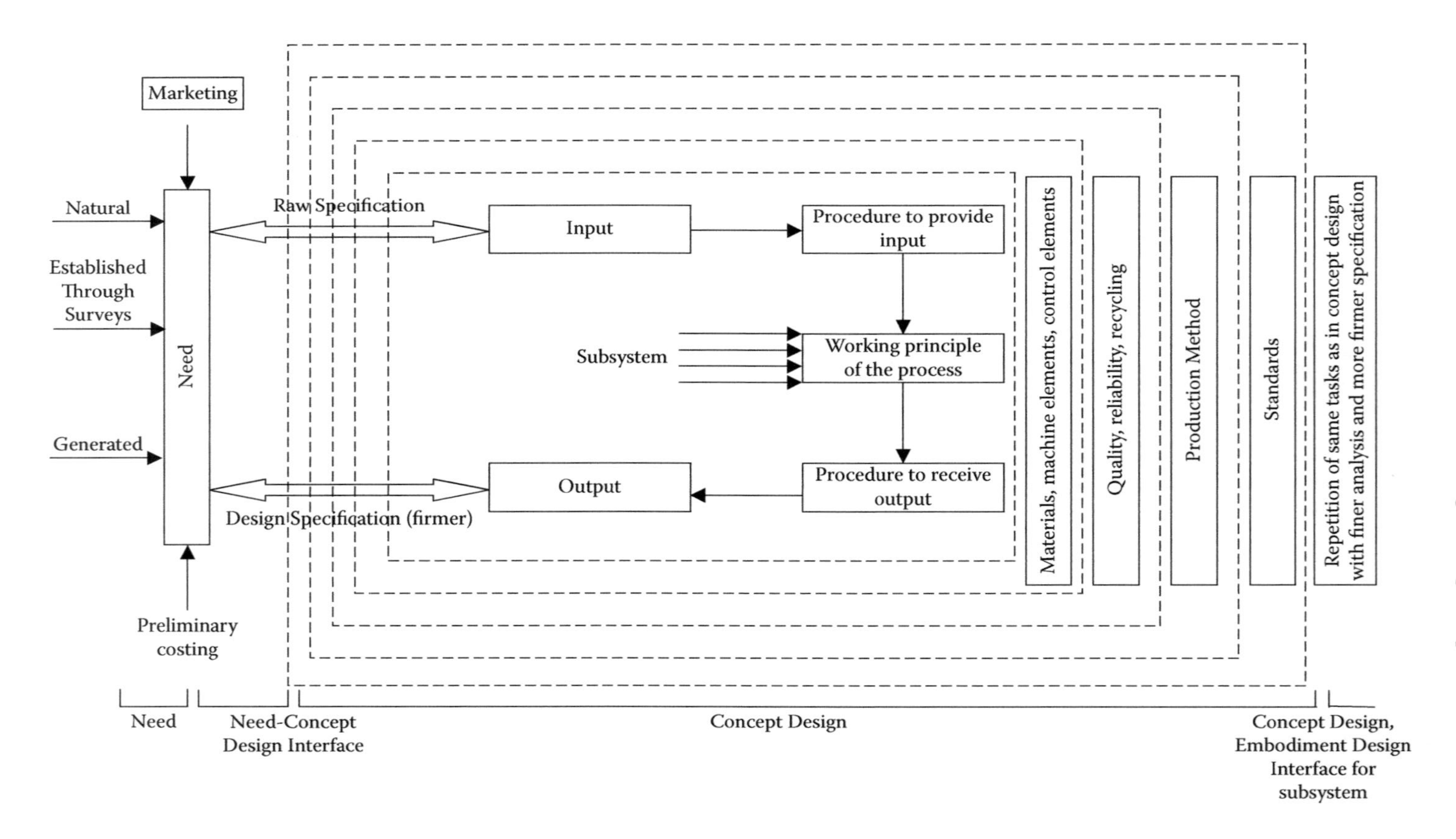

FIGURE 5.2 The need–concept design interface for a subsystem of a discrete product.

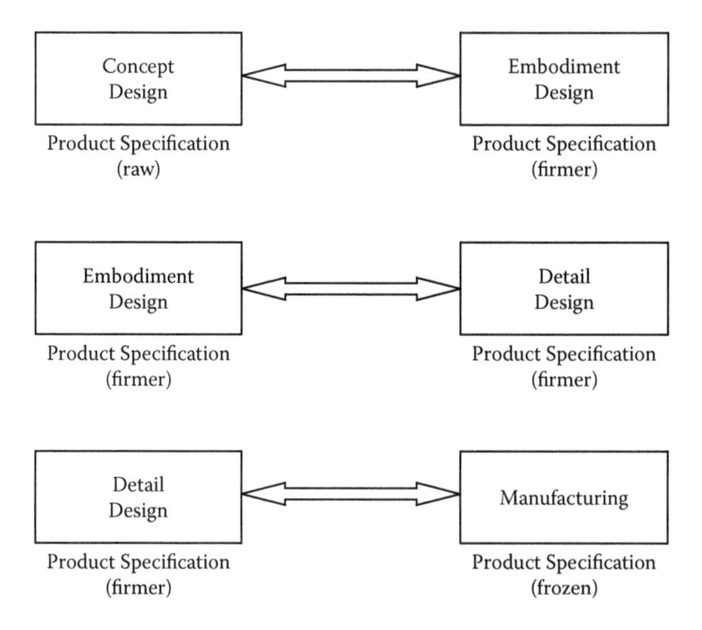

FIGURE 5.3 Interfaces between various subfunctions in the system approach to the design cycle.

parameter or group of parameters, or experimental determination of the parameter through testing a prototype or model of the physical element to which the parameter belongs. Sometimes more than one method is adopted to evaluate the parameter to bring more certainty to the design.

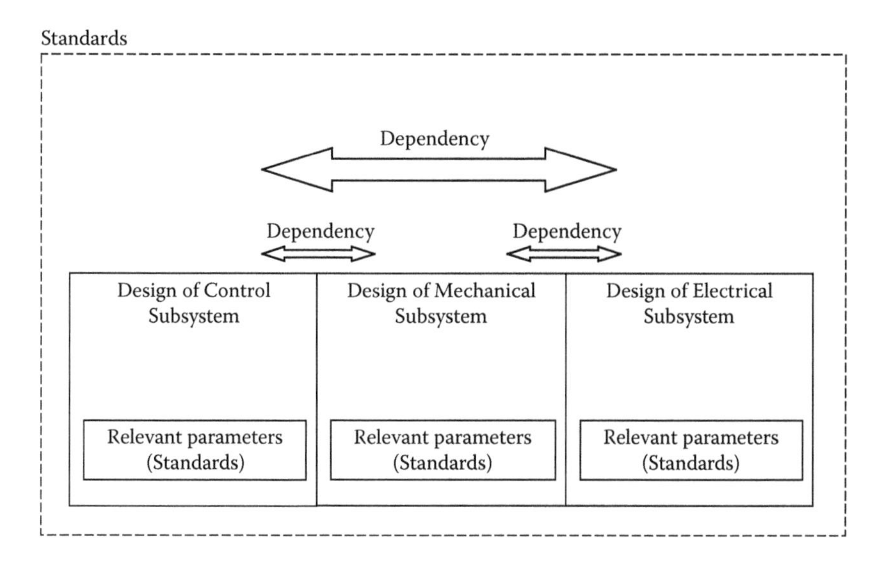

FIGURE 5.4 Concurrency in the design of discrete products.

Once the discrete product has been considered as a system and further divided into subsystems, the identification of the field and types of analysis and determination of required parameters become comparatively easier.

The following example, derived from Shigley and Mischke (1996),* demonstrates the use of analytical governing equations for deflection caused by bending and the use of pertinent standards. These equations are:

$$q/EI = d^4y/dx^4 \tag{5.1}$$

$$V/EI = d^3y/dx^3 \tag{5.2}$$

$$M/EI = d^2y/dx^2 \tag{5.3}$$

$$\theta = dy/dx \tag{5.4}$$

$$y = f(x) \tag{5.5}$$

in which the variables are as follows:

E is Young's modulus
I is the second moment of area
M is the moment
V is the unit load
V is the shear force
x is the x-coordinate
y is the y-coordinate
θ is the slope, torsional deflection

These relations are illustrated by the beam in Figure 5.5. Analysis of a beam by use of the above governing equations results in the conclusion that the material properties represented by parameter E and geometrical properties represented by parameter I can be strategically utilized to control the deflection in the component by varying the cross section or materials. Standards organizations that work in this area provide values of E and I (for selected cross sections), and the governing equations help compute the magnitude of parameters such as maximum shear force, bending moment, and deflection. Mechanical and physical properties of gray cast iron as listed, for example, by DIN EN1561 can be used to finalize the design of the beam. Similar use of standards applies to examples given in Sections 5.2.1 and 5.2.2.

5.2.1 Combined Loading on the Power Screw—An Example

As an extension of concept presented in Section 5.2, the following example provides a method for developing the governing equations for the combined loading on a machine element known as a power screw. Subsequently, references

* Reproduced from *Standard Handbook of Machine Design* by Shigley and Mischke with the permission of McGraw-Hill.

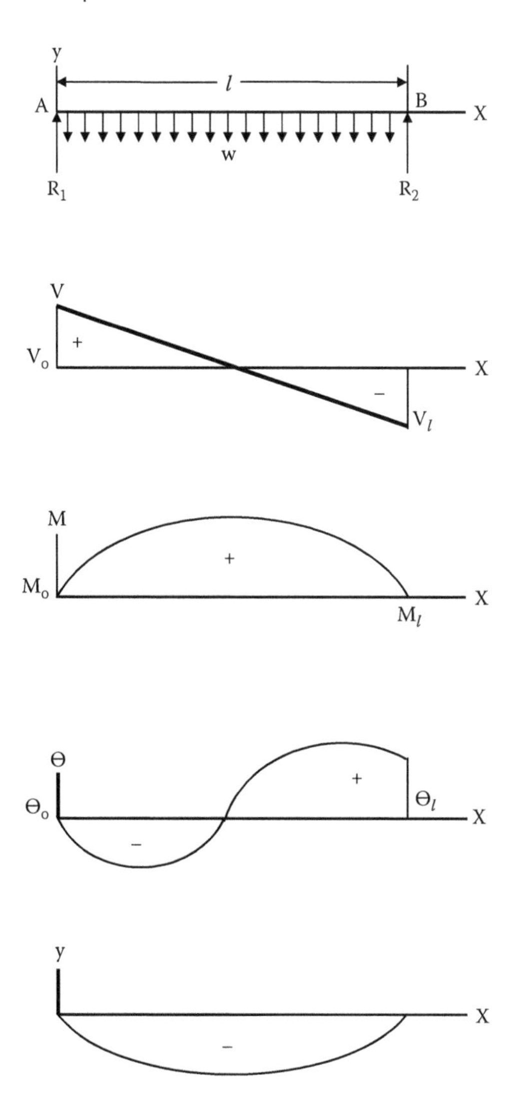

FIGURE 5.5 (a) Loading diagram of beam supported at A and B, with uniform load w having units of force per unit length, $R_1 = R_2 = w\ l/2$. (b) Shear-force diagram of end conditions. (c) Moment diagram. (d) Slope diagram. (e) Deflection diagram.

are provided for use of standard values in parametric governing equations for shear force and banding moment to find an optimum solution for their permissible values in the cross sections considered.

A power screw is a mechanical device that converts rotary motion of either the nut or the screw to relatively slow linear motion of a mating member for transmitting power along the screw axis. Power screws are also called linear actuators or translation screws. The main applications of a power screw are as follows:

To raise the load (e.g., as in screw jack)
To obtain accurate motion in machining operations (e.g., as in lathe machine)
To clamp a work piece (e.g., as in vise)
To exert large forces (e.g., as in presses and tensile testing machines)

The main advantage of power screws is their large load-carrying capacity despite small overall dimensions. Power screws are simple to design, easy to manufacture, and give smooth and noiseless service. They provide large mechanical advantage and highly accurate motion. Their main drawback is poor efficiency because of large frictional losses.

Combined loading is a combination of following different types of loading:

1. Loading with respect to time
   a. Static loading
   b. Sustained loading
   c. Impact loading
   d. Repeated loading
2. Loading with respect to area over which it is applied
   a. Concentrated loading
   b. Distributed loading
3. Loading with respect to location and method of application
   a. Centric loading
   b. Torsional loading
   c. Bending and flexural loading
4. Any other nonmechanical loading or exposure

Four types of loads are considered to be acting on the power screws:

Distributed load caused by the weight of the table
Torsion caused by the excessive resistive forces
Bending caused by load wa
Weight of the power screw

The details of the generalized calculations with appropriate examples are given below. The effects caused by the weight of the power screw are neglected. Figure 5.6 shows the combined loading on the power screw.

The first case considered is that of the load at any position x of the power screw (Figure 5.7). The load on the power screw is the distributed load of the table top. The distributed load is spread over variable length a, which makes the net load wa at any portion on the power screw. This load is moving, and its

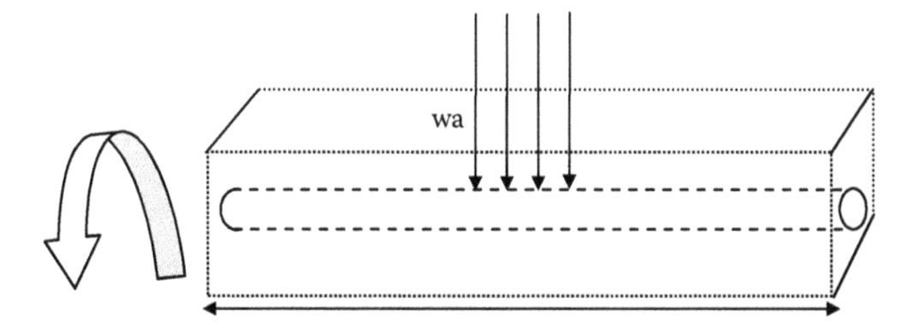

FIGURE 5.6 Combined loading on the power screw.

position can be considered to be x units away from the left corner. The reactions at the supports, the shearing forces, and the bending moments are calculated, considering the variable quantities in parametric form.

$$+\Sigma F_y = 0$$

$$R_A + R_B = wa \tag{5.6}$$

$$+\Sigma M_A = 0;$$

$$R_B l - wax = 0$$

$$R_B = \frac{wax}{l}$$

Substituting the value of R_B in equation (5.6),

$$R_A + \frac{wax}{l} = wax$$

$$R_A = wa\frac{(l-x)}{l}$$

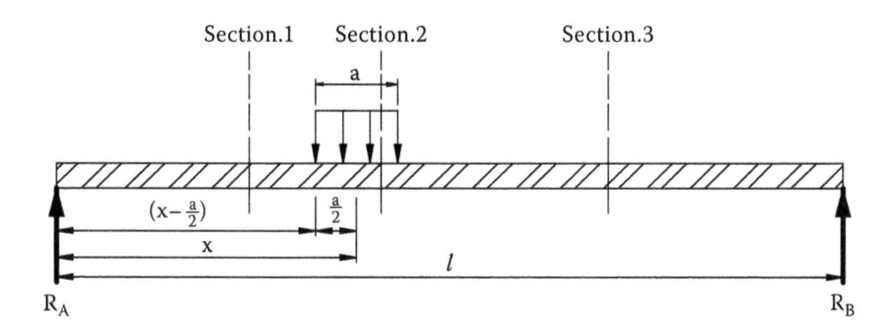

FIGURE 5.7 Loading on the power screw.

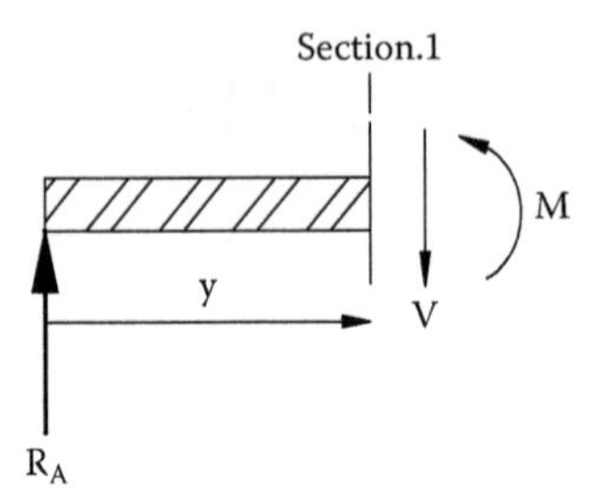

FIGURE 5.8 Section 1 of the power screw.

Considering section 1 of the power screw (Figure 5.8): $0 \le y \le (x - \frac{a}{2})$

$$V = R_A = wa\frac{(l-x)}{l}$$

$$M = R_A y = wa\frac{(l-x)}{l}y$$

at $y = 0$:

$$V = wa\frac{(l-x)}{l}$$

$$M = 0$$

at $y = \left(x - \frac{a}{2}\right)$

$$V = wa\frac{(l-x)}{l}$$

$$M = wa\frac{(l-x)}{l}\left(x - \frac{a}{2}\right)$$

Considering section 2 of the power screw (Figure 5.9): $(x - \frac{a}{2}) \le y \le (x - \frac{a}{2})$

$$V = R_A - w\left(y - x + \frac{a}{2}\right)$$

$$V = wa\frac{(l-x)}{l} - w\left(y - x + \frac{a}{2}\right)$$

$$M = R_A y - w\left(y - x + \frac{a}{2}\right)\frac{\left(y - x + \frac{a}{2}\right)}{2}$$

$$M = wa\frac{(l-x)}{l} - \frac{w}{2}\left(y - x + \frac{a}{2}\right)^2$$

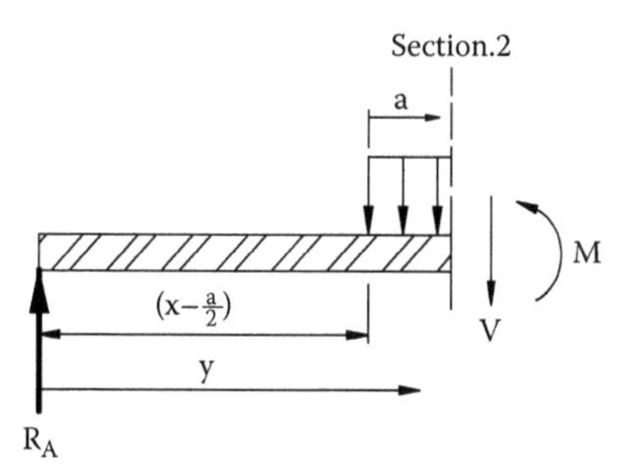

FIGURE 5.9 Section 2 of the power screw.

$$\text{at} \quad y=\left(x-\frac{a}{2}\right)$$

$$V = wa\frac{(l-x)}{l} - w\left(x-\frac{a}{2}-x+\frac{a}{2}\right) = wa\frac{(l-x)}{l}$$

$$M = wa\frac{(l-x)}{l}\left(x-\frac{a}{2}\right)$$

$$\text{at} \quad y=\left(x+\frac{a}{2}\right)$$

$$V = wa\frac{(l-x)}{l} - w\left(x+\frac{a}{2}-x+\frac{a}{2}\right) = -\frac{wax}{l}$$

$$M = wa\frac{(l-x)}{l}\left(x+\frac{a}{2}\right) - \frac{wa^2}{2}$$

Considering section 3 of the power screw (Figure 5.10): $(x+\frac{a}{2}) \le y \le l$

$$V = R_A - wa = wa\frac{(l-x)}{l} - wa = -\frac{wax}{l}$$

$$M = R_A y - wa(y-x) = wa\frac{(l-x)}{l}y - wa(y-x)$$

$$\text{at} \quad y=\left(x+\frac{a}{2}\right)$$

$$M = wa\frac{(l-x)}{l}\left(x+\frac{a}{2}\right) - \frac{wa^2}{2}$$

$$\text{at} \quad y = l$$

$$M = 0$$

From the above calculations, the moment at the start ($y = 0$) and at the end ($y = 1$) of the power screw is equal to 0. Figure 5.11 illustrates the shear force

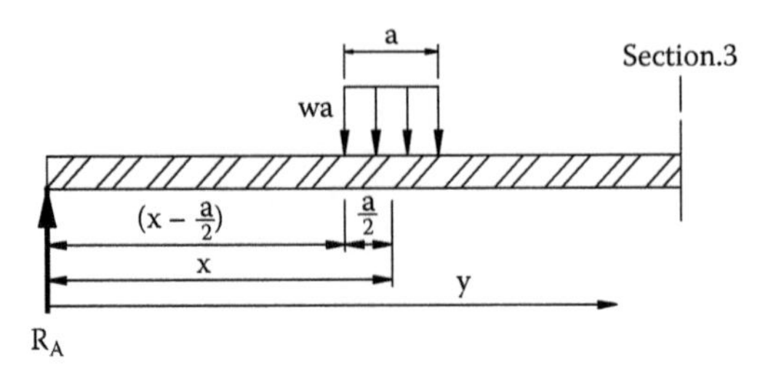

FIGURE 5.10 Section 3 of the power screw.

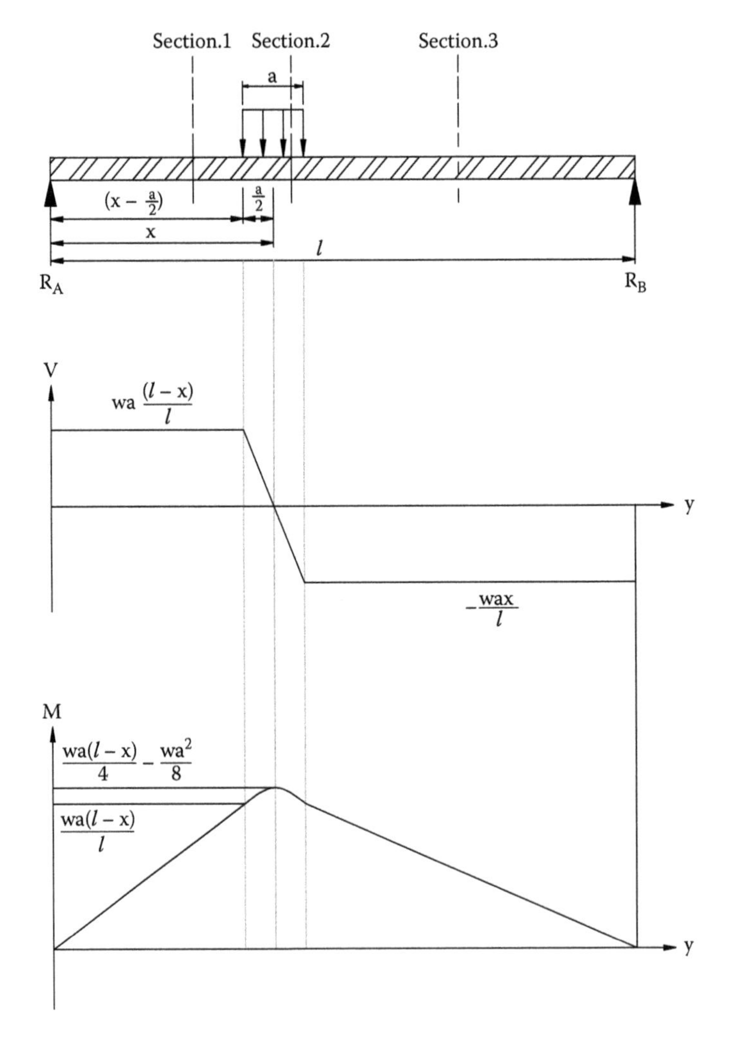

FIGURE 5.11 Shear force and bending moment diagram of the power screw.

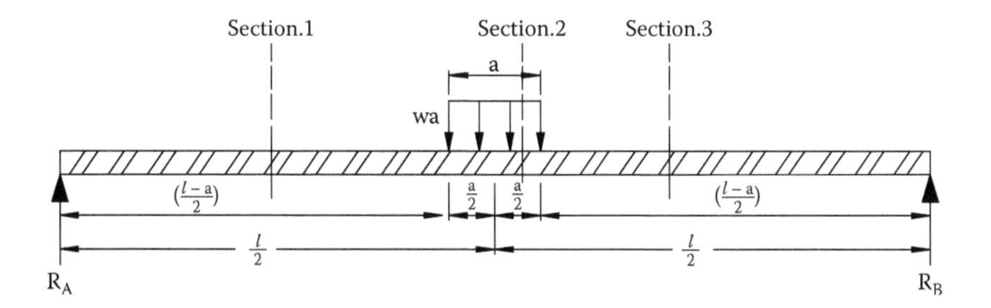

FIGURE 5.12 Loading at the center of the power screw.

and bending moment diagram for the case. Now consider the example of distributed load first at the center of the screw and draw its shear force and bending moment diagrams. Then consider another example of the same load when it is at one of the edges.

In the second case, the load *wa* is at the center of the power screw (Figure 5.12). Replace the value of x for the center position, which will be $x = l/2$.

Considering section 1 of the power screw (Figure 5.13):

$$V = R_A = \frac{wa}{2}$$

$$M = R_A y = \frac{way}{2}$$

$$\text{at} \quad y = 0:$$

$$M = 0$$

$$\text{at} \quad y = \left(\frac{l-a}{2}\right):$$

$$M = \frac{wa}{4}(l-a)$$

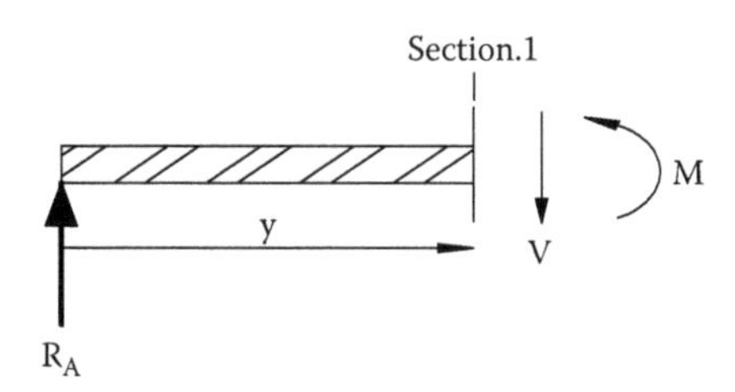

FIGURE 5.13 Section 1 of the power screw.

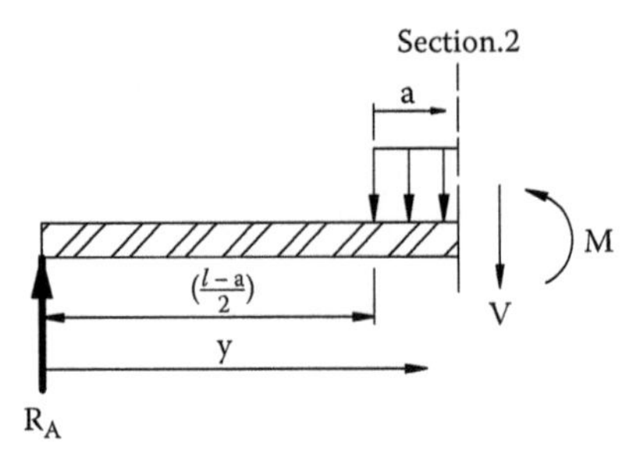

FIGURE 5.14 Section 2 of the power screw.

Considering section 2 of the power screw (Figure 5.14):

$$V = R_A - w\left(y - \frac{l-a}{2}\right) = \frac{wl}{2} - wy$$

$$M = R_A y - w\left(y - \frac{l-a}{2}\right)\frac{\left(y - \frac{l-a}{2}\right)}{2}$$

$$M = \frac{way}{2} - \frac{w}{2}\left(y - \frac{l-a}{2}\right)^2$$

at $y = \left(\frac{l-a}{2}\right)$:

$$V = \frac{wa}{2}$$

$$M = \frac{alw - a^2 w}{4}$$

at $y = \left(\frac{l+a}{2}\right)$:

$$V = -\frac{wa}{2}$$

$$M = \frac{alw - a^2 w}{4}$$

Considering section 3 of the power screw (Figure 5.15):

$$V = R_A - wa = -\frac{wa}{2}$$

$$M = R_A y - wa\left(y - \frac{l}{2}\right)$$

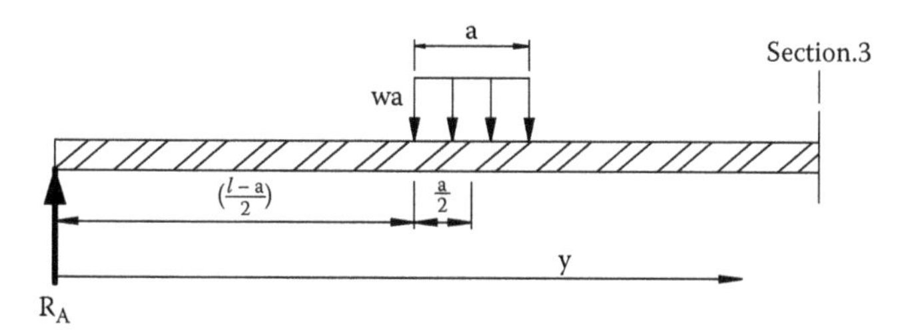

FIGURE 5.15 Section 3 of the power screw.

$$\text{at}\quad y=\left(\frac{l+a}{2}\right):$$

$$M=\frac{alw-a^2w}{4}$$

$$\text{at}\quad y=l:$$

$$M=0$$

The maximum bending moment will be at the center of the power screw, which can be evaluated by use of the parametric equation of section 2.

$$M=\frac{way}{2}-\frac{w}{2}\left(y-\frac{l-a}{2}\right)^2$$

$$\text{Put}\quad y=\frac{l}{2}$$

$$M_{\max}=\frac{wal}{4}-\frac{wa^2}{8}$$

The shear force and bending moment diagram of the power screw when the load is at the center is shown in Figure 5.16.

In the third case, the load is at the edge of the power screw (Figure 5.17). Consider the distributed load *wa* to be at the left end of the power screw. Calculating the most general equations valid for the reactions, shearing forces, and bending moments in the required two sections of the screw:

$$+\Sigma\, F_y=0$$

$$R_A+R_B-wa=0$$

$$+\Sigma\, M_B=0$$

$$wa\left(l-\frac{a}{2}\right)-R_A\,l=0$$

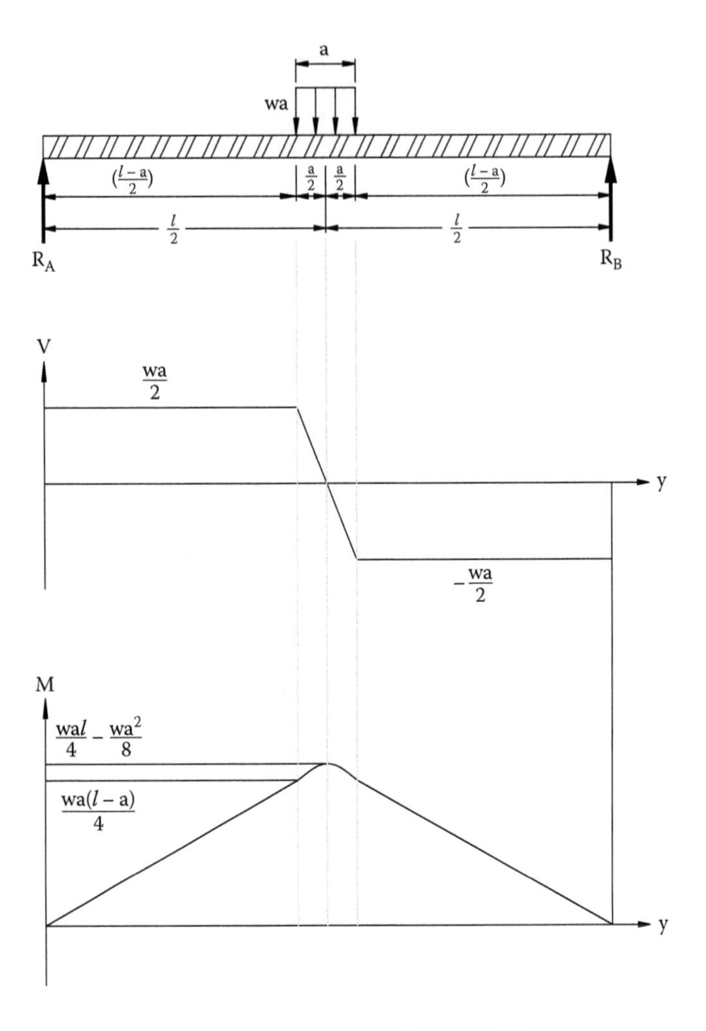

FIGURE 5.16 Shear force and bending-moment diagram of the power screw.

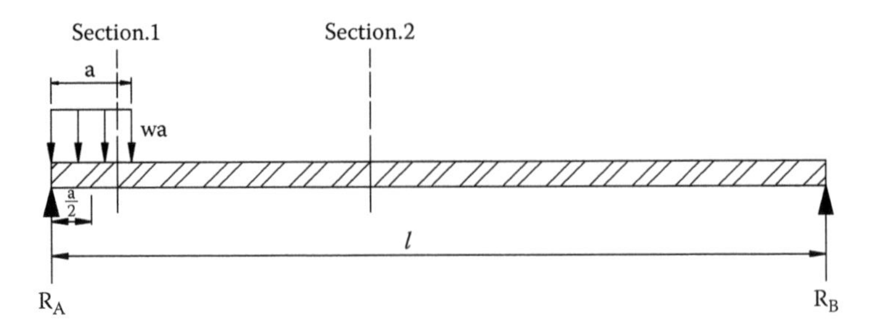

FIGURE 5.17 Power screw loaded at the edge.

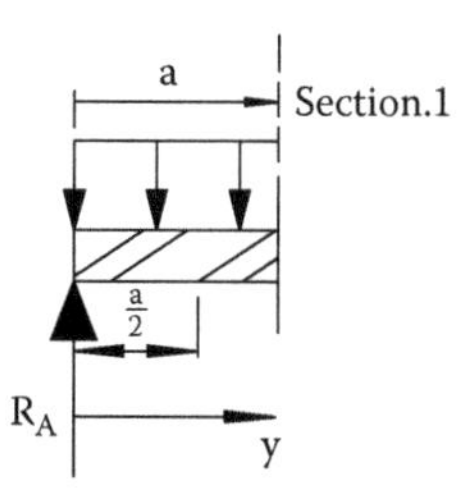

FIGURE 5.18 Section 1 of the power screw.

Solving above equations, we get

$$R_A = \frac{wa}{l}\left(l - \frac{a}{2}\right)$$

$$R_B = wa - \frac{wa}{l}\left(l - \frac{a}{2}\right) = \frac{wa^2}{2l}$$

Considering section 1 of the power screw (Figure 5.18):

$$V = R_A - wy = \frac{wa}{l}\left(l - \frac{a}{2}\right) - wy$$

$$M = R_A y - wy\frac{y}{2} = \frac{wa}{l}\left(l - \frac{a}{2}\right)y - \frac{wy^2}{2}$$

at $y = 0$:

$$V = R_A = \frac{wa^2}{2l}$$

$$M = 0$$

at $y = a$:

$$V = -\frac{wa^2}{2l}$$

$$M = \frac{wa^2}{2}\left(\frac{l-a}{l}\right)$$

Considering section 2 of the power screw (Figure 5.19):

$$V = R_A - wa = \frac{wa}{l}\left(l - \frac{a}{2}\right) - wa = -\frac{wa^2}{2l}$$

$$M = R_A y - wa\left(y - \frac{a}{2}\right) = \frac{wa}{l}\left(l - \frac{a}{2}\right)y - wa\left(y - \frac{a}{2}\right)$$

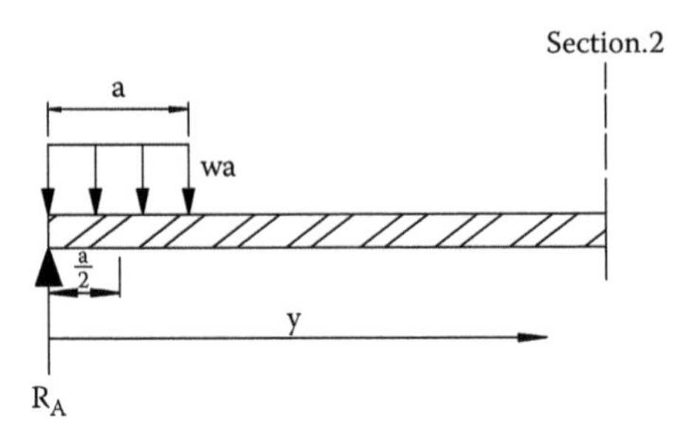

FIGURE 5.19 Section 2 of the power screw.

at $y = a:$

$$M = \frac{wa^2}{l}\left(l - \frac{a}{2}\right) - \frac{wa^2}{2}$$

at $y = l:$

$$M = wa\left(l - \frac{a}{2}\right) - wa\left(l - \frac{a}{2}\right) = 0$$

Share force and bending moment diagrams for this case are presented in Figure 5.20.

Now consider the torsional effect on the power screw. Torsion is the analysis of stresses and strains in a member of the circular cross section subjected to twisting couples, or torques, T and T' of common magnitude, opposite senses, represented generally by curved arrows.

The resistive force caused by the frictional effect of the distributed load *wa* produces torsion in the power screw. The direction of the resistive torque is opposite the direction of rotation, as show in Figure 5.21.

The supports at the edges also exert some frictional force, which is negligible compared with frictional effect of the tabletop. The effect of torsion on the power screw is calculated by use of the following variables:

Length of screw from the rotating end to the point of application of load *wa* is l

c is radius of power screw

f is resistive force caused by friction

$T = fc$ is torque produced

τ is shearing stress

γ is shearing strain

G is shearing modulus

φ is angle of twist

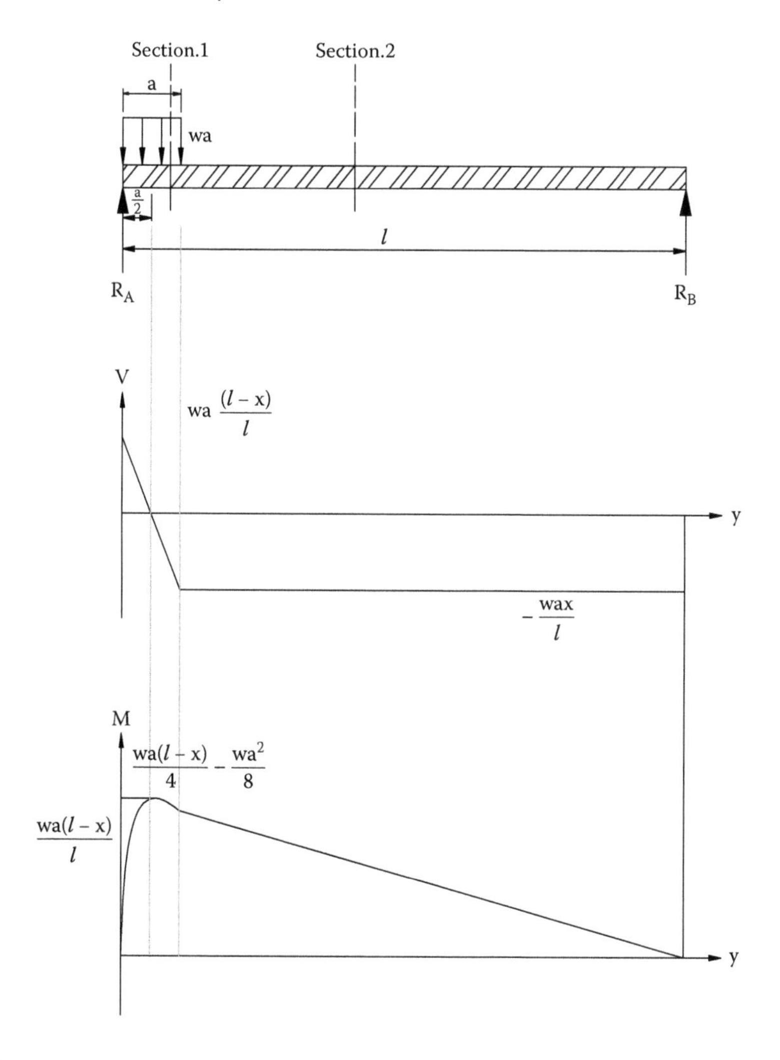

FIGURE 5.20 Shear force and bending-moment diagram of the power screw.

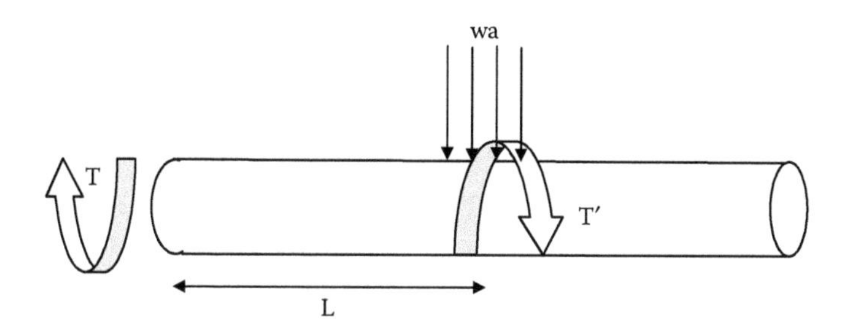

FIGURE 5.21 Torsional loading of the power screw.

Stresses and strains on the screw within the elastic limit are given as follows:

$$\tau_{max} = \frac{Tc}{J}, \quad \text{where} \quad J = \frac{\pi}{2}c^4$$

$$\tau_{max} = \frac{Tc}{\frac{\pi}{2}c^4} = \frac{2f}{\pi c^2} = \frac{2f}{A}$$

Angle of twist ϕ is proportional to the torque T applied to it and is expressed in radians.

$$\phi = \frac{TL}{JG} = \frac{fcL}{\frac{\pi}{2}c^4G} = \frac{2fL}{\pi c^3G}$$

$$\gamma = \frac{c\phi}{L} = \frac{2f}{\pi c^2G}$$

These equations give the effect of the torsion on the power screw.

Considering the stresses caused by combined loads at an arbitrary section NN' and two points K and L that lie on the circular cross section of the screw (the rest of the points on that particular cross section are the same because of symmetry), passing the section NN', the two points K and L are obtained as in the Figure 5.22.

Consider stresses on the first point, K, as in Figure 5.23. Normal stress σ_x is caused by the moment M_Z, which is

$$\sigma_x = \frac{M_Zc}{I_Z}$$

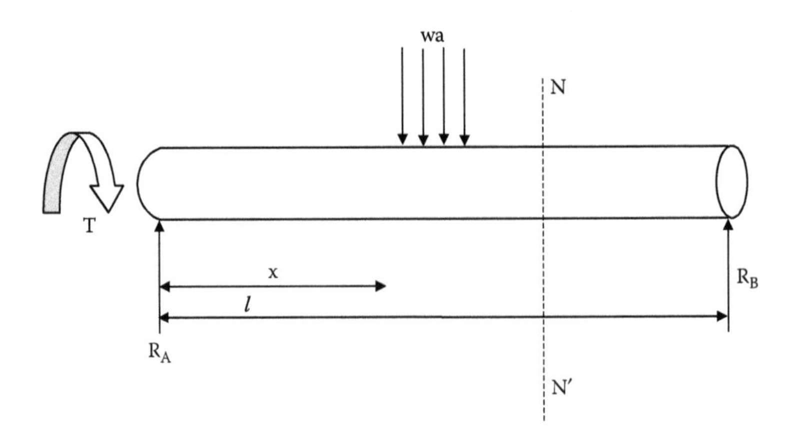

FIGURE 5.22 Torsional loading at section NN'.

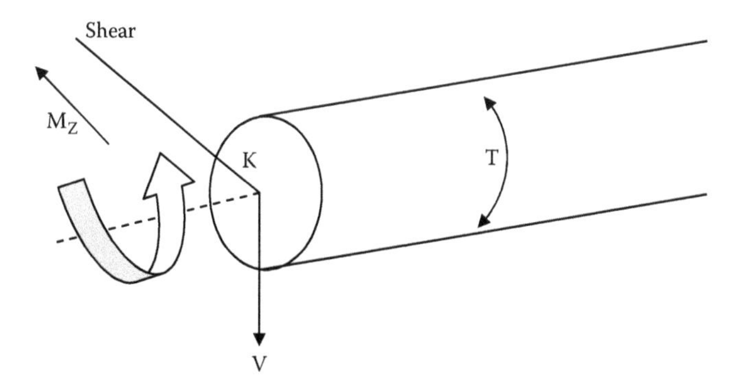

FIGURE 5.23 Stresses at point *K*.

Since

$$I_Z = \frac{\pi c^4}{4} \quad \text{and} \quad M_Z = Fx,$$

therefore,

$$\sigma_x = \frac{M_Z c}{\frac{\pi c^4}{4}} = \frac{4M_Z}{\pi c^3} = \frac{4Fx}{\pi c^3}$$

At this point, the shearing force τ_{xy} is caused by the twist or the torsion effect *T;* that is, the direction of torque depends upon the direction of rotation of power screw.

$$\tau_{x,y(Twist)} = \frac{Tc}{J}$$

From the previous calculation of torsion

$$\tau_{max} = \frac{2f}{A}$$

Therefore, for this case $\tau_{max} = \tau_{x,y(Twist)}$

So

$$\tau_{x,y(Twist)} = \frac{2f}{A}$$

Now consider the second point, L, as in Figure 5.24. At this point, the normal stress $\sigma_x = 0$, because no moment or normal force acts on this point. The only stresses at this point are the shearing stresses caused by *T* and the shearing force *V*,

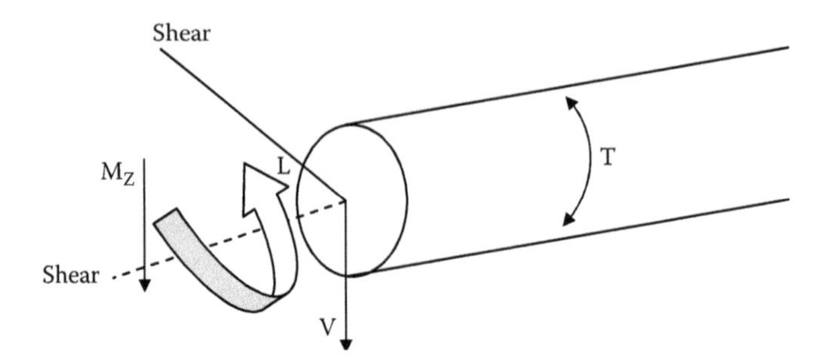

FIGURE 5.24 Stresses at point L.

which can vary for different positions of the load, depending upon the corresponding reactions. Therefore,

$$\tau_{(x,y)v} = \frac{VQ}{It},$$

where

$$Q = A\bar{y}, \quad I = \frac{\pi c^4}{4}, \quad t = 2c$$

$$\tau_{(x,y)v} = \frac{VA\bar{y}}{It} = \frac{V\left(\frac{\pi c^2}{2}\right)\bar{y}}{\left(\frac{\pi c^4}{4}\right)(2c)} = V\frac{\pi c^2}{2}\frac{4}{\pi c^4}\frac{\bar{y}}{2c} = \frac{V\bar{y}}{c^3}$$

The analytical expressions developed in this example can be used to compute the values of the critical parameters of deflection, shear force, bending moment, shearing stress, and shearing strain when standard values associated with physical and mechanical properties of the system are introduced. Several organizations provide standards test data for such an application (e.g., DIN and ASTM). For other mechanical loading conditions, a parametric expression can be developed by use of a similar analytical process, or an existing equation suitable for the loading conditions. Numerical and experimental procedures can also be adopted, either as the only method available for the analysis of given loading conditions or to verify the results of analytical computation. Again, use of standard values from various standards organizations play an important role. Combination of loading conditions and various types of exposures complicate the analysis task whether performed by use of analytical, numerical, or experimental procedures. A systematic analysis scheme, along with use of standards, proves useful for the design and analysis of multidisciplinary discrete products.

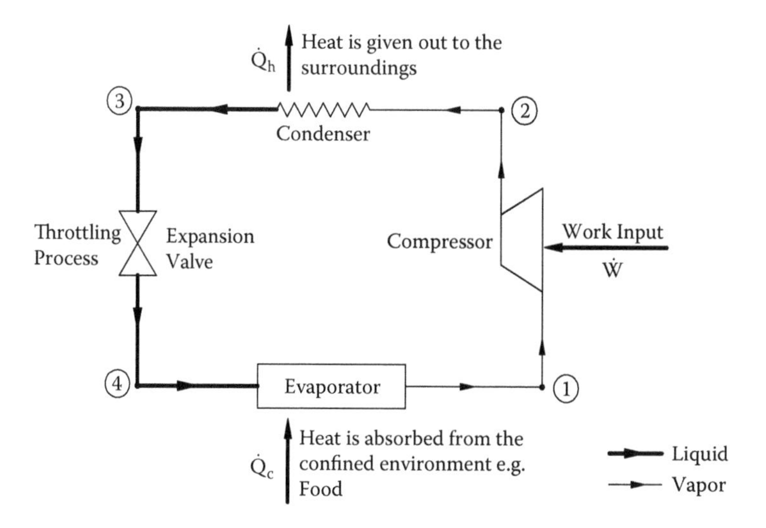

FIGURE 5.25 Black diagram of the refrigeration cycle.

5.2.2 Coefficient of Performance—An Example

This section provides an example pertaining to the design of processes. Refrigeration is considered to demonstrate use of governing equations in this area. Maintaining the temperature of a confined space below the temperature of its surroundings is called refrigeration. Following are some major applications of refrigeration:

1. Food industry
   a. Preparation
   b. Storage and preparation
   c. Distribution
2. Chemical and process industry
   a. Separation of gases
   b. Condensation of gases
   c. Dehumidification of air
   d. Solidification of solute from solvent
   e. Low-pressure storage in liquid form
   f. Removal of heat of reaction
3. Manufacturing industry
   a. Cold treatment of metals

In the refrigeration cycle, heat flows from lower temperature to higher temperature. Evaporators absorb heat from the lower-temperature environment, and

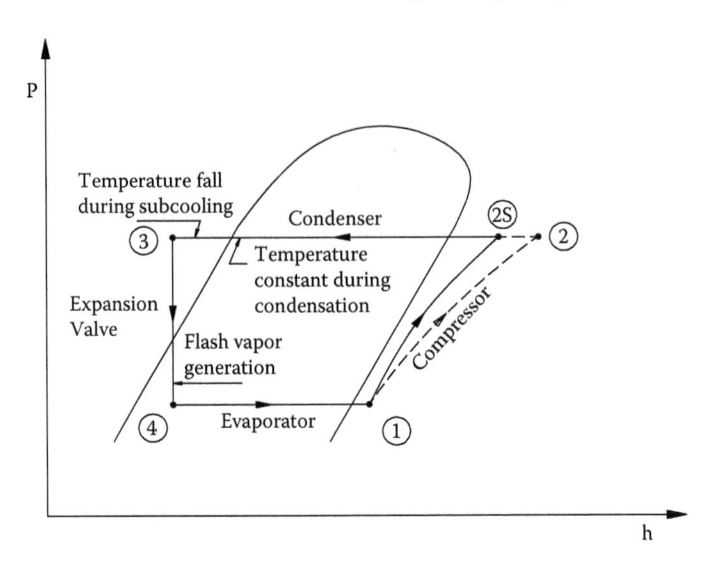

FIGURE 5.26 The refrigeration cycle.

condensers expel this heat into the higher-temperature environment. The refrigeration cycle is shown in Figure 5.25 and is explained below:

- State 1 to state 2: Compression. Starting with the superheated vapor, the temperature and enthalpy increases with the increase in pressure. However, the ideal compression is isentropic (i.e., constant entropy process), which is shown by 1 to 2S in Figure 5.26.
- State 2 to state 3: Condensation. The temperature remains constant during condensation and drops during subcooling, and the heat energy is expelled during the phase change from vapor to liquid. The whole process is assumed to take place at constant pressure.
- State 3 to state 4: Expansion. In expansion device enthalpy remains constant while pressure and temperature drops appreciably. Flash vapors are generated during expansion, which do not contribute to the refrigeration effect.
- State 4 to state 1: Cooling. When the liquid refrigerant enters into the evaporator at low pressure, it absorbs heat from the surroundings (e.g., food) and becomes vapor. During this phase change, the refrigerant produces the refrigeration effect.

If the heat energy absorbed by the refrigerant from the confined space of the evaporator is Q_c, the heat is absorbed during the phase change of the refrigerant from liquid state to vapor state at constant temperature. W is the work input to the compressor to increase the pressure and temperature of the refrigerant in the vapor form from state 1 to state 2. Q_h is the heat energy expelled by the refrigerant to the surrounding environment during its phase change from vapor state to liquid state (condensation). This heat is added to the refrigerant in the evaporator and compressor. $\dot{m}$ is the mass flow rate of refrigerant.

The coefficient of performance (COP) is always greater than 1 because it is the ratio of two inputs: the heat absorbed by the refrigerant (which is an output in terms of refrigeration) and the work input to the compressor.

$$\text{Refrigeration effect:} \quad \dot{Q}_c = \dot{m}\,(h_1 - h_4)$$

$$\text{Power Input:} \quad \dot{W} = \dot{m}\,(h_2 - h_1)$$

$$\text{Heat effect:} \quad \dot{Q}_h = \dot{m}\,(h_1 - h_3)$$

$$\text{Coeffiecent of Performance:} \quad COP = \frac{Ouput}{Input}$$

For refrigeration $COP_{ref} = \dfrac{\dot{Q}_c}{\dot{W}}$, and for heat pump $COP_{hp} = \dfrac{\dot{Q}_h}{\dot{W}}$.

According to the Energy Balance equation:

Energy input to the system = Energy output from the system

From Figure 5.26, we have

Energy out from the condenser = Energy input in the evaporator + Energy input in the compressor

$$\dot{Q}_h = \dot{Q}_c + \dot{W}$$

Dividing both sides by $\dot{W}$,

$$\frac{\dot{Q}_c}{\dot{W}} = \frac{\dot{Q}_h}{\dot{W}} + \frac{\dot{W}}{\dot{W}}$$

$$\frac{\dot{Q}_c}{\dot{W}} = \frac{\dot{Q}_h}{\dot{W}} + 1$$

$$COP_{hp} = COP_{ref} + 1$$

In a standard vapor compression cycle (Figure 5.27), production of A tons of refrigerant requires a condenser temperature of X °F and an evaporator temperature of Y °F. To calculate:

The refrigeration effect
Mass flow rate of refrigerant
Power required by the compressor
Volume flow rate

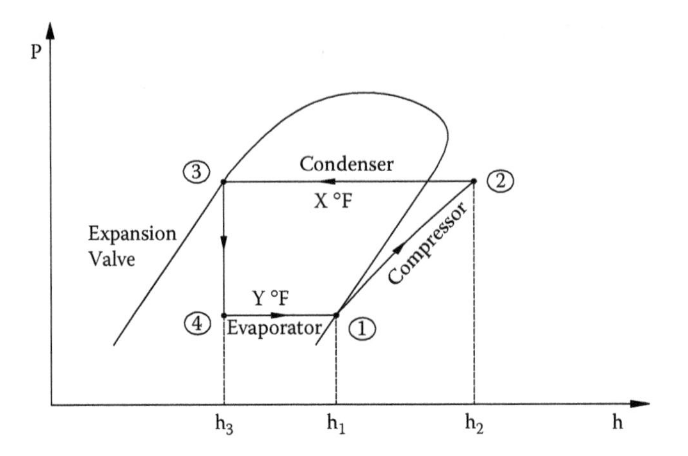

FIGURE 5.27 A standard vapor compression cycle.

Refer to a table of refrigerants, such as the ICI KLEA 134a pressure–enthalpy diagram. The enthalpy of the process is

$$\text{Refrigeration effect} = h_1 - h_4$$

$$\text{Refrigeration capacity} = \dot{m}\ (\text{Refrigeration effect})$$

$$\dot{m} = \frac{(A \text{ tons})\left(200 \frac{\text{BTU}}{\text{min ton}}\right)}{\left(B \frac{\text{BTU}}{\text{lb}}\right)} = C \frac{\text{lb}}{\text{min}}$$

$$\text{Compressor Power} = \dot{m}\ (h_2 - h_1)$$

$$\text{Compressor Power} = C \frac{\text{lb}}{\text{min}} (h_2 - h_1) \frac{\text{BTU}}{\text{lb}}$$

$$\text{Compressor Power} = D \frac{\text{BTU}}{\text{min}}$$

$$\because \quad 778 \text{ ft lb} = 1 \text{ BTU}$$

$$\text{or} \quad 33000 \text{ ft lb} = 1 \text{ hp}$$

$$\therefore \quad \text{Compressor Power} = \frac{D \frac{\text{BTU}}{\text{min}}\ 778 \frac{\text{ft lb}}{\text{BTU}}}{33000 \frac{\text{ft lb}}{\text{min hp}}}$$

$$\text{Compressor Power} = E \text{ hp}$$

From the table of refrigerants as given by ICI, KLEA134a, the specific volume v of the refrigerant is

$$\therefore \quad \text{Volume flow rate} = \dot{m}\, v$$

$$\text{Volume flow rate} = C\,\frac{\text{lb}}{\text{min}}\; v\,\frac{\text{ft}^3}{\text{lb}}$$

$$\text{Volume flow rate per ton} = \frac{C\,\dfrac{\text{lb}}{\text{min}}\; v\,\dfrac{\text{ft}^3}{\text{lb}}}{A\ \text{tons}}$$

$$\text{Volume flow rate per ton} = F\,\frac{\text{ft}^3}{\text{min ton}}$$

5.2.3 Engineering Properties and Standards

The discrete product based on the input-process-output principle uses several engineering properties in each domain of engineering. Awareness of the optimal values of these properties, initially deduced through various types and ranges of analysis and, subsequently, verified through appropriate standards, is necessary. Several national, international, and professional standards-developing organizations are listed in Chapter 2 that sufficiently cover these properties. Whether mechanical properties, physical properties, or process parameters, use of standards values of various engineering properties in the design and manufacture of discrete products gives the benefits listed in Section 1.1.

5.3 MULTIDISCIPLINARY ANALYSIS FOR DISCRETE PRODUCTS

The discrete products that comprise multidisciplinary features require analysis for all the features. If the discrete products are considered as systems and divided into subsystems at subassembly and component levels, this task can be performed independently. Design by use of basic analytical, numerical, or experimental technologies for multidisciplinary aspect then becomes possible. As a basic example, the analytical steps described in Figure 5.4 provide connection between various types of analyses required for multidisciplinary discrete products. Figure 5.28 provides a simple example of a discrete product under analysis for strength. Figure 5.29 provides the possible results that may be deduced from such an analysis.

One of the important aspects of Figure 5.29 is the size (diameter) of the stress raiser on the left hand side of the diagram to be used for placing a bearing

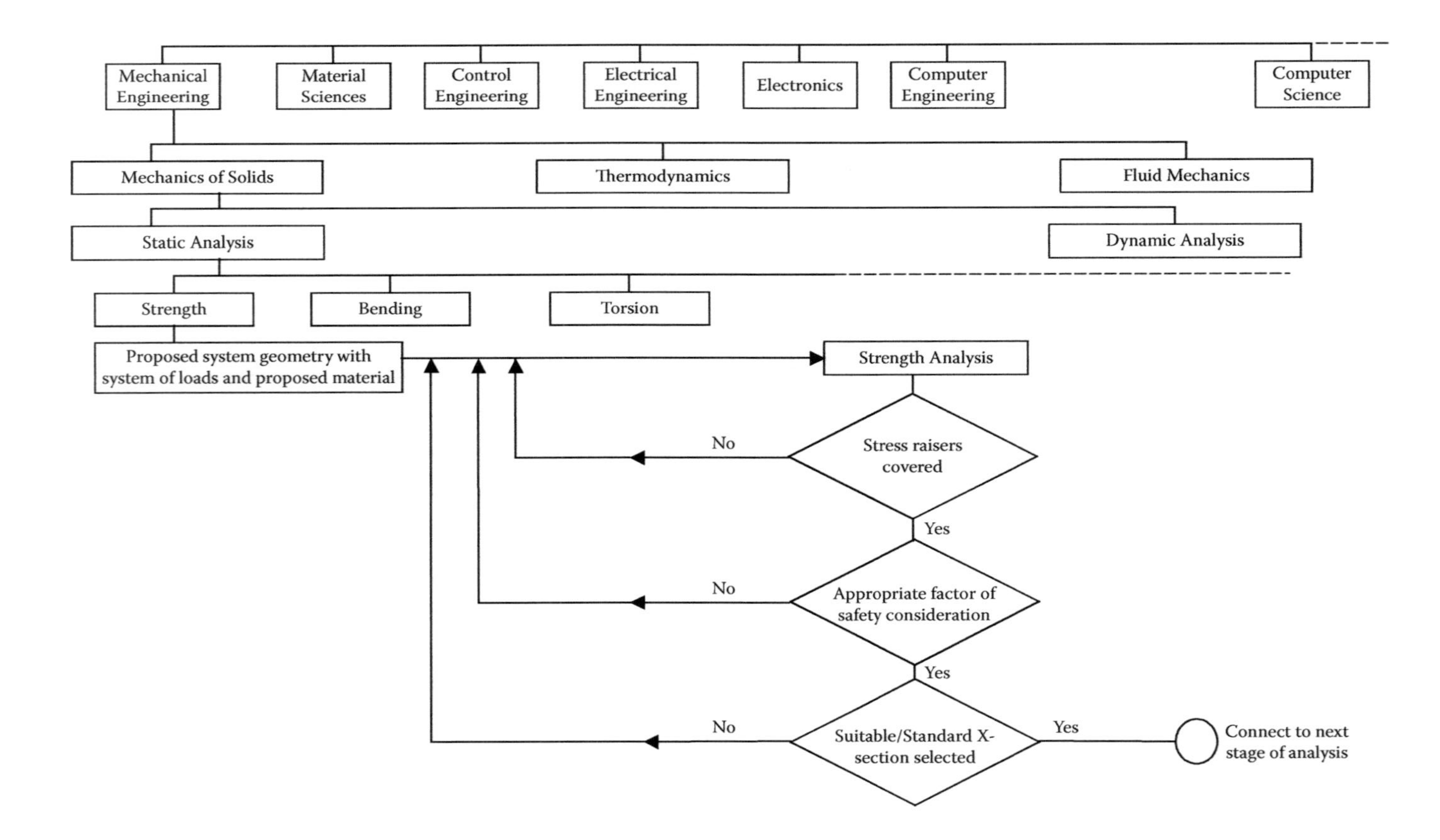

FIGURE 5.28 Strength analysis for discrete products.

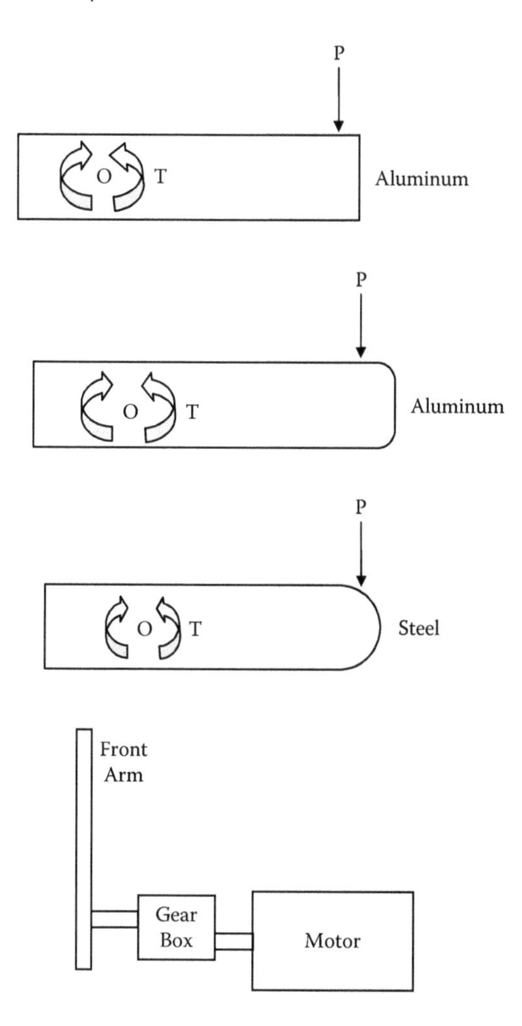

FIGURE 5.29 Design for strength. Revision for component geometry and material of robot arm.

(a machine element). Such features bring the element of selection to the design process and that of bridging parameters between two technologies. The allowable size of a bearing to be placed involves comparison between the bridging parameter of deduced size through analytical or numerical methods and the available size and type from the manufacturer's standard product. At the same spot, another bridging parameter is involved that requires a comparison between the torque necessary to rotate the lever clockwise or counterclockwise and the standard holding torque of a stepper motor. Figure 5.30 provides an insight into this aspect.

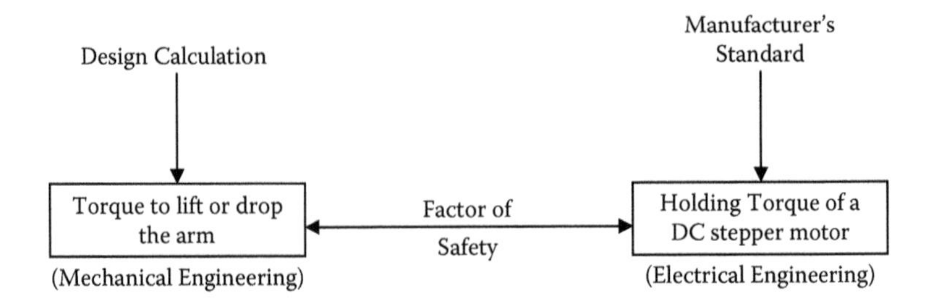

FIGURE 5.30 Bridging parameter comparison in design of multidisciplinary products.

5.4 SELECTION OF STANDARD COMPONENTS

A large variety of mechanical, electrical, electronics, and control engineering components are available from manufacturers. A major characteristic of these components is the use of standards related to size, shape, material, input parameters, process parameters, and output parameters. These components are normally designed on the basis of a collection of standards from international standards organizations, national standards organizations, professional standards organizations, and company standards. The standards are normally widely accepted. The procedure to select these components is provided in the online catalogs or paper-based catalogs of these components. Refer to various appendices for catalogs of mechanical components, machine elements, and control elements.

BIBLIOGRAPHY

Ferrrari, G., Evolution of new design standards, *Printed Circuit Design*, 14(5), 11–16, 1997.

Hicks T.G., *Handbook of Mechanical Engineering Calculations*, McGraw-Hill, New York, 1997.

Incropera, F.P. and Dewitt, D.P., *Fundamentals of Heat and Mass Transfer*, 5th ed., John Wiley & Sons, New York, 2002.

McQuiston, F.C., Parker, J.D., and Spitler, J.D., *Heating, Ventilating, and Air-conditioning —Analysis and Design*, 5th ed., John Wiley & Sons, New York, 2000.

O'Connell, L., Design Automation Standards Need Integration, Proceedings—Design Automation Conference, 562, 1987.

Mangonon, P.L., *The Principles of Materials Selection for Engineering Design*, 1st ed., Prentice Hall, New York, 1998.

Riesgo, T., de la Torre, E., Torroja, Y., and Uceda, J., Use of standards in electronic design, *IECON Proc.* 1, 407–412, 1996.

Shigley, J.E. and Mischke, C.R., *Standard Handbook of Machine Design*, 2nd ed., McGraw-Hill, New York, 1996.

Sonntag, R.E., Borgnakke, C., and Van Wylen, G.J., *Fundamentals of Thermodynamics*, 5th ed., John Wiley and Sons, New York, 1998.

Sweet, J., Design quality: standards and liability, *Pressure Vessel Piping Technol.*, 941–949, 1985.

Toms, P., Standards and the design process, *Design Stud.*, 9(2), 115–122, 1988.

Witt, J.W. and Johnson, R.L., Design standards/public liability, *ASCE*, 192–197, 1985.

Part 3

Standards in Manufacturing of Discrete Products

6 Review of Standards for Manufacturing—Equipment Perspective

6.1 MANUFACTURING PROCESSES

An enormous amount of manufacturing equipment exists for implementation of basic discrete manufacturing processes in production machinery. This manufacturing equipment consists of discrete products capable of manufacturing components, assemblies, structures, mechanisms, and machines. Figure 6.1 provides a list of processes available under different categories of discrete manufacturing that are used in actual production. The equipment under each category of discrete manufacturing is further classified according to the following features:

Type
Size
Accessories
Operation parameters
Type of control

6.2 PRODUCTION EQUIPMENT

Discrete manufacturing processes are developed as mechanical artifacts to perform production. These machines produce a variety of components, structures, assemblies, mechanisms, and machines. Each manufacturing process implemented into mechanical artifacts has three distinct features:

The input to the equipment (e.g., raw material type, form, and feeding mechanism, final dimension of the product, energy source, and other auxiliaries).

The process implementation allows transformation of raw material into the required size, shape, and surface finish by use of tools (e.g., metal-cutting tools, high-energy beams, or various types of jets) and utilization of tool-holding devices, work-holding devices, measuring devices, and manufacturing instructions. In-process supplies such as lubricating oil and coolants may also be used.

The output from the equipment comprises a component (the building block of structure, mechanisms, or machines) and scrap.

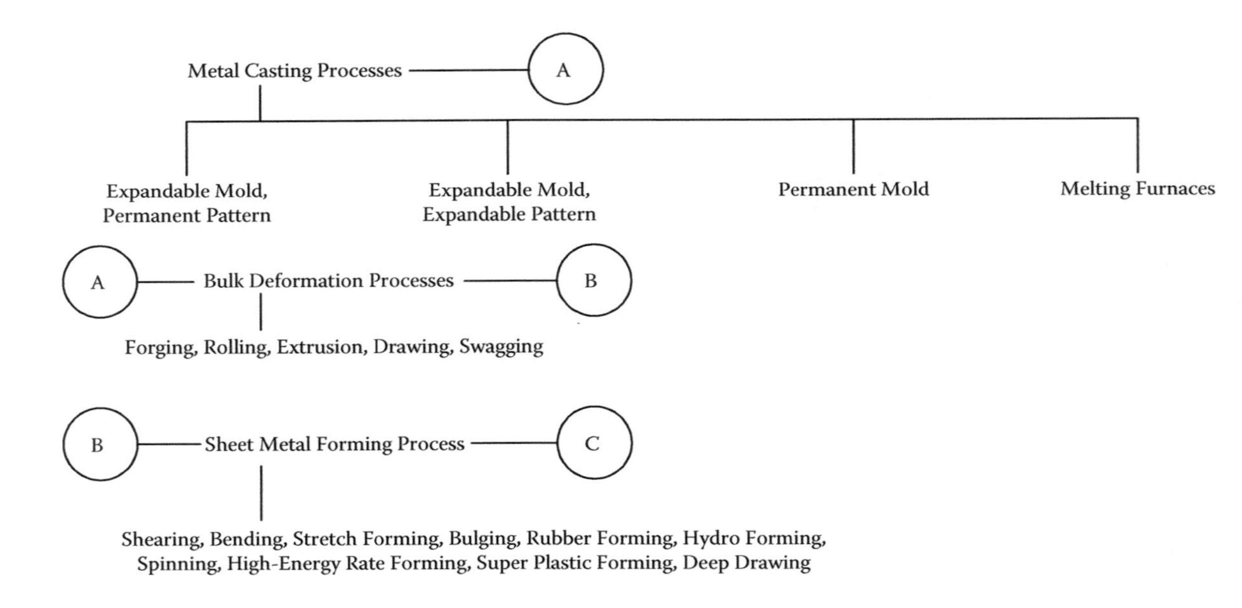

FIGURE 6.1 Process used in discrete manufacturing (from *Manufacturing Processes for Engineering Materials* by S. Kalpakjian and S.R. Schmid, used with permission from Pearson Education, Inc.). (*Continued*)

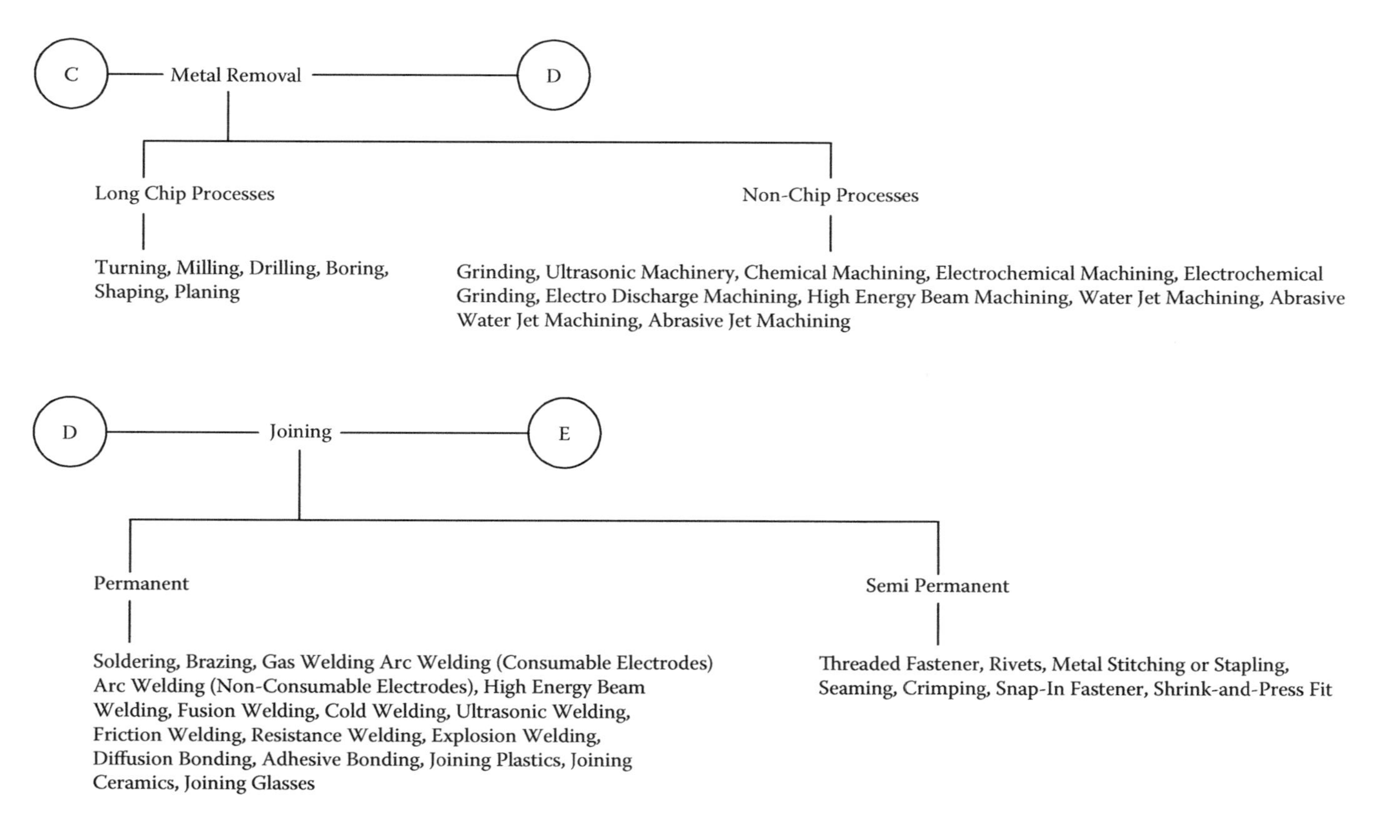

FIGURE 6.1 *(Continued)*

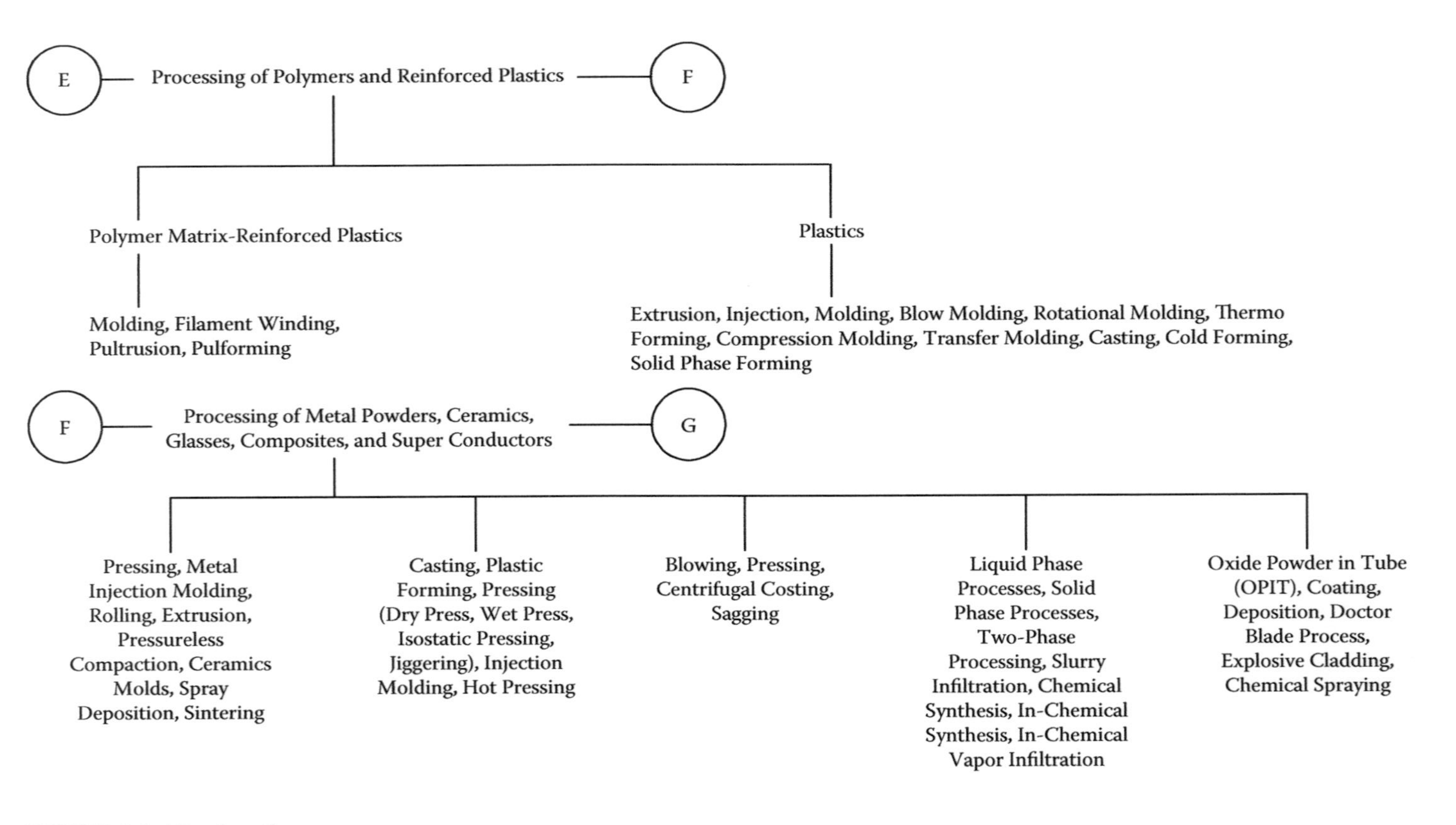

FIGURE 6.1 (*Continued*)

Several features may be present at the production machinery to make the task of manufacturing simpler and easy to control and may result in a high production rate. The design of manufacturing equipment relies on the system approach to design and considers some or all of the following special design requirements:

1. Structural consideration for machine frame and assemblies
2. Thermal effects on the equipment
3. Noise emission from the equipment
4. Vibration in the machinery
5. Environmental effects on the equipment
6. Geometric and kinematics behavior of the manufacturing equipment
7. Static and dynamic behavior of the equipment
8. Availability of computer numerical control (CNC) or process control by use of programmable logic controllers (PLC)
9. Electronic circuitry for implementing control features
10. Foundation and installation requirements

Like any other discrete product, the manufacturing equipment commonly utilizes standard mechanical components, machine elements, control elements, electrical and electronic components, and software components. Special assemblies and other accessories utilized in the construction of manufacturing equipment may also include the following:

1. Tool (mechanical tool, beams, jets, etc.)
2. Tool-holding devices
3. Work-holding devices
4. Lubricating oil pump assembly
5. Coolant circulation pump assembly
6. Material-handling equipment
7. Scrap-handling equipment

An illustration of the implementation of a manufacturing process into production equipment is presented in Figure 6.2.

Today, special consideration is given to the automation and the use of computer numerical control and programmable logic controllers have taken root. Figure 6.3 and Figure 6.4 provide block diagrams of the working principle of open-loop and closed-loop computer numerical control systems. Figure 6.5 and Figure 6.6 provide operation of the programmable logic controllers performing axes control and process control, respectively.

6.3 ELEMENTS OF MANUFACTURING SYSTEMS[1]

A manufacturing system comprises manufacturing equipment arranged in a certain fashion. Manufacturing systems have a physical layout, whereas intangible production control operates on production philosophies as listed in Section 1.3.7.

---

[1] Reproduced from Chryssolovris, G., Manufacturing Systems: *Theory and Practice*, with permission of Springer Verlag.

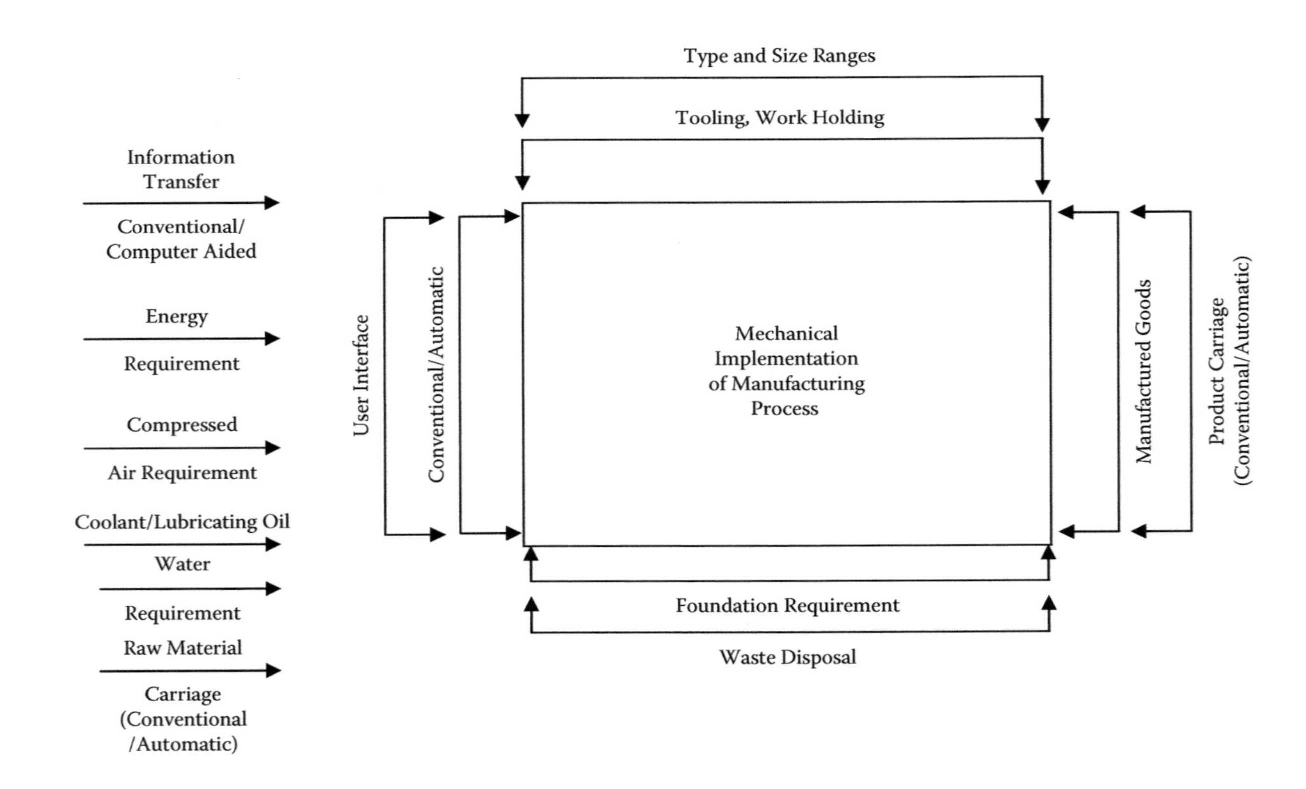

FIGURE 6.2 Manufacturing process implementation into equipment.

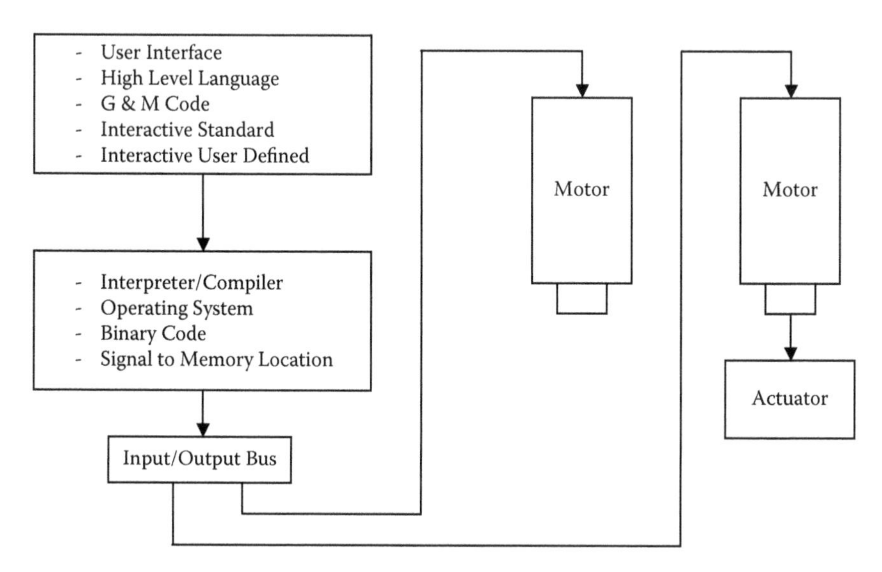

FIGURE 6.3 Open-loop control of CNC machines.

Other important elements of the manufacturing systems are methods of information and material transfer. The physical layout of the manufacturing systems is normally divided into two areas: the processing area and the assembly area.

In industrial practice, four general approaches are used in structuring the processing area for discrete manufacturing: the job shop, the project shop, the cellular system, and the flow line.

In a job shop (Figure 6.7), machines with the same or similar material processing capabilities are grouped together. The lathes form a turning work center, the milling machines form milling work center, and so forth.

In a project shop, a product's position remains fixed during manufacturing because of its size, weight, or both. Materials, people, and machines are brought to the product as needed. Facilities organized as project shops can be found in the aircraft and shipbuilding industries and in bridge and building construction. A schematic of project shop is presented in Figure 6.8.

In manufacturing systems organized according to the cellular plan, the equipment or machinery is grouped according to the process combinations that occur in families of parts. Each cell contains machines that can produce a certain family of parts. Figure 6.9 provides arrangement of the equipment in the cellular systems.

In the flow line, equipment is ordered according to the process sequences of the product to be manufactured. A flow line consists of a sequence of machines, which are typically dedicated to one particular part or, at most, to a few very similar parts. Only one part type is produced at a time (Figure 6.10).

In the actual manufacturing world, these standard system structures often occur in combinations or with slight changes. The choice of a manufacturing system depends on the design of the parts to be manufactured, the lot sizes of

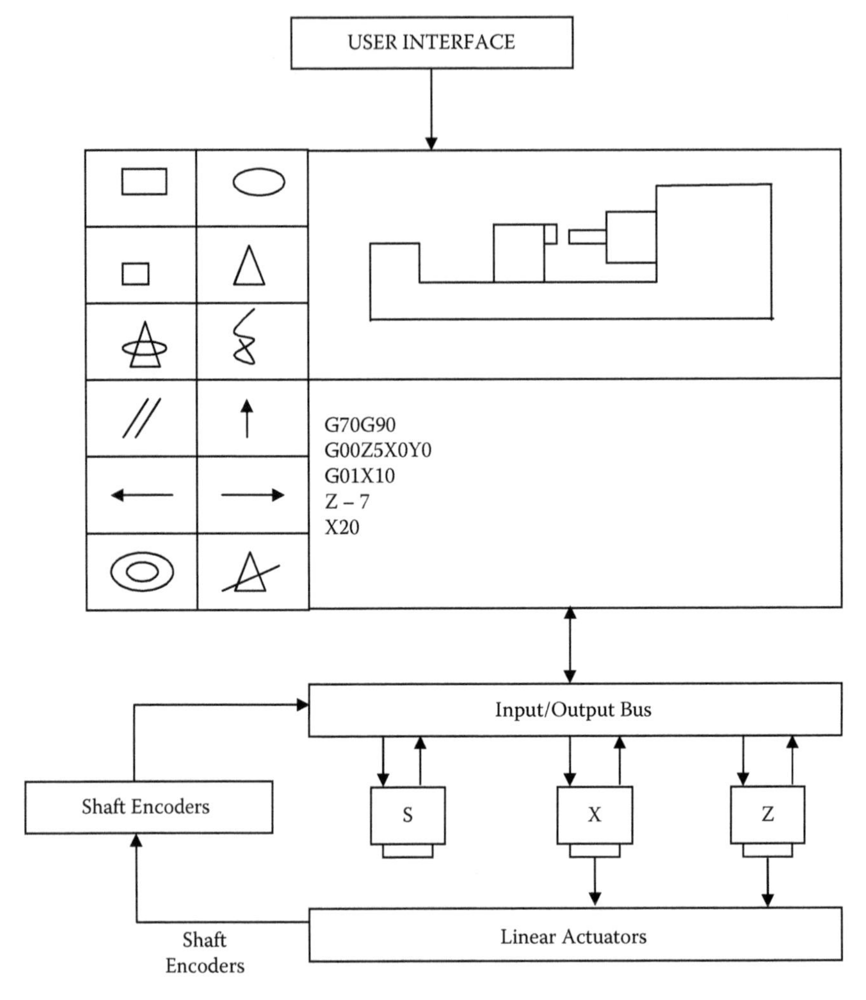

FIGURE 6.4 Closed-loop control of CNC turning machines.

the parts, and market factors such as the required responsiveness. A suitable lot size for each manufacturing system described above is presented in Figure 6.11.

A system that comprises a combination of the basic structures is the flexible manufacturing system (FMS). An FMS is a hybrid of job shop and a cellular manufacturing system. It provides great flexibility in terms of the types of parts and the process sequencing that can be used, a consequence of its high degree of automation. Material and information flow throughout the system is totally automated, and very little human intervention is required.

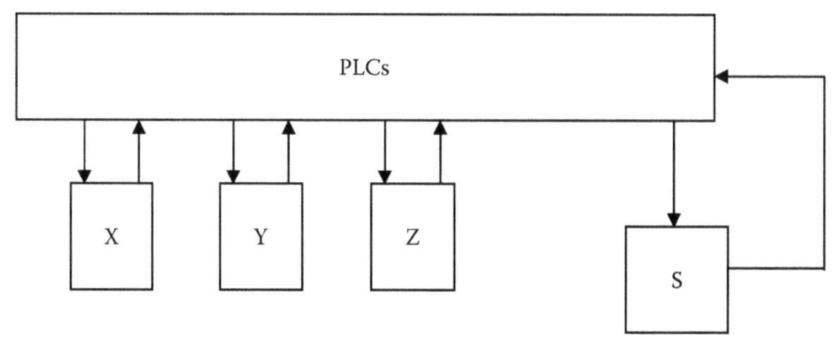

FIGURE 6.5 Axes control by use of a programmable logic controller (PLC).

Another substantial part of a manufacturing company's production facilities is the assembly system. Assembly systems can be categorized according to the motion of parts and workplaces. Stationary-part systems are usually employed for large assemblies, such as airplanes, which are difficult to move around. Moving-part systems can be divided into stationary workplace systems, in which parts are brought to stationery workplaces, and moving workplace system, in which the workplaces move along with the parts. Assembly systems with stationery parts tend to have higher floor-area requirements. They also tend to have more work at each workplace than do moving-part systems. Moving-part systems are generally more expensive because complicated material-handling equipment is required to move parts quickly from workplace to workplace.

In moving-part, stationary-workplace systems, the assembly operations at each workplace are usually short in duration and are highly repetitive. Moving-part, moving-workplace systems allow workers to work on each assembly for a longer period of time, which makes the assembly work less repetitive. The

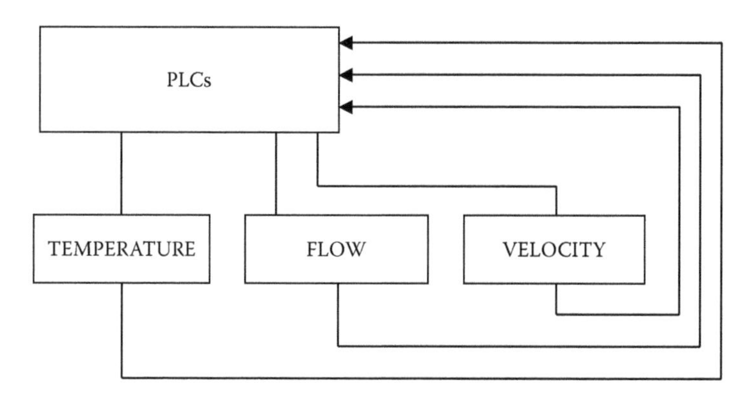

FIGURE 6.6 Process parameter control by use of a programmable logic controller (PLC).

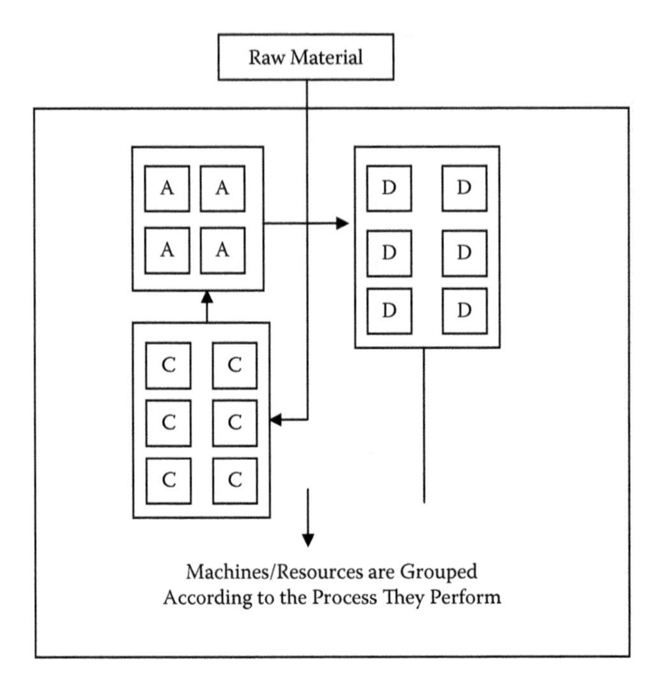

FIGURE 6.7 Schematic diagram of a job shop (used with permission from Chryssolouris).

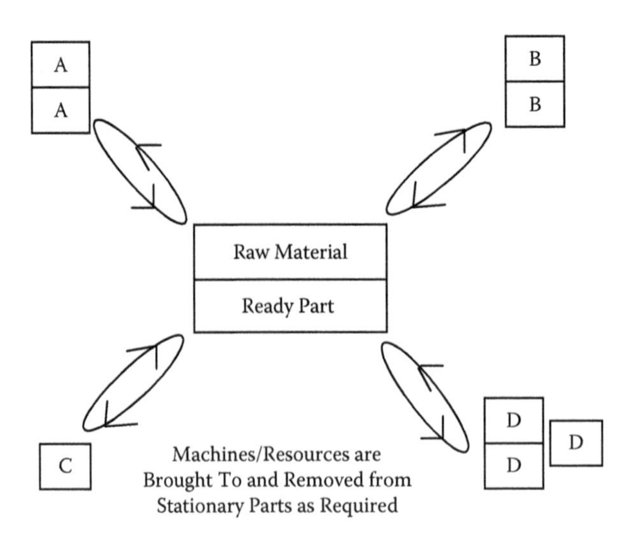

FIGURE 6.8 A project shop (used with permission from Chryssolouris).

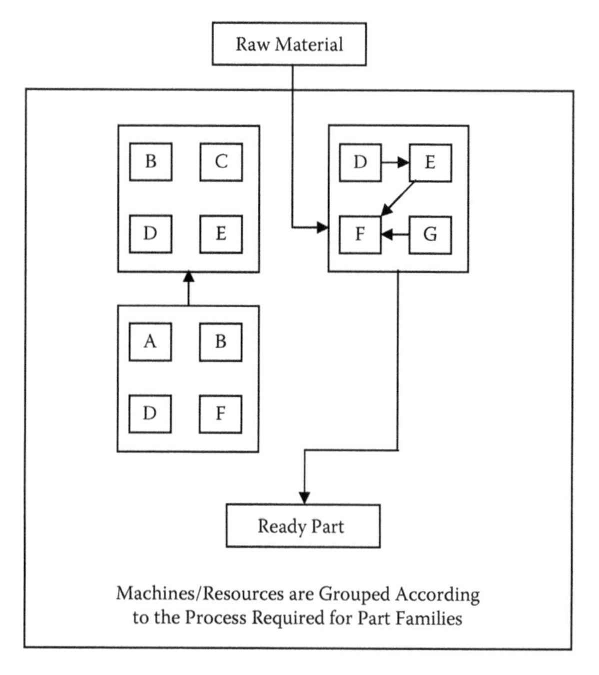

FIGURE 6.9 Schematic diagram of a cellular manufacturing system (used with permission from Chryssolouris).

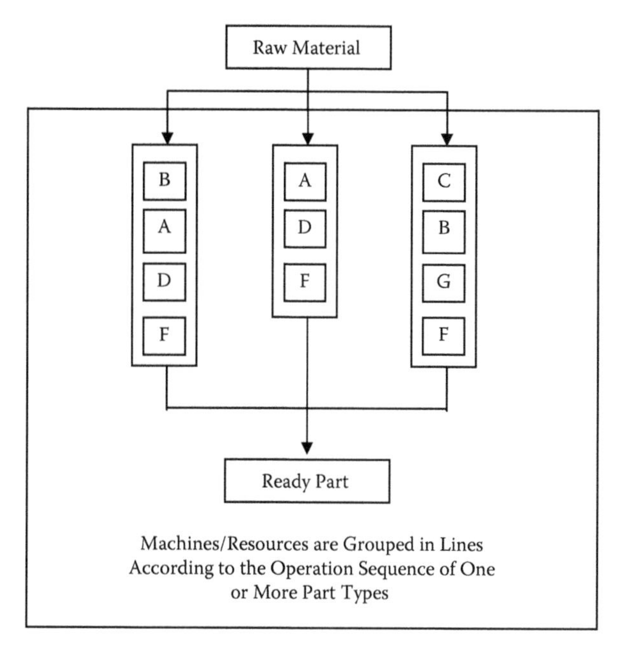

FIGURE 6.10 Schematic diagram of a flow line (used with permission from Chryssolouris).

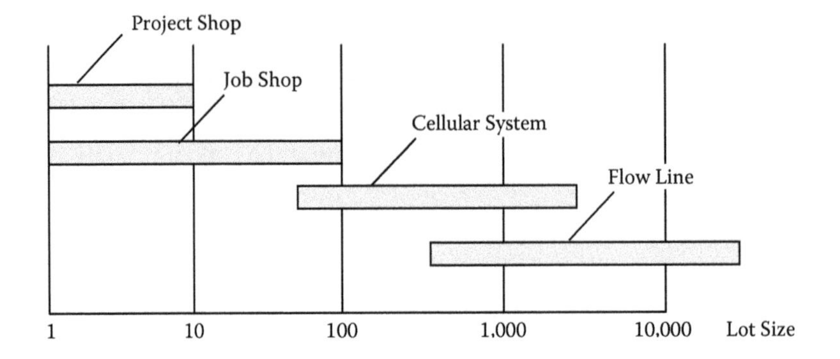

FIGURE 6.11 Suitable manufacturing system types as a function of lot size (used with permission from Chryssolouris).

assembly systems are presented in Figure 6.12. The characteristics of different assembly systems are shown in Figure 6.13.

6.4 STANDARDS FOR MANUFACTURING EQUIPMENT

The number of manufacturing equipment and related parameters are so extensive that listing all existing standards in this book is not possible. Therefore, a domain specification, a target standards organization, and the search parameters are presented. All the headings of various sections in this chapter constitute a domain specification (Figure 6.14). As the American National Standards Institute (ANSI)

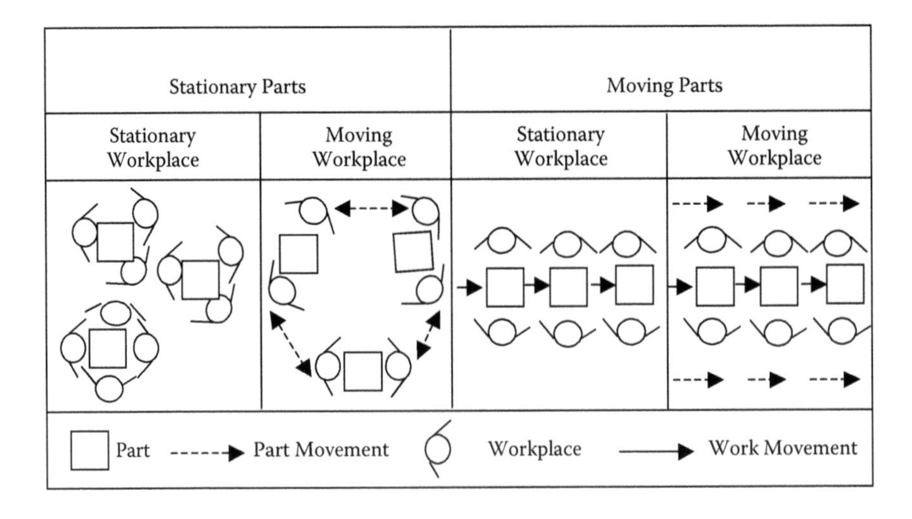

FIGURE 6.12 Types of assembly systems (used with permission from Chryssolouris).

	Stationary Parts		Moving Parts	
	Stationary Workplace	Moving Workplace	Stationary Workplace	Moving Workplace
Area Requirement	High	High	Low	Medium
Work Contents at Each Workplace	High	Medium	Low	Medium
Cost of System	Low	Medium	High	High

FIGURE 6.13 Characteristics of different assembly systems (used with permission from Chryssolouris).

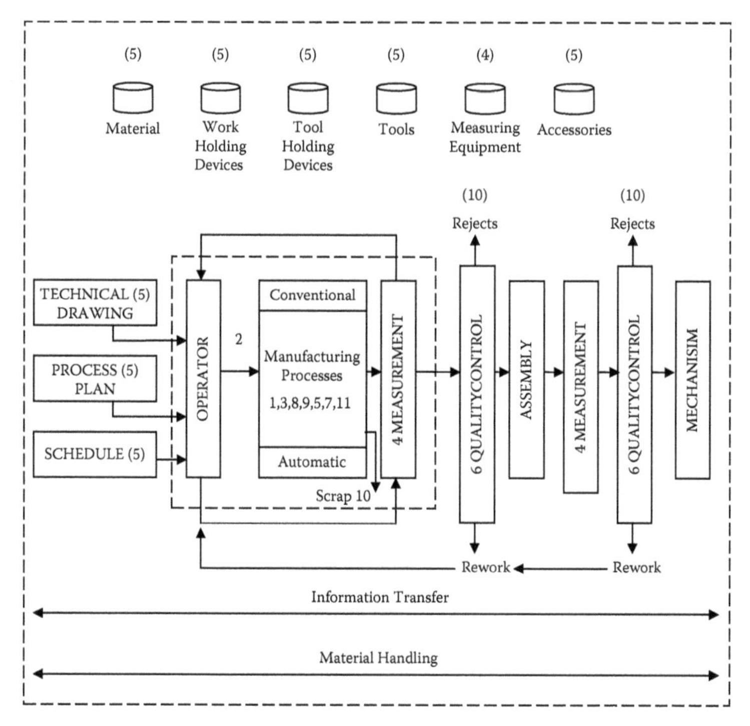

1. Standards for design of equipment
2. Man Machine Interface Standards
   Human Computer Interaction Standards
3. Standards Related to operation of equipment
4. Standards for measuring equipment precision
5. Standards used in manufacturing operation
6. Standards related to quality control
7. Safety Standards related to use of equipment
8. Reliability Standards related to use of equipment
9. Maintainability Standards related to use of equipment
10. Waste disposal standards
11. Equipment Life Cycle standards
12. Standards related to information transfer
13. Standards related to material handling

FIGURE 6.14 General layout for use of standards in production.

is the target standards organization in Chapter 4, this chapter considers the British Standards Institute (BSI) as the target standards organization. A search of the compact disc version of the catalog of the standards organization was performed. For each domain specification, alternative standards organizations are also listed. The filtered listing of standards, normally diverse, is presented as the achievement of the whole exercise. Sometimes this information may be useful to the reader, whereas at the other time, information about the alternative standards organization may be useful. The reader may also refer to the information provided in Chapter 2, about standards, standards organization, and the search for standards.

The objective of this section is to underline the quantum and diversity in discrete manufacturing and the practices it involves. On the basis of the information provided earlier, the more difficult task of selecting the standards for a discrete manufacturing facility shall be attempted. In most of the cases finding all the standards needed for a discrete manufacturing facility from a single source may be difficult. The search shall then examine the committee work of the target organization to find out what direction it is taking, register for the new developments at the target organization, or explore the alternate standards organization.

Although the main searches in a target organization are carried out for a demonstration of the first step in the standardization process of a discrete manufacturing facility, the case study presented in Chapter 7 may consider any standard organization. It, however, considers the modalities of taking such a step.

6.5 STANDARDS FOR DESIGN OF MANUFACTURING EQUIPMENT

Design of the diverse production equipment is performed by use of the system approach to design, with special consideration of the requirements of the production process. The domain specification for this aspect of production equipment is provided in Figure 6.15. Although many parameters in manufacturing processes have been standardized, a large number still remains as the proprietary jurisdiction of the source company. Workpiece materials, measuring equipment and their precision, and accessories are independent parameters and are well covered by various standards organization. On the other hand, the design parameter (such as the structure of the

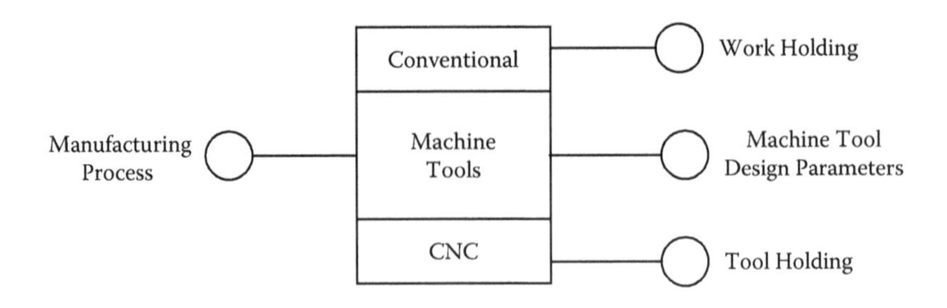

FIGURE 6.15 Domain of standards for the design of production equipment.

equipment), tools, and types of work-holding and tool-holding equipment are dependent parameters for the manufacturing equipment. In this domain specification, several standards organizations are active. These organizations include the following:

1. The American Bearing Manufacturers Association (ABMA)
2. The American Gear Manufacturers Association (AGMA)
3. The American Society for Testing and Materials (ASTM)
4. The American Society of Mechanical Engineers (ASME)
5. The American Welding Society (AWS)
6. The Consortium for Advanced Manufacturing—International (CAM-I)
7. The Electronic Industries Alliance (EIA)
8. The Institute of Electrical and Electronics Engineers (IEEE)
9. The European Committee for Standardization (CEN)
10. The International Electrotechnical Commission (IEC)
11. The International Organization for Standardization (ISO)
12. The Deutsches Institut für Normung (DIN)
13. The Japanese Industrial Standards Committee (JISC)
14. The British Standards Institution (BSI)
15. The American National Standards Institute (ANSI)

The British Standards Institution has been taken as the target organization in the case of Standards for Design of Production Equipment. The following search parameters are used in this domain:

Design
Design of production equipment
Production equipment
Production machinery
Machine tool
Machine tool design

Some typical results are given in Table 6.1. The search may be performed by use of the current electronic catalog or the online database.

6.6 STANDARDS RELATED TO MMI AND HCI

Man–machine interaction (MMI) is a subject that has posed complexities with each evolution or introduction of new manufacturing equipment. The contemporary use of the computer at the user interface of the equipment has given it a new dimension. Figure 6.16 provides a schematic detailing MMI and human–computer interaction (HCI) at the manufacturing equipment. The following standards organizations are recommended for the search of standards related to MMI and HCI in the area of manufacturing equipment:

1. The Human Factors and Ergonomics Society (HFES)
2. The Institute of Electrical and Electronics Engineers (IEEE)

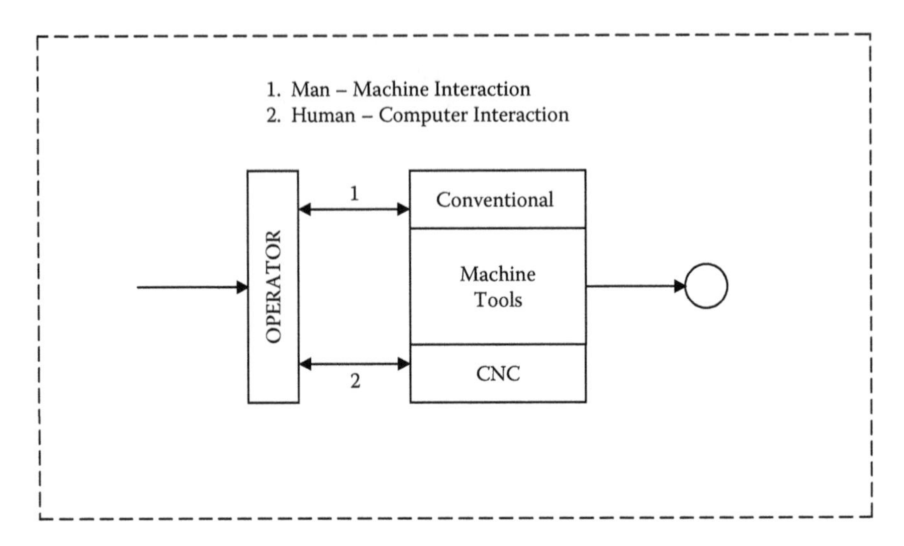

FIGURE 6.16 Domain of standards for interaction with manufacturing equipment.

3. Deutsches Institut für Normung (DIN)
4. The Japanese Industrial Standards Committee (JISC)
5. The British Standards Institution (BSI)
6. The American National Standards Institute (ANSI)
7. The International Organization for Standardization (ISO)

The following search parameters are used in this domain:

Man–machine interaction
MMI
Human–computer interaction
HCI

The search does not result in any standard directly related to domain specification.

6.7 STANDARDS RELATED TO OPERATION OF THE EQUIPMENT

The standards related to the operation of the equipment have a two-tier approach for the existing equipment. The first tier, normally general in nature, is defined by the standards organization. The second tier, more specific in nature, is defined by the manufacturers in the form of the operations manual that may contain references to standards from multiple organizations for components, assemblies, and mechanisms used in the manufacturing equipment. For manufacturing equipment that is novel and proprietary, standards organization may not provide detailed standards for the operation of the manufacturing process, and the manufacturers may

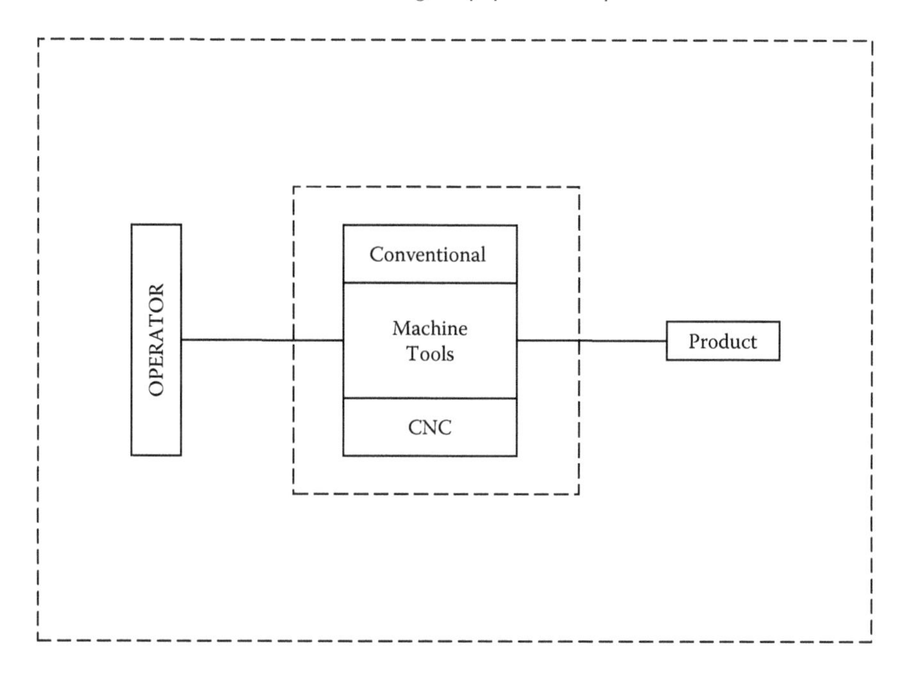

FIGURE 6.17 Domain of standards related to the operation of equipment.

not give the details of standards they are procuring from various standards organizations. Figure 6.17 shows the domain of standards that falls into this category. The following standards organizations provide standards related to the operation of several categories of manufacturing equipment on the basis of processes described in Figure 6.1:

1. The American Welding Society (AWS)
2. The American Society of Mechanical Engineers (ASME)
3. The Consortium for Advanced Manufacturing–International (CAM-I)
4. The Institute of Electrical and Electronics Engineers (IEEE)
5. The National Electrical Manufacturers Association (NEMA)
6. The National Fluid Power Association (NFPA)
7. The Resistance Welder Manufacturers' Association (RWMA)
8. The Society of Automotive Engineers (SAE)
9. The Association Française de Normalisation (AFNOR)
10. Deutsches Institut für Normung (DIN)
11. The Japanese Industrial Standards Committee (JISC)
12. The British Standards Institution (BSI)
13. The American National Standards Institute (ANSI)
14. The European Committee for Standardization (CEN)
15. The International Electrotechnical Commission (IEC)
16. The International Organization for Standardization (ISO)

TABLE 6.1
Standards for Design of Manufacturing Equipment

Pub Id: BS 4185-1:1967

Status: Confirmed, current

Title: Machine Tool Components. Locknuts ('C' Type, Socket Set Screw Locking)

Description: Materials, dimensions, and tolerances; finish, designation, and marking of separate millimeter and inch series; the millimeter series suitable for threaded components of diameter 10 mm to 125 mm; the inch series suitable for threaded components of diameter 1/2 in to 5 in.

Confirmation Date: 15/03/2002

ICS Classification: 21.060.20;25.060.99

Pub Id: BS 4185-2:1967

Status: Proposed for confirmation, current

Title: Machine Tool Components. Collars

Description: Materials, dimensions, and tolerances; finish, designation, and marking of separate millimeter and inch series; the millimeter series suitable for shafts of diameter 6 mm to 100 mm; the inch series for shafts of diameter 1/4 in to 4 in.

Confirmation Date: 15/07/1991

ICS Classification: 21.060.60;25.060.99

Pub Id: BS 4185-7:1973

Status: Confirmed, current

Title: Machine Tool Components. Wide Range Collets

Description: Specifies three distinct types and gives dimensions for interchangeability and accuracy requirements; collets may be used in either lead length or draw-in collet chucks of suitable design, and may be used in various makes of capstan and turret lathes and on single-spindle automatic lathes

Confirmation Date: 15/03/2002

ICS Classification: 25.060.20

TABLE 6.1 (Continued)
Standards for Design of Manufacturing Equipment

Pub Id: BS 4185-10:1977

Status: Confirmed, current

Title: Machine Tool Components. Trapezoidal Threads for Lead and Feed Screw Assemblies

Description: Range of nominal diameters (10 to 100 mm), nominal pitches (2 to 20 mm), and tolerances; does not include mounting criteria, length, and material

Confirmation Date: 15/03/2002

Int Relationships: ISO 2901 NEQ;ISO 2902 NEQ

ICS Classification: 21.040.10;21.040.30

Pub Id: BS 4185-11:1983

Status: Confirmed, current

Title: Machine Tool Components. Recommendations for Accuracy Grades of Gears

Description: Gives recommended grades of accuracy, selected from BS 436-2 for straight spur and single helical gears

Confirmation Date: 15/03/2002

Int Relationships: ISO 1328 NEQ

ICS Classification: 21.200; 25.060.01

Pub Id: BS 4185-13:1985

Status: Confirmed, current

Title: Machine Tool Components. Specification for Dimensions of Counter Bored Holes

Description: Gives dimensions of counter bored holes for use with hexagon socket head cap screws complying with BS 4186 of size ranges from M2.5 to M30

Confirmation Date: 15/03/2002

ICS Classification: 21.060.99;25.060.99

(*Continued*)

TABLE 6.1 (Continued)
Standards for Design of Manufacturing Equipment

Pub Id: BS 4185-14:1987

Status: Confirmed, current

Title: Machine Tool Components. Method for Determination of the Pilot Diameter of 60° Center Holes

Description: Applies to certain types of center holes described in BS 328-2. Calculations apply to work pieces that are to be machined on centers, provided they satisfy specified conditions.

Confirmation Date: 15/03/2002

Int Relationships: DIN 332:Part 7;DIN 333

ICS Classification: 21.060.99; 25.060.99

Pub Id: BS 5063:1992

Status: Work in hand, current

Title: Classification for Rationalized Range of Lubricants for Machine Tool Application

Description: Twelve classes of mineral lubricating oils and one class of grease; oils categorized by viscosity grading

Confirmation Date: 15/12/1997

Int Relationships: ISO/TR 3498:1986 NEQ

ICS Classification: 75.100

Pub Id: BS 6101-1:1981

Status: Confirmed, current

Title: Machine Tool Ball Screws. Methods of Calculating Dynamic and Static Load and Life Ratings

Description: Applies to recirculating ball-screw assemblies that have the shaft and nuts manufactured from hardened steel

Confirmation Date: 15/03/2002

ICS Classification: 25.060.10

TABLE 6.1 (Continued)
Standards for Design of Manufacturing Equipment

Pub Id: BS 6101-2:1991

Status: Confirmed, current

Title: Machine Tool Ball Screws. Specification for Accuracy, Including Geometrical Tests

Description: Requirements for assemblies that have shafts and nuts manufactured from hardened steel and before their mounting in machine tools

Confirmation Date: 15/03/2002

Int Relationships: ISO/DIS 3408-3:1987 NEQ

ICS Classification: 25.060.99

Pub Id: BS 6101-3:1992

Status: Current

Title: Machine Tool Ball Screws. Glossary of Terms

Int Relationships: ISO 3408-1:1991 IDT

ICS Classification: 01.040.25; 25.060.99

Committee Ref: MTE/1

ISBN: 0–580–21201–7

Replaces: BS 6101:Part 3:1984

Form: A4

Pages: 16

Pub Id: BS 6101-4:1987

Status: Confirmed, work in hand, current

Title: Machine Tool Ball Screws. Method of Calculation of Static Axial Rigidity

(*Continued*)

TABLE 6.1 (Continued)
Standards for Design of Manufacturing Equipment

Confirmation Date: 15/03/2002

ICS Classification: 25.060.10

Committee Ref: MTE/1

ISBN: 0–580–15441–6

Amended By: AMD 6264, November 1989(£0.00(M), £0.00(NM)):R

Form: A4

Pages: 26

Pub Id: BS 6101-5:1992

Status: Current

Title: Machine Tool Ball Screws. Specification for Nominal Diameters and Nominal Leads

Int Relationships: ISO 3408-2:1991 IDT

ICS Classification: 25.060.99

Committee Ref: MTE/1

ISBN: 0–580–21200–9

Replaces: BS 6101:Part 5:1984

Form: A4

Pages: 8

Pub Id: BS 6101-6:1990

Status: Current

Title: Machine Tool Ball Screws. Specification for Ball Nuts and Principal Dimensions of Ball Screws

Description: Includes load and life ratings for ball nuts; applies to ball nuts with both internal and external return systems; gives marking requirements and ordering instructions

TABLE 6.1 (Continued)
Standards for Design of Manufacturing Equipment

ICS Classification: 25.060.10

Committee Ref: MTE/15

ISBN: 0–580–18332–7

Form: A4

Pages: 16

Pub Id: BS 2573-1:1983

Status: Current

Title: Rules for The Design of Cranes. Specification for Classification, Stress Calculations and Design Criteria for Structures

Description: Gives a base for computing stresses in crane structures and permissible stresses for steels to BS 4360; covers classification based on severity of intended use of load to be considered, selection of steel, minimum thickness of plates and sections, working stresses, basic stresses in joint fastenings, web plates and stiffeners, and fluctuating loads; appendices give classification, use of steels of higher tensile strength than those to BS 4360, effective lengths of jibs, basic formula, design checks, and fatigue strength of structural components

Int Relationships: ISO 4301-1 EQV

Pub Id: BS 2573-2:1980

Status: Current

Title: Rules for the Design of Cranes. Specification for Classification, Stress Calculations, and Design of Mechanisms

Description: Classification of mechanisms for all types of cranes based on severity of intended use, loads and load combinations, calculation of stresses, stresses in components and the selection and verification of proprietary items; appendices give classifications normally associated with different types of cranes, assessment of torque, and determination of fatigue reference stresses

Int Relationships: ISO 4301-1 EQV

(Continued)

TABLE 6.1 (Continued)
Standards for Design of Manufacturing Equipment

Pub Id: BS 6548-2:1992

Status: Confirmed, current

Title: Maintainability of Equipment. Guide to Maintainability Studies During the Design Phase

Description: Guidance on studies that should be carried out during the design phase of a project, relating to maintainability and maintenance support tasks; the studies assist design decision making, prediction of quantitative maintainability characteristics of an item, and evaluation of alternative design options

Confirmation Date: 15/03/2002

The following search parameters are used in this domain:

Operation
<manufacturing equipment> Operation
Machine tool
<machine tool> Operation

Some selected results are listed in Table 6.2.

6.8 STANDARDS RELATED TO MEASUREMENTS AND MEASURING EQUIPMENT PRECISION

Measuring equipment and its precision plays an important role in the manufacturing operation. Both manual and computer-aided techniques are employed to gauge different parameters of different manufacturing equipment. Figure 6.18 presents a schematic for this domain specification. The following standards organizations are recommended for searches related to measurements of various parameters and equipment used in this process:

1. Deutsches Institut für Normung (DIN)
2. The Japanese Industrial Standards Committee (JISC)
3. The Association Française de Normalisation (AFNOR)
4. The British Standards Institution (BSI)
5. The American National Standards Institute (ANSI)
6. The International Organization for Standardization (ISO)

TABLE 6.2
Standards Related to Operation of Equipment

Pub Id: BS 1983-5:1989

Status: Confirmed, current

Title: Chucks for Machine Tools and Portable Power Tools. Code of Practice for the Safe Operation of Work Holding Chucks Used on Lathes

Description: Safe practices for design and operation of power and manually operated work-holding chucks used on turning machines

Confirmation Date: 15/03/2002

Int Relationships: ISO/TR 13618 EQV

Pub Id: BS EN ISO 14877:2002

Status: Current

Title: Protective Clothing for Abrasive Blasting Operations Using Granular Abrasives

Int Relationships: EN ISO 14877:2002 IDT;ISO 14877:2002 IDT

Pub Id: BS EN 1825-2:2002

Status: Current

Title: Grease Separators. Selection of Nominal Size, Installation, Operation, and Maintenance

Int Relationships: EN 1825-2:2002 IDT

The following search parameters are used in this domain:

Measurement
<measuring equipment>
<parameter> Measurement

Some selected standards are listed in Table 6.3.

6.9 USE OF STANDARDS IN MANUFACTURING OPERATIONS

The term manufacturing operation is used to define tangible processes in production that are acted upon on a workpiece (solid, granular, or powder). It is supported by intangible processes, and the functions are collectively known as the production function. Figure 6.19 shows a manufacturing operation domain supported by intangible processes of the production function.

Standards used in manufacturing operations include standards related to the operation of the equipment, standards related to MMI and HCI, standards

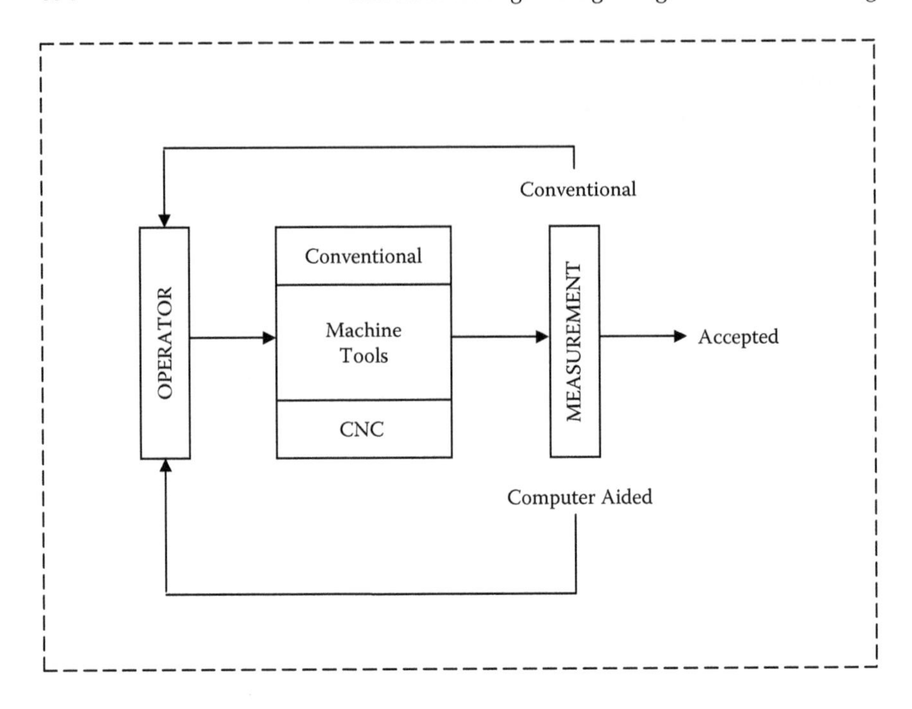

FIGURE 6.18 Domain of standards related to measurement and measuring equipment precision.

related to measurement and measuring equipment precision, and standards related to tools, tool holding, and workpiece holding. The support standards from design function, that is, the standards for technical drawing, standards for workpiece material, and standards for practices related to company operation such as process planning and scheduling, help achieve production targets. The following organizations are recommended for searches in this domain specification:

1. The Aerospace Industries Association (AIA)
2. The American Institute of Aeronautics and Astronautics (AIAA)
3. The American Society of Mechanical Engineers (ASME)
4. The American Welding Society (AWS)
5. The Consortium for Advanced Manufacturing–International (CAM-I)
6. The Society of Automotive Engineers (SAE)
7. The International Organization for Standardization (ISO)
8. The China State Bureau of Quality and Technical Supervision (CSBTS)
9. The Association Française de Normalisation (AFNOR)
10. Deutsches Institut für Normung (DIN)
11. The Bureau of Indian Standards (BIS)

TABLE 6.3
Standards Related to Measurements and Measuring Equipment Precision

Pub Id: BS 1597:1985

Status: Confirmed, current

Title: Symbolic Representation for Process Measurement Control Functions and Instrumentation. Basic Requirements

Description: Intended for use at all stages of design, manufacture, installation, and operation; depicts measurement techniques embodied in a particular instrument or means of activation; not intended to replace graphic symbols for electrical equipment

Int Relationships: ISO 3511-1:1977 IDT

ICS Classification: 01.080.40;25.040.40

Pub Id: BS 1747-1:1969

Status: Confirmed, current

Title: Methods for the Measurement of Air Pollution. Deposit Gauges

Description: Construction, installation, use for collection and measurement of atmospheric impurities deposited by their own weight or with assistance of rain, and estimate of rainfall

Confirmation Date: 15/06/1991

Pub Id: BS 1747-2:1969

Status: Confirmed, current

Title: Methods for the Measurement of Air Pollution. Determination of Concentration of Suspended Matter

Description: Construction, use of apparatus for determination of fine suspended particles (smoke) by comparison with arbitrary standard and by absolute (weighing) method

Confirmation Date: 15/06/1991

Pub Id: BS 1747-3:1969

Status: Confirmed, current

Title: Methods for Measurement of Air Pollution. Determination of Sulfur Dioxide

(Continued)

TABLE 6.3 (Continued)
Standards Related to Measurements and Measuring Equipment Precision

Description: Construction, use of apparatus

Confirmation Date: 15/06/1991

Pub Id: BS 1747-5:1972

Status: Confirmed, current

Title: Methods for the Measurement of Air Pollution. Directional Dust Gauges

Description: Construction, installation, use for collection and measurement of atmospheric impurities, precipitated as a liquid or as fine dust

Confirmation Date: 15/06/1991

Pub Id: BS 1902-3.11:1983

Status: Confirmed, current

Title: Specification for Glass and Reference Electrodes for the Measurement of pH

Description: Specifies for each type of electrode for general laboratory use, performance requirements, methods of test, maximum permitted errors, and other essential characteristics and information to be supplied by the manufacturer

Confirmation Date: 15/02/1999

Pub Id: BS ISO 10012-2:1997

Status: Work in hand, current

Title: Quality Assurance for Measuring Equipment. Guidelines for Control of Measurement Processes

Int Relationships: ISO 10012-2:1997 IDT

Pub Id: BS EN ISO 14253-1:1999

Status: Current

Title: Geometrical Product Specifications (GPS). Inspection by Measurement of Workpieces and Measuring Equipment. Decision Rules for Proving Conformance or Nonconformance with Specifications

Int Relationships: EN ISO 14253-1:1998 IDT;ISO 14253-1:1998 IDT

TABLE 6.3 (Continued)
Standards Related to Measurements and Measuring Equipment Precision

Pub Id: BS EN 30012-1:1994

Status: Work in hand, current

Title: Quality Assurance Requirements for Measuring Equipment. Metrological Confirmation System for Measuring Equipment

Description: Gives requirements and guidance to organizations that operate a quality system or provide a service in which measurement results are used to demonstrate compliance with specified requirements

Int Relationships: EN 30012-1:1993 IDT;ISO 10012-1:1992 IDT

Pub Id: BS EN 61187:1996

Status: Current

Title: Electrical and Electronic Measuring Equipment. Documentation

Description: Specifies technical documentation to be supplied with electrical and electronic measuring equipment for use in laboratories

Int Relationships: EN 61187:1994 IDT;IEC 61187:1993 MOD

Pub Id: BS EN 61557-10:2001

Status: Current

Title: Electrical Safety in Low Voltage Distribution Systems Up to 1000 V A.C. and 1500 V D.C. Equipment for Testing, Measuring or Monitoring of Protective Measures. Combined Measuring Equipment for Testing, Measuring or Monitoring of Protective Measures

Description: To be read in conjunction with BS EN 61557-1:1997

Int Relationships: EN 61557-10:2001 IDT;IEC 61557-10:2000 IDT

Pub Id: 98/262470 DC

Status: Drafts for comment, current

Title: IEC 61987. Ed.1. Terms and Structures of Measuring Equipment. IEC/65B/349/CD

Int Relationships: IEC Document 65B/349/CD IDT

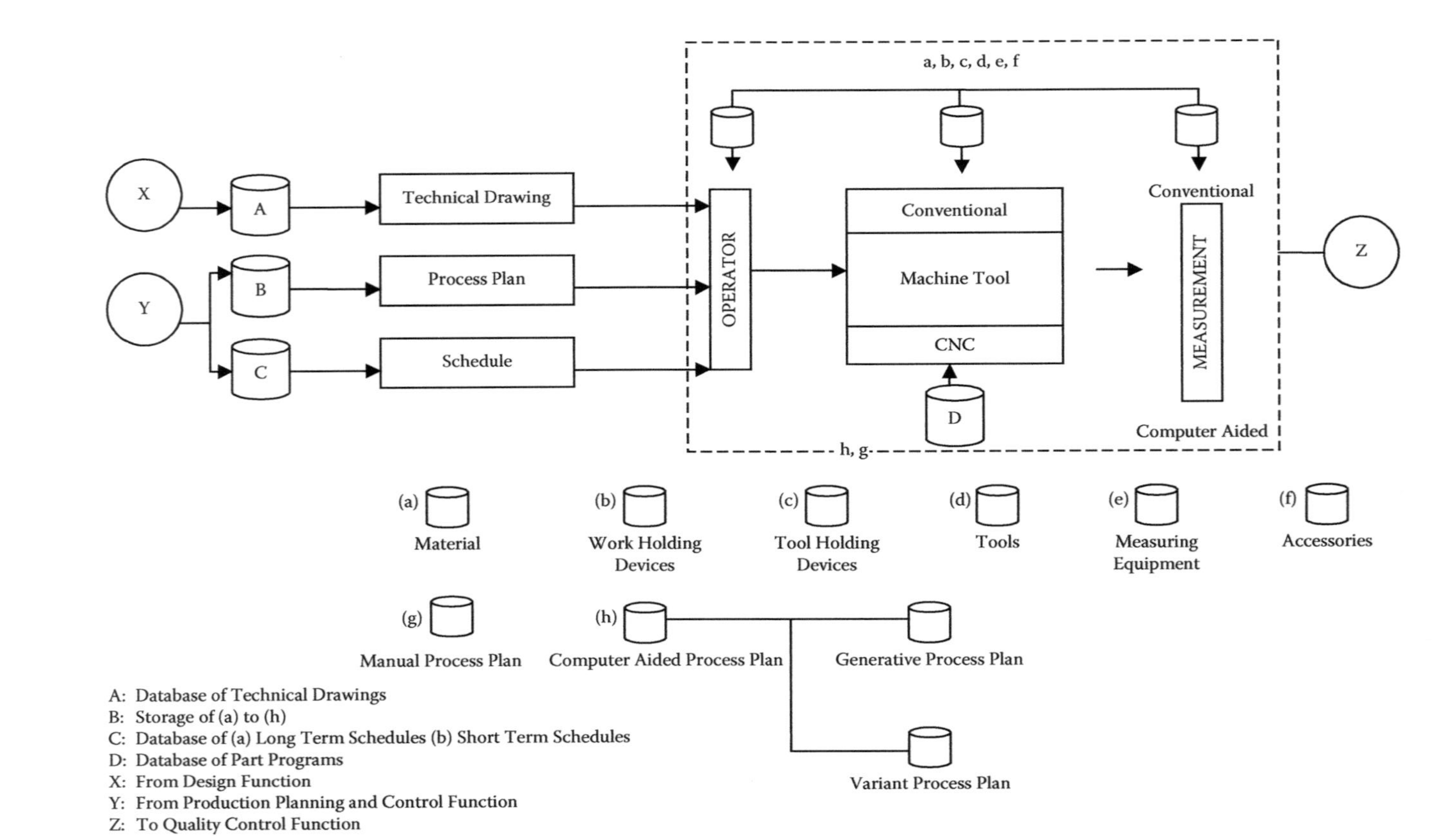

FIGURE 6.19 Standards and functions used in the manufacturing operation.

12. The Japanese Industrial Standards Committee (JISC)
13. The Department of Standards Malaysia (DSM)
14. The State Committee of the Russian Federation for Standardization, Metrology and Certification (GOST-R)
15. SPRING Singapore (Standards, Productivity and Innovation Board)
16. The British Standards Institution (BSI)
17. The American National Standards Institute (ANSI)

Considering the diverse nature of manufacturing, a small number of standards in this domain specification are presented from the target organization. The following search parameters are used in this domain:

Manufacturing
Manufacturing operation
<process>
Operation of <process>
Manufacturing system operation

The typical results of the search are given in Table 6.4.

6.10 STANDARDS RELATED TO QUALITY CONTROL

Quality control is an aggregate activity of design analysis and inspection for defects that is meant to ensure adequate quality in manufactured products. Quality control domain specification is provided in Figure 6.20. The following major standards organization involved in issuing standards for quality control:

1. Deutsches Institut für Normung (DIN)
2. The Japanese Industrial Standards Committee (JISC)
3. The Association Française de Normalisation (AFNOR)
4. The British Standards Institution (BSI)
5. The American National Standards Institute (ANSI)
6. The International Organization for Standardization (ISO)

Some of the typical results obtained from the target organization are in Table 6.5. The search parameter used in this domain is quality control.

6.11 SAFETY STANDARDS RELATED TO USE OF EQUIPMENT

Standards available for the safety of equipment are as important as those required for safety in the workplace. In this domain, specifications for both general and specific standards are available. General standards cater for the manufacturing system, whereas specific standards are applicable to individual manufacturing equipment. Figure 6.21 provides an illustration of domain specifications under consideration. The following organizations are the most suited for finding standards

TABLE 6.4
Use of Standards in Manufacturing Operations

Pub Id: BS 949-4:1990

Status: Current

Title: Screwing Taps. Specification for Metric Ground Thread Taps: Manufacturing Tolerances

Description: Manufacturing tolerances on the threaded portion of taps for producing ISO metric threads of tolerance classes 4H to 8H and 4G to 6G; valid for short taps specified in ISO/R 529, as well as any other kind of ground thread taps with the same diameters and pitches

Int Relationships: EN 22857:1989 IDT;ISO 2857:1973 IDT

Pub Id: DD 194:1990

Status: Current

Title: Computer Integrated Manufacturing (CIM): CIM Systems Architecture Framework for Modeling

Description: Structure for identifying and coordinating standards development for computer-based modeling of enterprises; emphasis is on discrete parts manufacturing and is relevant to the implementation of computer-executable models

Int Relationships: ENV 40003:1990 IDT

Pub Id: BS IEC 60748-23-3:2002

Status: Current

Title: Semiconductor Devices. Integrated Circuits. Hybrid Integrated Circuits and Film Structures. Manufacturing Line Certification. Manufacturers' Self-Audit Checklist and Report

Int Relationships: IEC 60748-23-3:2002 IDT

Pub Id: BS EN 2078:2002

Status: Current

Title: Aerospace Series. Metallic Materials. Manufacturing Schedule, Inspection Schedule, Inspection and Test Report. Definition, General Principles, Preparation and Approval

Int Relationships: EN 2078:2001 IDT

TABLE 6.4 (Continued)
Use of Standards in Manufacturing Operations

Pub Id: BS EN 61943:1999

Status: Current

Title: Integrated Circuits. Manufacturing Line Approval Application Guideline

Description: How to apply the principles and requirements given in IEC 61739 to manufacturers of monolithic integrated circuits applying for manufacturing line approval; to be read in conjunction with IEC 61739:1996

Int Relationships: EN 61943:1999 IDT;IEC 61943:1999 IDT;QC 211001:1999 IDT

Pub Id: 95/715775 DC

Status: Drafts for comment, current

Title: Footwear, Leather and Imitation Leather Goods Manufacturing Machines. Shoe and Leather Presses. Safety Requirements (Pren 12203)

Int Relationships: PREN 12203:1995 IDT

Pub Id: 98/121863 DC

Status: Drafts for comment, current

Title: Fibre-Reinforced Plastics. Test Plates Manufacturing Methods. Part 9. Moulding of GMT/STC. ISO/CD 1268-9

Int Relationships: ISO/CD 1268-9:1998 IDT

Pub Id: 98/125029 DC

Status: Drafts for comment, current

Title: ISO/CD 1268-4. Fibre-Reinforced Plastics. Test Plate Manufacturing Methods. Part 4. Molding of Preimpregnates (Using Autoclave, Pressclave, Hydraulic Press and Vacuum Bag Equipment)

Int Relationships: ISO/CD 1268-4:1998 IDT

Pub Id: 99/614084 DC

Status: Drafts for comment, current

Title: ISO 15531-1. Industrial Automation Systems and Integration. Industrial Manufacturing Management Data. Part 1. General Overview

Int Relationships: ISO/DIS 15531-1:1999 IDT

(*Continued*)

TABLE 6.4 (Continued)
Use of Standards in Manufacturing Operations

Pub Id: 00/123691 DC

Status: Drafts for Comment, current

Title: ISO/CD 1268-10 A. Fibre Reinforced Plastics. Test Plates Manufacturing Methods. Part 10. Injection Moulding of Long Fibre-Reinforced Molding Compounds (BMC, DMC and SMC). A. General Principles and Molding of Multipurpose Test Specimens

Int Relationships: ISO/CD 1268-10 A:2000 IDT

Pub Id: 00/123692 DC

Status: Drafts for comment, current

Title: ISO/CD 1268-10 B. Fibre Reinforced Plastics. Test Plates Manufacturing Methods. Part 10. Injection Molding of Long Fibre-Reinforced Molding Compounds (BMC, DMC and SMC). B. Small Plates

Int Relationships: ISO/CD 1268-10 B:2000 IDT

Pub Id: 01/120080 DC

Status: Drafts for comment, current

Title: BS EN 13923. Filament Wound FRP Pressure Vessels. Materials, Design, Calculation, Manufacturing and Testing

Int Relationships: PREN 13923:2000 IDT

Pub Id: 02/615002 DC

Status: Drafts for comment, current

Title: ISO/DIS 16100-2. Industrial Automation Systems and Integration. Manufacturing Software Capability Profiling. Part 2: Profiling Methodology

Int Relationships: ISO/DIS 16100-2:2002 IDT

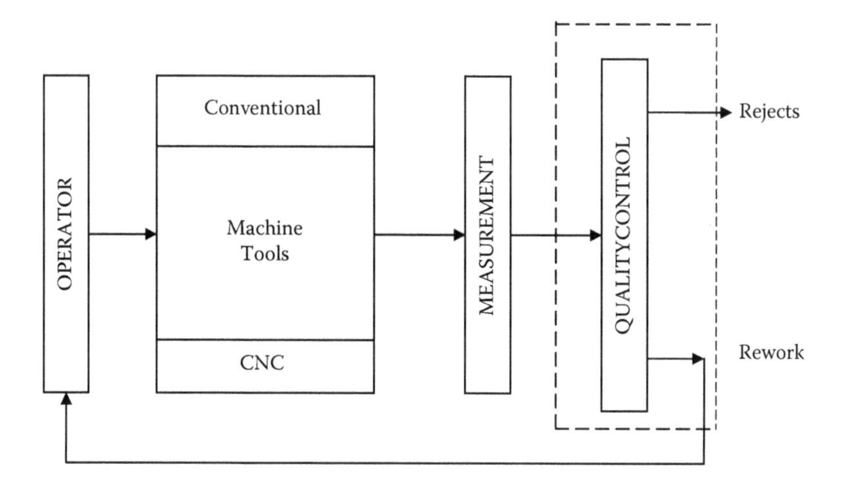

FIGURE 6.20 Standards related to quality control.

relevant to safety at specific equipment and safety standards related to the whole of the manufacturing system:

1. Deutsches Institut für Normung (DIN)
2. The Japanese Industrial Standards Committee (JISC)
3. The Association Française de Normalisation (AFNOR)
4. The British Standards Institution (BSI)
5. The American National Standards Institute (ANSI)
6. The International Organization for Standardization (ISO)
7. The American Society of Safety Engineers (ASSE)
8. The Occupational Safety and Health Administration (OSHA)

The following search parameters are used in this domain:

Safety
Safety at manufacturing equipment
<manufacturing equipment> Safety
Occupational hazards

Some of the selected results are given in Table 6.6.

6.12 RELIABILITY STANDARDS RELATED TO THE USE OF EQUIPMENT

Reliability is defined as the quality or state of being reliable. This definition applies to the individual equipment as well as to the whole manufacturing system. Reliability standards domain specification is provided in Figure 6.22. The recommended

TABLE 6.5
Standards Related to Quality Control

Pub Id: BS 600:2000

Status: Current

Title: A Guide to the Application of Statistical Methods to Quality and Standardization

Pub Id: BS 835:1973

Status: Confirmed, current

Pub Id: BS 4778-2:1991

Status: Current

Title: Quality Vocabulary. Quality Concepts and Related Definitions

Description: Gives concepts and defines terms in five sectors: quality, management, control, inspection general, and inspection statistical; terms defined in previous publications, now omitted, found in Part 3; related dictionary terms now in the Appendix

Pub Id: BS 4778-3.1:1991

Status: Proposed for confirmation, current

Title: Quality Vocabulary. Availability, Reliability and Maintainability Terms. Guide to Concepts and Related Definitions

Description: Complements and extends the internationally agreed definitions contained in Section 3.2 by providing conceptual explanations of many of these terms and by including additional terms

Int Relationships: IEC 60050-191:1990 NEQ

Pub Id: BS 4778-3.2:1991

Status: Confirmed, current

Title: Quality Vocabulary. Availability, Reliability and Maintainability Terms. Glossary of International Terms

Description: Includes faults and failures, performance measures, and various processes such as testing, analysis, design, and improvement; also provides a vocabulary of quality of service in telecommunication

Confirmation Date: 15/03/2002

Int Relationships: IEC 60050-191:1990 IDT

TABLE 6.5 (Continued)
Standards Related to Quality Control

Pub Id: BS 5703-2:1980

Status: Work in hand, current

Title: Guide to Data Analysis and Quality Control Using Cusum Techniques. Decision Rules and Statistical Tests for Cusum Charts and Tabulations

Description: Introduces simple decision rules for monitoring, control, and retrospective analysis; statistical tests, examples, and use of extended Cusum techniques

Confirmation Date: 15/03/1992

Pub Id: BS 5750-8:1991

Status: Confirmed, current

Title: Quality Systems. Guide to Quality Management and Quality Systems Elements for Services

Description: Guidance on the quality system elements of organizations that provide services; closely linked with service needs and terminology and enables management in this field to obtain and provide quality assurance

Confirmation Date: 15/10/1997

Int Relationships: EN 29004-2:1993 IDT;ISO 9004-2:1991 IDT

Pub Id: BS 5750-14:1993

Status: Confirmed, Work in hand, current

Title: Quality Systems. Guide to Dependability Program Management

Description: Outlines the essential features of a dependability program, planned and managed to produce products that will be reliable and maintainable

Confirmation Date: 15/03/2002

standards organizations in this domain specification are the following:

1. Deutsches Institut für Normung (DIN)
2. The Japanese Industrial Standards Committee (JISC)
3. The Association Française de Normalisation (AFNOR)
4. The British Standards Institution (BSI)
5. The American National Standards Institute (ANSI)
6. The International Organization for Standardization (ISO)

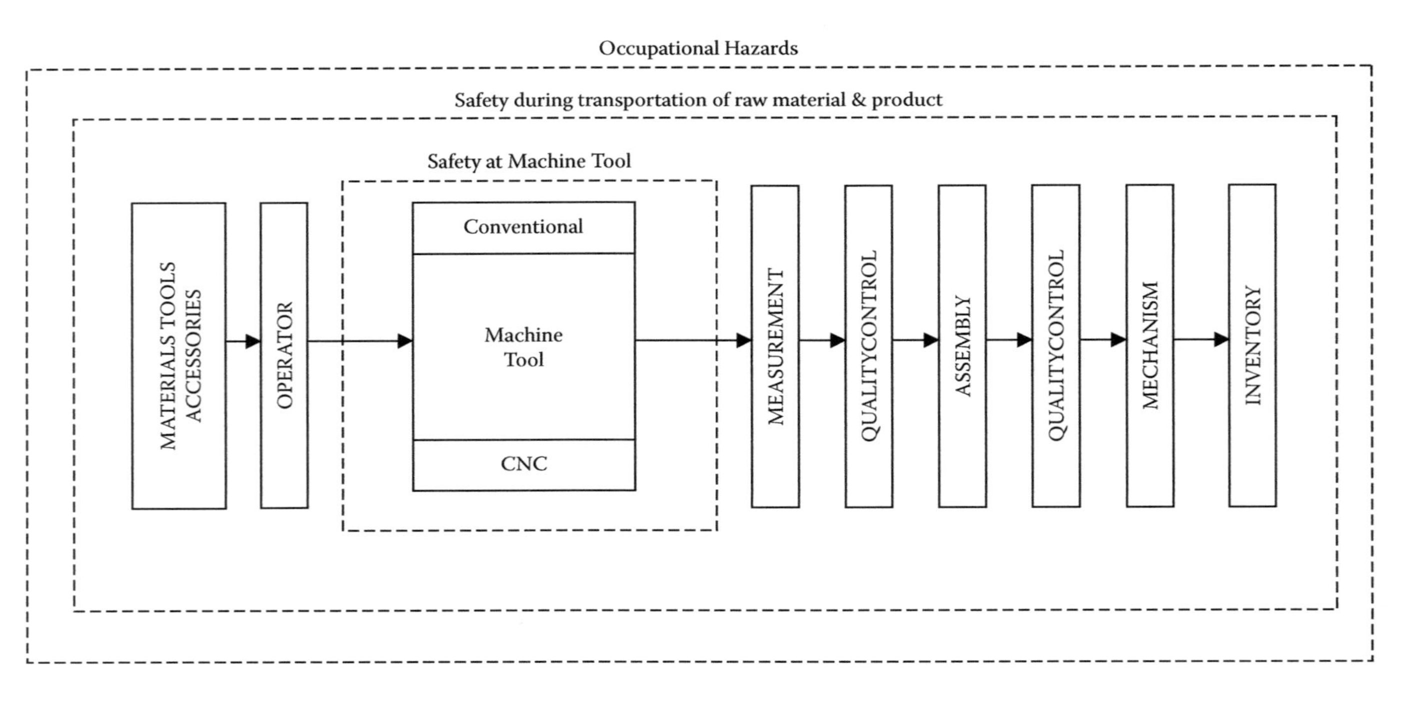

FIGURE 6.21 Domain of standards related to safety at discrete manufacturing facilities.

TABLE 6.6
Safety Standards Related to Use of Equipment

Pub Id: BS 5499-3:1990

Status: Current

Title: Fire Safety Signs, Notices and Graphic Symbols. Specification for Internally-Illuminated Fire Safety Signs

Description: Specifies the construction requirements for internally illuminated signs by relating them to the appropriate British standard for luminaires; appropriate tests also included; specific requirements for cinemas and theatres given separately; to be read in conjunction with BS 4533-101

Pub Id: BS 5499-4:2000

Status: Current

Title: Safety Signs, Including Fire Safety Signs. Code of Practice for Escape Route Signing

Pub Id: BS 5499-5:2002

Status: Current

Title: Graphical Symbols and Signs. Safety Signs, Including Fire Safety Signs. Signs with Specific Safety Meanings

Pub Id: BS 5499-6:2002

Status: Current

Title: Graphical Symbols and Signs. Safety Signs, Including Fire Safety Signs. Creation and Design of Graphical Symbols for Use in Safety Signs. Requirements

Description: to be read in conjunction with BS 5499-1:2002

Pub Id: BS 5499-11:2002

Status: Current

Title: Graphical Symbols and Signs. Safety Signs, Including Fire Safety Signs. Water Safety Signs

ICS Classification: 01.080.10;97.220.40

(*Continued*)

TABLE 6.6 (Continued)
Safety Standards Related to Use of Equipment

Pub Id: BS 5667-2:1979

Status: Proposed for withdrawal, current

Title: Specification for Continuous Mechanical Handling Equipment—Safety Requirements. Loose Bulk Materials: Pneumatic Handling Installations

Pub Id: BS 5667-3:1979

Status: Proposed for withdrawal, current

Title: Specification for Continuous Mechanical Handling Equipment—Safety Requirements. Loose Bulk Materials: Storage Equipment Fed by a Pneumatic Handling System

Pub Id: BS 5667-4:1979

Status: Proposed for withdrawal, current

Title: Specification for Continuous Mechanical Handling Equipment—Safety Requirements. Loose Bulk Materials: Mobile Suction Pipes Suspended from Derrick Jibs Used in Pneumatic Handling

Pub Id: BS 5667-5:1979

Status: Proposed for withdrawal, current

Title: Specification for Continuous Mechanical Handling Equipment—Safety Requirements. Loose Bulk Materials: Couplings and Hose Components Used in Pneumatic Handling

Pub Id: BS 5667-6:1979

Status: Proposed for withdrawal, current

Title: Specification for Continuous Mechanical Handling Equipment—Safety Requirements. Loose Bulk Materials: Rotary Feeders Used in Pneumatic Handling

Pub Id: BS 5667-7:1979

Status: Proposed for withdrawal, current

Title: Specification for Continuous Mechanical Handling Equipment—Safety Requirements. Loose Bulk Materials: Rotary Drum Feeders and Rotary Vane Feeders

Pub Id: BS 5667-8:1979

Status: Proposed for withdrawal, current

Title: Specification for Continuous Mechanical Handling Equipment—Safety Requirements. Loose Bulk Materials: Hand-Operated Power Shovels

TABLE 6.6 (Continued)
Safety Standards Related to Use of Equipment

Pub Id: BS 5667-9:1979

Status: Proposed for withdrawal, current

Title: Specification for Continuous Mechanical Handling Equipment—Safety Requirements. Loose Bulk Materials: Bulk Throwing Machines

Pub Id: BS 5667-10:1979

Status: Proposed for withdrawal, current

Title: Specification for Continuous Mechanical Handling Equipment—Safety Requirements. Loose Bulk Materials: Vertical Screw Conveyors

Pub Id: BS 5667-11:1979

Status: Proposed for withdrawal, current

Title: Specification for Continuous Mechanical Handling Equipment—Safety Requirements. Unit Loads: Fixed Slat Conveyors (Metal or Wood) with Horizontal Shafts

Pub Id: BS 5667-12:1979

Status: Proposed for withdrawal, current

Title: Specification for Continuous Mechanical Handling Equipment—Safety Requirements. Unit Loads: Mobile Slat Conveyors (Metal or Wood) with Horizontal Shafts

Pub Id: BS 5667-13:1979

Status: Proposed for withdrawal, current

Title: Specification for Continuous Mechanical Handling Equipment—Safety Requirements. Unit Loads: Arm Elevators and Push Bar Conveyors

Pub Id: BS 5667-14:1979

Status: Proposed for withdrawal, current

Title: Specification for Continuous Mechanical Handling Equipment—Safety Requirements. Unit Loads: Live Roller Conveyors (with Positive or Friction Drive)

Pub Id: BS 5667-15:1979

Status: Proposed for withdrawal, current

(*Continued*)

TABLE 6.6 (Continued)
Safety Standards Related to Use of Equipment

Title: Specification for Continuous Mechanical Handling Equipment—Safety Requirements. Unit Loads: Crate-Carrying Chain Conveyors Having Biplanar Chains for Flat-Bottomed Unit Loads

Pub Id: BS 5667-16:1979

Status: Proposed for withdrawal, current

Title: Specification for Continuous Mechanical Handling Equipment—Safety Requirements. Unit Loads: Slat Band Chain Conveyors

Pub Id: BS 5667-17:1979

Status: Proposed for withdrawal, current

Title: Specification for Continuous Mechanical Handling Equipment—Safety Requirements. Unit Loads: Continuous Plate Conveyors (Horizontal)

Pub Id: BS 5667-18:1979

Status: Proposed for withdrawal, current

Title: Specification for Continuous Mechanical Handling Equipment—Safety Requirements. Conveyors and Elevators with Chain-Elements—Examples for Guarding of Nip Points

Pub Id: BS 5667-19:1980

Status: Proposed for withdrawal, current

Title: Specification for Continuous Mechanical Handling Equipment—Safety Requirements. Belt Conveyors—Examples for Guarding of Nip Points

Pub Id: BS EN 61310-3:1999

Status: Current

Title: Safety of Machinery. Indication, Marking and Actuation. Requirements for the Location and Operation of Actuators

Int Relationships:
EN 61310-3:1999 IDT;IEC 61310-3:1999 IDT

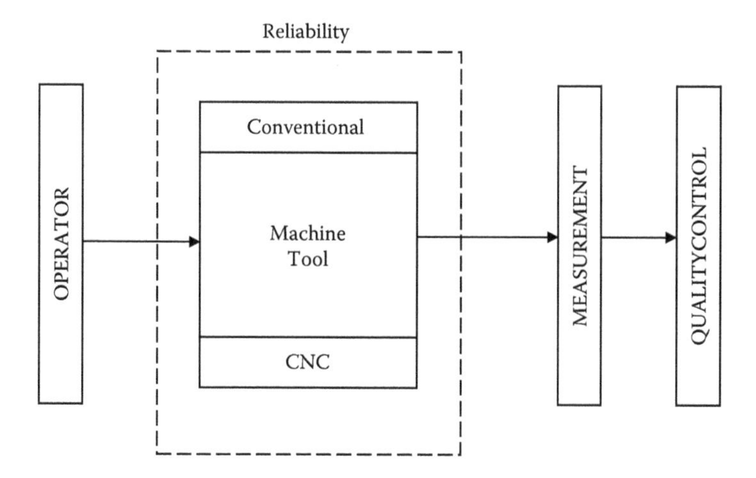

FIGURE 6.22 Domain of standards related to reliability of manufacturing equipment.

The following search parameters are used in this domain:

Reliability
<manufacturing system> Reliability
<manufacturing equipment> Reliability

A selection of the result is given in Table 6.7.

6.13 MAINTAINABILITY STANDARDS RELATED TO USE OF EQUIPMENT

Maintainability is a parameter that plays an important role from the design stage to operation of the equipment. The domain specification in this case is presented in Figure 6.23. The recommended standards organizations for search related to maintainability are the following:

1. Deutsches Institut für Normung (DIN)
2. The Japanese Industrial Standards Committee (JISC)
3. The Association Française de Normalisation (AFNOR)
4. The British Standards Institution (BSI)
5. The American National Standards Institute (ANSI)
6. The International Organization for Standardization (ISO)

The following search parameters are used in this domain:

Maintainability
<equipment> Maintainability
<manufacturing system> Maintainability

Some selected results are given in Table 6.8.

TABLE 6.7
Reliability Standards Related to the Use of Equipment

Pub Id: BS 4778-3.1:1991

Status: Proposed for confirmation, current

Title: Quality Vocabulary. Availability, Reliability and Maintainability Terms. Guide to Concepts and Related Definitions

Description: Complements and extends the internationally agreed definitions contained in Section 3.2 by providing conceptual explanations of many of these terms and by including additional terms

Int Relationships: IEC 60050-191:1990 NEQ

Pub Id: BS 4778-3.2:1991

Status: Confirmed, current

Title: Quality Vocabulary. Availability, Reliability and Maintainability Terms. Glossary of International Terms

Description: Includes faults and failures, performance measures, and various processes such as testing, analysis, design and improvement; also provides a vocabulary of quality of service in telecommunication

Confirmation Date: 15/03/2002

Int Relationships: IEC 60050-191:1990 IDT

Pub Id: BS 5760-0:1986

Status: Confirmed, current

Title: Reliability of Systems, Equipment and Components. Introductory Guide to Reliability

Description: Aimed at directors of companies looking for overall advantages, engineers not trained in quality and reliability to show how reliability can help in their technical decision making, and middle management not specialized in engineering to demonstrate how reliability should be dovetailed with other disciplines to give the best result

Confirmation Date: 15/08/1993

(*Continued*)

TABLE 6.7 (Continued)
Reliability Standards Related to the Use of Equipment

Pub Id: BS 5760-1:1996

Status: Confirmed, work in hand, current

Title: Reliability of Systems, Equipment and Components. Dependability Program Elements and Tasks

Confirmation Date: 15/03/2002

Int Relationships: EN 60300-2:1996 IDT;IEC 60300-2:1995 IDT

Pub Id: BS 5760-2:1994

Status: Confirmed, current

Title: Reliability of Systems, Equipment and Components. Guide to the Assessment of Reliability

Description: Includes human reliability and software reliability assessment techniques

Confirmation Date: 15/07/2002

Pub Id: BS 5760-3:1982

Status: Proposed for withdrawal, current

Title: Reliability of Systems, Equipment and Components. Guide to Reliability Practices: Examples

Description: Shows, by way of examples, how some of the principles described in Parts 1 and 2 are applied in industry; each is a practical example of the use of reliability techniques, although names of the organizations are omitted; where appropriate, reference is made to the clauses in Parts 1 and 2 of BS 5760 that the examples illustrate

Confirmation Date: 15/08/1993

6.14 WASTE-DISPOSAL STANDARDS

Figure 6.24 details the domain specification. Diverse standards are available in the target organization for the general waste-disposal category. The recommended standards organizations for searches related to waste disposal are the following:

1. Deutsches Institut für Normung (DIN)
2. The Japanese Industrial Standards Committee (JISC)
3. The Association Française de Normalisation (AFNOR)
4. The British Standards Institution (BSI)

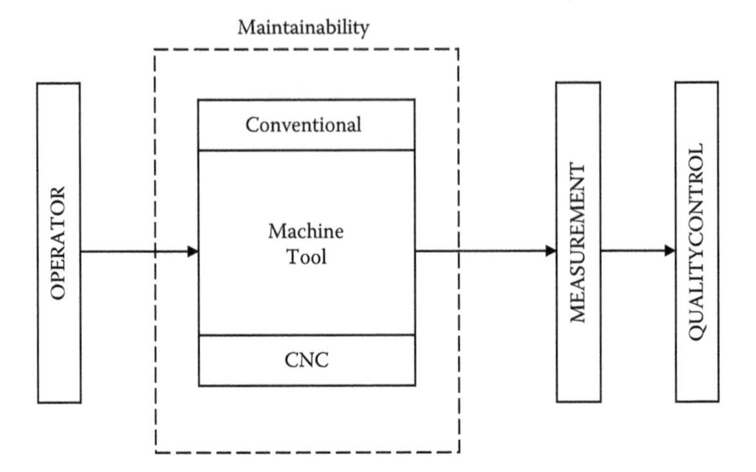

FIGURE 6.23 Domain of standards related to maintainability of manufacturing equipment.

5. The American National Standards Institute (ANSI)
6. The International Organization for Standardization (ISO)

The following search parameters are used in this domain:

Waste
<equipment> Waste
<manufacturing system> Waste
Mechanical waste

Some of the selected standards are given in Table 6.9.

6.15 EQUIPMENT LIFE CYCLE STANDARD

Figure 6.25 details the domain specification. The life cycle of the equipment is a composite parameter that compromises the life of the maintainable components and the life of the complete machine before it is salvaged. Diverse standards are available in the target organization for the general-equipment life cycle category. The recommended standards organizations for searches related to life cycle are the following:

1. Deutsches Institut für Normung (DIN)
2. The Japanese Industrial Standards Committee (JISC)
3. The Association Française de Normalisation (AFNOR)
4. The British Standards Institution (BSI)
5. The American National Standards Institute (ANSI)
6. The International Organization for Standardization (ISO)

TABLE 6.8
Maintainability Standards Related to Use of Equipment

Pub Id: BS 5760-12:1993

Status: Confirmed, current

Title: Reliability of Systems, Equipment and Components. Guide to the Presentation of Reliability, Maintainability and Availability Predictions

Confirmation Date: 15/03/2002

Int Relationships: IEC 60863:1986 IDT

Pub Id: BS 6548-1:1984

Status: Confirmed, partially replaced, current

Title: Maintainability of Equipment. Guide to Specifying and Contracting for Maintainability

Description: Gives an introduction to the concept of maintainability with guidance on specifying and contracting; a maintainability program description is also included

Confirmation Date: 15/08/1993

Int Relationships: IEC 60706-1:1982 IDT

Pub Id: BS 6548-2:1992

Status: Confirmed, current

Title: Maintainability of Equipment. Guide to Maintainability Studies During the Design Phase

Description: Guidance on studies that should be carried out during the design phase of a project, relating to maintainability and maintenance support tasks; the studies assist design decision making, prediction of quantitative maintainability characteristics of an item and evaluation of alternative design options.

Confirmation Date: 15/03/2002

Int Relationships: IEC 60706-2:1990 IDT

Pub Id: BS 6548-3:1992

Status: Confirmed, current

(*Continued*)

TABLE 6.8 (Continued)
Maintainability Standards Related to Use of Equipment

Title: Maintainability of Equipment. Guide to Maintainability, Verification and the Collection, Analysis and Presentation of Maintainability Data

Description: Provides guidance on various aspects of verification necessary to ensure that specified maintainability requirements have been met and indicates relevant procedures and test methods; outlines major considerations

Confirmation Date: 15/03/2002

Int Relationships: IEC 60706-3:1987 IDT

Pub Id: BS 6548-4:1993

Status: Confirmed, current

Title: Maintainability of Equipment. Guide to the Planning of Maintenance and Maintenance Support

Description: Provides guidance on tasks to be performed during the system-acquisition phase to ensure that availability objectives are met in the operational phase; describes development of maintenance policy, planning of maintenance tasks, and determination of maintenance support resources including personnel, documentation, test equipment, and spares

Confirmation Date: 15/03/2002

Int Relationships: IEC 60706-4:1992 IDT

Pub Id: BS 6548-5:1995

Status: Confirmed, current

Title: Maintainability of Equipment. Guide to Diagnostic Testing

Confirmation Date: 15/03/2002

Int Relationships: IEC 60706-5:1994 IDT

Pub Id: BS 6548-6:1995

Status: Confirmed, current

Title: Maintainability of Equipment. Guide to Statistical Methods in Maintainability Evaluation

TABLE 6.8 (Continued)
Maintainability Standards Related to Use of Equipment

Confirmation Date: 15/03/2002

Int Relationships: IEC 60706-6:1994 IDT

Pub Id: BS IEC 60300-3-10:2001

Status: Current

Title: Dependability Management. Application Guide. Maintainability

Description: Can be used to implement a maintainability program that covers the initiation, development, and in-service phases of a product; to be read in conjunction with IEC 60300-2:1995

Int Relationships: IEC 60300-3-10:2001 IDT

Pub Id: BS EN 50126:1999

Status: Current

Title: Railway Applications. The Specification and Demonstration of Reliability, Availability, Maintainability and Safety (RAMS)

Int Relationships: EN 50126:1999 IDT

Pub Id: BS EN 61703:2002

Status: Current

Title: Mathematical Expressions for Reliability, Availability, Maintainability and Maintenance Support Terms

Int Relationships: EN 61703:2002 IDT; IEC 61703:2001 IDT

Pub Id: 96/704522 DC

Status: Drafts for comment, current

Title: Gas Turbines. Procurement. Part 11. Reliability, Availability, Maintainability and Safety (ISO/DIS 3977-11)

Int Relationships: ISO/DIS 3977-11:1996 IDT

(*Continued*)

TABLE 6.8 (Continued)
Maintainability Standards Related to Use of Equipment

Pub Id: 97/204722 DC

Status: Drafts for comment, current

Title: IEC 61732. Measuring Relays. Predicted Availability, Reliability and Maintainability of Static Protection Equipment (95/51/CD)

Int Relationships: IEC Document 95/51/CD IDT

Pub Id: 99/105620 DC

Status: Drafts for comment, current

Title: BS 6913. New Part. Operation and Maintenance of Earth-Moving Machinery. Part XX. Maintainability Guidelines

Int Relationships: ISO/DIS 12510:1999 IDT

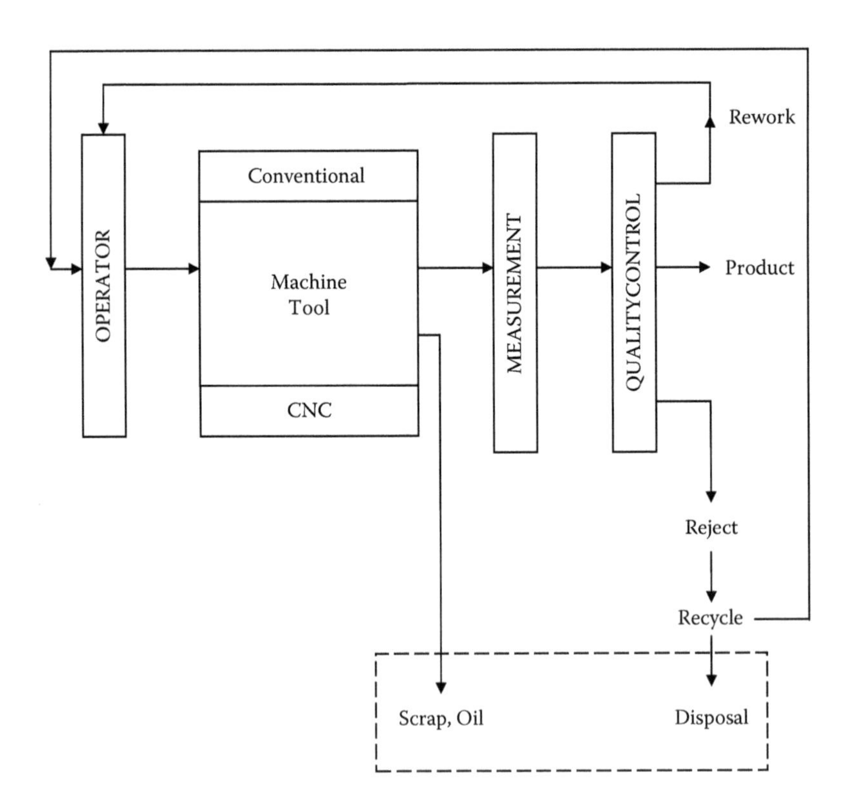

FIGURE 6.24 Waste-disposal standards domain.

TABLE 6.9
Waste Disposal Standards

Pub Id: BS 1577:1949

Status: Withdrawn

Title: Mild Steel Refuse or Food Waste Containers Title: Mild Steel Refuse or Food Waste Containers

Pub Id: BS 7033-2:1989

Status: Confirmed, current

Title: Packaging. Requirements for the Use of European Standards in the Field of Packaging and Packaging Waste

Pub Id: 98/716650 DC

Status: Drafts for comment, current

Title: Chemical Durability Test. Soxhlet-Mode Chemical Durability Test. Application to Vitrified Matrixes for High-Level Radioactive Waste (ISO/CD 16797)

Pub Id: 02/101520 DC

Status: Drafts for comment, current

Title: BS EN 14346. Characterization of Waste. Calculation of Dry Matter by Determination of Dry Residue or Water Content

Pub Id: 02/102047 DC

Status: Drafts for comment, current

Title: BS EN 14345. Characterization of Waste. Determination of Hydrocarbon Content By Gravimetry

The following search parameters are used in this domain:

Life cycle
<equipment> Life cycle

Some of the selected standards are given in Table 6.10.

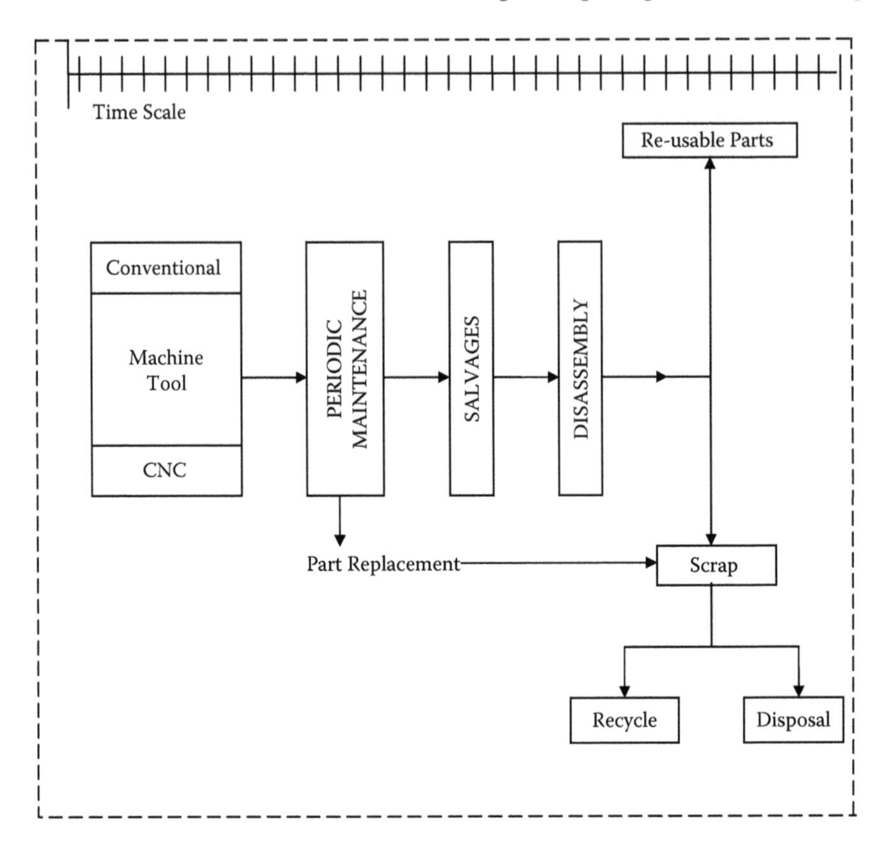

FIGURE 6.25 Equipment life cycle domain.

6.16 STANDARDS RELATED TO INFORMATION TRANSFER

The management of a manufacturing process strongly relies on the right information available at the right place and at the right time. Forms and memos are normally used to convey the information from one node to another in a manual system. With the advent of digital computers and networks, multiple standards have become available for information transfer in a partially or fully computer-integrated environment. The standards most commonly quoted in this category are ISO open-system interconnection for computer integrated manufacturing. More details on digital-information–transfer standards are provided in Chapter 10.

6.17 STANDARDS RELATED TO MATERIAL-HANDLING EQUIPMENT

The material-handling equipment is an essential part of the manufacturing operation. Several classes of material-handling equipment are used to augment

TABLE 6.10
Equipment Life Cycle Standard

Pub Id: BS ISO/IEC 15288:2002

Status: Current

Title: Systems Engineering. System Life Cycle Processes

Int Relationships: ISO/IEC 15288:2002 IDT

Pub Id: DD ISO/TS 14048:2002

Status: Current

Title: Environmental Management. Life Cycle Assessment. Data Documentation Format

Int Relationships: ISO/TS 14048:2002 IDT

Pub Id: 96/703122 DC

Status: Drafts for comment, current

Title: Technical Documentation in the Product Life Cycle. Part 1. Definition of a Life Cycle Model and Allocation of Technical Documents to Phases and Activities (ISO/CD 15226)

Int Relationships: ISO/CD 15226:1996 IDT

Pub Id: 02/401005 DC

Status: Drafts for comment, current

Title: IEC 60300-3-3. Ed.2. Dependability Management. Part 3-3: Life Cycle Costing Analysis. Application Guide

Int Relationships: IEC Document 56/826/CDV IDT

productivity in manufacturing systems. Major organizations that provide standards related to this category are the following:

1. Deutsches Institut für Normung (DIN)
2. The Japanese Industrial Standards Committee (JISC)
3. The Association Française de Normalisation (AFNOR)
4. The British Standards Institution (BSI)
5. The American National Standards Institute (ANSI)
6. The International Organization for Standardization (ISO)

The following search parameters are used in this domain:

Handling
Material handling
<handling equipment>

Some of the selected standards are given in Table 6.11.

6.18 FUTURE DIRECTIONS IN THE STANDARDIZATION OF MANUFACTURING EQUIPMENT[2]

The standardization of manufacturing equipment is an uphill task because of the diversity of existing manufacturing processes, emerging procedures related to new and old manufacturing technologies, accessories used with existing and new manufacturing machineries, and control systems associated with these machines.

The state-of-the-art machine tools, in terms of use of multidisciplinary features, mainly belong to metal-cutting processes such as turning, milling, and drilling machines. These processes, along with the grinding process, make most of the mechanical artifacts used to produce discrete components (products).

The machine tool industry in the United States is a relatively small group. This multibillion-dollar industry benefits from factors related to the machine tools, such as gain in productivity, decline in inventory requirements, and product improvements for price, quality, and energy efficiency.

The Manufacturing Engineering Laboratory (MEL) of the National Institute of Science and Technology (NIST) is currently working on a program to investigate enhancement in critical areas of a machine tool as a system and its integration in manufacturing environment. This program, known as the Smart Machine Tools program, has the following goals:

> In collaboration with Industry, develop, validate and demonstrate the enabling metrology, smart sensor technology, and standards needed to characterize, monitor, and improve the accuracy and reliability of machining and turning centers, leading to the realization of autonomous smart machine tool that machine the first and every subsequent part to specification without unscheduled delays.

The Smart Machine Tool program at the MEL was initiated in 2003 and is expected to be completed in 2005. The program objectives, technical outputs, and anticipated impacts are listed below.

The first objective of this program is to enhance measurement methods, models, and parameters for the performance characterization of milling and turning centers, with a focus on practical, intuitive, yet comprehensive, performance parameters and associated validation tests; address new developments in

[2] Summary reproduced with the permission of National Institute of Science and Technology (NIST).

TABLE 6.11
Standards Related to Material-Handling Equipment

Pub Id: BS 2629

Status: Superseded, withdrawn

Title: Pallets for Materials Handling for Through Transit

Pub Id: BS 3810-1:1964

Status: Current

Title: Glossary of Terms Used in Materials Handling. Terms Used in Connection with Pallets, Stillages, Hand and Powered Trucks

Pub Id: BS 3810-2:1965

Status: Current

Title: Glossary of Terms Used in Materials Handling. Terms Used in Connection with Conveyors and Elevators (Excluding Pneumatic and Hydraulic Handling)

Pub Id: BS 3810-3:1967

Status: Current

Title: Glossary of Terms Used in Materials Handling. Terms Used in Connection with Pneumatic and Hydraulic Handling

Pub Id: BS 3810-4:1968

Status: Current

Title: Glossary of Terms Used in Materials Handling. Terms Used in Connection with Cranes

Pub Id: BS 3810-5:1971

Status: Current

Title: Glossary of Terms Used in Materials Handling. Terms Used in Connection with Lifting Tackle

Pub Id: BS 3810-6:1973

Status: Current

(*Continued*)

TABLE 6.11 (Continued)
Standards Related to Material-Handling Equipment

Title: Glossary of Terms Used in Materials Handling. Terms Used in Connection with Pulley Blocks

Pub Id: BS 3810-7:1973

Status: Current

Title: Glossary of Terms Used in Materials Handling. Terms Used in Connection with Aerial Ropeways and Cableways

Pub Id: BS 3810-8:1975

Status: Current

Title: Glossary of Terms Used in Materials Handling. Terms Used in Connection with Lifts, Lifting Platforms and Inclined Haulages

machine tools and test metrology; and provide leadership to facilitate the development of the respective national and international standards.

The technical outputs related to the first objective are the following:

1. Draft international standard ISO 2307-7 "Axis of Rotation."
2. Make recommendation to ISO on updates to ISO 230-3 "Thermal Effects of Machine Tools."
3. Report to ASME B5.54 on dynamic measurements of geometric errors.
4. Report to project sponsor on procedures and results of an extensive baseline performance evaluation of a tested turning center.
5. Complete study on performance measures for high-speed contouring and procedures to facilitate tuning of controller parameters.
6. Validate (virtual machining) procedures to translate generic performance data available in industrial environments into expected tolerances of parts, providing graded tradeoffs between uncertainties of the predictions and the required detail in analysis and data.
7. Develop harmonized and unbiased U.S. and ISO standards on machine tool performance evaluation that incorporate the latest science in machine tool technology and test methods.
8. Restructure U.S. standards into smaller generic and machine-specific sections.
9. Represent of U.S. interests in ISO by providing the Secretariat of ISO/TC39/SC2 "Machine Tool Test Conditions."

The second objective of this program is to develop unambiguous standardized Extensible Markup Language (XML) data formats for machine tool performance data; provide facilities to demonstrate performance and test conformance of the emerging standards and technology for interoperability; and develop machine tool self-identification capability enabled by an internal database of performance data that can be queried and updated through standard protocols.

The technical outputs related to second objective are the following:

1. Mature drafts, reference XML schemas, and comprehensive collection of XML examples for ASME B5.59-1 "Data Specification for Machine Tool Performance Tests" and B5.59-2 "Data Specification for the Properties of Machining and Turning Centers."
2. Smart machine tool database with Web interface that provides self-knowledge of turning-center test bed as defined by data elements in B5.59-1 and B5.59-2.

The third objective of the smart machine tool program is to develop standards for smart sensor interfaces, wireless sensor connectivity, and sensor network capability relevant to manufacturing; provide facilities to demonstrate, promote, test conformance, and validate emerging standards and technology for interoperability; and apply generic, standardized, smart transducer connectivity infrastructure to machine tools, with real-time processing of sensor data by embedded systems.

The technical outputs related to the third objective are the following:

1. An open-source collaborative framework that allows industry, government, and academia to participate in the development of a unified implementation of IEEE 1451.1.
2. An open-source–based IEEE 1451–embedded NCAP that demonstrates a distributed measurement and control system for machine tool monitoring and control.
3. A balloted IEEE standard for a smart transducer interface for sensor and actuator mixed-mode communication protocols and Transducer Electronic Data Sheet (TEDS) formats, IEEE standard P1451.4.
4. Ballot-ready IEEE 1451.5 wireless smart transducer interface specification.

The fourth objective of this program is to develop smart sensor systems and procedures to monitor and adjust machine parameters that affect form and size of parts manufactured on the smart machine-tool test bed, including use of process-intermittent and postprocess inspection data and techniques for *in situ* inspection.

The technical outputs related to the fourth objective are the following:

1. Procedures, validated on test-bed, to monitor machine tool error sources from postprocess and process-intermittent inspection data.
2. Exploratory report on metrology concepts that enable machine tools to self-assess errors and provide *in situ* inspection of mechanical parts.
3. Implementation and evaluation of various error-reduction techniques on test bed.

The fifth objective of the program is to identify main failure modes of key machine components and develop procedures and sensor applications to detect and (remotely) diagnose abnormal machine behavior and heightened probability of failure.

The technical outputs for the fifth objective are the following:

1. Report on the robust application of sensors and signal-processing techniques to monitor the condition of machine tool spindles that emphasizes a physics-based understanding.
2. Report on methods and potential standards for the specification and estimation of the MTBF of machine tools.

Anticipated impacts of the program include the following:

1. Improvements in productivity, time to market, and quality through smart machine tool concepts that facilitate the harmonization of part design with machine capability, machining the first and every subsequent part to specification, and machine maintenance.
2. Harmonized machine tool performance standards that better address realistic production capabilities.
3. Efficient exchange and application of machine tool data through standardize unambiguous data formats.
4. Simplified and reliable transducer application in products and manufacturing through the network and enterprise connectivity and plug-and-play integration provided by IEEE 1451, leading to cost-effective integration of embedded intelligence and sensor technology.
5. New market opportunities through standards, and new product capabilities for manufacturers of machine tools, machine tool test equipment, manufacturing software, and transducer systems.

BIBLIOGRAPHY

Anjard, S.R. and Ronald, P., Computer integrated manufacturing: A dream becoming a reality, *Industrial Manage. Data Syst.* 95(1), 3–4, 1995.

Baker, L., Manufacturing Motors in a New World Economy. Equipment Standards, Proceedings of the Electrical/Electronics Insulation Conference, 2001, 63–67.

Blyth, G., PCB design and manufacture—The need for integration standards, *Computer-Aided Eng. J.,* 3(6), 225–228, 1986.

Chan, S.C.F., Mak, H., and Cloutier, N., Architecture for Standards-Based Manufacturing Data Integration, SME Technical Paper (Series) MM, 1993, 1–10.

Dowding, D., Hildebrant, A., Chindamo, D., and Rearick, J., Collaboration through industry standards for manufacturing success, *Proc. IEEE/CPMT/SEMI* 29, 259–262, 2004.

Drach, B., Use manufacturing standards to drive continuous cost improvement, *Prod. Inventory Manage. J.* 35(1), 20–25, 1994.

DeGarmo, E.P., Black J.T., and Kohser, R.A., *Materials and Processes in Manufacturing*, 9th ed., John Wiley & Sons, New York, 2002.

Eichener, Impact of technical standards on the diffusion of anthropocentric production systems, *International Journal of Human Factors in Manufacturing,* 6(2),131–145, 1996.

Endelman L.L., Effect of standards on new equipment design by new international standards and industry restraints, *Proc. SPIE,* 1346, 90–92, 1990.

Fusaro, D., Standardizing manufacturing skills, *Control*, 14(5), 13, 2001.

Schrader, G.F. and Elshennawy, A.K., *Manufacturing Processes and Materials,* 4th ed., Society of Manufacturing Engineers, 2000.

Hardwick, M. and Loffredo, D., The STEP international data standard is becoming a manufacturing tool, *Manuf. Eng.,* 126(1), 38–50, 2001.

Hillman, G., Christ, M., Barnard, A.J., Jr., Nelson, R., and English, R.T., Development and use of standards in the semiconductor manufacturing industry, *Electrochem. Soc. Extended Abstr.* 85(1), 148, 1985.

Holmes, T., Review of the Automated Manufacturing Research Facility at the National Bureau of Standards—Federal Government Involvement in Manufacturing Innovation, SAE Technical Paper Series, 1986.

Kemelhor, R.E., Advanced manufacturing technology—Computers, open systems interconnections, international standards, and the Japanese, *Johns Hopkins APL Technical Dig.,* 9(4), 383–387, 1988.

Klein, M.T., IEEE 802. Four Standards for Factory Communications: An Overview, Wescon Conference Record, 2(1),1986.

Krepchin, I.P., Can standards ease automation confusion? *Mod. Mater. Handling*, 42(1), 97–100, 1987.

Leaver, E.W., Process and manufacturing automation—Past, present, and future, *ISA Trans.*, 26(2), 45–50, 1987.

Cornelius, L.T., *Computer Aided and Integrated Manufacturing Systems: Manufacturing Processes*, World Scientific Publishing Company, 2003.

Lorincz, J.A., Setting standards of manufacturing excellence, *Tooling & Singapore Prod.*, 59(4), 44–46, 1993.

Mikell, P.G., *Fundamentals of Modern Manufacturing: Materials, Processes, and Systems*, 2nd ed., John Wiley & Sons, New York, 2001.

Miller, J.C., Ten essential elements of an effective engineering standards program, *Agric. Eng.*, 69(7), 11–13, 1988.

Ming, X.G., Mak, K.L., and Yan, J.Q., PDES/STEP-based information model for computer-aided process planning, *Robotics Computer-Integrated Manuf.*, 14(5–6), 347–361, 1998.

Mufti, A.A., Morris, M.L., and Spencer, W.B., Data exchange standards for computer-aided engineering and manufacturing, *Int. J. Comput. Appl. Technol.*, 3(2), 70–80, 1990.

Phillip F.O. and Muñoz, J., *Manufacturing Processes and Systems,* 9th ed., John Wiley & Sons, New York, 1997.

Qiao, L.-H., Zhang, C., Liu, T.-H., Wang, H.-P.B., and Fischer, G.W., PDES/STEP-based product data preparation procedure for computer-aided process planning, *Comput. Ind.* 21(1), 11–22, 1993.

Ramirez-Valdivia, M.T., Christian, P., Govande, V., and Zimmers, E.W., Jr., Design and implementation of a cellular manufacturing process: A simulation modeling approach, *Int. J. Industrial Eng. Theory Appl. Pract.,* 7(4), 281–285, 2000.

Todd, R.H., Allen, D.K., and Alting L., *Manufacturing Processes Reference Guide,* 1st ed., Industrial Press, New York, 1994.

Askin, R.G. and Charles, R.S., *Modeling and Analysis of Manufacturing Systems,* 1st ed., John Wiley & Sons, New York, 1993.

Schey, J.A., *Introduction to Manufacturing Processes,* 3rd ed., McGraw-Hill, New York, 1999.

Shang, H., Zhao, Z., and Thorn, R., Implementing manufacturing message specifications (MMS) within collaborative virtual environments over the Internet, *Int. J. Computer-Integrated Manuf.,* 16(2), 112–127, 2003.

Smith, R., Consistency of manufacturing standards, *N.Z. Eng.* 50(8), 30, 1995.

Stefanac, D.R. and Klager, J.R., Evolution of process controls toward factory automation, *Industrial Heat.,* 53(7), 30–34, 1986.

Stewart, D.T., Application of standards in the integrated CAD/CAM environment,*ASTM Stand. News*, 15(1), 52–56, 1987.

Sylla, C. and Toraskar, K., Linking standardization and quality: Some manufacturing implications, *Manufacturing Rev.* 3(1), 6–15, 1990.

Thomas, E.V., Berry, W.L., and Whybark, D.C., *Manufacturing Planning and Control Systems*, 4th ed., McGraw-Hill, New York, 1997.

Wechsler, J., Safety and manufacturing standards, *Pharm. Technol.*, 25(10), 18–26, 2001.

Whiteside, R.A., Pancerella, C.M., and Klevgard, P.A., CORBA-based manufacturing environment, *Proc. Hawaii Int. Conf. Syst. Sci. Software Technol. Architecture*, 1, 34–43, 1997.

Zha, X.F. and Ji, P. Assembly process sequence planning for STEP based mechanical products: An integrated model and system implementation, *Int. J. Industrial Eng. Theory Appl. Pract.,* 10(3), 279–288, 2003.

7 Process Planning for a Revolute Robot Using Pertinent Standards—A Case Study

7.1 PROCESS PLANNING

Process planning is an activity commonly performed in manufacturing organizations to define all the steps required to convert raw material into a finished product. The following are the types of process planning:

Manual process planning
Computer-aided process planning (CAPP)
  Generative process planning
  Variant process planning

In manual process planning, all the tasks are defined without reference to any other process plans. The experience of the planner plays the most important role. Computer-aided process planning involves the extensive use of computers to perform the task. It allows utilization of databases of geometry of parts, databases of types of machine tools and facilities available at manufacturing organizations, databases of manufacturing auxiliaries such as tools, coolants, lubricating oils, and measuring equipment. It also allows access to materials databases, along with their available cross sections.

Generative process planning makes use of all the facilities available in a computerized format; however, it does not refer to any previous process plan for the generation of a new process plan. Variant process planning makes use of a special process-plans database that contains previous process plans. It initially selects a process plan from the process-plans database and modifies it to make it suitable for the current task.

The use of standards in process planning is paramount. Standards catalogs, whether print versions for manual process planning or online versions for computer-aided process planning, play an important role in selecting various parameters.

7.2 MANUFACTURING FACILITIES AND AVAILABLE RESOURCES

Each discrete manufacturing facility differs from others in some respect, although they may have the same manufacturing system definition, such as job shop or project shop. Process planning in different organizations takes into account maximum use of available facilities to produce a part or product to reduce the chances of outsourcing. The process plans produced are normally different for differing facilities that produce the same product. One reason is that planners at different facilities think differently.

7.3 USE OF STANDARDS IN PROCESS PLANNING

Manufacturing operations are normally well covered by company, professional, national, regional, or international standards. Practices that are not covered by standards normally become a benchmark for a manufacturing organization.

The major imports to process planning from the design functions are technical drawings of parts and products. The process planner decides how and where each component should be manufactured, according to specified tolerances, and later assembled to give the final products. The technical drawings are normally created according to specified standards, and any further elaboration of part or assembly during process planning is also made according to a standard.

The operations performed on computers or during assembly may use processes as shown in Figure 6.1. Most of these processes and their accessories (such as work holding) usually conform to standards.

7.4 PROCESS PLANNING FOR A REVOLUTE ROBOT

Manual process planning for a model revolute robot is performed as a case study to demonstrate the use of standards in this function. The revolute robot under study is shown in Figure 7.1. The robot comprises five main assemblies:

Gripper
Front arm
Back arm
Column
Base

In subsequent sections, each of these subassemblies of the robot is deconstructed into individual parts, and a procedure is set (on a procedure sheet) by which to manufacture the component in a reference manufacturing facility. The reference facility in this case is a small job shop normally associated with various kinds of discrete manufacturing units.

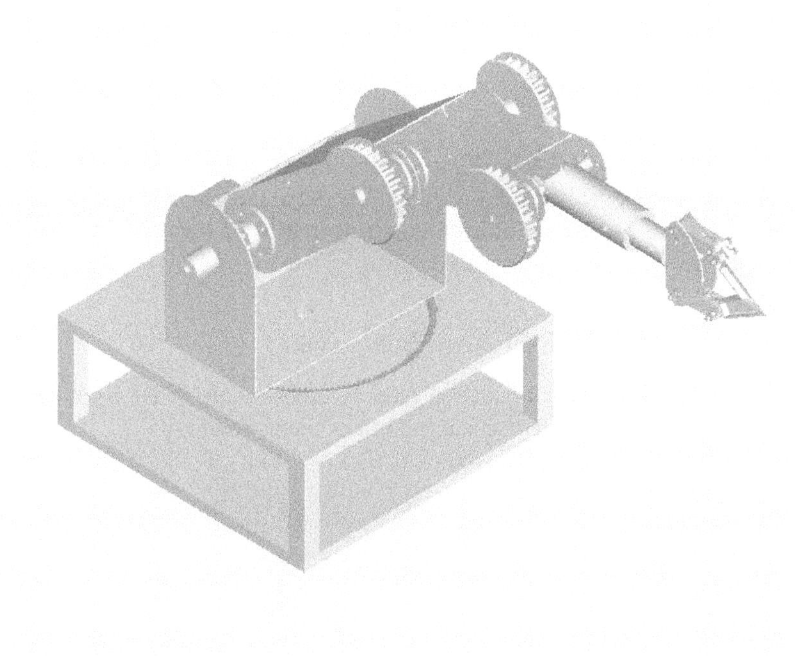

FIGURE 7.1 Model of the revolute robot.

During preparation of the procedure sheet for components, assemblies, and the product, the aim is to utilize as many standards as possible. However, the first problem that a real-world manufacturing organization faces is that all the standards it requires are not available from a single standards organization. Therefore, multiple standard organizations have to be contacted. Furthermore, much of the information required in producing a procedure sheet is not fully covered by professional, national, or international standards organizations. In such cases, company standards are commonly used. In other cases, certain solutions are not standardized at all, so the operator's skill is used to resolve these situations. An example is the use of custom-made tools to cut gear teeth for a standard module. The HSS bars for this purpose are produced and marketed, which indicates a need for such an item and such a solution.

7.4.1 Process Planning for Parts in Subassemblies

This section provides details on developing procedure sheets for all the parts of each assembly. The part drawing and the procedure sheet for each part corresponds to the following sequence:

Gripper
Front arm
Back arm
Column
Base

Figure 7.2 to Figure 7.40 provide the drawing according to pertinent standards for all the components. Table 7.1 to Table 7.39 detail the manufacturing processes required for the production of each component. Use of standards is taken as a prime measure.

7.4.2 Process Planning for the Subassemblies

The procedure sheets for production of each subassembly of the revolute robot are given in this section. These subassemblies are as follows:

Gripper
Front arm
Back arm
Column
Base

Figure 7.41 to Figure 7.45 detail the assembly drawing in standard format. Table 7.40 to Table 7.44 provide the assembly procedure using the components.

7.4.3 Process Planning for the Robot Assembly

This section provides the procedure sheets for assembly of the complete robot.

Figure 7.46 provides the assembly drawing or the revolute robot, and Table 7.45 details construction of robot using the major assemblies.

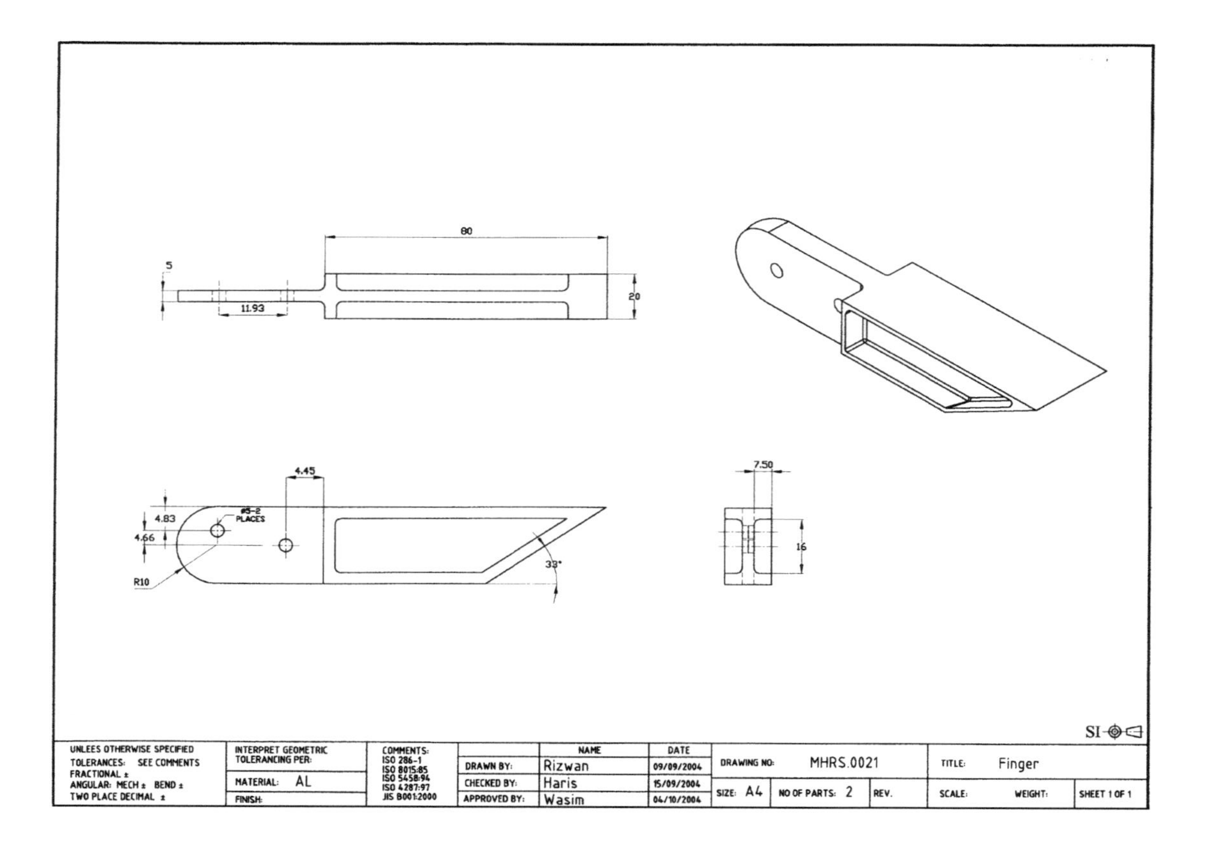

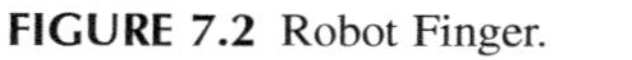
FIGURE 7.2 Robot Finger.

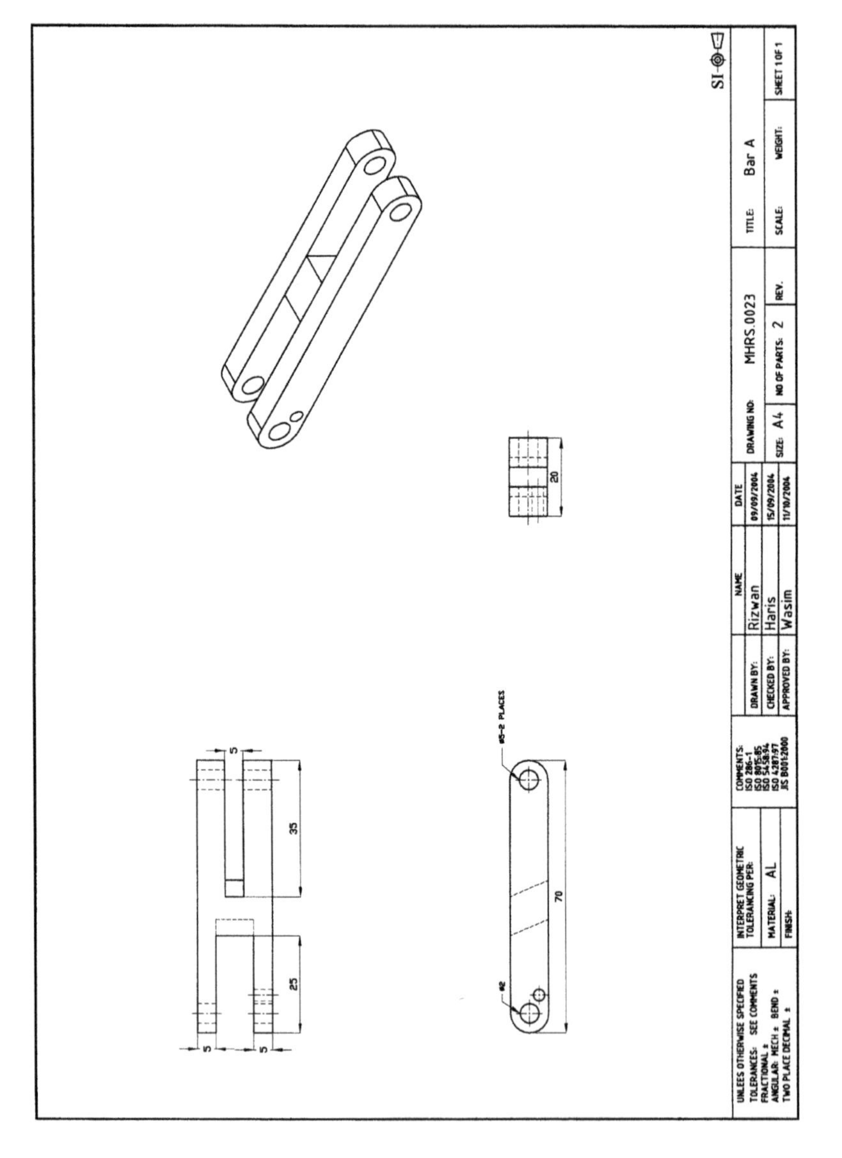

FIGURE 7.3 Bar A for the gripper assembly.

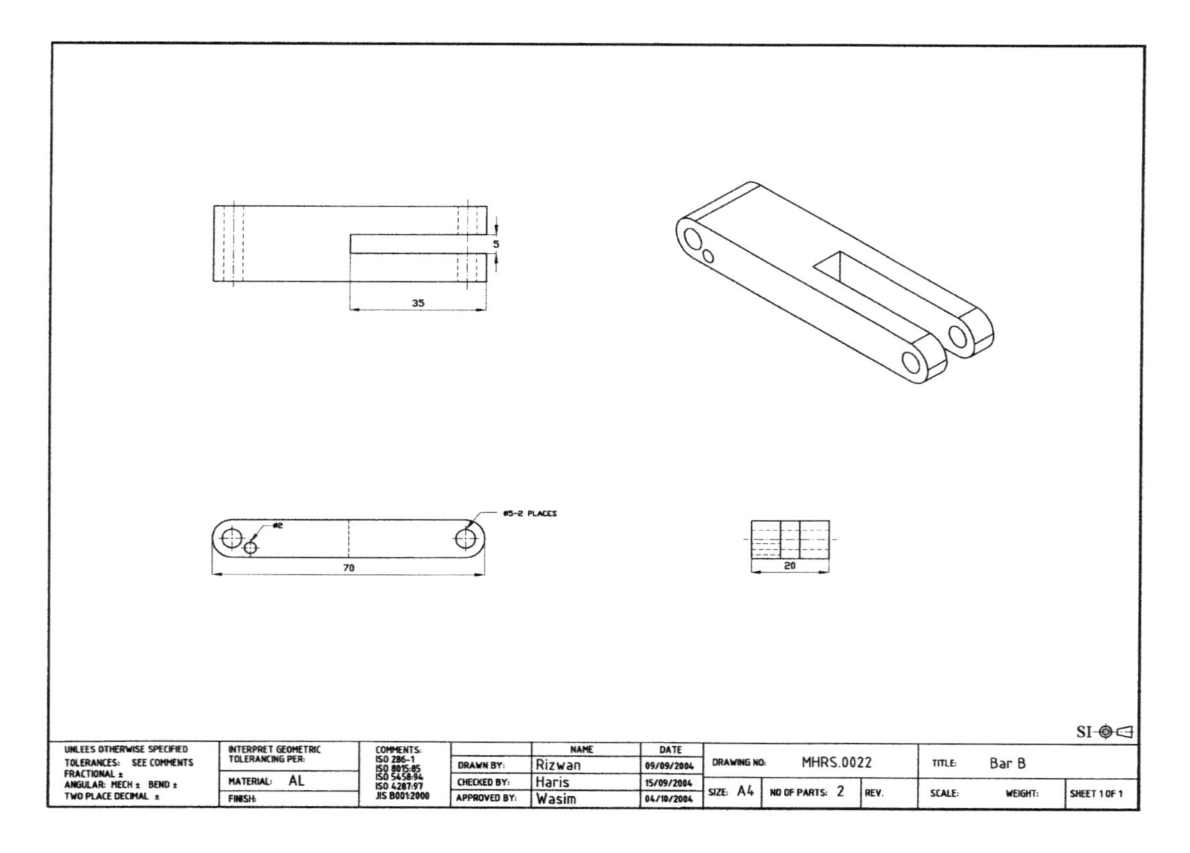

FIGURE 7.4 Bar B for the gripper assembly.

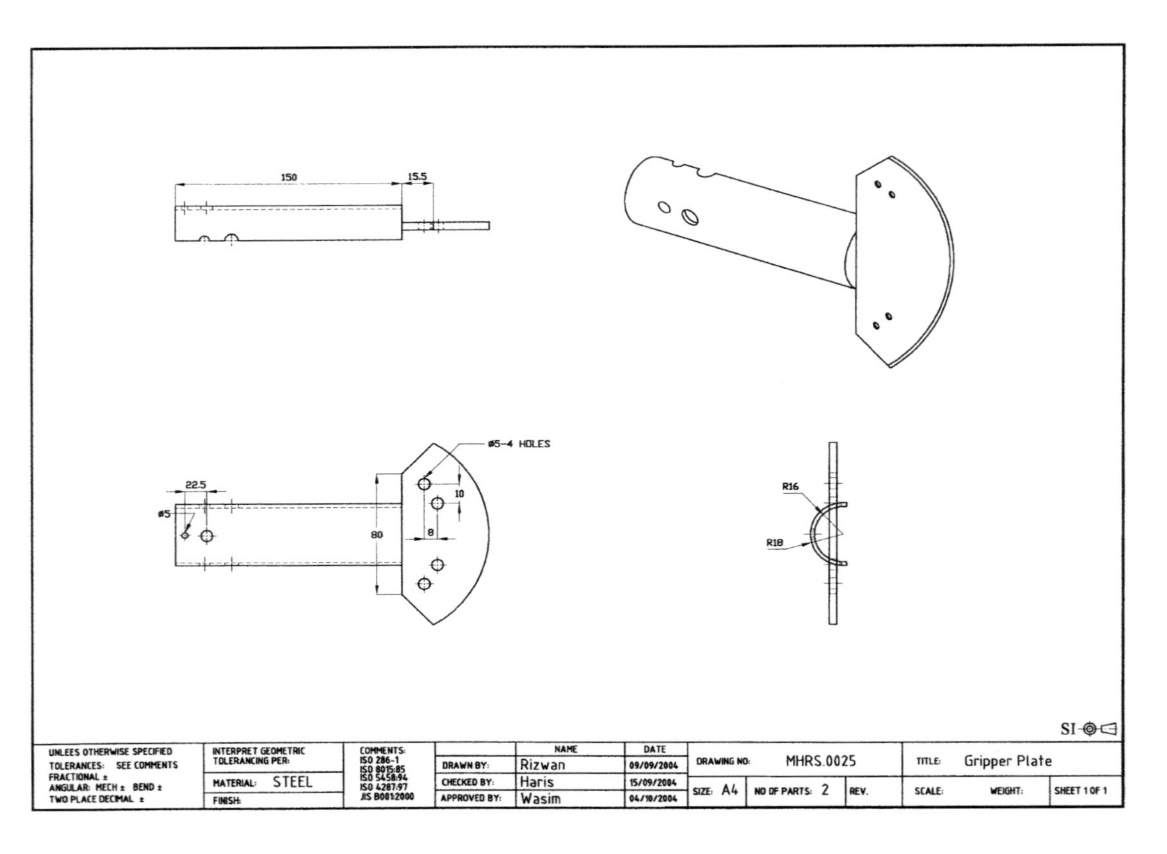

FIGURE 7.5 Gripper plate for gripper assembly.

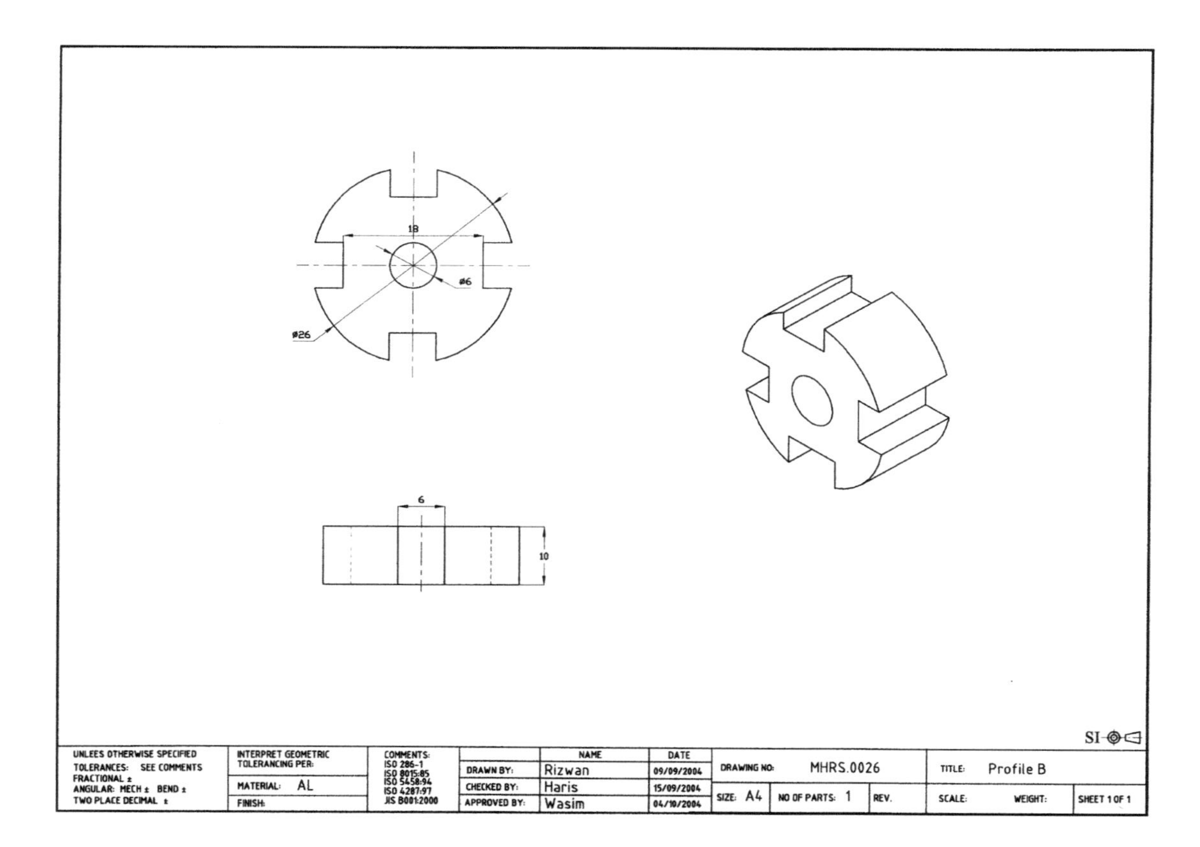

FIGURE 7.6 Profile B for the gripper assembly.

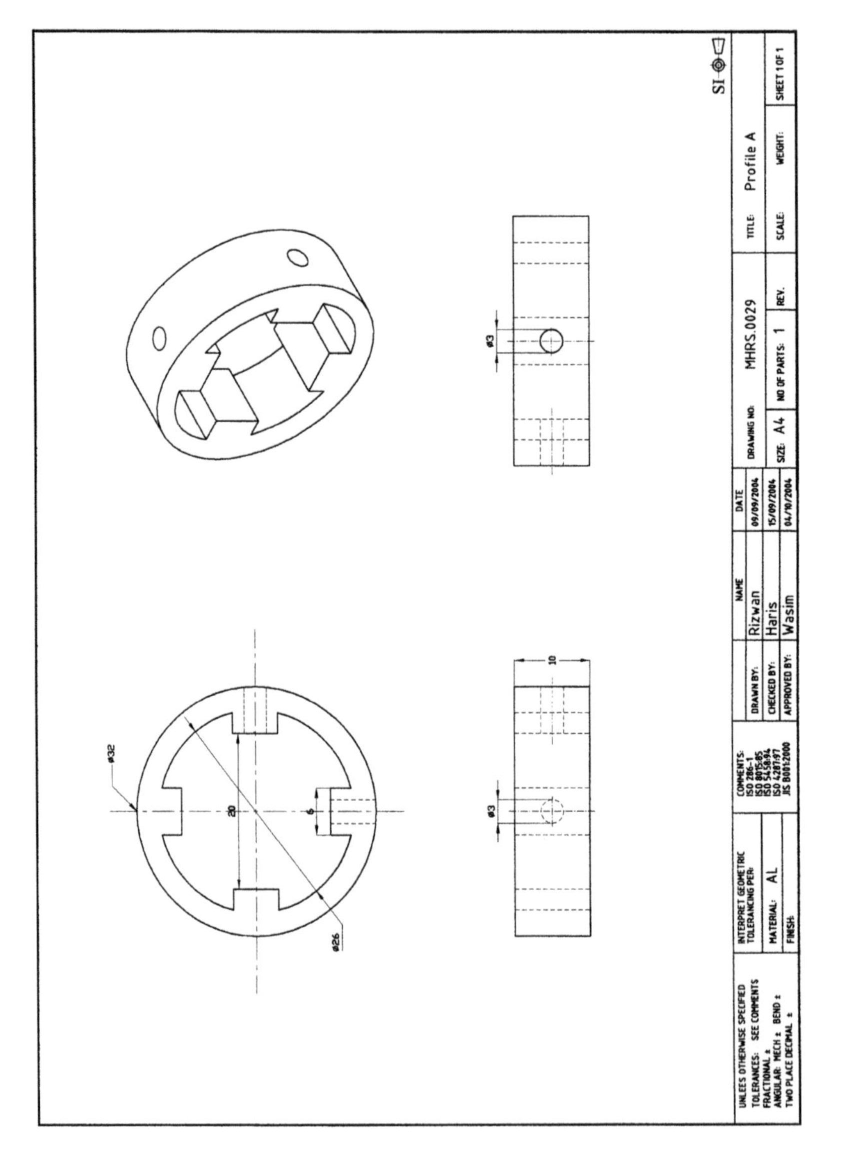

FIGURE 7.7 Profile A for the gripper assembly.

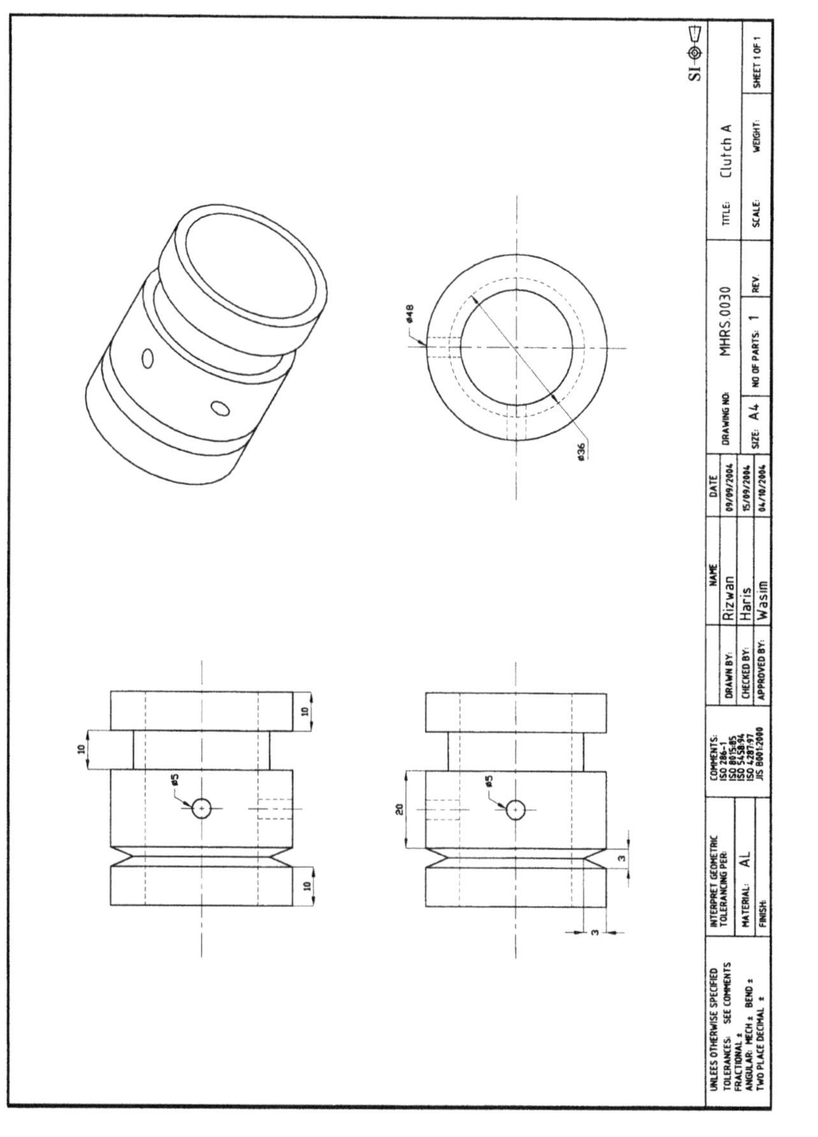

FIGURE 7.8 Clutch A for the gripper assembly.

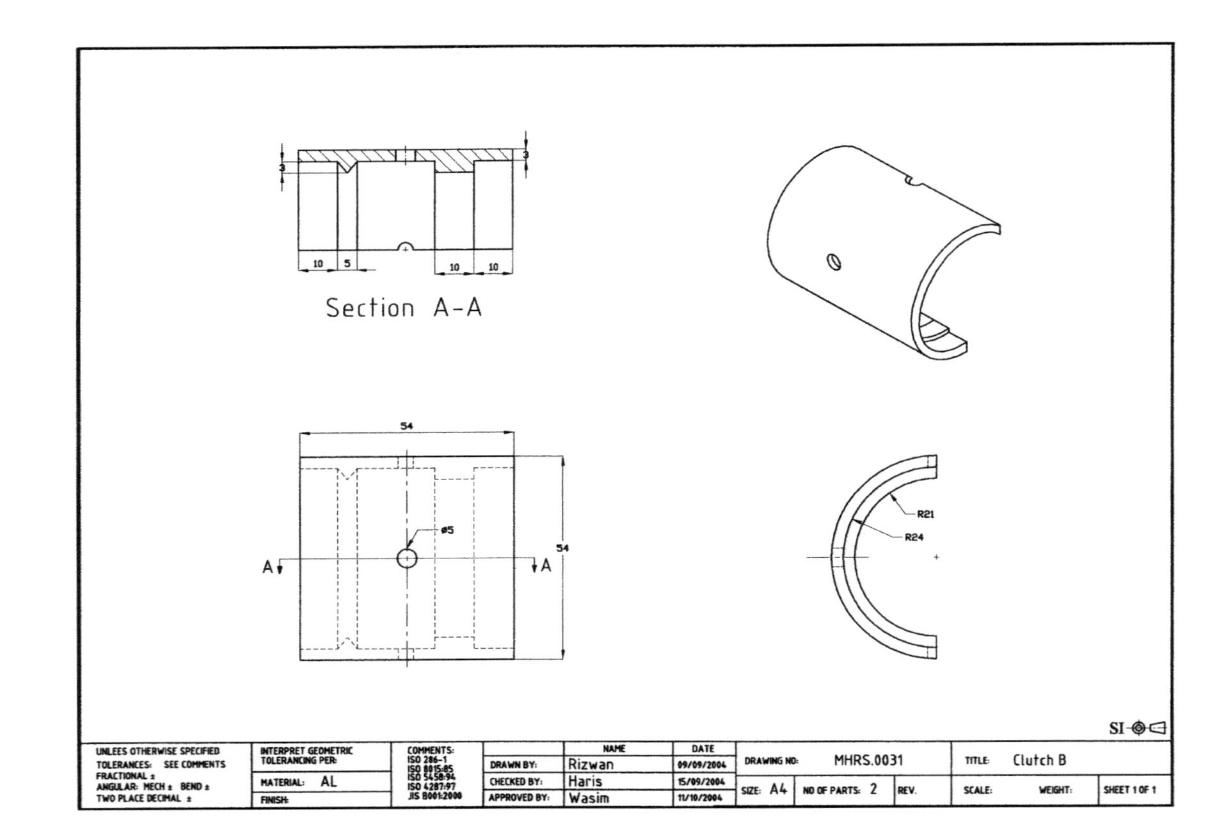

FIGURE 7.9 Clutch B for the gripper assembly.

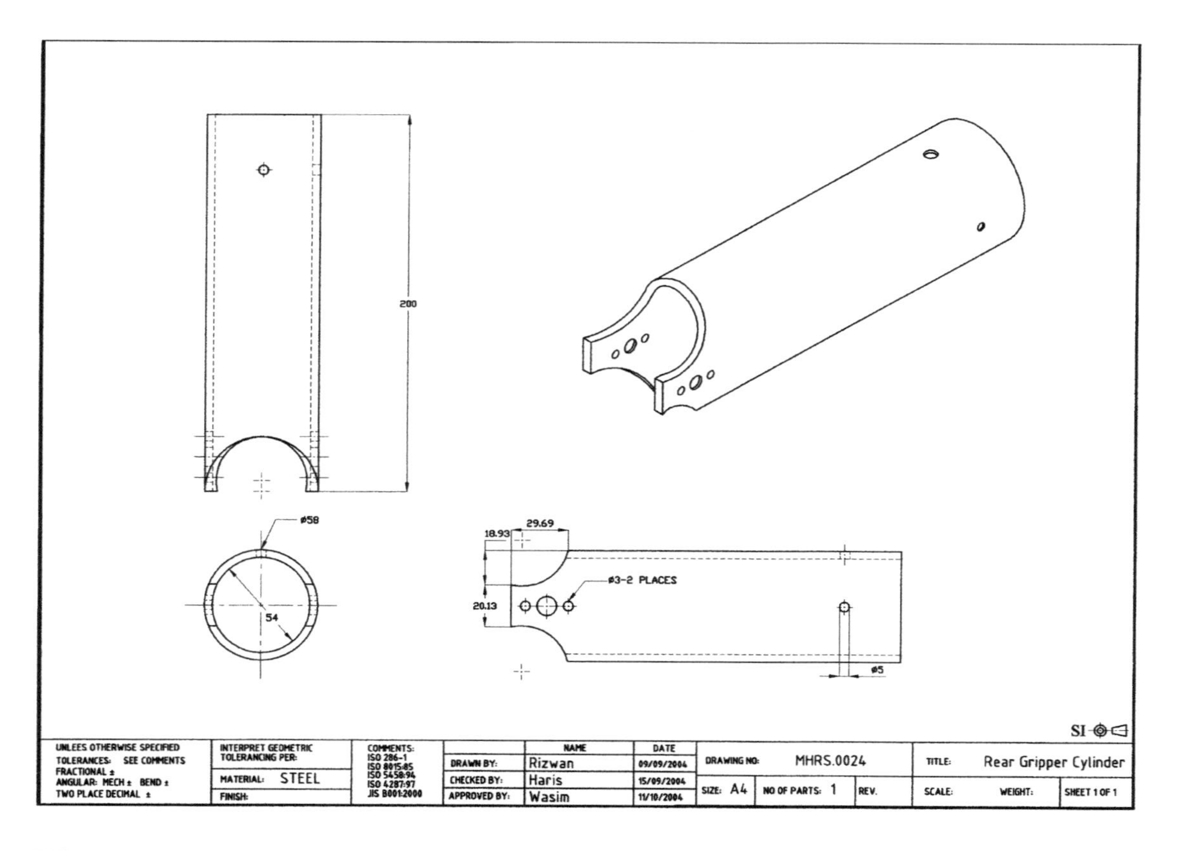

FIGURE 7.10 Rear gripper cylinder for the gripper assembly.

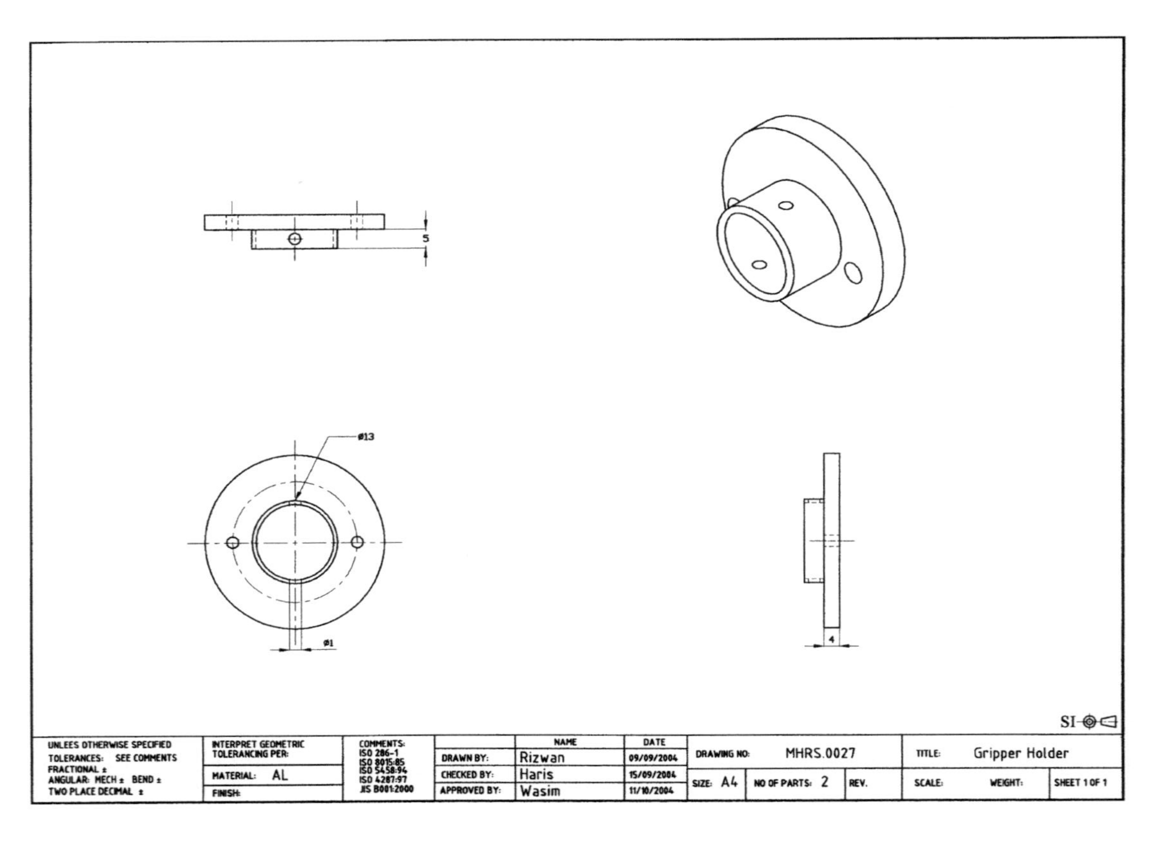

FIGURE 7.11 Gripper holder for the gripper assembly.

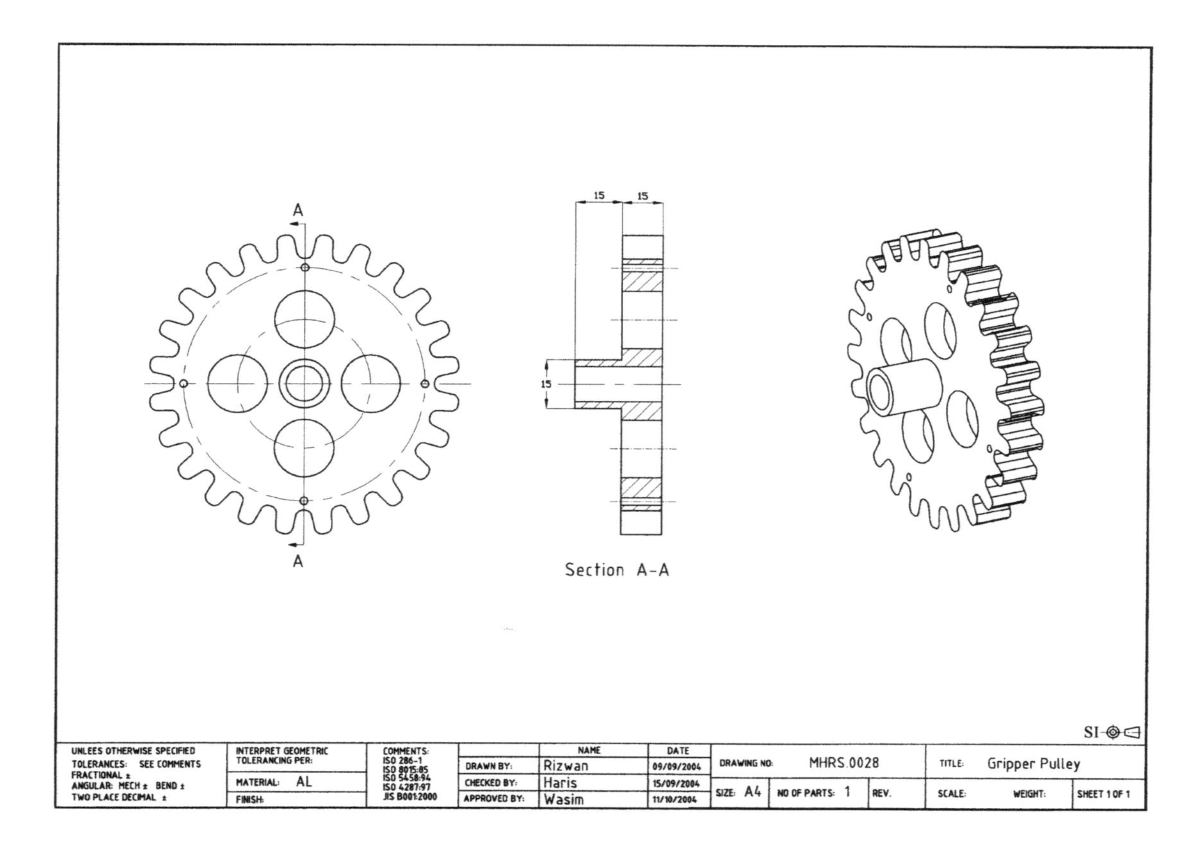

FIGURE 7.12 Gripper pulley for the gripper assembly.

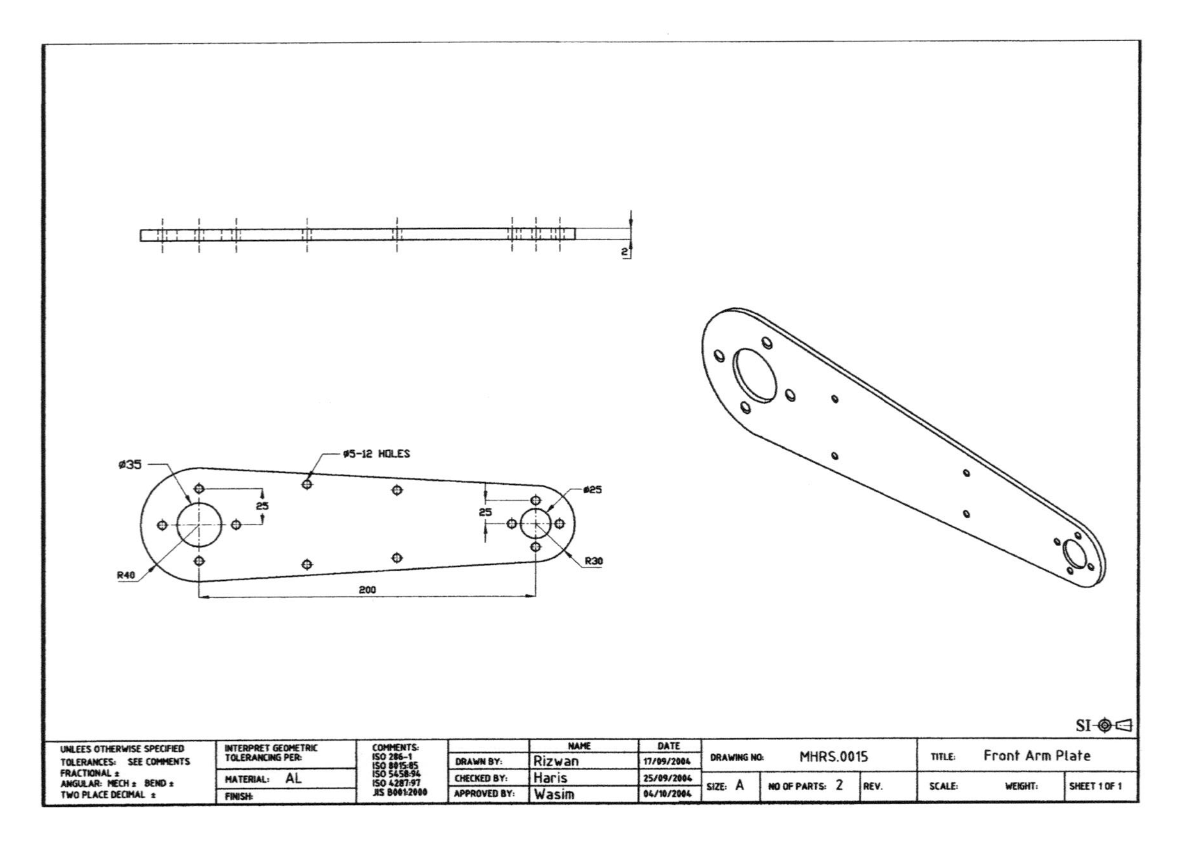

FIGURE 7.13 Front arm plate.

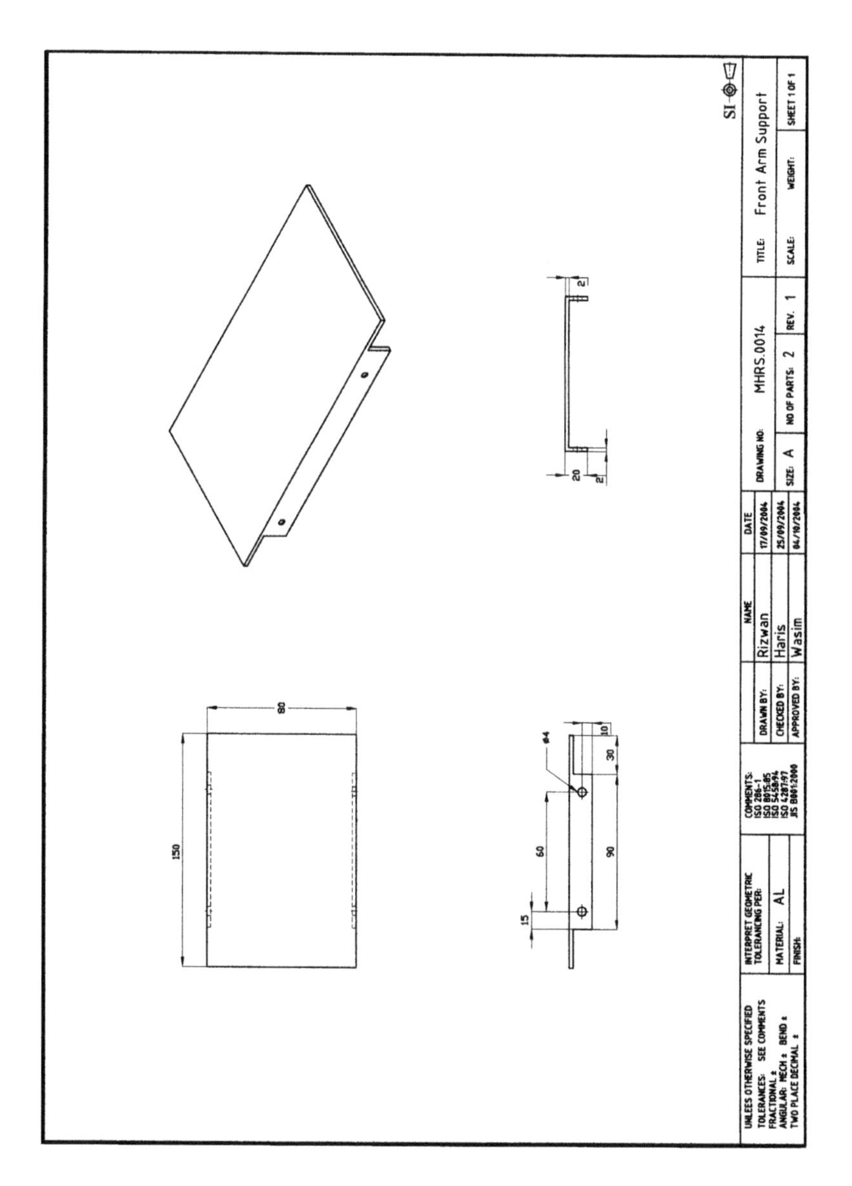

FIGURE 7.14 Front arm support.

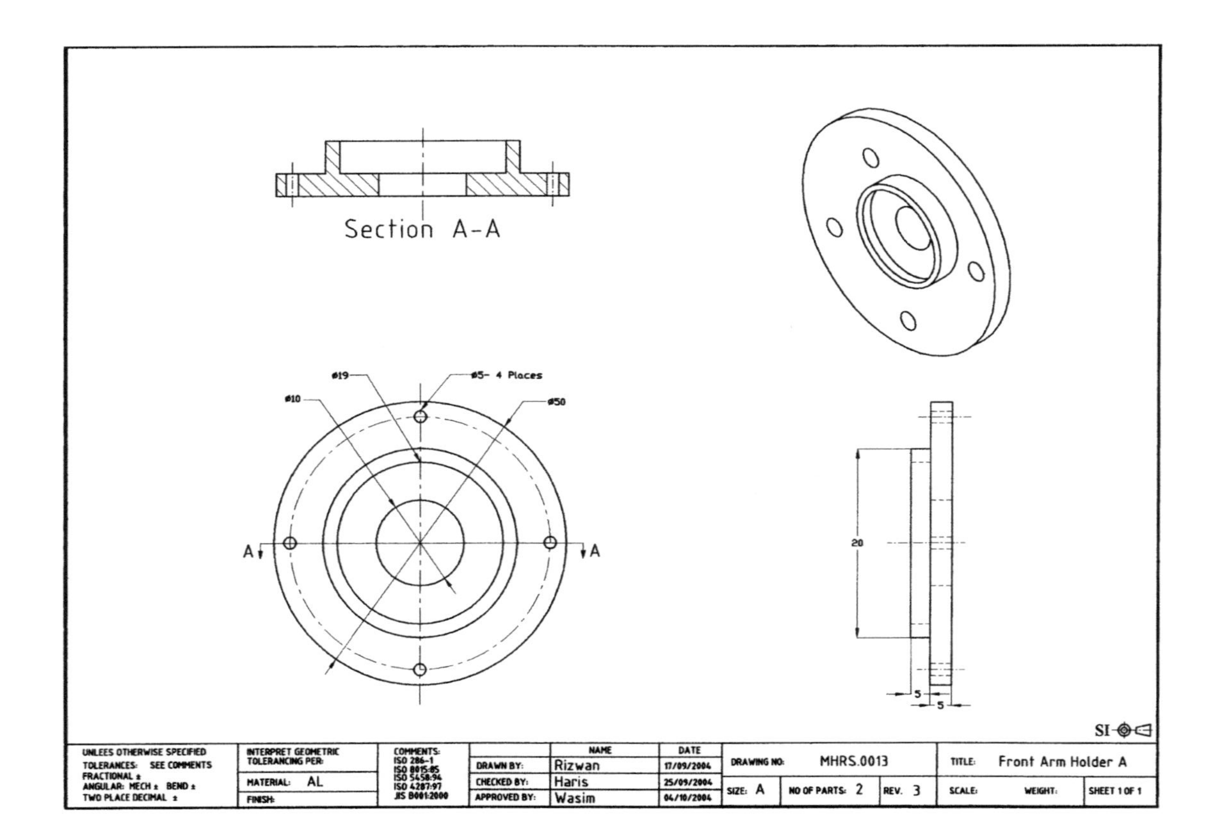

FIGURE 7.15 Front arm holder A.

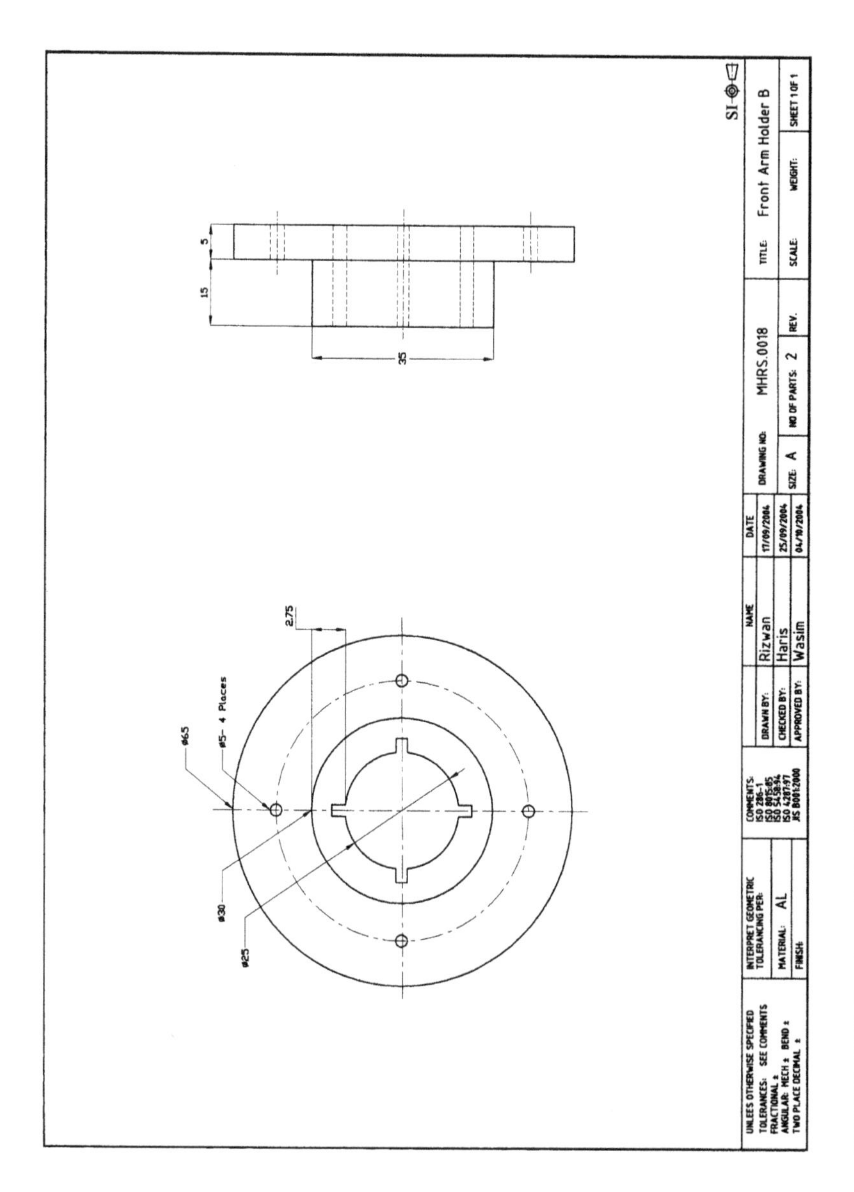

FIGURE 7.16 Front arm holder B.

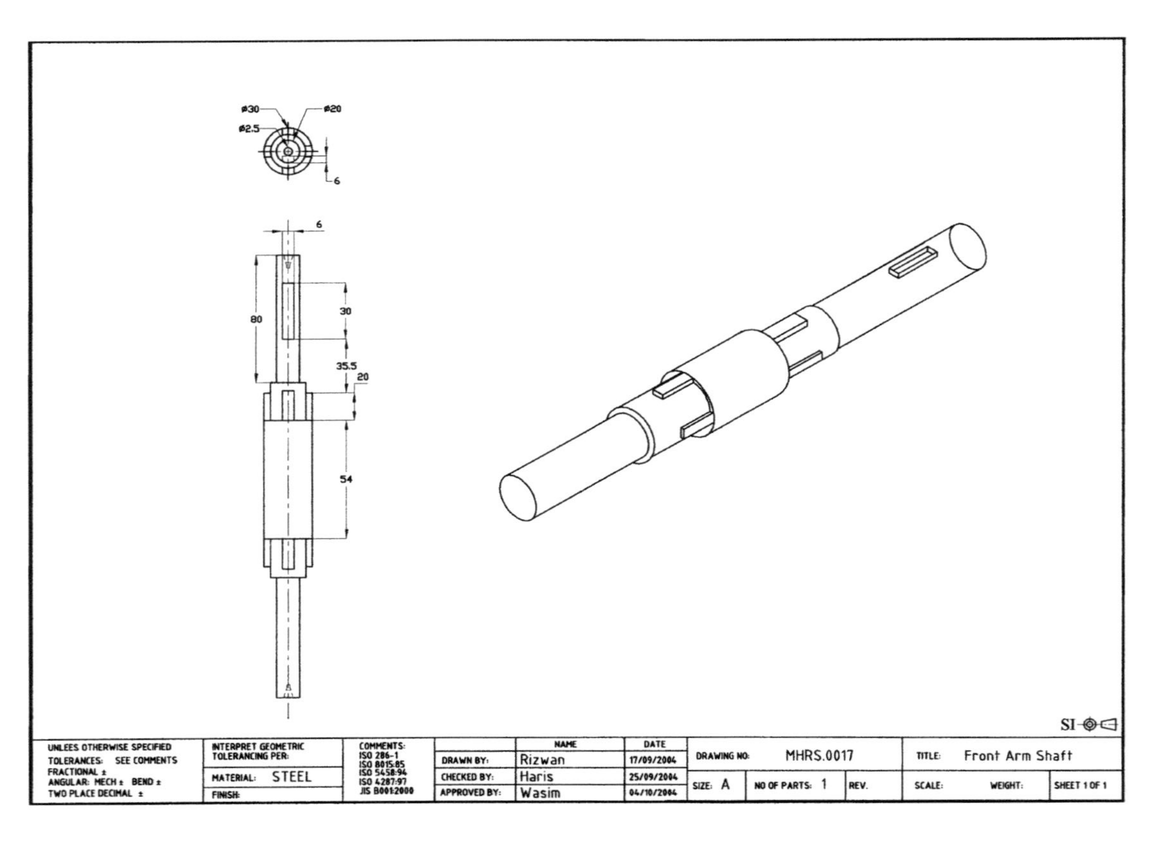

FIGURE 7.17 Front arm shaft.

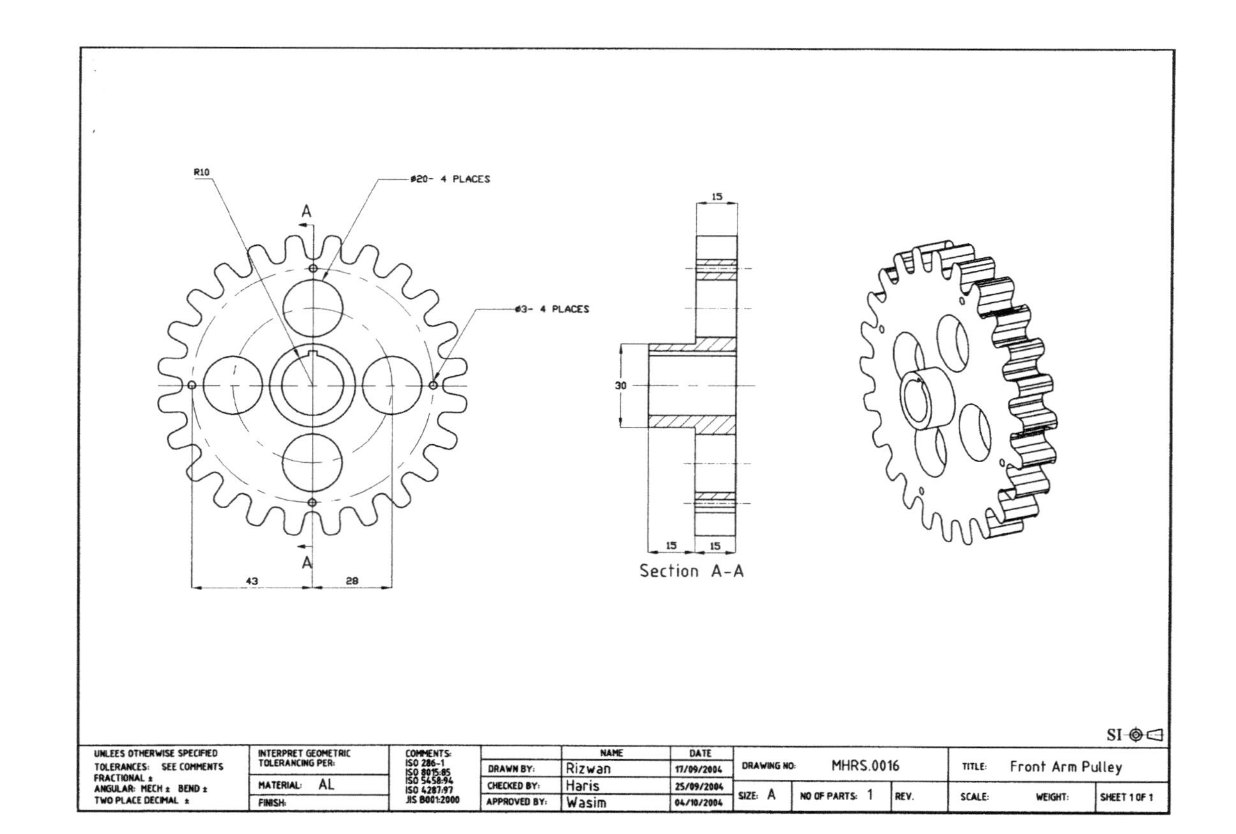

FIGURE 7.18 Front arm pulley.

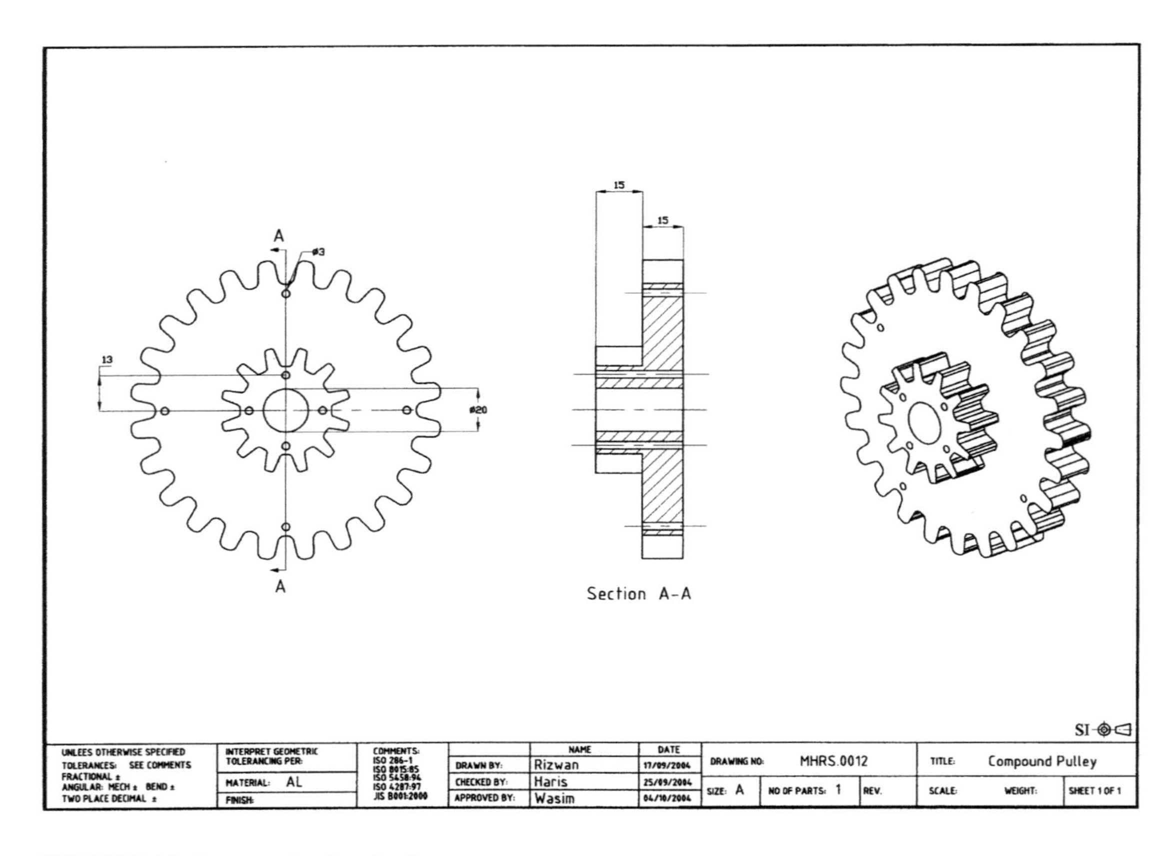

FIGURE 7.19 Compound pulley for front arm.

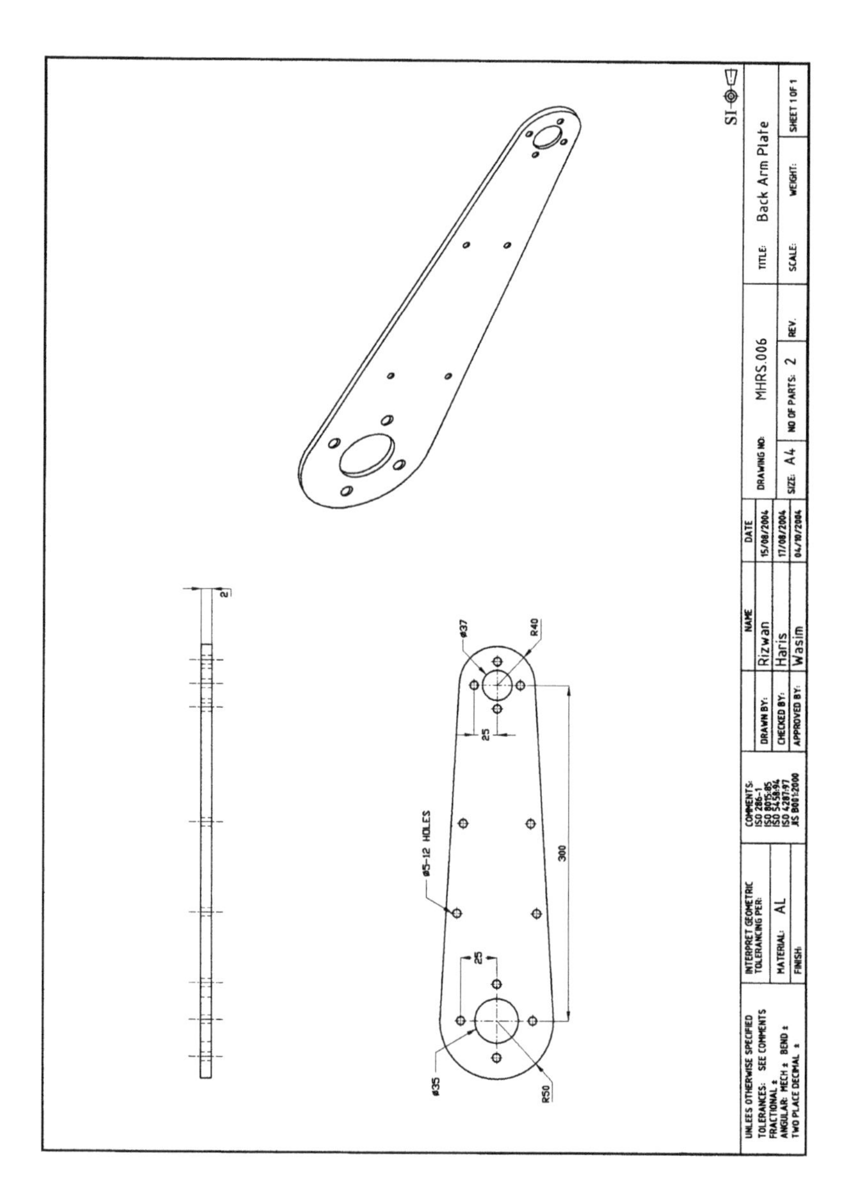

FIGURE 7.20 Back arm plate.

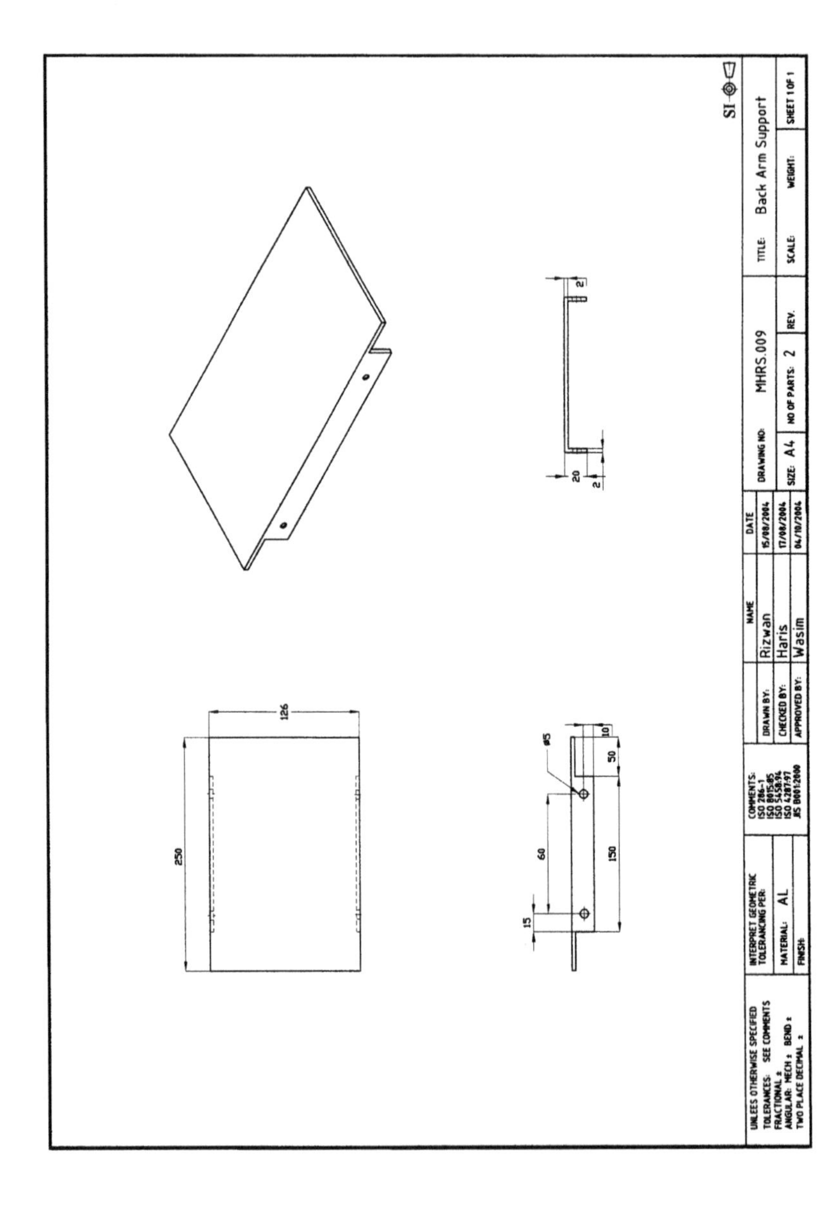

FIGURE 7.21 Back arm support.

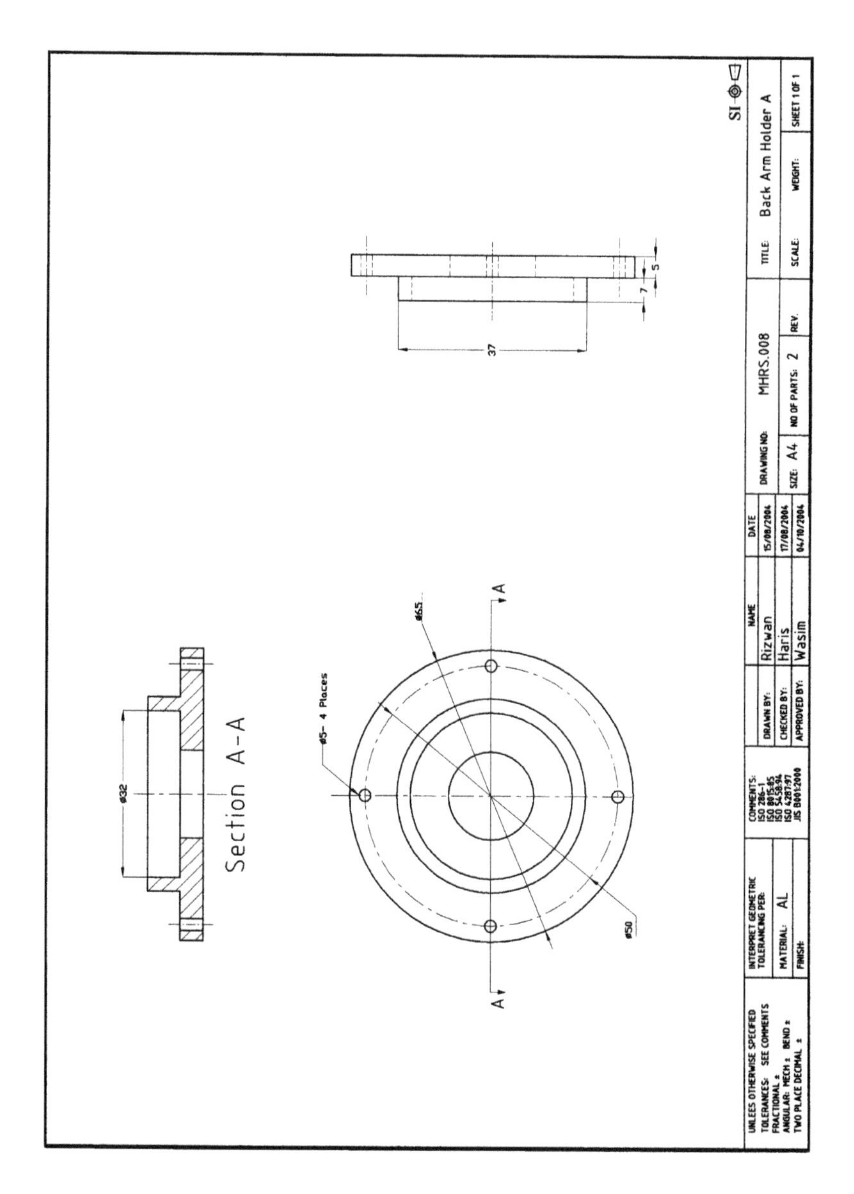

FIGURE 7.22 Back arm holder A.

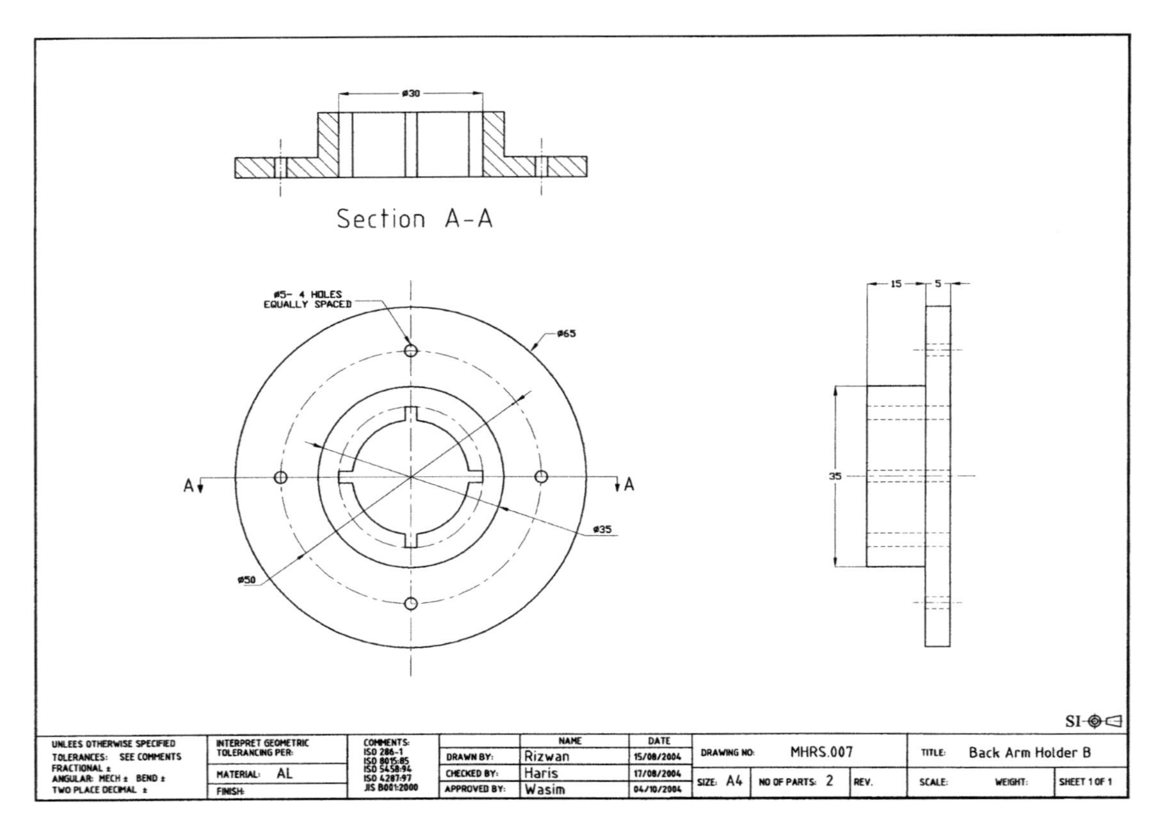

FIGURE 7.23 Back arm holder B.

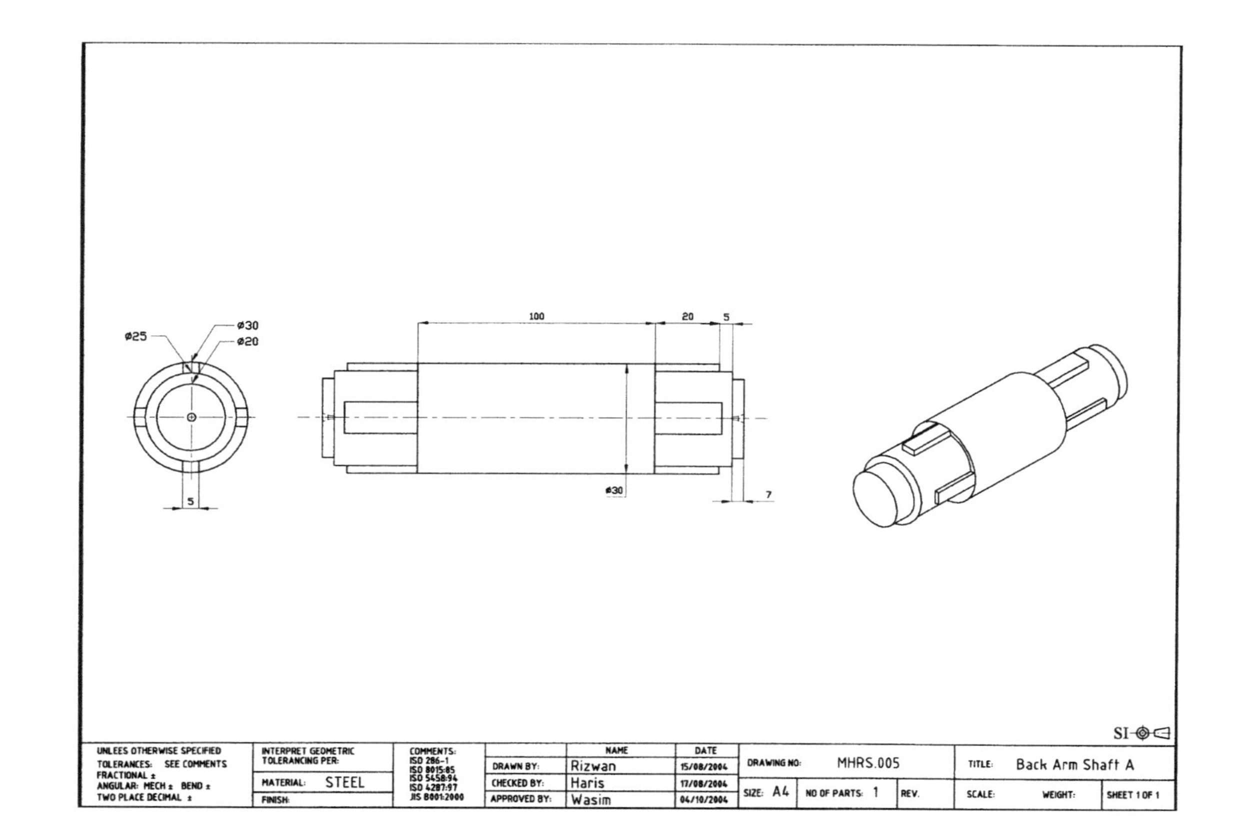

FIGURE 7.24 Back arm shaft A.

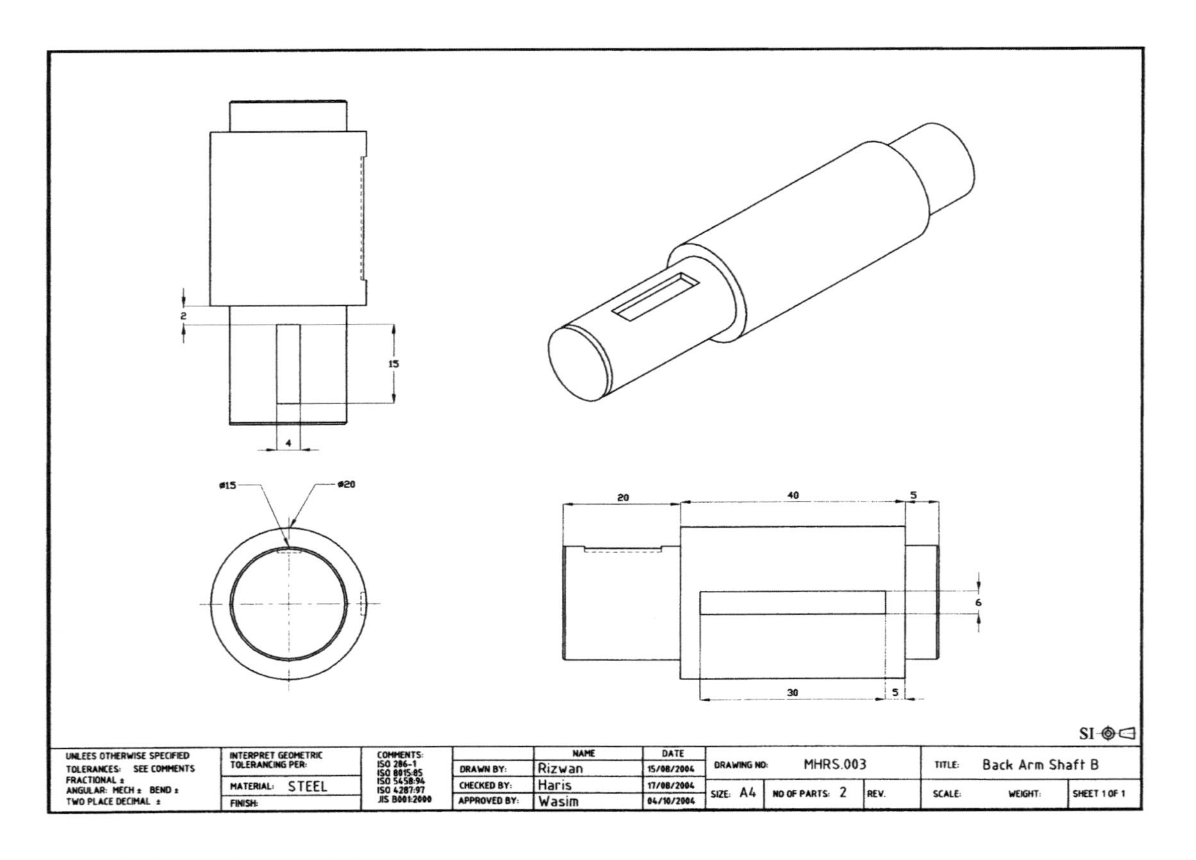

FIGURE 7.25 Back arm shaft B.

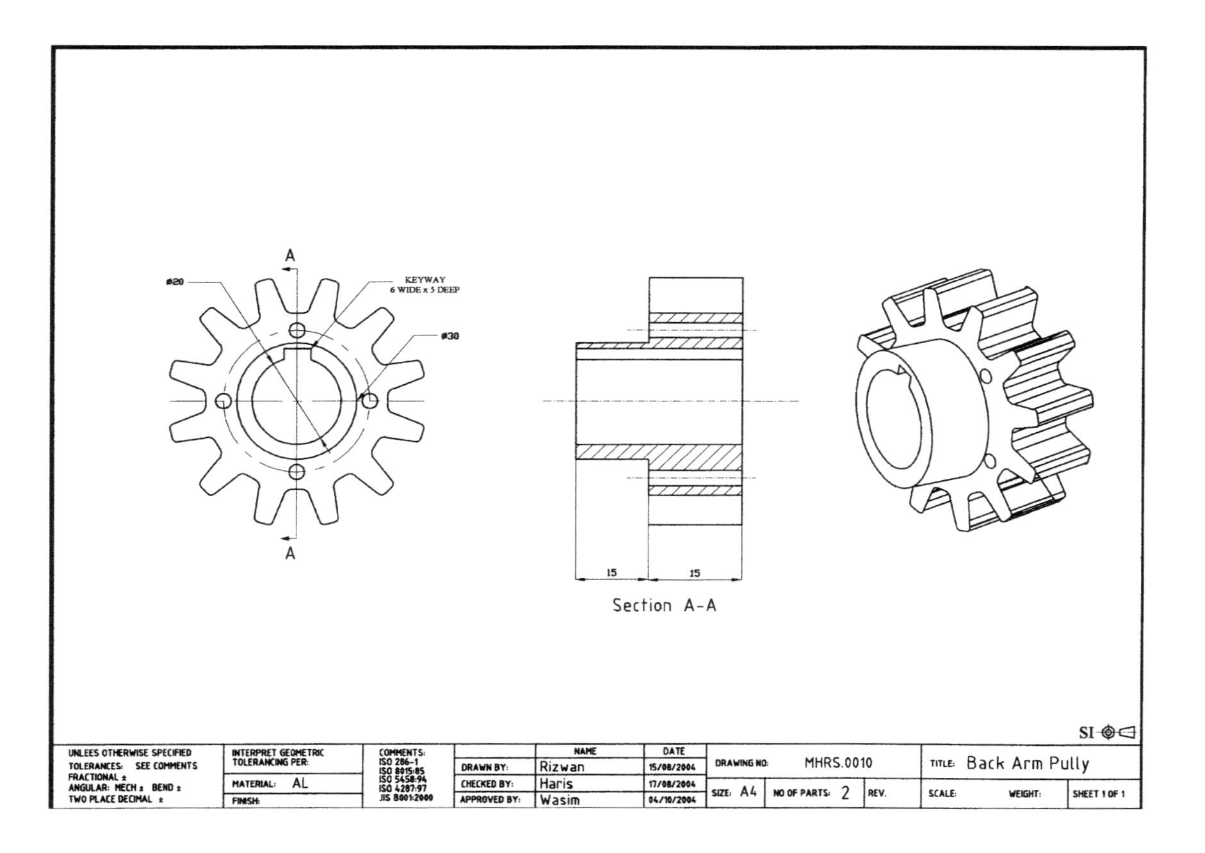

FIGURE 7.26 Back arm pulley.

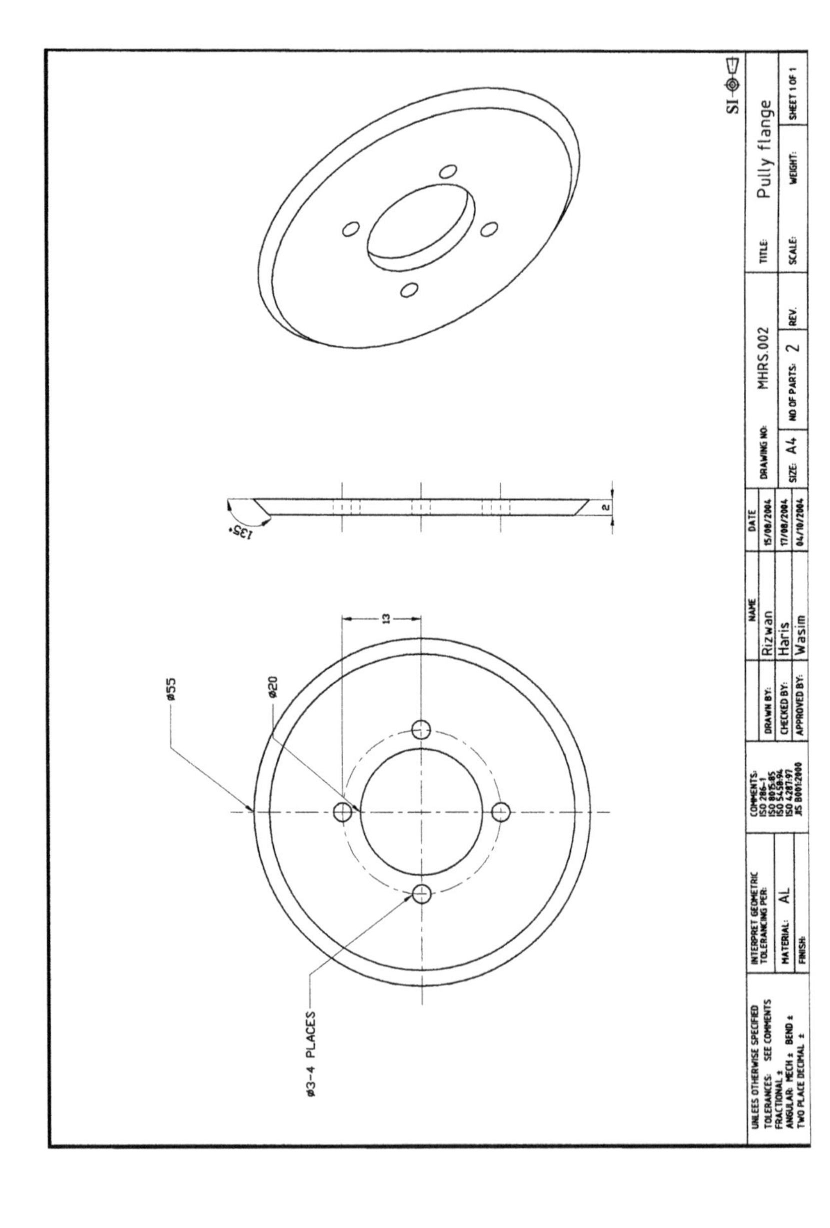

FIGURE 7.27 Pulley flange for back arm.

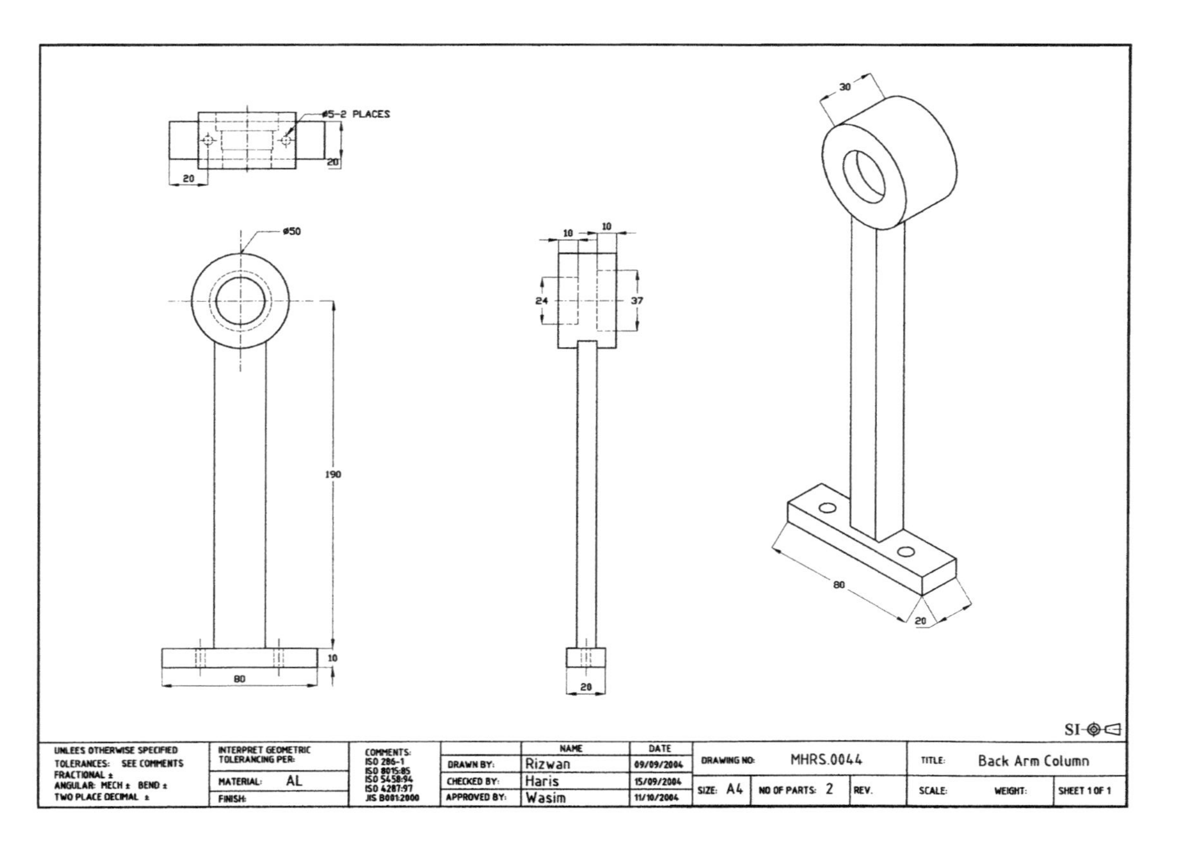

FIGURE 7.28 Back arm column.

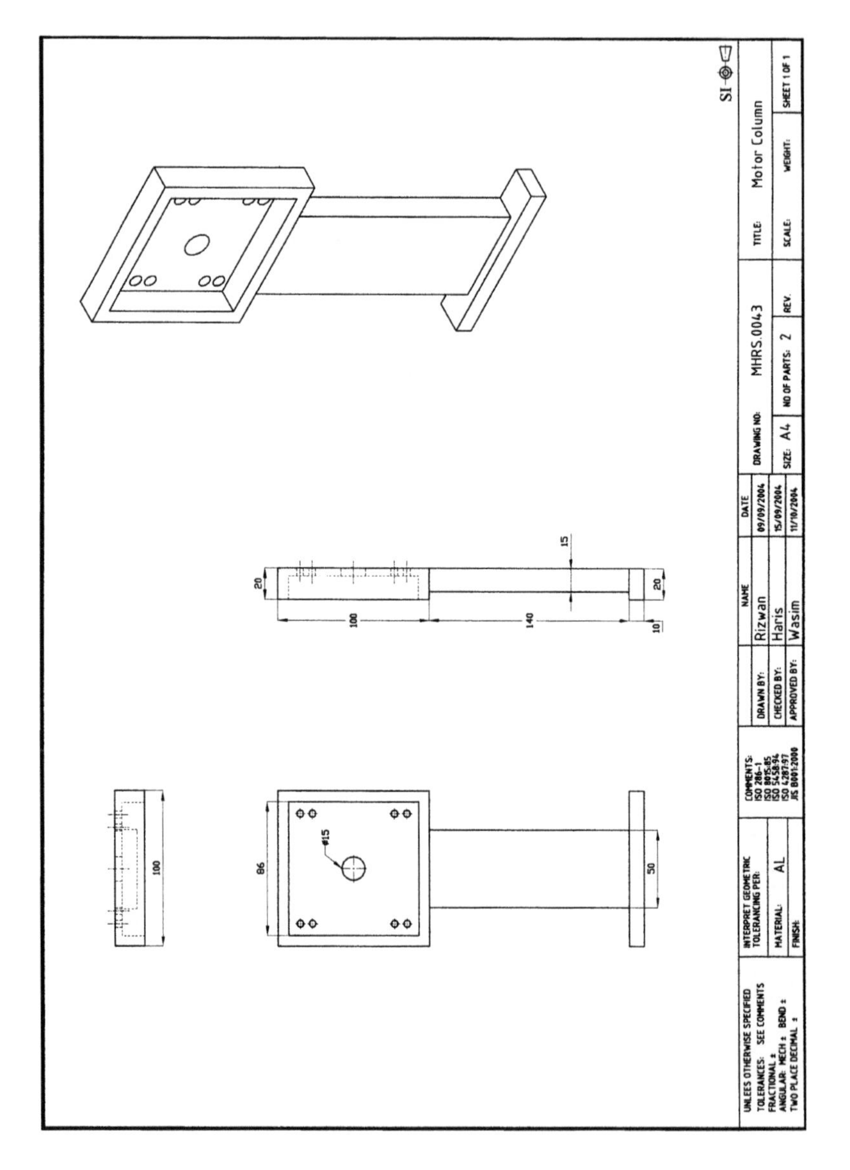

FIGURE 7.29 Motor column.

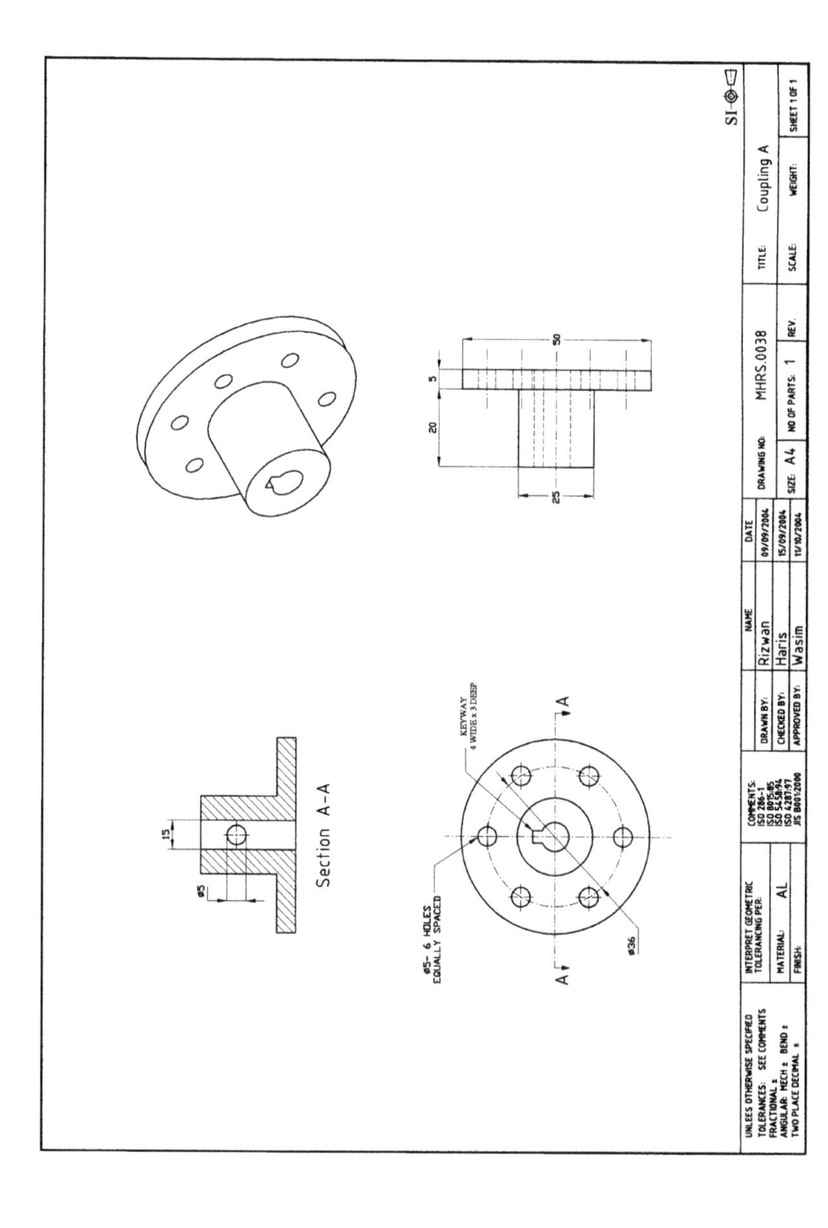

FIGURE 7.30 Coupling A for motor column.

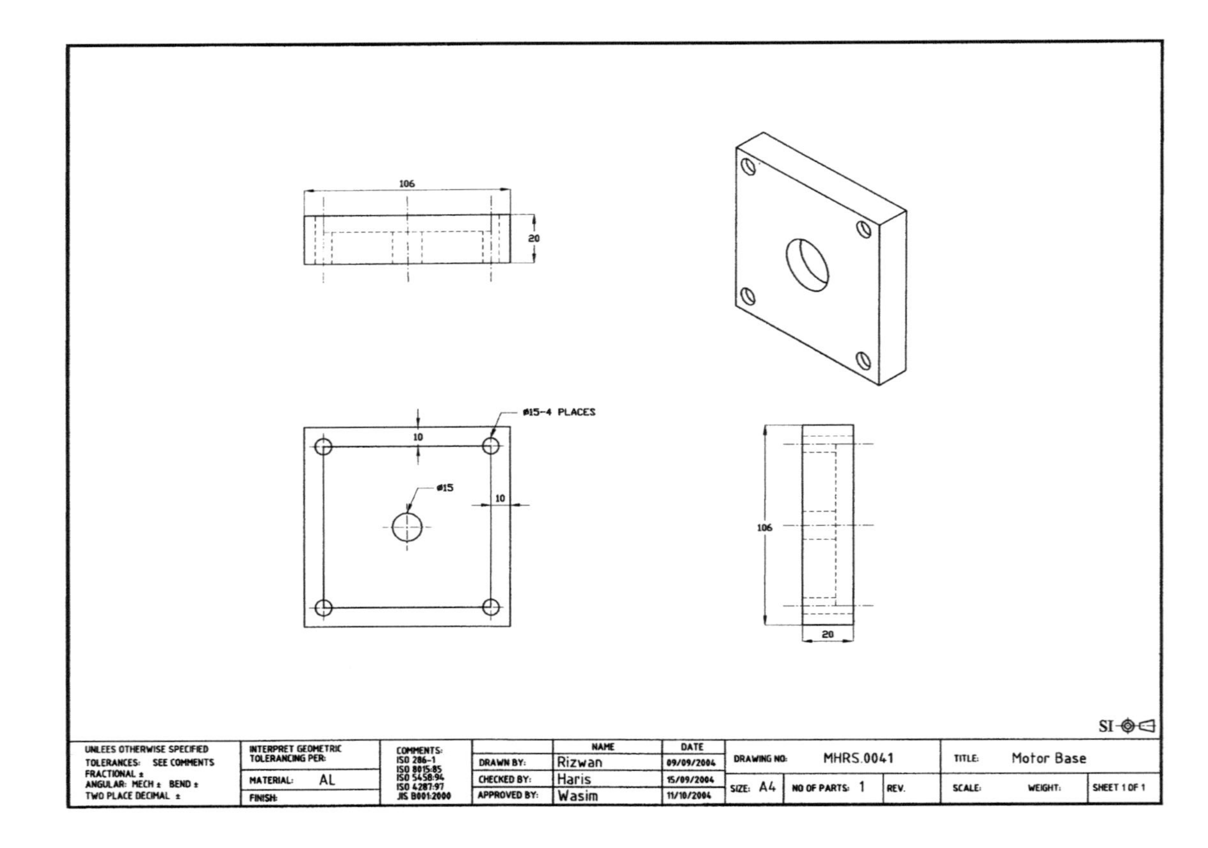

FIGURE 7.31 Motor base.

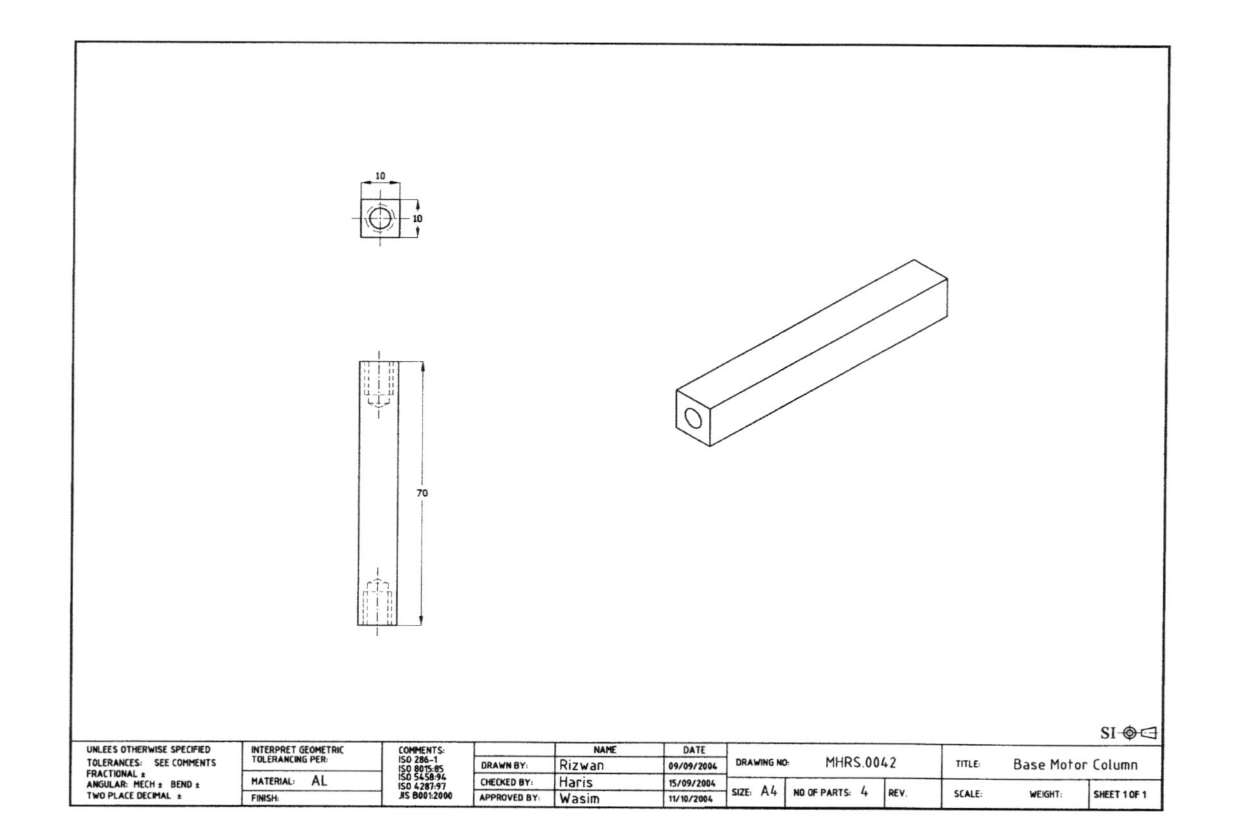

FIGURE 7.32 Base motor column.

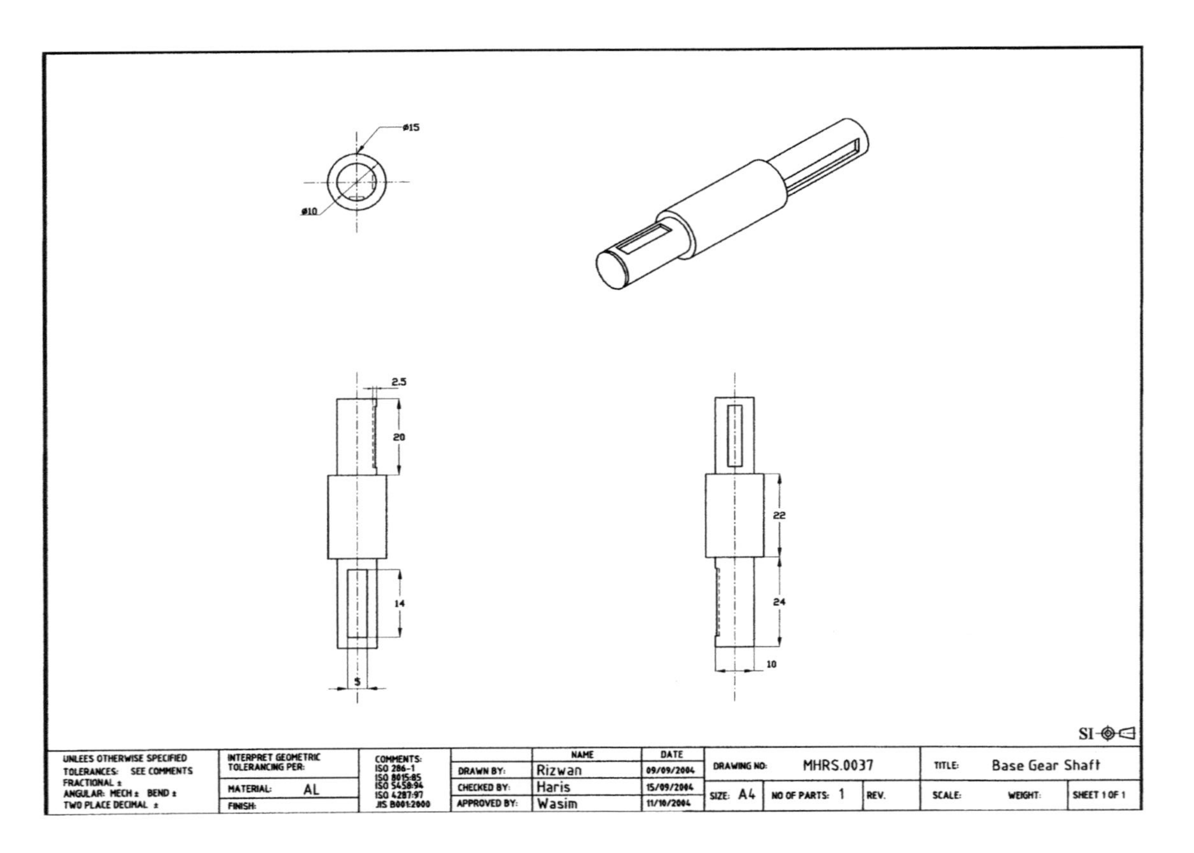

FIGURE 7.33 Base gear shaft.

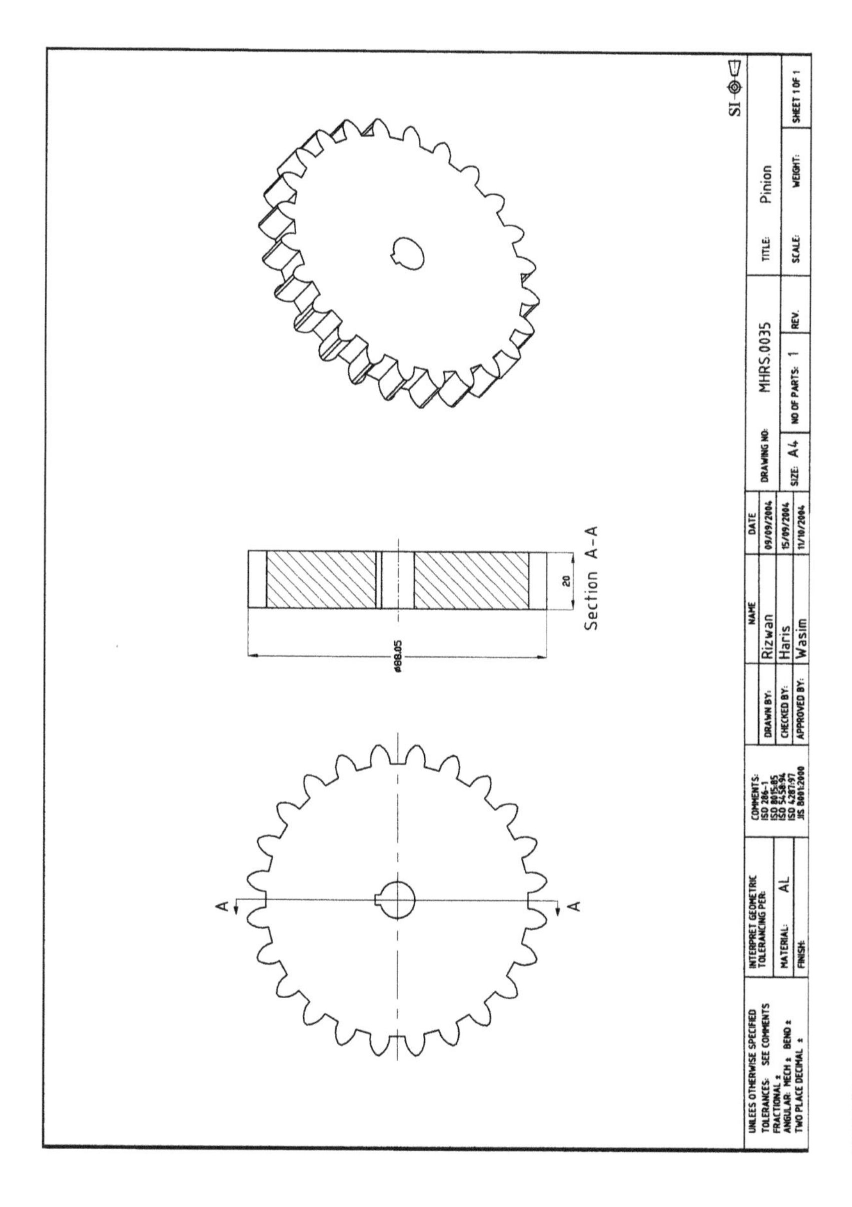

FIGURE 7.34 Pinion.

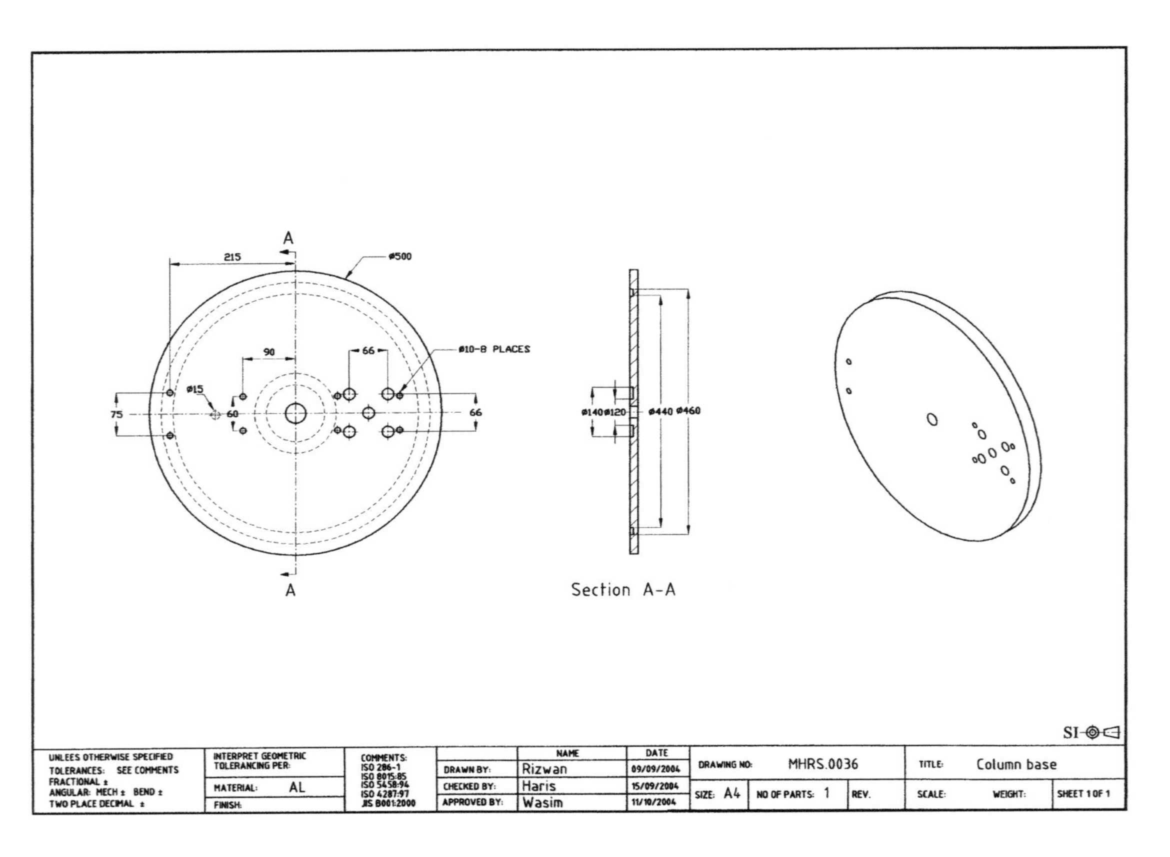

FIGURE 7.35 Column base.

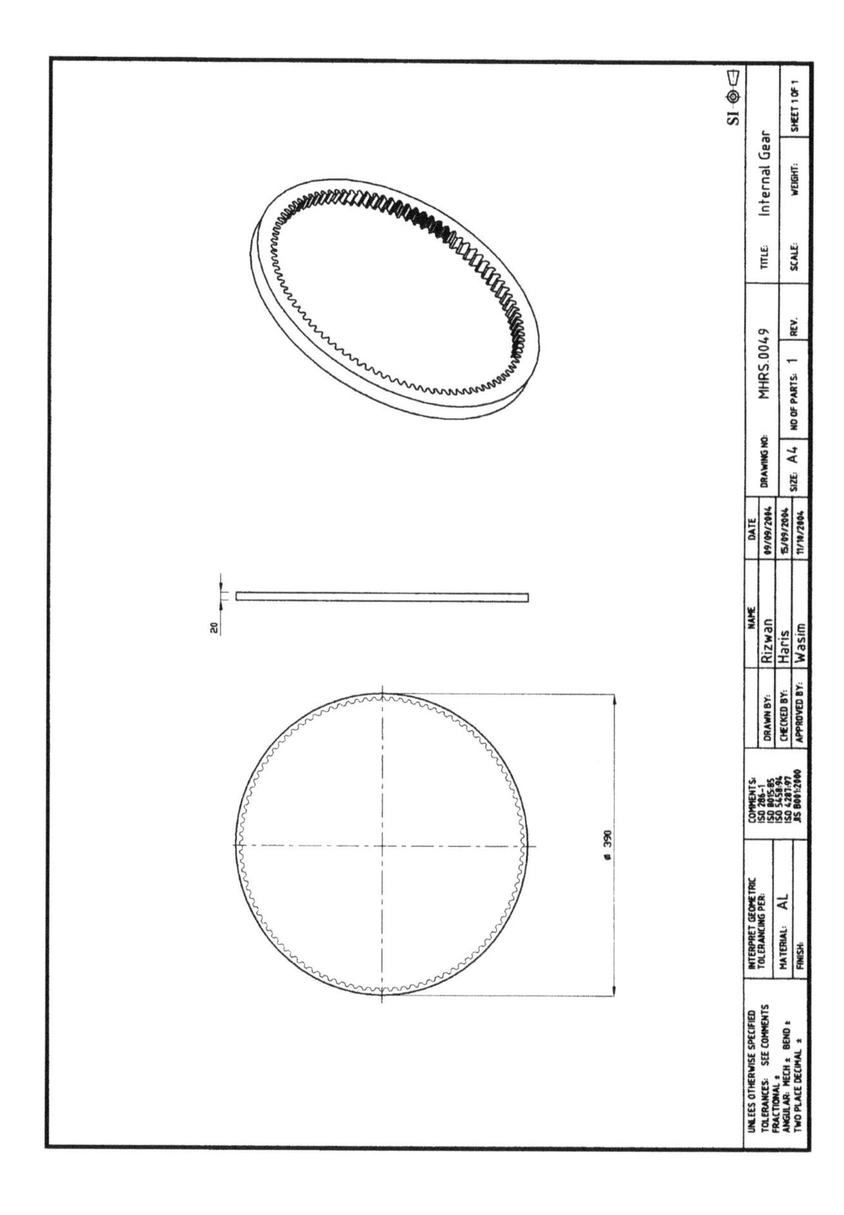

FIGURE 7.36 Internal gear for the base assembly.

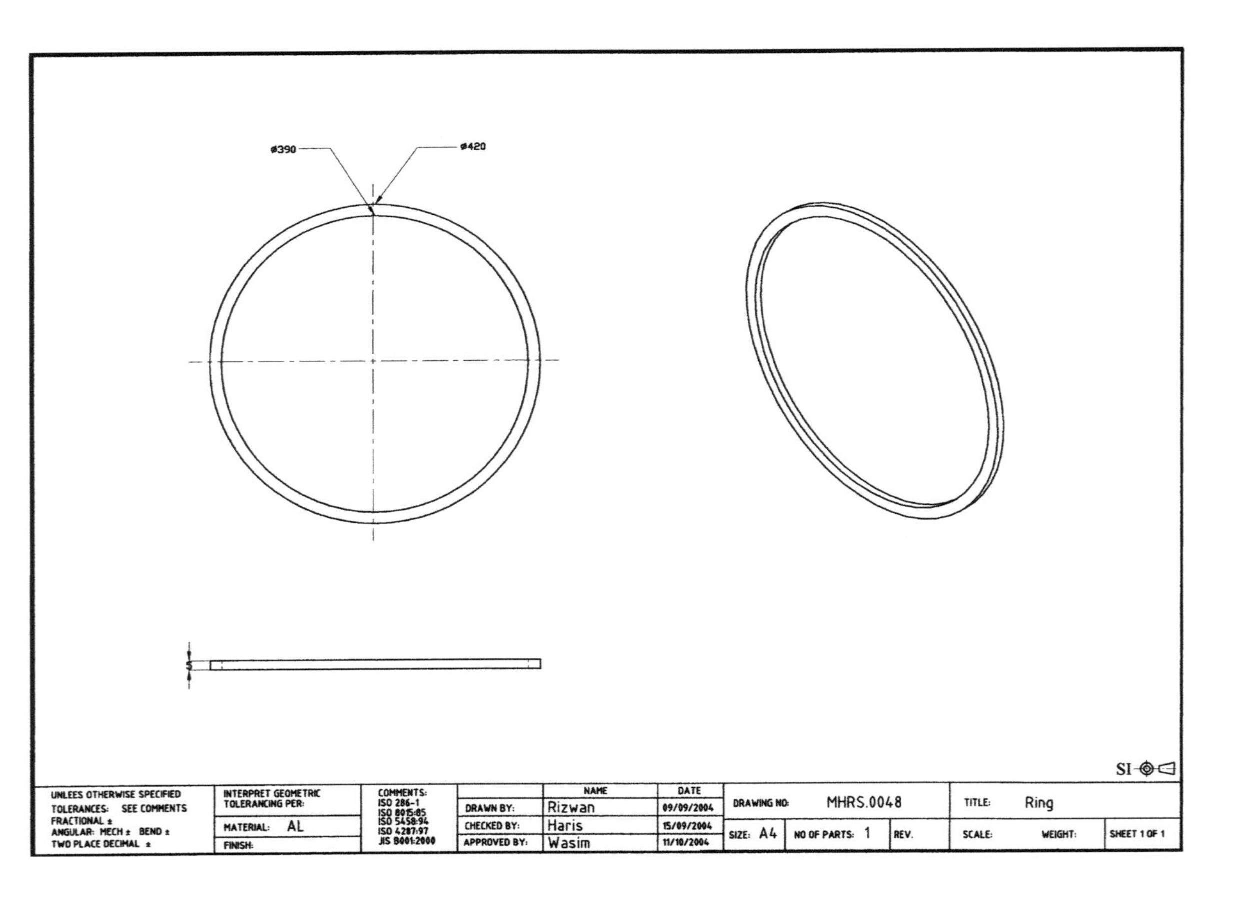

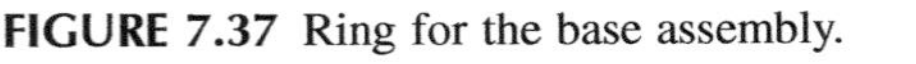

FIGURE 7.37 Ring for the base assembly.

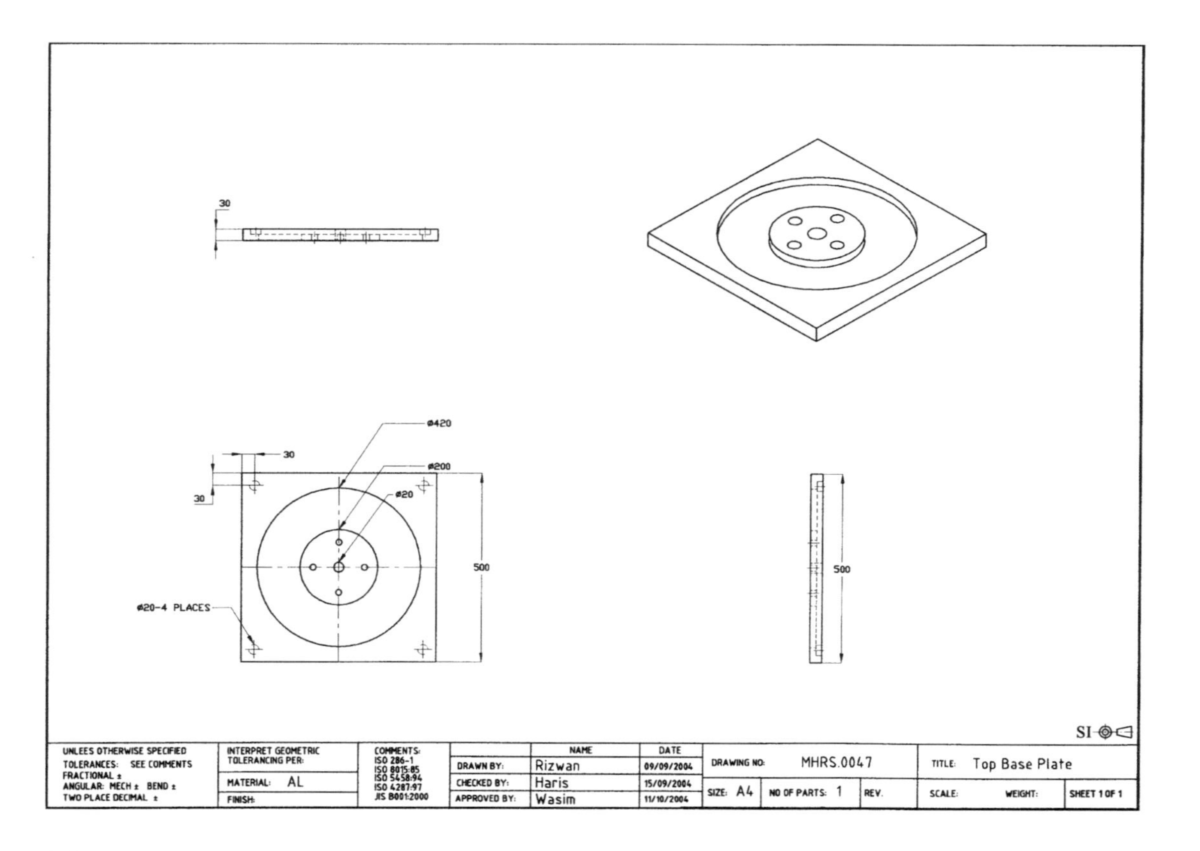

FIGURE 7.38 Top base plate.

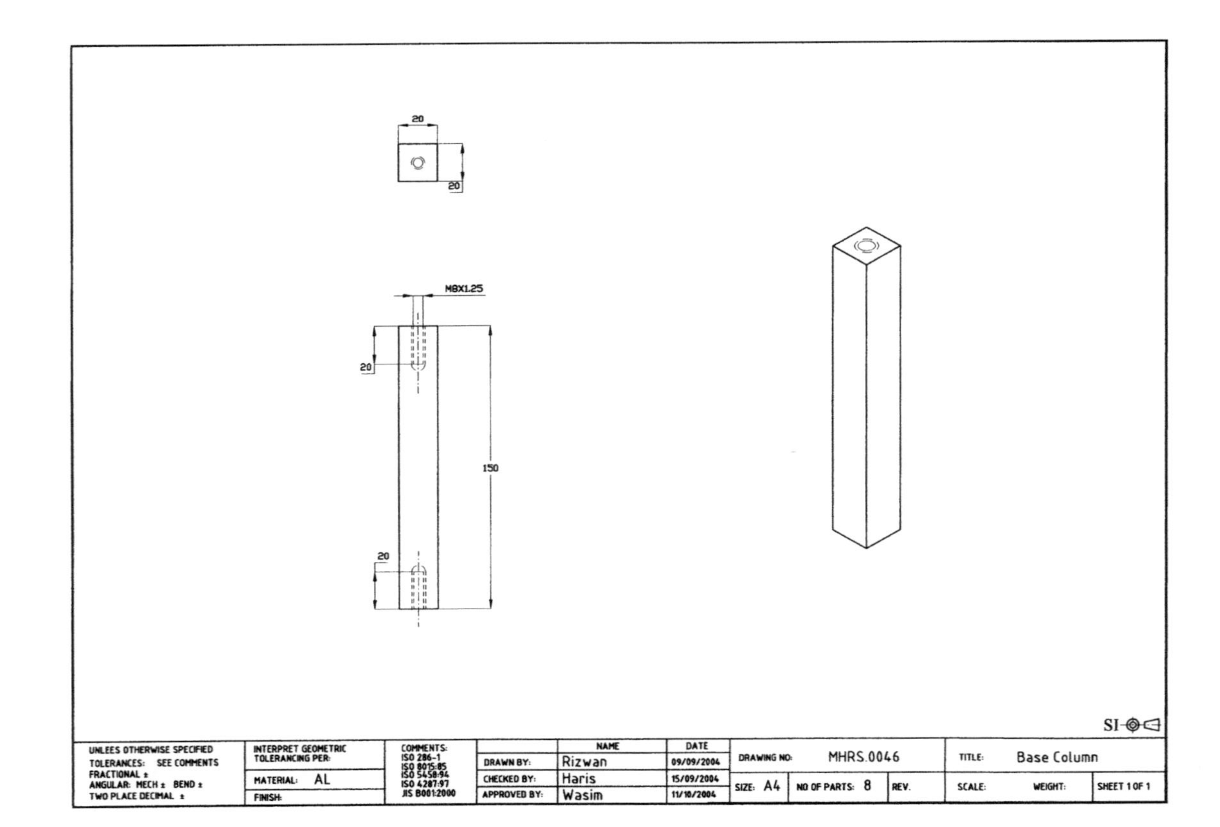

FIGURE 7.39 Base column.

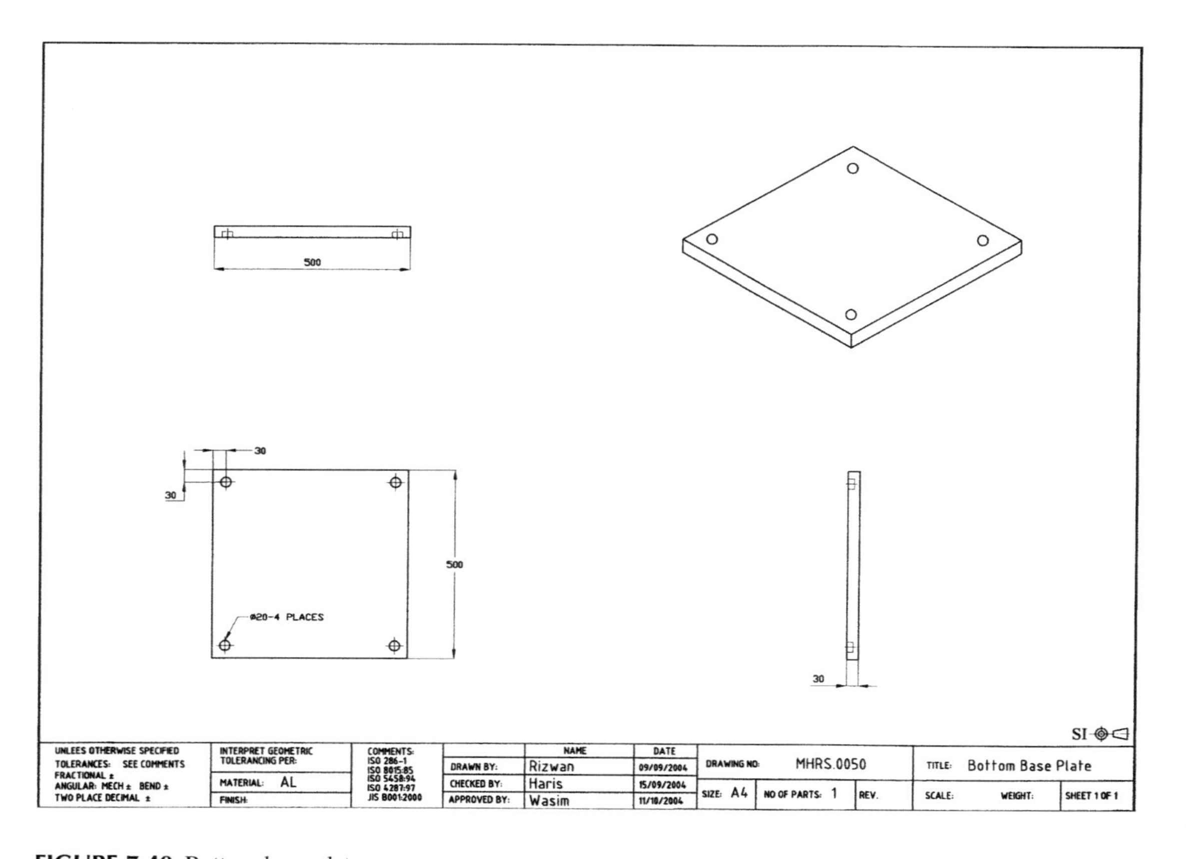

FIGURE 7.40 Bottom base plate.

TABLE 7.1
Procedure Sheet for Robot Finger

PROCEDURE SHEET					
SUB ASSEMBLY NAME	GRIPPER ASSY.	No. OF PART	2		
PART NAME	FINGER	PART No.	1		
DRAWING No.	MHRS.0021	MATERIAL	Al		

S.No.	OPERATION NAME	SKETCH OF OPERATION	TOOL	TOOL CLASSIFICATION	MACHINES
1	RAW MATERIAL SQUARE 20x20X113L	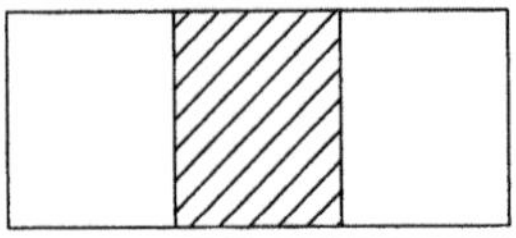			
2	END MILLING BOTH SIDE A & B	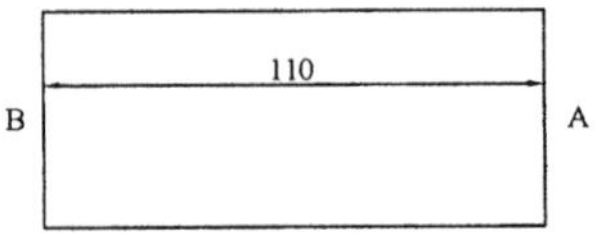 B 110 A	ENDMILL	ø 20mm BS122/4	VERTICAL MILLING MACHINE
3	ENDMILL LEFT SIDE30mm LONG AND 7.5mm DEPTH	 7.5	ENDMILL	ø 15mm BS122/4	VERTICAL MILLING MACHINE
4	ENDMILL THE OPPOSITE SIDE AS 3	 7.5 7.5	ENDMILL	ø 15mm BS122/4	VERTICAL MILLING MACHINE
5	ENDMILL AT 7.5mm DEPTH	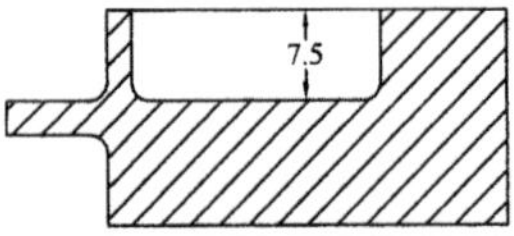 7.5	ENDMILL	ø 5mm BS122/4	VERTICAL MILLING MACHINE
6	ENDMILL THE OPPOSITE SIDE	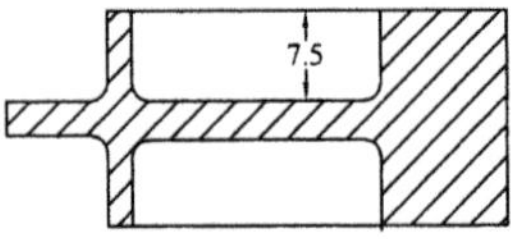 7.5	ENDMILL	ø 5mm BS122/4	VERTICAL MILLING MACHINE
7	FINISH THE CONTOUR	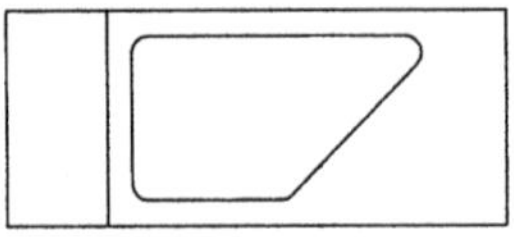	ENDMILL	ø 5mm BS122/4	VERTICAL MILLING MACHINE
8	DO THE OPPOSITE SIDE	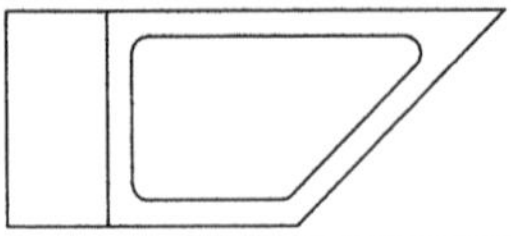	ENDMILL	ø 15mm BS122/4	VERTICAL MILLING MACHINE

(Continued)

TABLE 7.1 (Continued)
Procedure Sheet for Robot Finger

PROCEDURE SHEET

SUB ASSEMBLY NAME GRIPPER ASSY. No. OF PART 2
PART NAME FINGER PART No. 1
DRAWING No. MHRS.0021 MATERIAL Al

S.No.	OPERATION NAME	SKETCH OF OPERATION	TOOL	TOOL CLASSIFICATION	MACHINES
9	DRILLING ø5mm 2 HOLES	Ø5mm Ø5mm	TWIST DRILL	ø 5mm DIN/1897 RN	DRILLING MACHINE
10	FITTING ROUND THE END OF THE FINGER		FILE	SANDVIK 6' MILL SMOOTH FILE	MANUAL

TABLE 7.2
Procedure Sheet for Bar A

PROCEDURE SHEET

SUB ASSEMBLY NAME	GRIPPER ASSY.	No. OF PART	2
PART NAME	BAR A	PART No.	3
DRAWING No.	MHRS.0023	MATERIAL	Al

S.No.	OPERATION NAME	SKETCH OF OPERATION	TOOL	TOOL CLASSIFICATION	MACHINES
1	RAW MATERIAL SQUARE 20x20x75mm	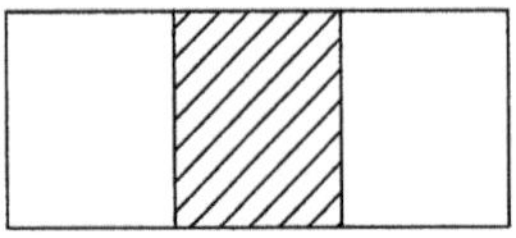			
2	ENDMILL BOTH SIDE A & B	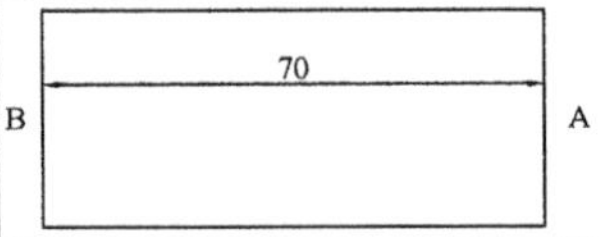 B 70 A	ENDMILL	ø 20mm BS122/4	MILLING MACHINE
3	ENDMILL SIDE TO GET 10mm THICKNESS	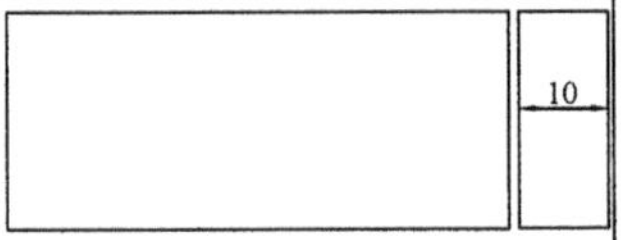 10	ENDMILL	ø 20mm BS122/4	MILLING MACHINE
4	ENDMILL THE GROOVE OF 5mm WIDTH & 35mm LONG	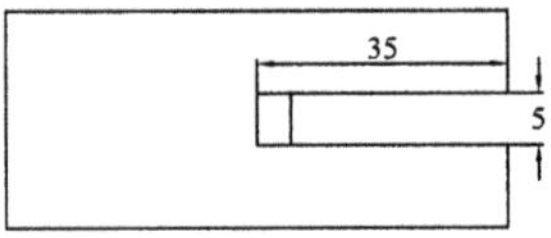 35 5	ENDMILL	ø 20mm BS122/4	MILLING MACHINE
5	ENDMILL THE GROOVE OF 10mm WIDTH & 25mm LONG	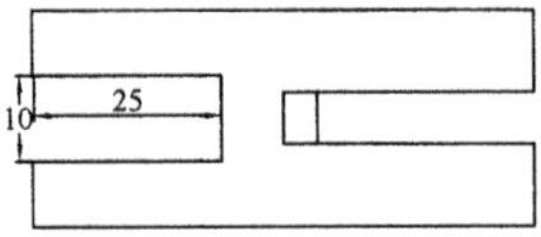 10 25	ENDMILL	ø 10mm BS122/4	MILLING MACHINE
6	DRILLING TWO HOLE ø5mm & ø2 mm DIAMETER	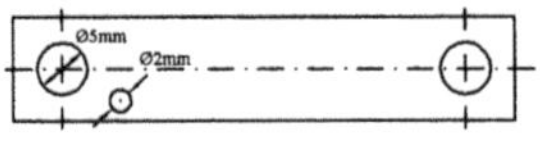 Ø5mm Ø2mm	TWIST DRILL	ø 5mm ø 2mm DIN1897 RN	DRILLING MACHINE
7	FITTING ROUND BOTH ENDS	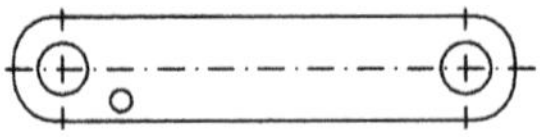	FILE	SANDVIK 6' MILL SMOOTH FILE	MANUAL

TABLE 7.3
Procedure Sheet for Bar B

PROCEDURE SHEET

SUB ASSEMBLY NAME	GRIPPER ASSY.	No. OF PART	2
PART NAME	BAR B	PART No.	2
DRAWING No.	MHRS.0022	MATERIAL	Al

S.No.	OPERATION NAME	SKETCH OF OPERATION	TOOL	TOOL CLASSIFICATION	MACHINES
1	RAW MATERIAL SQUARE 20x20X75mm LONG	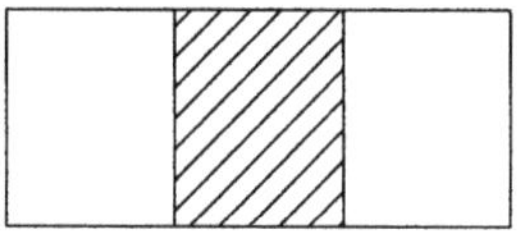			
2	ENDMILL BOTH SIDE A & B	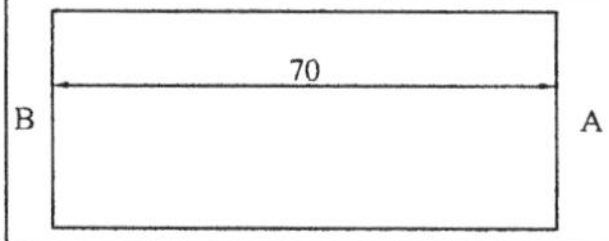 B 70 A	ENDMILL	ø 20mm BS122/4	MILLING MACHINE
3	ENDMILL ONE SIDE TO GET 10mm THICKNESS	 10	ENDMILL	ø 20mm BS122/4	MILLING MACHINE
4	ENDMILL GROOVE OF 5mm WIDTH & 35mm LONG	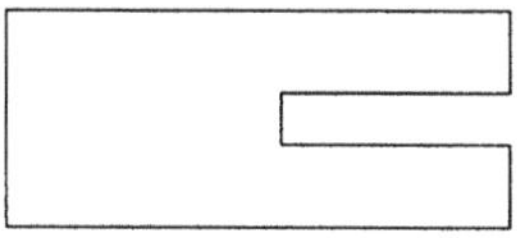	ENDMILL	ø 5mm BS122/4	MILLING MACHINE
5	DRILLING BOTH SIDE TO ø5 mm HOLES	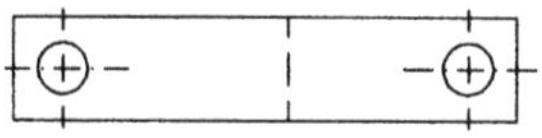	TWIST DRILL	ø 5mm DIN1897 RN	DRILLING MACHINE
7	FITTING ROUND BOTH SIDES PLUS THE END OF THE GROOVE	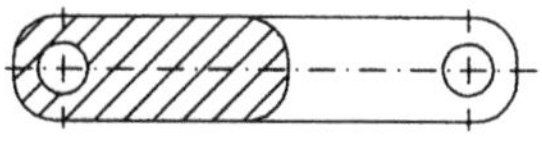	FILE	SANDVIK 6" MILL SMOOTH FILE	MANUAL

TABLE 7.4A
Procedure Sheet for Gripper Plate

PROCEDURE SHEET

SUB ASSEMBLY NAME	GRIPPER ASSY.	No. OF PART	2
PART NAME	GRIPPER PLATE	PART No.	4
DRAWING No.	MHRS.0025	MATERIAL	ST.

S.No.	OPERATION NAME	SKETCH OF OPERATION	TOOL	TOOL CLASSIFICATION	MACHINES
1	RAWMATERIAL,TUBE O.D ø36mm THICKNESS 2mm				
2	FACING A,B	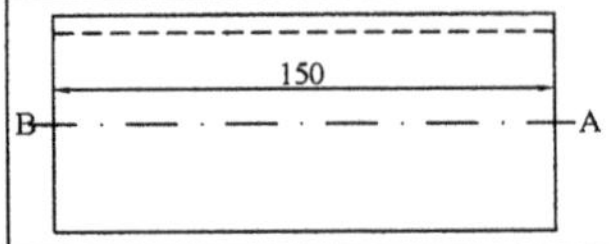 150; B; A	R.H.T.	SANDVIK T-MAX P	C.L.M.
3	CUT THETUBE INTO TWO HALF'S	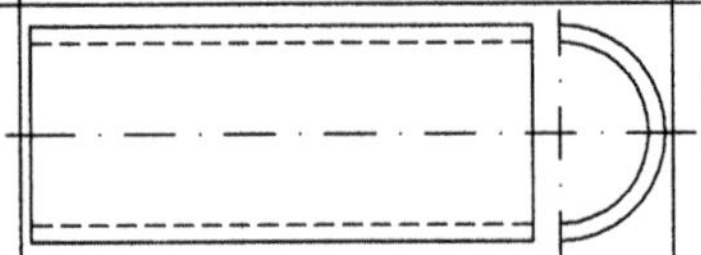	HACK SAW	SANDVIK BAHCO JUNIOR HACKSAW	MANUAL
4	MILL A 15mm BY 2mm GROOVE	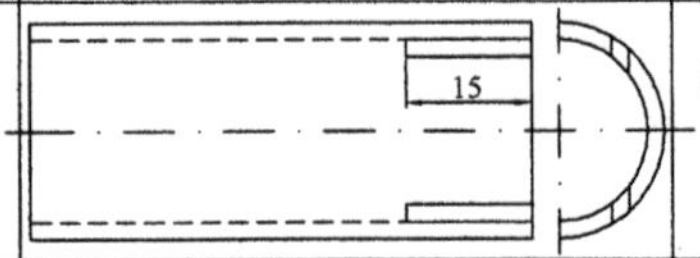 15	CUTTER	2 mm DIN1838	HORIZONTAL MILLING MACHINE
5	DRILLING AT ø 3mm AND ø5mm HOLES	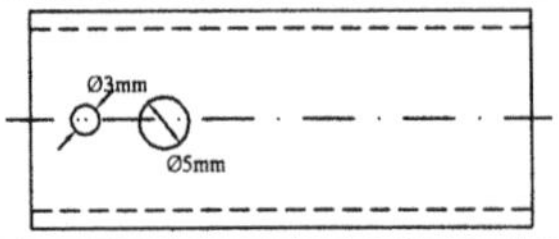 Ø3mm; Ø5mm	TWIST DRILL	ø 5mm ø 3mm DIN1897 RN	DRILLING MACHINE

C.L.M. CENTER LATHE MACHINE

(Continued)

TABLE 7.4B (Continued)
Procedure Sheet for Gripper Plate

PROCEDURE SHEET

SUB ASSEMBLY NAME GRIPPER ASSY. No. OF PART 2
PART NAME GRIPPER PLATE PART No. 4
DRAWING No. MHRS.0025 MATERIAL ST.

S.No.	OPERATION NAME	SKETCH OF OPERATION	TOOL	TOOL CLASSIFICATION	MACHINES
1	RAW MATERIAL SH. 2X110X70mm	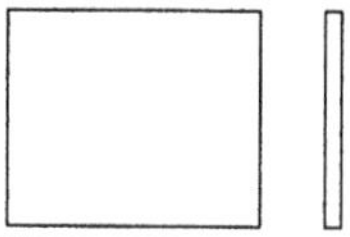			
2	MARKING	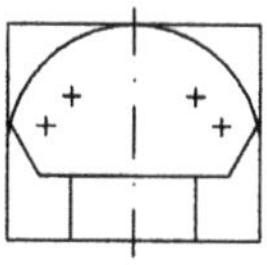	SCRIBOR, CENTER-PUNCH, RULER	SUTTON SINGLE POINT SCRIBER (MT007), SUTTON CENTER PUNCH (MT010), SUTTON 12'STAN-LESS STEEL RULER (MT035)	MANUAL
3	SHEAR OFF EXTRA MATERIAL	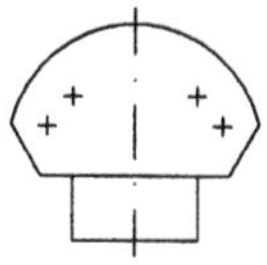	DIVIDER	SUTTON 6'MOORE &WRIGHT DIVIDER (CL007)	MANUAL
4	DRILLING 5mm	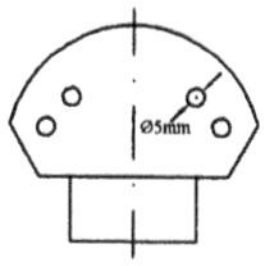 Ø5mm	TWIST DRILL	Ø 5mm DIN1897 RN	MILLING MACHINE
5	WELDING THE TWO PARTS	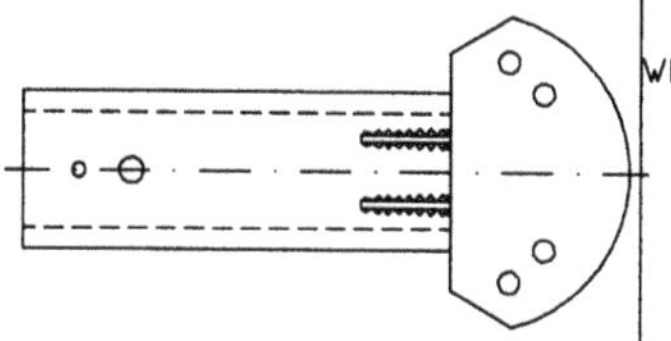	WELDING	ISO9692: 1992 ISO4063: 1990	WELDING

TABLE 7.5
Procedure Sheet for Profile B

PROCEDURE SHEET					
SUB ASSEMBLY NAME	GRIPPER ASSY.	No. OF PART	1		
PART NAME	PROFILE B	PART No.	7		
DRAWING No.	MHRS.0026	MATERIAL	Al		

S.No.	OPERATION NAME	SKETCH OF OPERATION	TOOL	TOOL CLASSIFICATION	MACHINES
1 2	FACING A FACING B	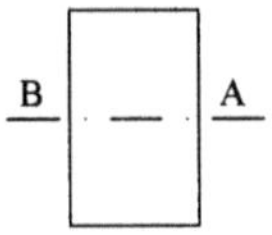 B A	R.H.T.	SANDVIK T-MAX P	C.L.M.
3	CENTRING	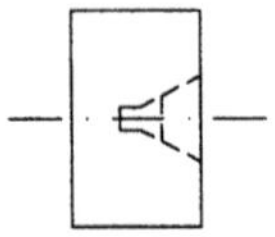	CENTER DRILL	Ø 2.5mm DIN333A	C.L.M.
4	DRILLING FOR HOLE Ø6mm	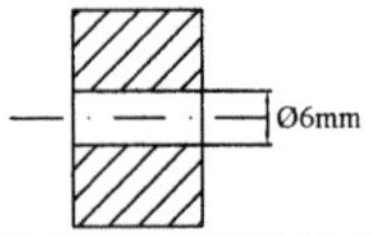 Ø6mm	TWIST DRILL	Ø 6mm DIN1897 RN	C.L.M.
5	EXTERNAL STRAIGHT TURNING FOR Ø26mm	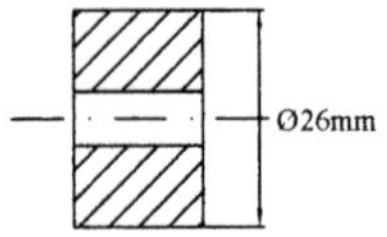 Ø26mm	R.H.T.	SANDVIK T-MAX P	C.L.M.
6	MILLING FOR (4) SLIDES	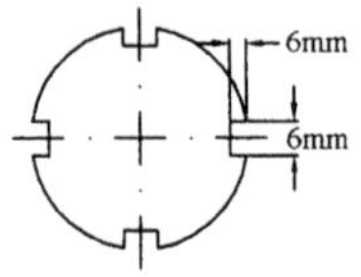 6mm 6mm	ENDMILL	Ø 6mm BS 122/4	VERTICAL MILLING MACHINE

C.L.M. CENTER LATHE MACHINE

TABLE 7.6
Procedure Sheet for Profile A

PROCEDURE SHEET

SUB ASSEMBLY NAME GRIPPER ASSY. No. OF PART 1
PART NAME PROFILE A PART No. 6
DRAWING No. MHRS.0029 MATERIAL Al

S.No.	OPERATION NAME	SKETCH OF OPERATION	TOOL	TOOL CLASSIFICATION	MACHINES
1 2	FACING A FACING B	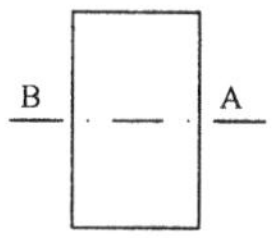 B A	R.H.T.	SANDVIK T-MAX P	C.L.M.
3	CENTRING	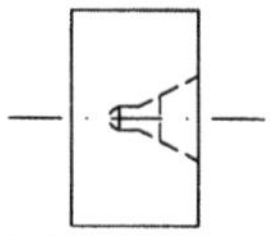	CENTER DRILL	ø 2.5mm DIN333A	C.L.M.
4	DRILLING HOLE ø15mm	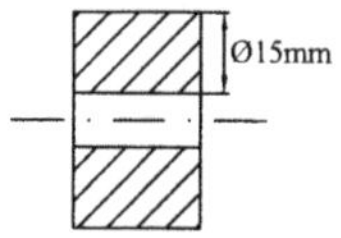 Ø15mm	TWIST DRILL	ø 15mm DIN1897 RN	C.L.M.
5	INTERNAL STRAIGHT TURNING FOR HOLE ø20mm (BORING)	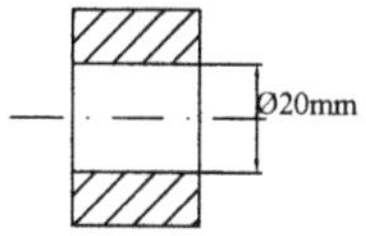 Ø20mm	BORING TOOL	SANDVIK CORОTURN 107	C.L.M.
6	EXTERNAL STRAIGHT TURNING FOR ø32mm	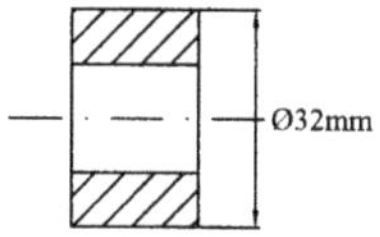 Ø32mm	R.H.T.	SANDVIK T-MAX P	C.L.M.
7	VERTICAL SHAPING (4) INTERNAL SLIDES.	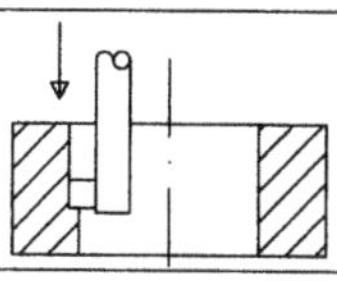	GROOVING TOOL	CUSTOM MADE TOOL FROM LEVICKS HYPEAK 202 HSS BAR	VERTICAL SHAPING MACHINE
8	DRILLING FOR (4) HOLE ø3mm	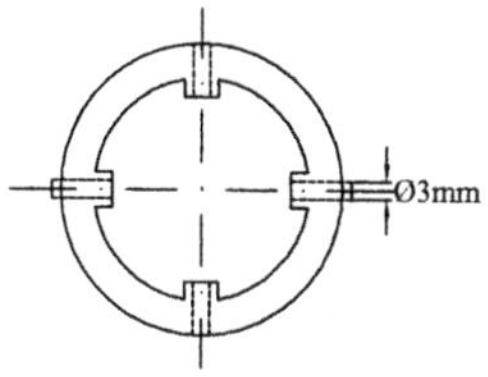 Ø3mm	TWIST DRILL	ø 3mm DIN1897 RN	DRILLING MACHINE

C.L.M. CENTER LATHE MACHINE

TABLE 7.7
Procedure Sheet for Clutch A

PROCEDURE SHEET					
SUB ASSEMBLY NAME: GRIPPER ASSY.			No. OF PART: 1		
PART NAME: CLUTCH A			PART No.: 5		
DRAWING No.: MHRS.0030			MATERIAL: Al		
S.No.	**OPERATION NAME**	**SKETCH OF OPERATION**	**TOOL**	**TOOL CLASSIFICATION**	**MACHINES**
1 2	FACING A FACING B	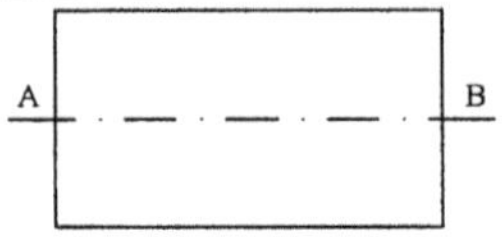 A B	R.H.T.	SANDVIK T-MAX P	C.L.M.
3	EXTERNAL STRAIGHT TURNING FOR OUTER DIA	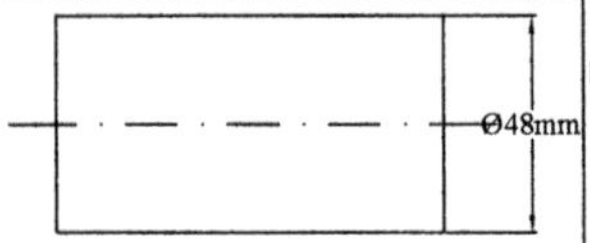 Ø48mm	R.H.T.	SANDVIK T-MAX P	C.L.M.
4	CENTRING	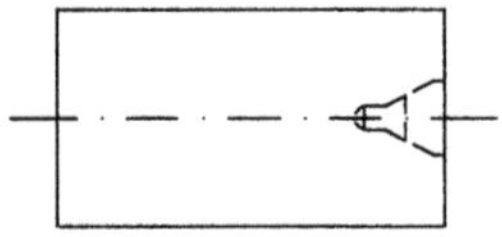	CENTER DRILL	ø 2.5mm DIN333A	C.L.M.
5	DRILLING ø15mm	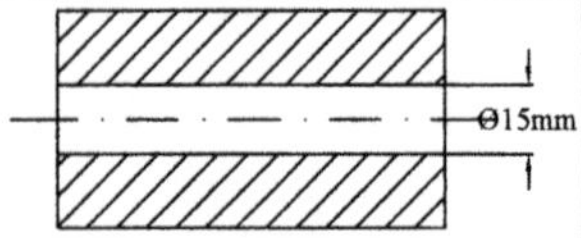 Ø15mm	TWIST DRILL	ø 15mm DIN1897 RN	C.L.M.
6	INTERNAL STRAIGHT TURNING (BORING) FOR ø36mm	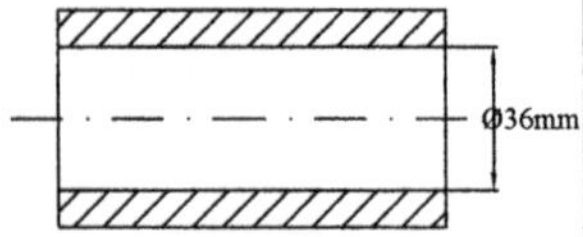 Ø36mm	BORING TOOL	SANDVIK COROTURN 107	C.L.M.
7	EXTERNAL GROOVING A	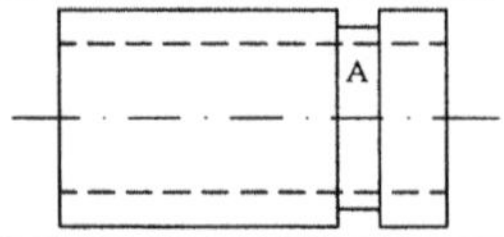 A	GROOVING TOOL	SANDVIK COROCUT 1	C.L.M.
8	EXTERNAL GROOVING B	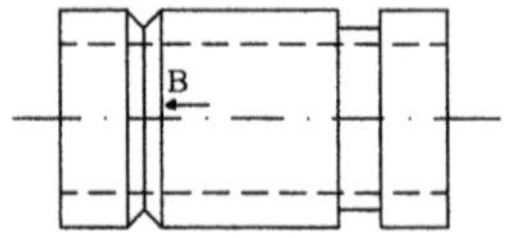 B	GROOVING TOOL	SANDVIK COROCUT 1	C.L.M.
9	DRILLING HOLE ø5mm, No. OF HOLES (4)	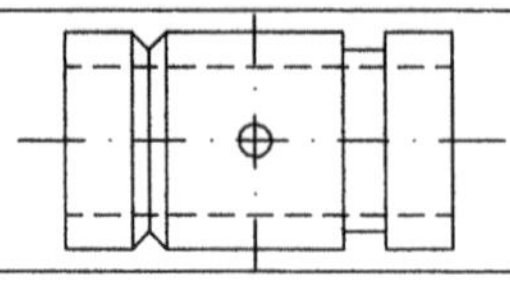	TWIST DRILL	ø 5mm DIN1897 RN	DRILLING MACHINE

C.L.M. CENTER LATHE MACHINE

TABLE 7.8A
Procedure Sheet for Clutch B

PROCEDURE SHEET

SUB ASSEMBLY NAME	GRIPPER ASSY.	No. OF PART	2
PART NAME	CLUTCH B	PART No.	8
DRAWING No.	MHRS.0031	MATERIAL	Al

S.No.	OPERATION NAME	SKETCH OF OPERATION	TOOL	TOOL CLASSIFICATION	MACHINES
1 2	FACING A FACING B	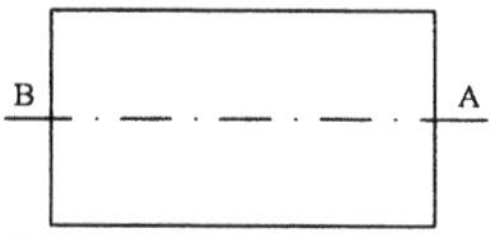 B A	R.H.T.	SANDVIK T-MAX P	C.L.M.
3	CENTRING	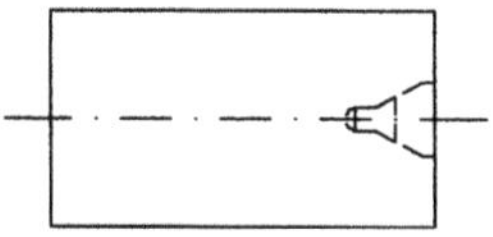	CENTER DRILL	ø 2.5mm DIN333A	C.L.M.
4	EXTERNAL STRAIGHT TURNING	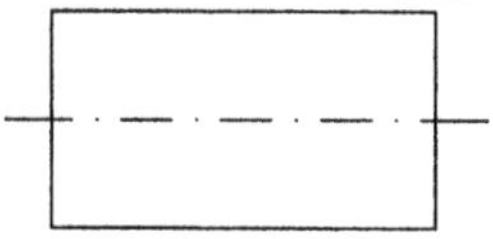	R.H.T.	SANDVIK T-MAX P	C.L.M.
5	DRILLING HOLE ø20mm	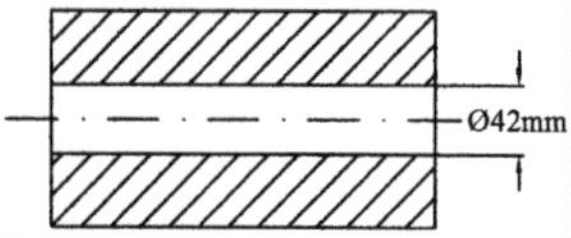 Ø42mm	TWIST DRILL	ø 20mm DIN1897 RN	C.L.M.
6	INTERNAL STRAIGHT TURNING (BORING) ø42mm	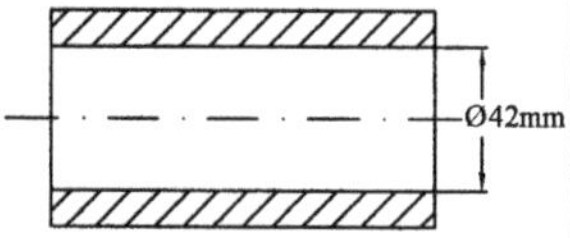 Ø42mm	BORING TOOL	SANDVIK COROTURN 107	C.L.M.
7	INTERNAL STRAIGHT TURNING (BORING) ø48mm	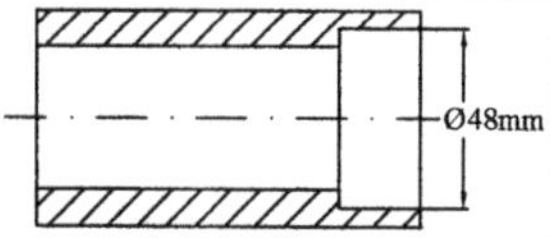 Ø48mm	BORING TOOL	SANDVIK COROTURN 107	C.L.M.
8	INTERNAL GROOVING	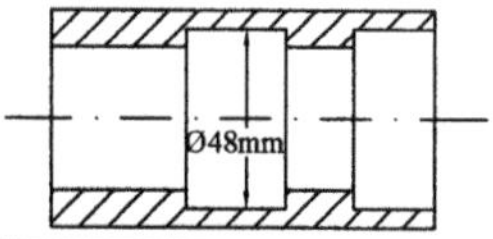 Ø48mm	GROOVING TOOL	SANDVIK COROTURN 107	C.L.M.
9	INTERNAL TAPERING	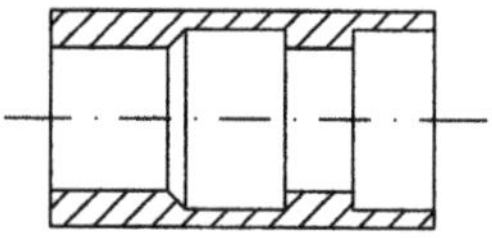	BORING TOOL	SANDVIK COROTURN 107	C.L.M.

C.L.M. CENTER LATHE MACHINE

(Continued)

TABLE 7.8B (continued)
Procedure Sheet for Clutch B

PROCEDURE SHEET

SUB ASSEMBLY NAME	GRIPPER ASSY.	No. OF PART	2
PART NAME	CLUTCH B	PART No.	8
DRAWING No.	MHRS.0031	MATERIAL	Al

S.No.	OPERATION NAME	SKETCH OF OPERATION	TOOL	TOOL CLASSIFICATION	MACHINES
10	INTERNAL STRAIGHT TURNING (BORING) ø48mm, ANOTHER SIDE	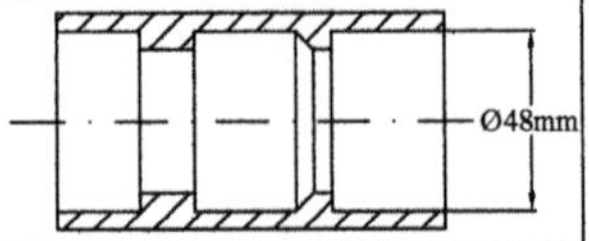 Ø48mm	BORING TOOL	SANDVIK COROTURN 107	C.L.M.
11	INTERNAL TAPERING	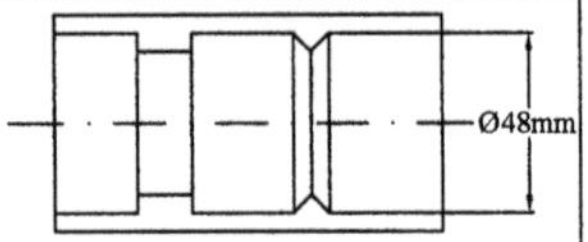 Ø48mm	BORING TOOL	SANDVIK COROTURN 107	C.L.M.
12	DRILLING FOR (4) HOLES ø5mm	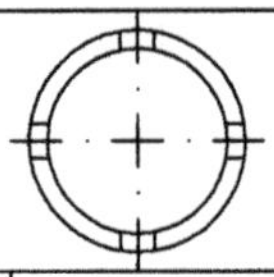	TWIST DRILL	ø 5mm DIN1897 RN	DRILLING MACHINE
13	CUTTING THE WORK PIECE TO TWO PARTS	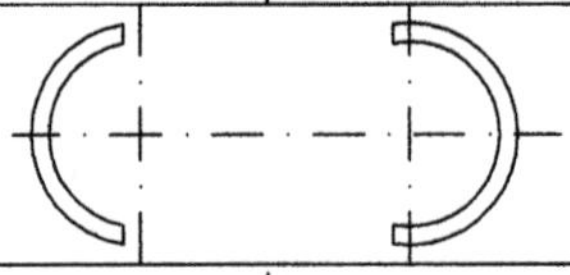	HACK SAW	SANDVIK BAHCO JUNIOR HACKSAW	MANUAL
14	REMOVE EXTRA MATERIAL AFTER PARTING OFF	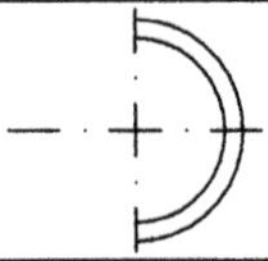	FILE	SANDVIK 6' MILL SMOOTH FILE	MANUAL

C.L.M. CENTER LATHE MACHINE

TABLE 7.9
Procedure Sheet for Rear Gripper Cylinder

PROCEDURE SHEET

SUB ASSEMBLY NAME GRIPPER ASSY. No. OF PART 1
PART NAME REAR GRIPPER CYLINDER PART No. 9
DRAWING No. MHRS.0024 MATERIAL ST

S.No.	OPERATION NAME	SKETCH OF OPERATION	TOOL	TOOL CLASSIFICATION	MACHINES
	RAW MATERIAL TUBE O.D. ø58mm THICKNESS 2mm				
1 2	FACING A FACING B		R.H.T.	SANDVIK T-MAX P	C.L.M.
3	MILLING	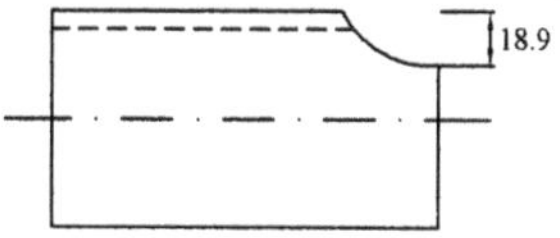 18.9	MILLING CUTTER	DIN1880	HORIZONTAL MILLING MACHINE
4	MILLING ANOTHER SIDE	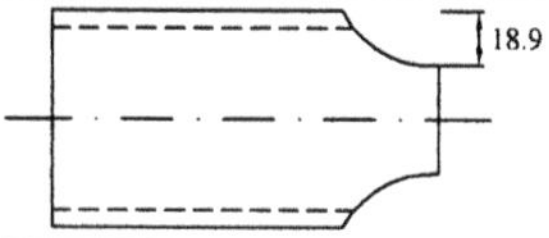 18.9	MILLING CUTTER	DIN1880	HORIZONTAL MILLING MACHINE
5	DRILLING HOLES ø10mm, ø3mm AND ø5mm	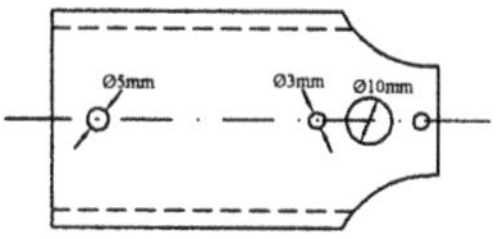 ø5mm ø3mm ø10mm	TWIST DRILL	ø 10mm ø 3mm DIN1897 RN	DRILLING MACHINE

C.L.M. CENTER LATHE MACHINE

TABLE 7.10
Procedure Sheet for Gripper Holder

PROCEDURE SHEET

SUB ASSEMBLY NAME	GRIPPER ASSY.	No. OF PART	2
PART NAME	GRIPPER HOLDER	PART No.	-
DRAWING No.	MHRS.0027	MATERIAL	Al

S.No.	OPERATION NAME	SKETCH OF OPERATION	TOOL	TOOL CLASSIFICATION	MACHINES
1 2	FACING A FACING B	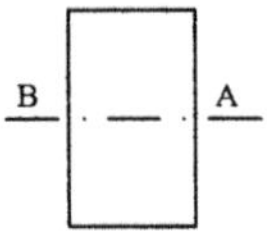 B A	R.H.T.	SANDVIK T-MAX P	C.L.M.
3	EXTERNAL STRAIGHT TURNING FOR DIA. ø23mm	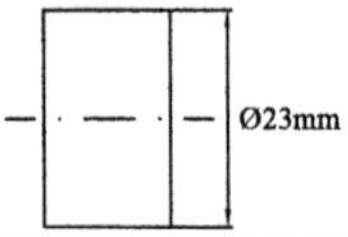 Ø23mm	R.H.T.	SANDVIK T-MAX P	C.L.M.
4	EXTERNAL STRAIGHT TURNING FOR DIA. ø13mm	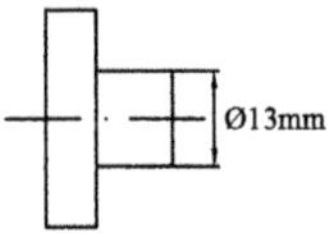 Ø13mm	R.H.T.	SANDVIK T-MAX P	C.L.M.
5	CENTRING	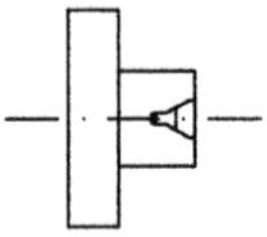	CENTER DRILL	ø 2.5mm DIN333A	C.L.M.
6	DRILLING 10ømm	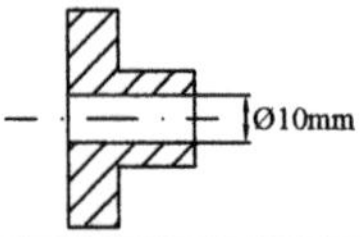 Ø10mm	TWIST DRILL	ø 10mm DIN1897 RN	DRILLING MACHINE
7	DRILLING (2) HOLES ø3mm AND (2) HOLE ø1mm	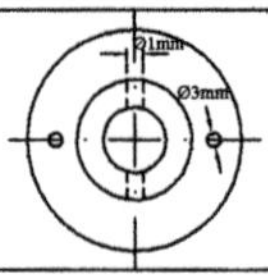 1mm Ø3mm	TWIST DRILL	ø 3mm ø 1mm DIN1897 RN	DRILLING MACHINE

C.L.M. CENTER LATHE MACHINE

TABLE 7.11
Procedure Sheet for Gripper Pulley

PROCEDURE SHEET

SUB ASSEMBLY NAME GRIPPER ASSY. No. OF PART 1
PART NAME GRIPPER PULLEY PART No. -
DRAWING No. MHRS.0028 MATERIAL Al

S.No.	OPERATION NAME	SKETCH OF OPERATION	TOOL	TOOL CLASSIFICATION	MACHINES
1 2	FACING A FACING B	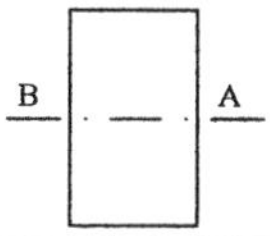 B A	R.H.T.	SANDVIK T-MAX P	C.L.M.
3	EXTERNAL STRAIGHT TURNING FOR DIA Ø95mm	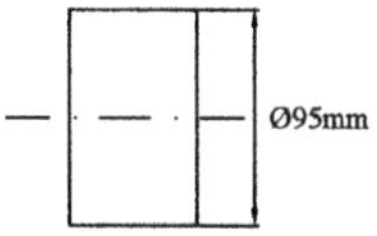 Ø95mm	R.H.T.	SANDVIK T-MAX P	C.L.M.
4	EXTERNAL STRAIGHT TURNING FOR DIA Ø15mm	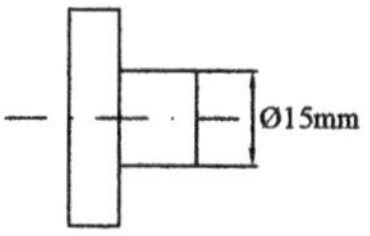 Ø15mm	R.H.T.	SANDVIK T-MAX P	C.L.M.
5	CENTRING	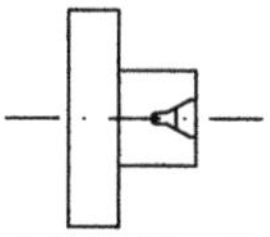	CENTER DRILL	Ø 2.5mm DIN333A	C.L.M.
6	DRILLING FOR HOLE Ø10mm	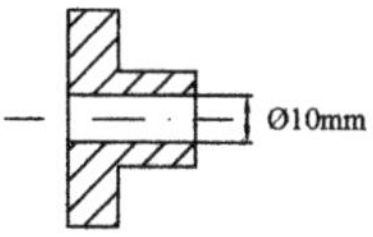 Ø10mm	TWIST DRILL	Ø 10mm DIN1897 RN	C.L.M.
7	DRILLING FOR (4) HOLES Ø25mm AND (4) HOLE Ø3mm	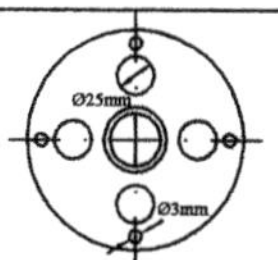 Ø25mm Ø3mm	TWIST DRILL	Ø 25mm Ø 3mm DIN1897 RN	DRILLING MACHINE
8	MILLING FOR SLIDES	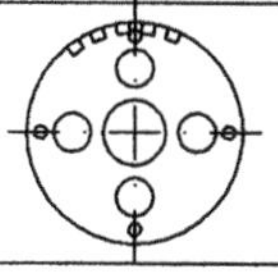	ENDMILL	BS 122/4	VERTICAL MILLING MACHINE

C.L.M. CENTER LATHE MACHINE

TABLE 7.12
Procedure Sheet for Front Arm Plate

PROCEDURE SHEET					
SUB ASSEMBLY NAME: FRONT ARM			No. OF PART: 2		
PART NAME: FRONT ARM PLATE			PART No.: 1		
DRAWING No.: MHRS.0015			MATERIAL: Al		
S.No.	**OPERATION NAME**	**SKETCH OF OPERATION**	**TOOL**	**TOOL CLASSIFICATION**	**MACHINES**
1	ALUMINUM SHEET 270x80X2mm				
2	MARKING	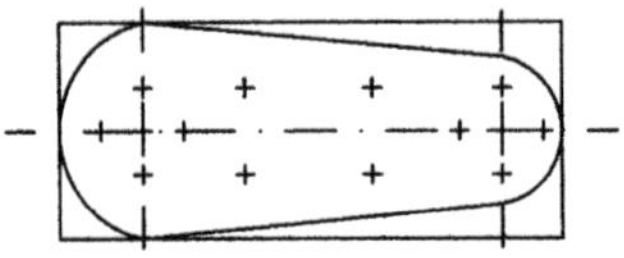	SCRIBER, CENTER, PUNCH, RULER, DIVIDER	SUTTON SINGLE POINT SCRIBER (MT007), SUTTON CENTER PUNCH (MT010), SUTTON12" STAINLESS STEEL RULER (MT035) SUTTON 6"MOORE &WRIGHT DIVIDER (CL007)	
3	SHEAR OFF ROUND SHAPE	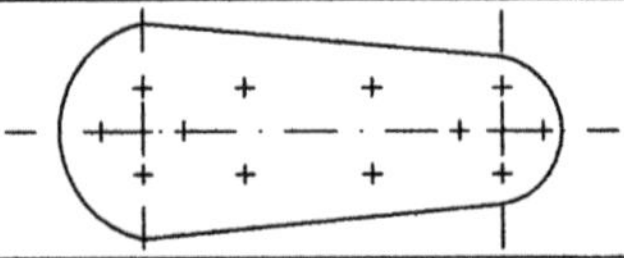	SHEAR	SHEAR TOOL	NIT MANUAL SHEARING MACHINE
4	DRILLING SMALL HOLES 4&5mm DIA	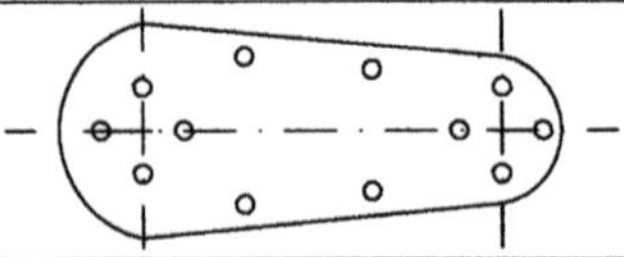	TWIST DRILL	ø 4mm ø 5mm DIN1897 RN	DRILLING MACHINE
5	DRILLING ENLARGE TO ø25mm	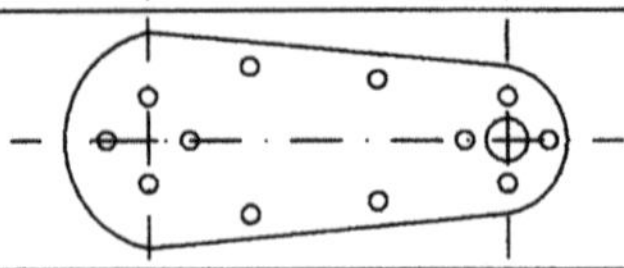	TWIST DRILL	ø 25mm DIN1897 RN	DRILLING MACHINE
6	DRILLING ENLARGE TO ø35mm	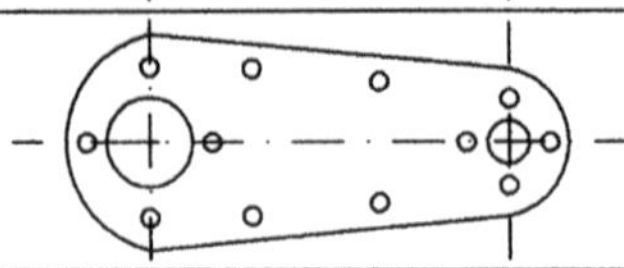	TWIST DRILL	ø 35mm DIN1897 RN	DRILLING MACHINE

TABLE 7.13
Procedure Sheet for Front Arm Support

PROCEDURE SHEET					
SUB ASSEMBLY NAME FRONT ARM			No. OF PART 2		
PART NAME FRONT ARM SUPPORT			PART No. 1		
DRAWING No. MHRS.0015			MATERIAL Al		
S.No.	OPERATION NAME	SKETCH OF OPERATION	TOOL	TOOL CLASSIFICATION	MACHINES
1	ALUMINUM SHEET 150x120X2mm	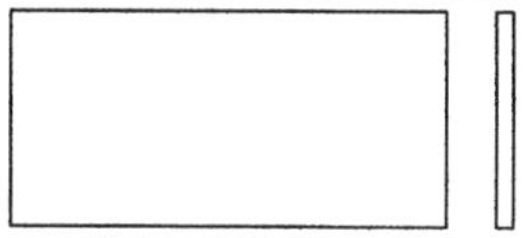			
2	MARKING	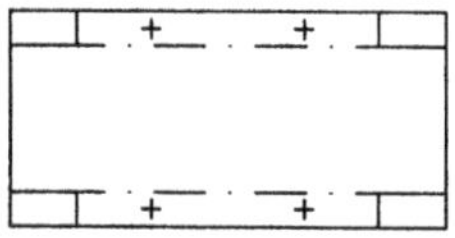	SCRIBER, CENTER PUNCH, RULER	SUTTON SINGLE POINT SCRIBER (MT007), SUTTON CENTER PUNCH (MT010), SUTTON 12"STAINLESS STEEL RULER (MT035)	MANUAL
3	SHEAR OFF	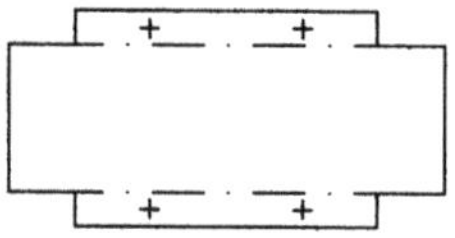	SHEAR	SHEAR TOOL	NIT MANUAL SHEARING MACHINE
4	DRILLING ø4mm	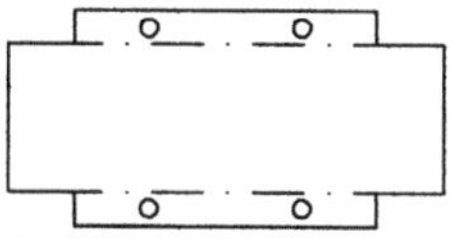	TWIST DRILL	ø 4mm DIN1897 RN	DRILLING MACHINE
5	BENDING				BENDING MACHINE

TABLE 7.14
Procedure Sheet for Front Arm Holder A

PROCEDURE SHEET

SUB ASSEMBLY NAME FRONT ARM No. OF PART 2
PART NAME FRONT ARM HOLDER A PART No. 4
DRAWING No. MHRS.0013 MATERIAL Al

S.No.	OPERATION NAME	SKETCH OF OPERATION	TOOL	TOOL CLASSIFICATION	MACHINES
1 2	FACING A FACING B	B A	R.H.T.	SANDVIK T-MAX P	C.L.M. C.L.M.
3	EXTERNAL STRAIGHT TURNING FOR ø25mm		R.H.T.	SANDVIK T-MAX P	C.L.M.
4	CENTRING		CENTER DRILL	ø 2.5mm DIN333A	C.L.M.
5	DRILLING HOLE ø10mm		TWIST DRILL	ø 10mm DIN1897 RN	C.L.M.
6	INTERNAL STRAIGHT TURNING (BORING) ø19mm		BORING TOOL	SANDVIK COROTURN 107	C.L.M.
7	DRILLING FOR HOLE ø5mm No. OF HOLES (4)		TWIST DRILL	ø 5mm DIN1897 RN	DRILLING MACHINE

TABLE 7.15
Procedure Sheet for Front Arm Holder B

PROCEDURE SHEET					
SUB ASSEMBLY NAME FRONT ARM			No. OF PART 2		
PART NAME FRONT ARM HOLDER B			PART No. 2		
DRAWING No. MHRS.0018			MATERIAL Al		
S.No.	OPERATION NAME	SKETCH OF OPERATION	TOOL	TOOL CLASSIFICATION	MACHINES
1 2	FACING A FACING B	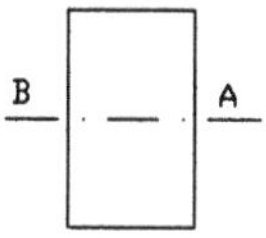 B A	R.H.T.	SANDVIK T-MAX P	C.L.M.
3	EXTERNAL STRAIGHT TURNING FOR ø35mm	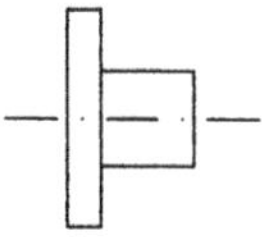	R.H.T.	SANDVIK T-MAX P	C.L.M.
4	CENTRING	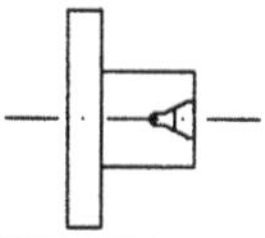	CENTER DRILL	ø 2.5mm DIN333A	C.L.M.
5	DRILLING HOLE ø15mm	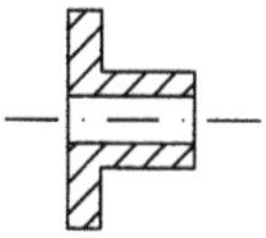	TWIST DRILL	15 ømm DIN1897 RN	C.L.M.
6	INTERNAL STRIGHT TURING (BORING) FOR ø25mm	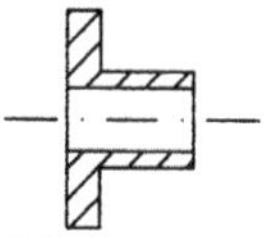	BORING TOOL	SANDVIK COROTURN 107	C.L.M.
7	DRILLING FOR HOLES ø5mm No. OFF HOLE (4)	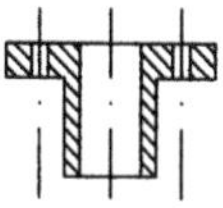	TWIST DRILL	ø 5mm DIN1897 RN	DRILLING MACHINE
8	VERTICAL SHAPING FOR (4) INTERNAL SLIDES	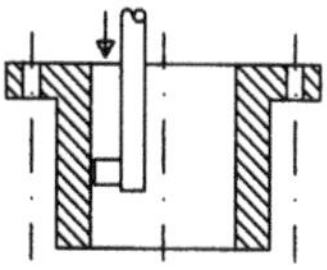	GROOVING TOOL	5mm CUSTOM MADE TOOL FROM LEVICKS HYPEAK 202 HSS BAR	VERTICAL SHAPING MACHINE

C.L.M. CENTER LATHE MACHINE

TABLE 7.16
Procedure Sheet for Front Arm Shaft

PROCEDURE SHEET

SUB ASSEMBLY NAME FRONT ARM No. OF PART 1
PART NAME FRONT ARM SHAFT PART No. 5
DRAWING No. MHRS.0017 MATERIAL ST. 37Ø 30mm

S.No.	OPERATION NAME	SKETCH OF OPERATION	TOOL	TOOL CLASSIFICATION	MACHINES
1	FACING A FACING B	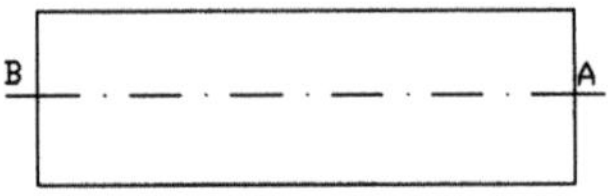 B A	R.H.T.	SANDVIK T-MAX P	C.L.M.
2	CENTERING A CENTERING B	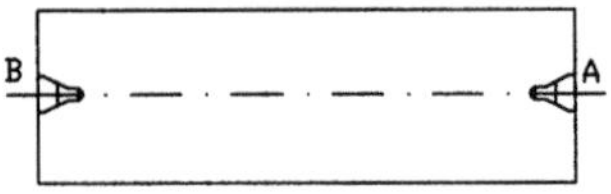 B A	CENTER DRILL	Ø 2.5mm DIN333A	C.L.M.
3	EXTERNAL STRAIGHT TURNING Ø25mm	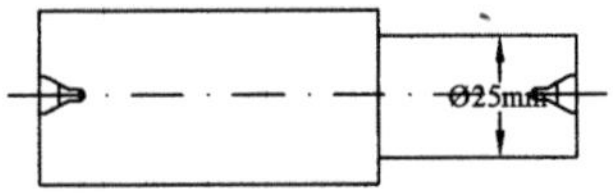 Ø25mm	R.H.T.	SANDVIK T-MAX P	C.L.M.
4	EXTERNAL STRAIGHT TURNING Ø20mm	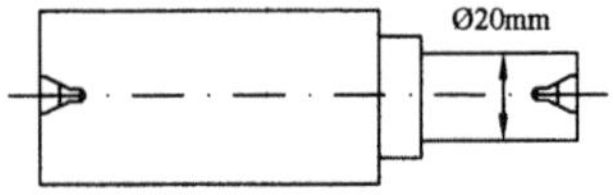 Ø20mm	R.H.T.	SANDVIK T-MAX P	C.L.M.
5	EXTERNAL STRAIGHT TURNING Ø25mm ANOTHER SIDE	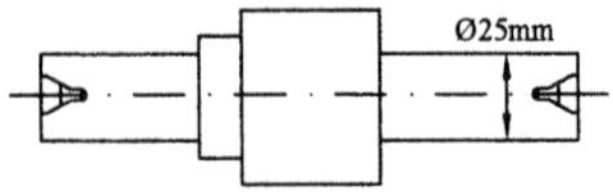 Ø25mm	R.H.T.	SANDVIK T-MAX P	C.L.M.
6	EXTERNAL STRAIGHT TURNING Ø20mm ANOTHER SIDE	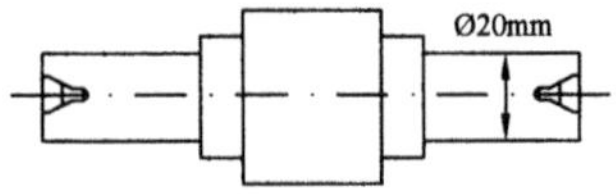 Ø20mm	R.H.T.	SANDVIK T-MAX P	C.L.M.
7	MAKE FORE EXTERNAL GROOVES AT Ø25mm, BOTH SIDES	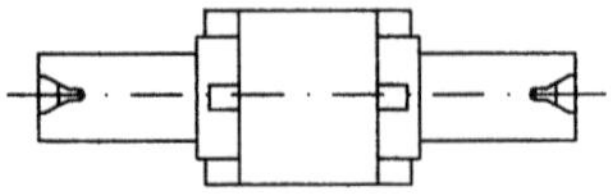	ENDMILL	Ø 5mm BS 112/4	VERTICAL MILLING MACHINE
8	MACHINE KEY WAY AT THE Ø20mm	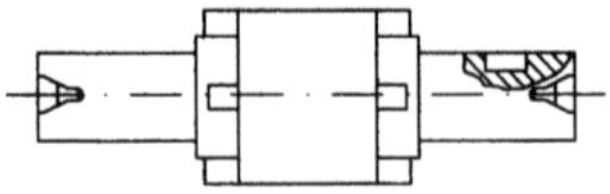	ENDMILL	Ø 6mm BS 112/4	VERTICAL MILLING MACHINE

C.L.M. CENTER LATHE MACHINE

TABLE 7.17
Procedure Sheet for Front Arm Pulley

PROCEDURE SHEET

SUB ASSEMBLY NAME	FRONT ARM	No. OF PART	1
PART NAME	FRONT ARM PULLEY	PART No.	9
DRAWING No.	MHRS.0016	MATERIAL	Al

S.No.	OPERATION NAME	SKETCH OF OPERATION	TOOL	TOOL CLASSIFICATION	MACHINES
1	FACING A FACING B	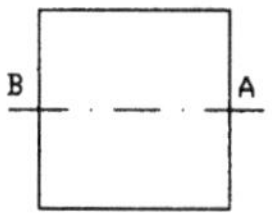 B A	R.H.T.	SANDVIK T-MAX P	C.L.M.
2	EXTERNAL STRAIGHT TURNING FOR ø30mm LONG ø15mm	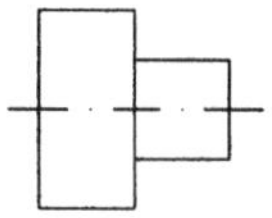	R.H.T.	SANDVIK T-MAX P	C.L.M.
3	CENTRING	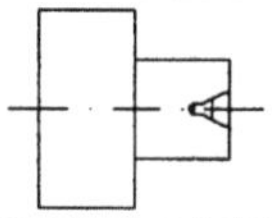	CENTER DRILL	ø 2.5mm DIN333A	C.L.M.
4	DRILLING FOR HOLE ø15mm	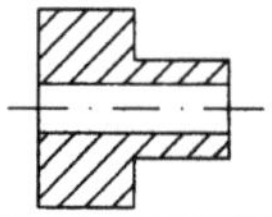	TWIST DRILL	ø 10mm DIN1897RN	C.L.M.
5	INTERNAL STRIGHT TURING FOR ø20mm (BORING)	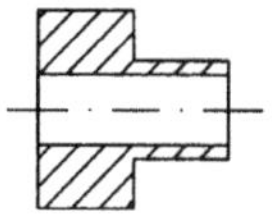	BORING TOOL	SANDVIK COROTURN 107	C.L.M.
6	DRILLING FOR HOLE ø10mm No. OF HOLE (4)	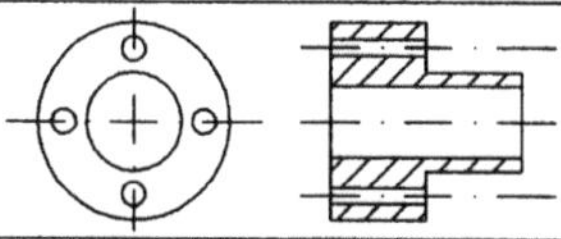	TWIST DRILL	ø 10mm DIN1897 RN	DRILLING MACHINE
7	ENLARGING TO ø20mm No. OF HOLE (4)	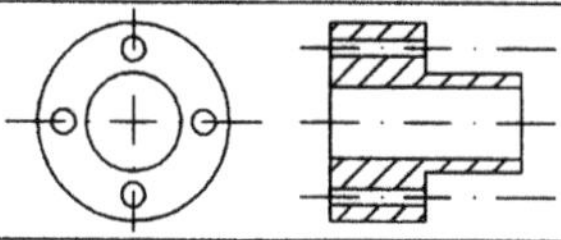	TWIST DRILL	ø 20mm DIN1897 RN	DRILLING MACHINE
8	VERTICAL SHAPING FOR KEY SLIDE	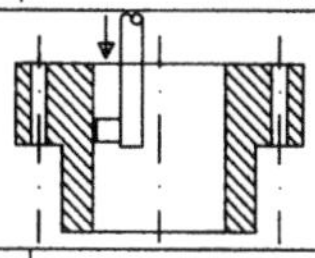	GROOVING TOOL	5mm CUSTOM MADETOOL FROM LEVICKS HYPEAK202 HSS BAR	VERTICAL SHAPING MACHINE
9	HORIZONTAL MILLING FOR TEETH	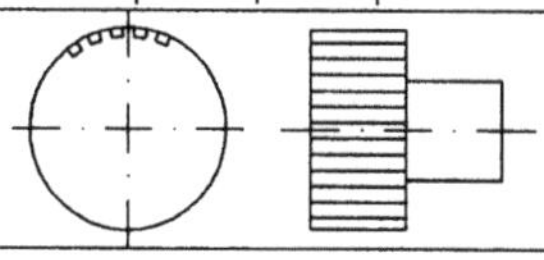	SIDE MILLING CUTTER	DIN885A	HORIZONTAL MILLING MACHINE

C.L.M. CENTER LATHE MACHINE

TABLE 7.18A
Procedure Sheet for Compound Pulley

PROCEDURE SHEET

SUB ASSEMBLY NAME	FRONT ARM	No. OF PART	1
PART NAME	COMPOUND PULLEY	PART No.	-
DRAWING No.	MHRS.0012	MATERIAL	Al

S.No.	OPERATION NAME	SKETCH OF OPERATION	TOOL	TOOL CLASSIFICATION	MACHINES
1 2	FACING A FACING B	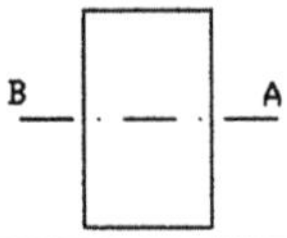 B A	R.H.T.	SANDVIK T-MAX P	C.L.M.
3	CENTRING	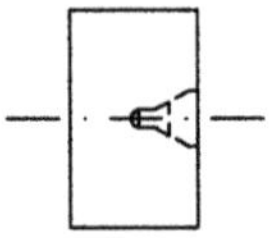	CENTER DRILL	ø 2.5mm DIN333A	C.L.M.
4	DRILLING HOLE ø15mm	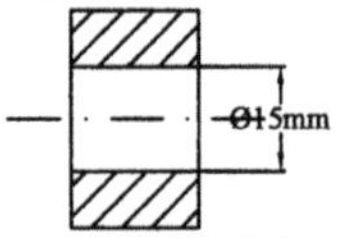 Ø15mm	TWIST DRILL	ø 15mm DIN1897 RN	C.L.M.
5	INTERNAL STRAIGHT TURING FOR ø30mm (BORING)	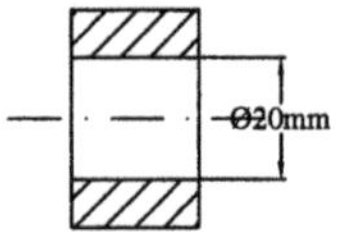 Ø20mm	BORING TOOL	SANDVIK COROTURN 107	C.L.M.
6	EXTERNAL STRAIGHT TURNING FOR OUTER DIAMETER	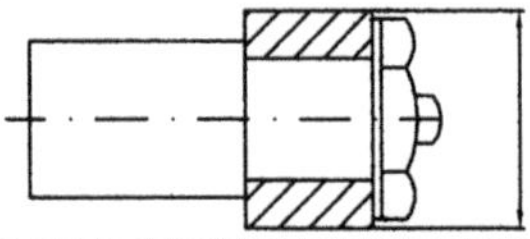	R.H.T.	SANDVIK T-MAX P	C.L.M.
7	DRILLING FOR HOLE ø3mm No. OF HOLE (4)	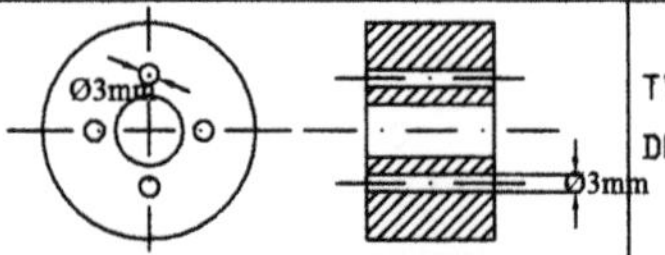 Ø3mm Ø3mm	TWIST DRILL	ø 3mm DIN1897 RN	DRILLING MACHINE
8	DRILLING FOR HOLE ø3mm No. OF HOLE (4)	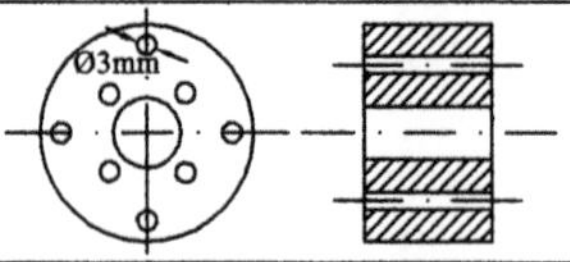 Ø3mm	TWIST DRILL	ø 3mm DIN1897 RN	DRILLING MACHINE
9	HORIZONTAL MILLING FOR TEETH	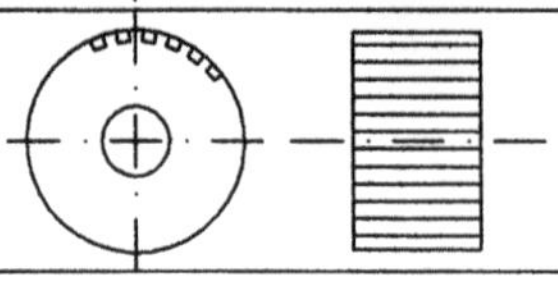	SIDE MILLING CUTTER	DIN885A	HORIZONTAL MILLING MACHINE

C.L.M. CENTER LATHE MACHINE

TABLE 7.18B (Continued)
Procedure Sheet for Compound Pulley

PROCEDURE SHEET

SUB ASSEMBLY NAME	FRONT ARM	No. OF PART	1
PART NAME	COMPOUND PULLEY	PART No.	-
DRAWING No.	MHRS.0012	MATERIAL	Al

S.No.	OPERATION NAME	SKETCH OF OPERATION	TOOL	TOOL CLASSIFICATION	MACHINES
1	FACING A FACING B	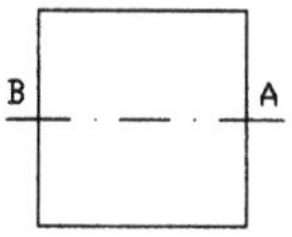 B A	R.H.T.	SANDVIK T-MAX P	C.L.M.
2	EXTERNAL STRAIGHT TURNING FOR ø30mm	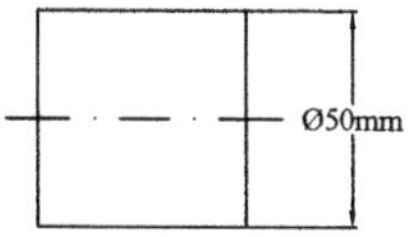 Ø50mm	R.H.T.	SANDVIK T-MAX P	C.L.M.
3	EXTERNAL STRAIGHT TURNING FOR ø30mm	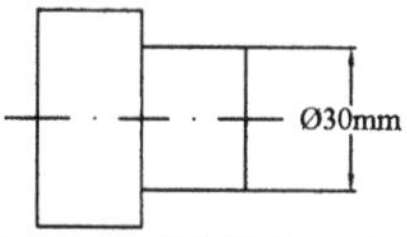 Ø30mm	R.H.T.	SANDVIK T-MAX P	C.L.M.
4	DRILLING FOR HOLE ø15mm	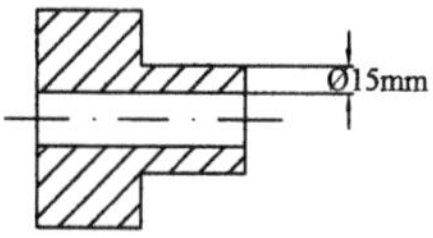 Ø15mm	TWIST DRILL	ø 15mm DIN1897 RN	C.L.M.
5	INTERNAL STRAIGHT TURING FOR ø20mm (BORING)	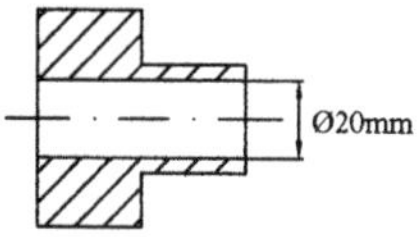 Ø20mm	BORING TOOL	SANDVIK COROTURN 107	C.L.M.
6	DRILLING FOR HOLE ø3mm No. OF HOLE (4)	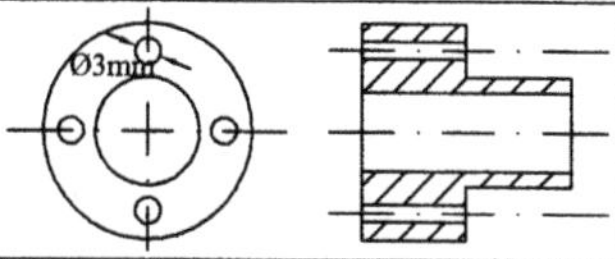 Ø3mm	TWIST DRILL	ø 3mm DIN1897 RN	DRILLING MACHINE
7	HORIZONTAL MILLING FOR TEETH	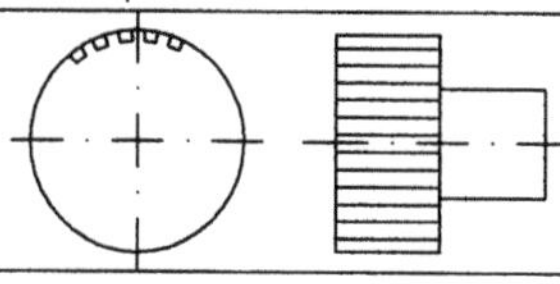	SIDE MILLING CUTTER	DIN885A	HORIZONTAL MILLING MACHINE

C.L.M. CENTER LATHE MACHINE

TABLE 7.19
Procedure Sheet for Back Arm

PROCEDURE SHEET

SUB ASSEMBLY NAME BACK ARM No. OF PART 2
PART NAME BACK ARM PLATE PART No. 1
DRAWING No. MHRS.0006 MATERIAL Al

S.No.	OPERATION NAME	SKETCH OF OPERATION	TOOL	TOOL CLASSIFICATION	MACHINES
1	RAW MATERIAL ALUMINUM SHEET 270x80X2mm				
2	MARKING	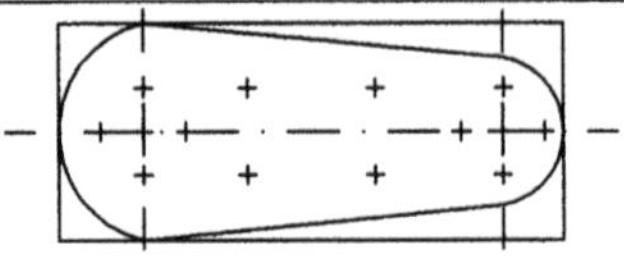	SCRIBER, CENTER, PUNCH, RULER, DIVIDER	SUTTON SINGLE POINT SCRIBER (MT007), SUTTON CENTER PUNCH (MT010), SUTTON12" STAINLE-SS STEEL RULER (MT035) SUTTON 6"MOORE &WRIGHT DIVIDER (CL007)	MANUAL
3	SHEAR OFF ROUND SHAPE	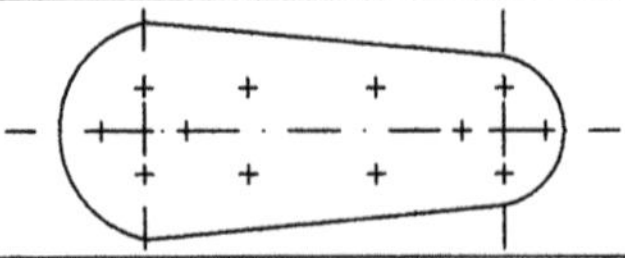	SHEAR	SHEAR TOOL	NIT MANUAL SHEARING MACHINE
4	DRILLING SMALL HOLES 4&5mm DIA	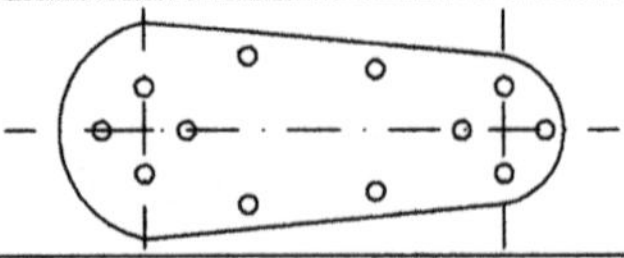	TWIST DRILL	ø 4mm ø 5mm DIN1897 RN	DRILLING MACHINE
5	DRILLING ENLARGE TO 25ø	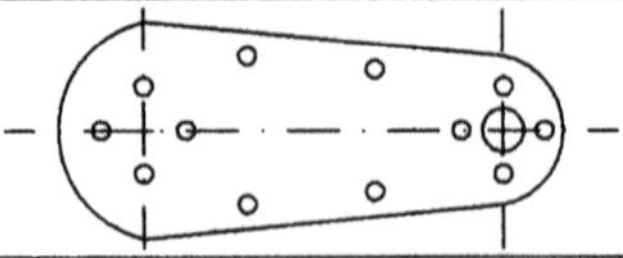	TWIST DRILL	ø 25mm DIN1897 RN	DRILLING MACHINE
6	DRILLING ENLARGE TO 35ø	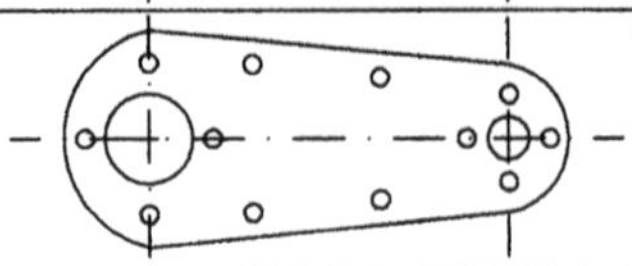	TWIST DRILL	ø 35mm DIN1897 RN	DRILLING MACHINE

TABLE 7.20
Procedure Sheet for Back Arm Support

PROCEDURE SHEET					
SUB ASSEMBLY NAME: BACK ARM	PART NAME: BACK ARM SUPPORT	DRAWING No.: MHRS.0009	No. OF PART: 2	PART No.: 3	MATERIAL: Al
S.No.	**OPERATION NAME**	**SKETCH OF OPERATION**	**TOOL**	**TOOL CLASSIFICATION**	**MACHINES**
1	RAW MATERIAL ALUMINUM SHEET 150x120X2mm	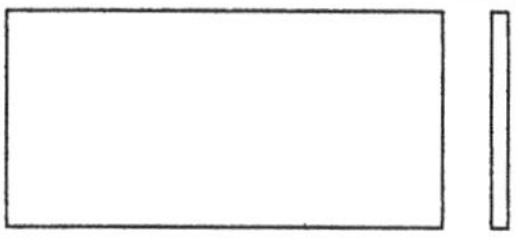			
2	MARKING	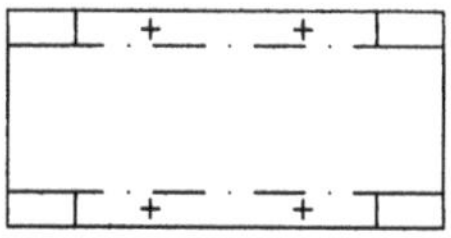	SCRIBER, CENTER, PUNCH, RULER,	SUTTON SINGLE POINT SCRIBER (MT007), SUTTON CENTER PUNCH (MT010), SUTTON 12"STAINLESS STEEL RULER (MT035)	MANUAL
3	SHEAR OFF	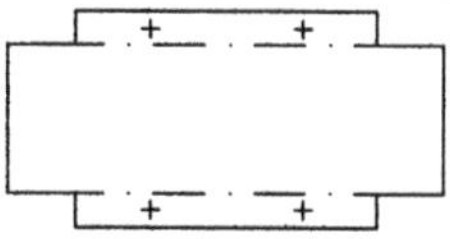	SHEAR	SHEAR TOOL	NIT MANUAL SHEARING MACHINE
4	DRILLING ø4mm	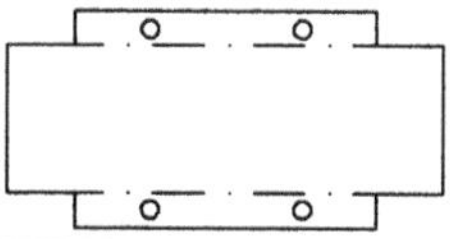	TWIST DRILL	ø 4mm DIN1897 RN	DRILLING MACHINE
5	BENDING	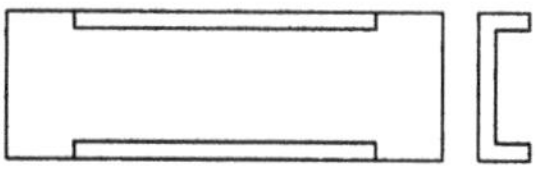	BENDING	BENDING	BENDING MACHINE

TABLE 7.21
Procedure Sheet for Back Arm Holder A

PROCEDURE SHEET					
SUB ASSEMBLY NAME: BACK ARM			No. OF PART: 2		
PART NAME: BACK ARM HOLDER A			PART No.: 4		
DRAWING No.: MHRS.0008			MATERIAL: Al		
S.No.	**OPERATION NAME**	**SKETCH OF OPERATION**	**TOOL**	**TOOL CLASSIFICATION**	**MACHINES**
1 2	FACING A FACING B	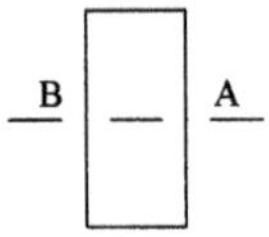 B A	R.H.T.	SANDVIK T-MAX P	C.L.M. C.L.M.
3	EXTERNAL STRAIGHT TURNING FOR ø25mm	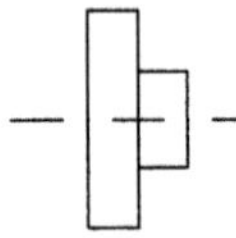	R.H.T.	SANDVIK T-MAX P	C.L.M.
4	CENTRING	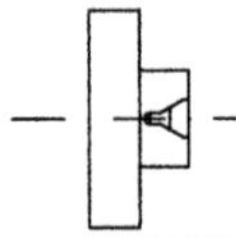	CENTER DRILL	ø 2.5mm DIN333A	C.L.M.
5	DRILLING FOR HOLE ø10mm	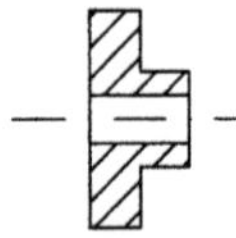	TWIST DRILL	ø 10mm DIN1897 RN	C.L.M.
6	INTERNAL STRAIGHT TURNING (BORING) FOR ø19mm	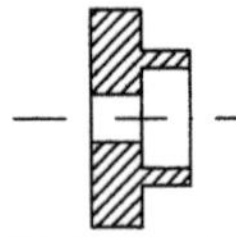	BORING TOOL	SANDVIK COROTURN 107	C.L.M.
7	DRILLING FOR HOLE ø5mm No. OF HOLES (4)	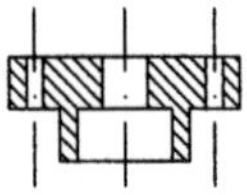	TWIST DRILL	ø 5mm DIN1897 RN	DRILLING MACHINE

C.L.M. CENTER LATHE MACHINE

TABLE 7.22
Procedure Sheet for Back Arm Holder B

PROCEDURE SHEET					
SUB ASSEMBLY NAME: BACK ARM			No. OF PART: 2		
PART NAME: BACK ARM HOLDER B			PART No.: 2		
DRAWING No.: MHRS.0002			MATERIAL: Al		
S.No.	**OPERATION NAME**	**SKETCH OF OPERATION**	**TOOL**	**TOOL CLASSIFICATION**	**MACHINES**
1 2	FACING A FACING B	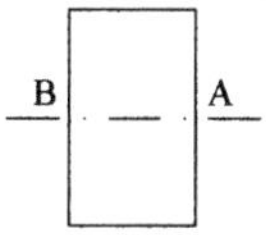 B A	R.H.T.	SANDVIK T-MAX P	C.L.M.
3	EXTERNAL STRAIGHT TURNING FOR ø35mm	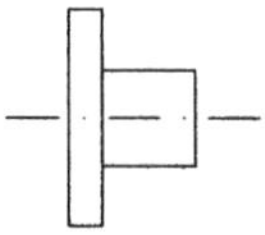	R.H.T.	SANDVIK T-MAX P	C.L.M.
4	CENTRING	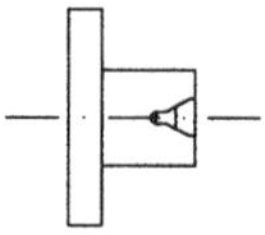	CENTER DRILL	ø 2.5mm DIN333A	C.L.M.
5	DRILLING HOLE ø15mm	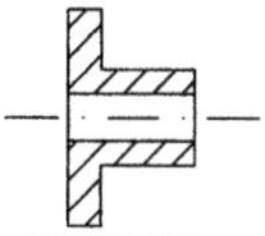	TWIST DRILL	ø 15mm DIN1897 RN	C.L.M.
6	INTERNAL STRIGHT TURING (BORING) FOR ø25mm	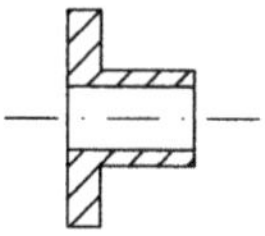	BORING TOOL	SANDVIK COROTURN 107	C.L.M.
7	DRILLING FOR HOLES ø5mm No. OFF HOLE (4)	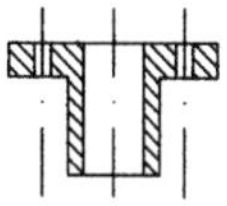	TWIST DRILL	ø 5mm DIN1897 RN	DRILLING MACHINE
8	VERTICAL SHAPING FOR (4) INTERNAL SLIDES	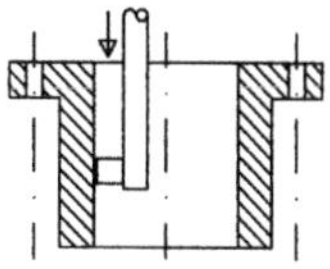	GROOVING TOOL	5mm CUSTOM MADETOOL FROM LEVICKS HYPEAK 202 HSS BAR	VERTICAL SHAPING MACHINE

C.L.M. CENTER LATHE MACHINE

TABLE 7.23
Procedure Sheet for Back Arm Shaft A

PROCEDURE SHEET

SUB ASSEMBLY NAME BACK ARM — No. OF PART 1
PART NAME BACK ARM SHAFT — PART No. 5
DRAWING No. MHRS.0004 — MATERIAL ST. 37

S.No.	OPERATION NAME	SKETCH OF OPERATION	TOOL	TOOL CLASSIFICATION	MACHINES
1	FACING A FACING B	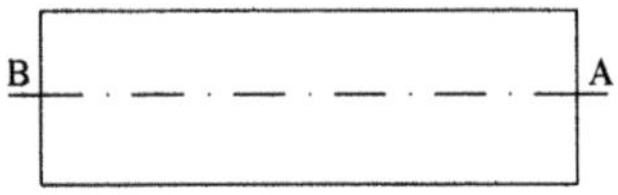 B A	R.H.T.	SANDVIK T-MAX P	C.L.M.
2	CENTERING A CENTERING B	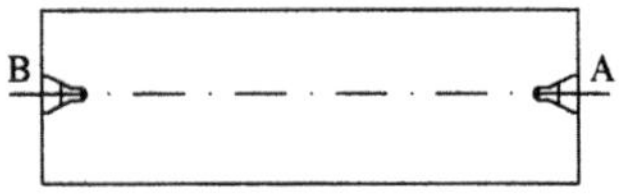 B A	CENTER DRILL	ø 2.5mm DIN333A	DRILLING MACHINE
3	EXTERNAL STRAIGHT TURNING ø20mm	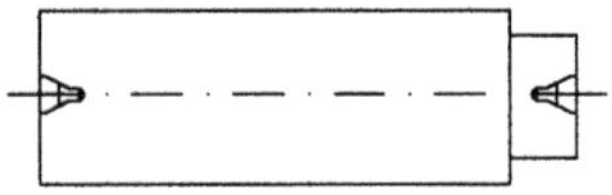	R.H.T.	SANDVIK T-MAX P	C.L.M.
4	EXTERNAL STRAIGHT TURNING ø20mm	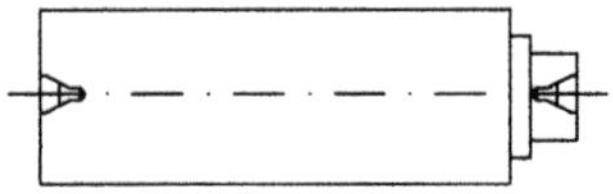	R.H.T.	SANDVIK T-MAX P	C.L.M.
5	EXTERNAL STRAIGHT TURNING ø25mm ANOTHER SIDE	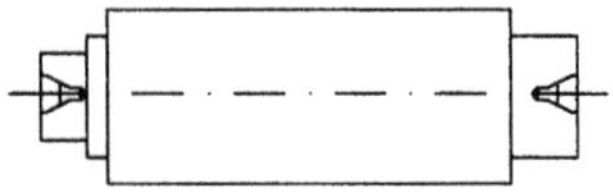	R.H.T.	SANDVIK T-MAX P	C.L.M.
6	EXTERNAL STRAIGHT TURNING ø20mm ANOTHER SIDE	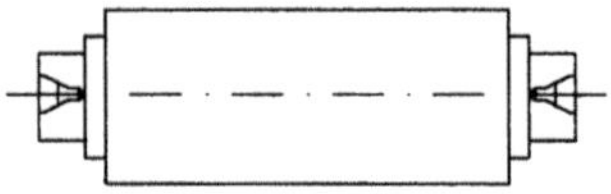	R.H.T.	SANDVIK T-MAX P	C.L.M.
7	MAKE FOUR EXTERNAL GROOVES AT THE ø25mm FOR BOTH SIDES	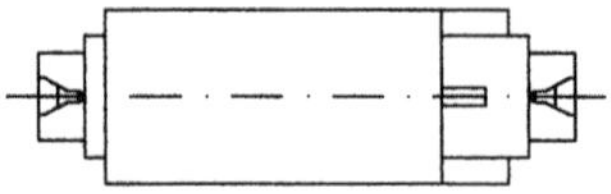	ENDMILL	ø 5mm BS 122/4	VERTICAL MILLING MACHINE
8	FOUR EXTERNAL GROOVES AT ø25mm BOTH SIDES	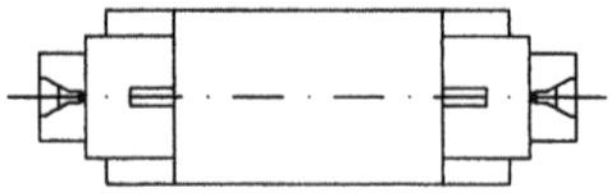	ENDMILL	ø 5mm BS 122/4	VERTICAL MILLING MACHINE

C.L.M. CENTER LATHE MACHINE

TABLE 7.24
Procedure Sheet for Back Arm Shaft B

PROCEDURE SHEET

SUB ASSEMBLY NAME	BACK ARM	No. OF PART	2
PART NAME	BACK ARM SHAFT B	PART No.	1
DRAWING No.	MHRS.0003	MATERIAL	ST 37

S.No.	OPERATION NAME	SKETCH OF OPERATION	TOOL	TOOL CLASSIFICATION	MACHINES
1	FACING A FACING B	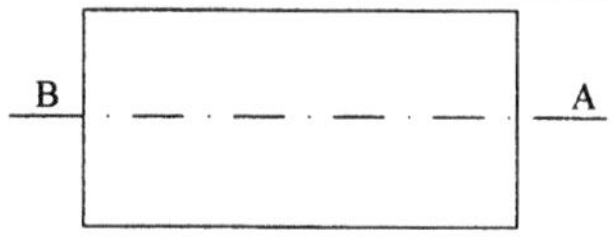 B A	R.H.T.	SANDVIK T-MAX P	C.L.M.
2	EXTERNAL STRAIGHT TURNING Ø15mm L=20mm	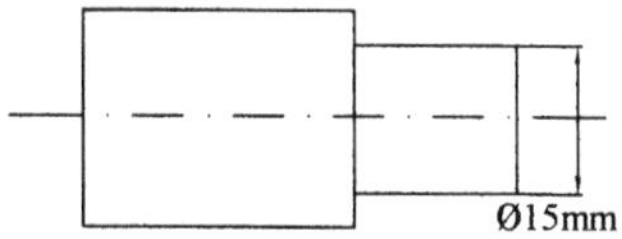 Ø15mm	R.H.T.	SANDVIK T-MAX P	C.L.M.
3	EXTERNAL STRAIGHT TURNING Ø15mm L=5mm	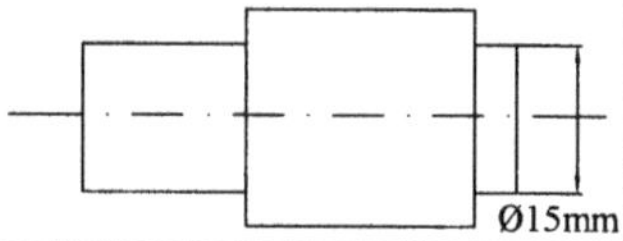 Ø15mm	R.H.T.	SANDVIK T-MAX P	C.L.M.
4	CHAMFERING A , B	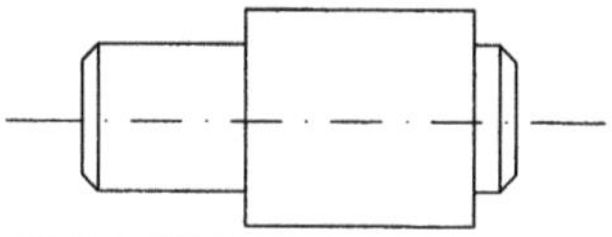	R.H.T.	SANDVIK T-MAX P	C.L.M.
5	KEY - WAY AT Ø20mm	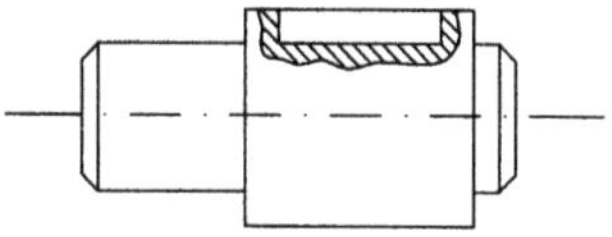	ENDMILL	Ø 4mm BS 122/4	VERTICAL MILLING MACHINE
6	KEY - WAY AT Ø15mm	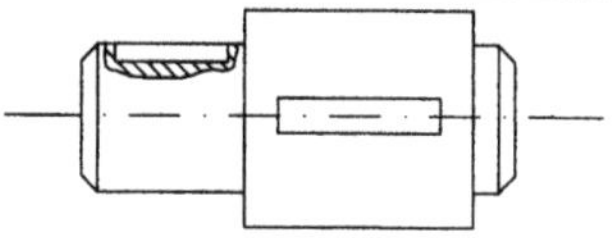	ENDMILL	Ø 4mm BS 122/4	VERTICAL MILLING MACHINE

C.L.M. CENTER LATHE MACHINE

TABLE 7.25
Procedure Sheet for Back Arm Pulley

PROCEDURE SHEET

SUB ASSEMBLY NAME	BACK ARM	No. OF PART	1
PART NAME	BACK ARM PULLEY	PART No.	9
DRAWING No.	MHRS.0010	MATERIAL	Al 95Ø mm

S.No.	OPERATION NAME	SKETCH OF OPERATION	TOOL	TOOL CLASSIFICATION	MACHINES
1	FACING A FACING B	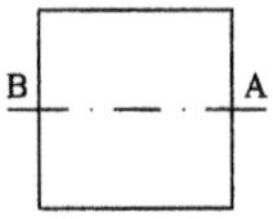 B A	R.H.T.	SANDVIK T-MAX P	C.L.M.
2	EXTERNAL STRAIGHT TURNING ø30mm LONG ø15mm	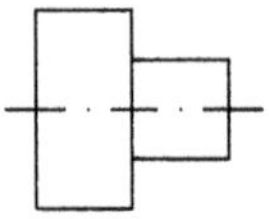	R.H.T.	SANDVIK T-MAX P	C.L.M.
3	CENTRING	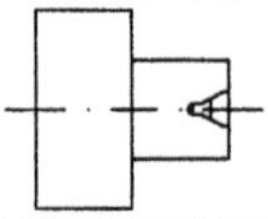	CENTER DRILL	ø 2.5mm DIN333A	C.L.M.
4	DRILLING HOLE ø15mm	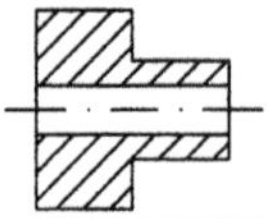	TWIST DRILL	ø 15mm DIN1897 RN	C.L.M.
5	INTERNAL STRIGHT TURNING ø20mm (BORING)	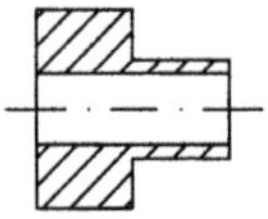	BORING TOOL	SANDVIK COROTURN 107	C.L.M.
6	DRILLING HOLE ø10mm No. OF HOLE (4)	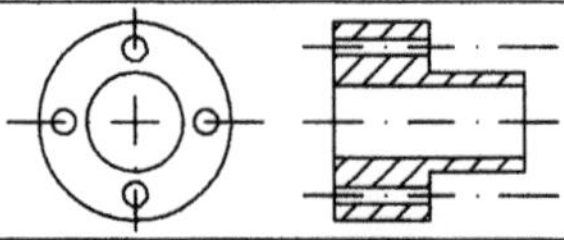	TWIST DRILL	ø 10mm DIN1897 RN	DRILLING MACHINE
7	ENLARGING TO ø20mm No. OF HOLE (4)	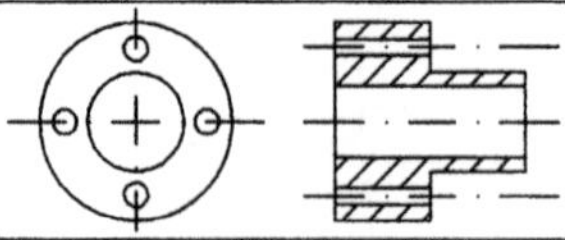	TWIST DRILL	ø 20mm DIN1897 RN	DRILLING MACHINE
8	VERTICAL SHAPING FOR KEY SLIDE	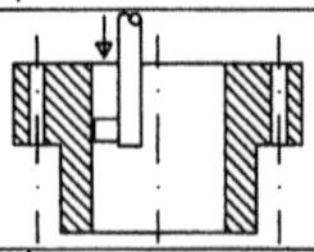	GROOVING TOOL	5mm CUSTOM MADETOOL FROM LEVICKS HYPEAK 202 HSS BAR	VERTICAL SHAPING MACHINE
9	HORIZONTAL MILLING FOR TEETH	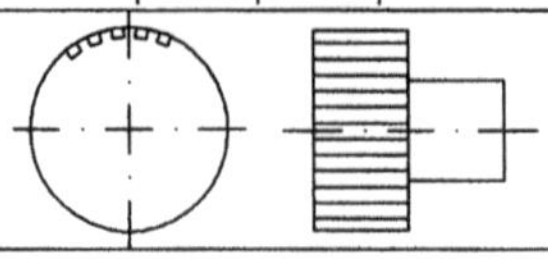	SIDE MILLING CUTTER	DIN885A	HORIZONTAL MILLING MACHINE

C.L.M. CENTER LATHE MACHINE

TABLE 7.26
Procedure Sheet for Pulley Flange

PROCEDURE SHEET

SUB ASSEMBLY NAME BACK ARM No. OF PART 2
PART NAME PULLEY FLANGE PART No. -
DRAWING No. MHRS.0002 MATERIAL Al

S.No.	OPERATION NAME	SKETCH OF OPERATION	TOOL	TOOL CLASSIFICATION	MACHINES
1	RAW MATERIAL Al SHEET 2x125X125mm	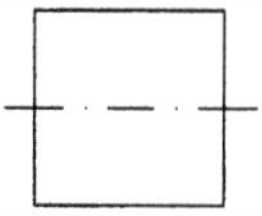			
2	MARKING		SCRIBER, CENTER, PUNCH, RULER,	SUTTON SINGLE POINT SCRIBER (MT007), SUTTON CENTER PUNCH (MT010), SUTTON 12"STAINLESS STEEL RULER (MT035)	MANUAL
3	SHEAR OFF	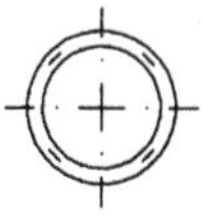	DIVIDER	SUTTON 6" MOORE & WRIGHT DIVIDER (CL007)	MANUAL
4	DRILLING ø10mm	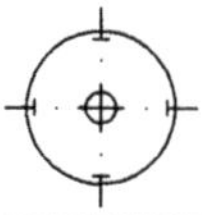	TWIST DRILL	ø 10mm DIN1897 RN	C.L.M.
5	EXTERNAL STRAIGHT TURNING OF OUTER DIAMETER	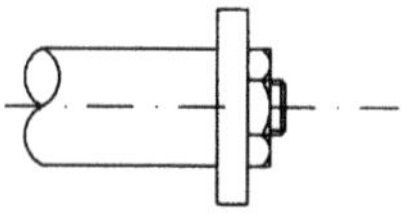	R.H.T.	SANDVIK T-MAX P	C.L.M.
6	CHAMFERING	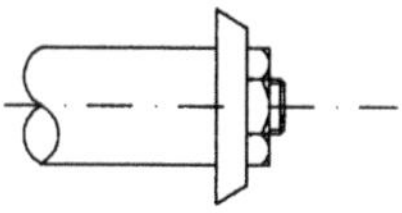	R.H.T.	SANDVIK T-MAX P	C.L.M.
7	DRILLING ø3mm No. OF HOLE (4)	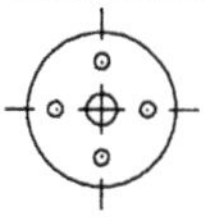	TWIST DRILL	ø3mm DIN1897 RN	DRILLING MACHINE
8	DRILLING	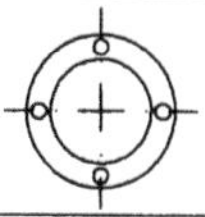	TWIST DRILL	DIN1897 RN	DRILLING MACHINE

C.L.M. CENTER LATHE MACHINE

TABLE 7.27A
Procedure Sheet for Back Arm Column

PROCEDURE SHEET

SUB ASSEMBLY NAME: COLUMN — No. OF PART: 2
PART NAME: BACK ARM COLUMN — PART No.: 1
DRAWING No.: MHRS.0044 — MATERIAL: Al

S.No.	OPERATION NAME	SKETCH OF OPERATION	TOOL	TOOL CLASSIFICATION	MACHINES
1	RAW MATERIAL 30x80X230mm	80, 30			
2	MILLING A	A, 5	ENDMILL	ø 20mm BS122/4	VERTICAL MILLING MACHINE
3	MILLING B	B	ENDMILL	ø 20mm BS122/4	VERTICAL MILLING MACHINE
4	MILLING C	C	ENDMILL	ø 20mm BS122/4	VERTICAL MILLING MACHINE
5	MILLING D	D, 20, 30	ENDMILL	ø 20mm BS122/4	VERTICAL MILLING MACHINE
6	MILLING E	E, 20	ENDMILL	ø 20mm BS122/4	VERTICAL MILLING MACHINE
7	MILLING F	F	ENDMILL	ø 20mm BS122/4	VERTICAL MILLING MACHINE
8	MILLING G	G	ENDMILL	ø 20mm BS122/4	VERTICAL MILLING MACHINE

TABLE 7.27B (Continued)
Procedure Sheet for Back Arm Column

PROCEDURE SHEET

SUB ASSEMBLY NAME COLUMN No. OF PART 2
PART NAME BACK ARM COLUMN PART No. 1
DRAWING No. MHRS.0044 MATERIAL Al

S.No.	OPERATION NAME	SKETCH OF OPERATION	TOOL	TOOL CLASSIFICATION	MACHINES
9	MILLING H	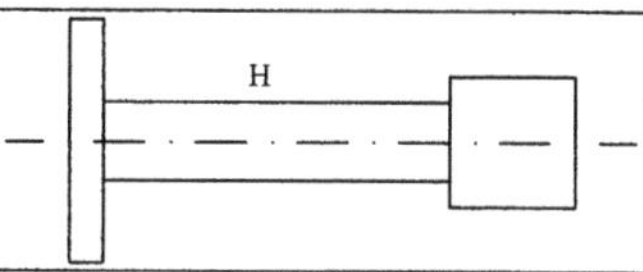 H	ENDMILL	Ø 20mm BS122/4	VERTICAL MILLING MACHINE
10	EXTERNAL STRAIGHT TURNING	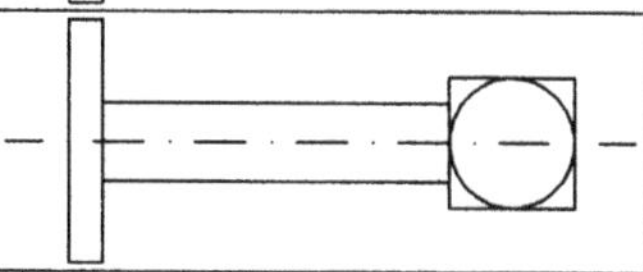	R.H.T.	SANDVIK T-MAX P	C.L.M.
11	CENTERING DRILLING THEN BORING	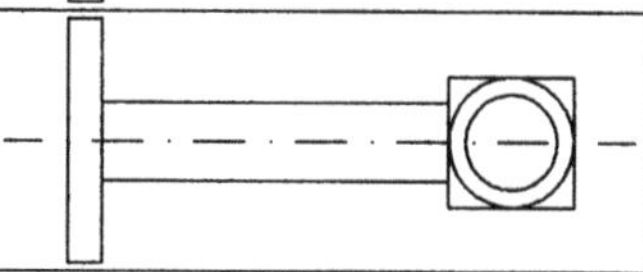	CENTER DRILL TWIST DRILL BORING TOOL	DIN333A, DIN1897 RN SANDVIK CORDTURN 107	C.L.M.
12	EXTERNAL STRAIGHT TURNING FOR ANOTHER SIDE	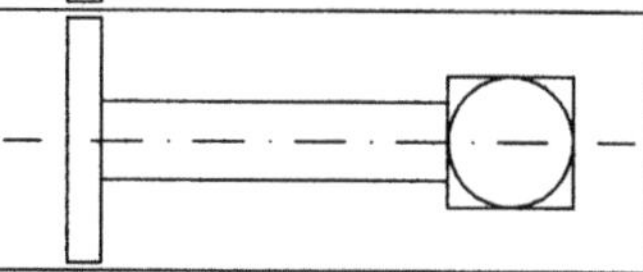	R.H.T.	SANDVIK T-MAX P	C.L.M.
13	CENTERING DRILLING THEN BORING	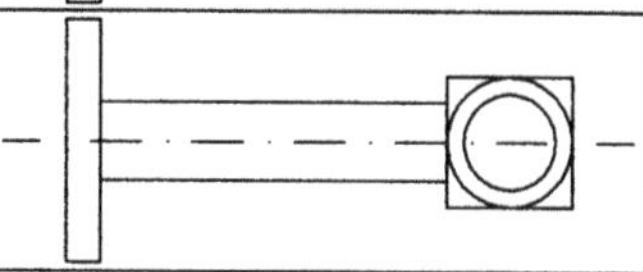	CENTER DRILL TWIST DRILL BORING TOOL	DIN333A, DIN1897 RN SANDVIK CORDTURN 107	C.L.M.
14	FITTING FOR ROUNDED SQUARE PART	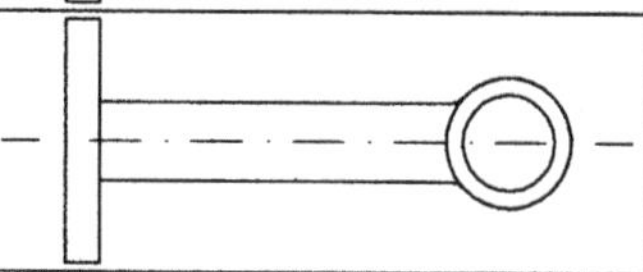	FILLE	SANDVIK 6" MILL SMOOTH FILE	MANUAL
15	DRILLING TWO HOLES IN THE BASE	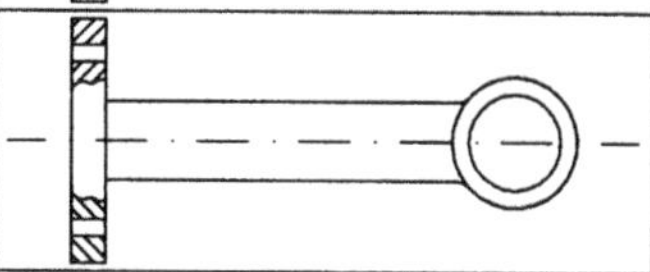	TWIST DRILL	DIN1897 RN	DRILLING MACHINE

C.L.M. CENTER LATHE MACHINE

TABLE 7.28
Procedure Sheet for Motor Column

PROCEDURE SHEET			
SUB ASSEMBLY NAME	COLUMN	No. OF PART	2
PART NAME	MOTOR COLUMN	PART No.	1
DRAWING No.	MHRS.0043	MATERIAL	Al

S.No.	OPERATION NAME	SKETCH OF OPERATION	TOOL	TOOL CLASSIFICATION	MACHINES
1	RAW MATERIAL PLATE: 105X255X20mm	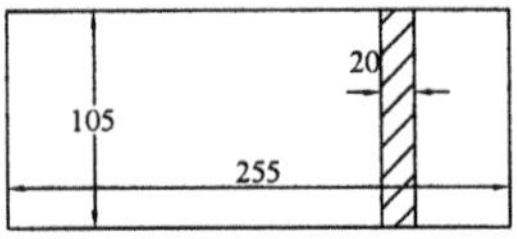 20, 105, 255		BS122/4	
2	ENDMILL ALL SIDE TO END UP WITH 100X250X20mm	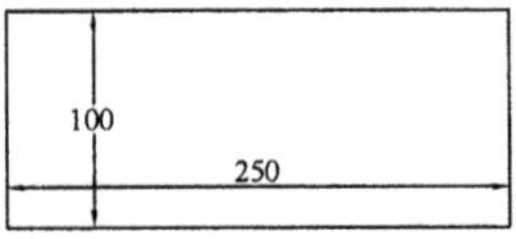 100, 250	ENDMILL	BS122/4	VERTICAL MILLING MACHINE
3	MILLING A	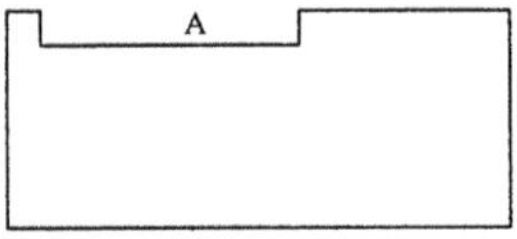 A	ENDMILL	BS122/4	VERTICAL MILLING MACHINE
4	MILLING B	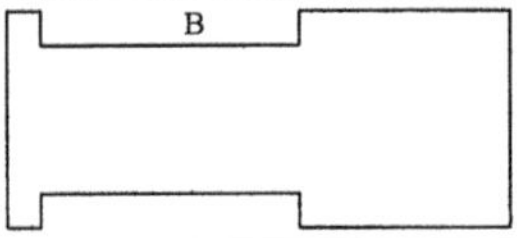 B	ENDMILL	BS122/4	VERTICAL MILLING MACHINE
5	MILLING C SQUARE SHAPE	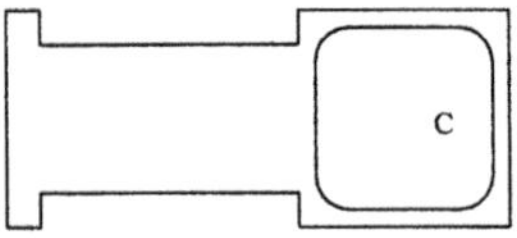 C	ENDMILL	BS122/4	VERTICAL MILLING MACHINE
6	MILLING D	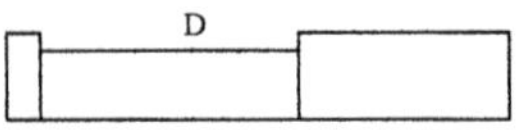 D	ENDMILL	BS122/4	VERTICAL MILLING MACHINE
7	DRILLING (1) HOLE ø15mm AND (8) HOLES ø6mm	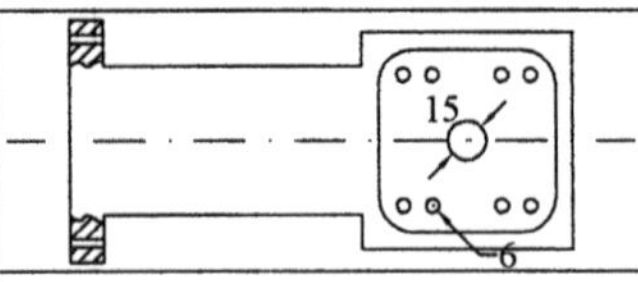 15, 6	TWIST DRILL	ø 6mm, ø 15mm DIN1897 RN	DRILLING MACHINE

TABLE 7.29
Procedure Sheet for Coupling A

PROCEDURE SHEET

SUB ASSEMBLY NAME	COLUMN	No. OF PART	1
PART NAME	COUPLING A	PART No.	-
DRAWING No.	MHRS.0038	MATERIAL	Al

S.No.	OPERATION NAME	SKETCH OF OPERATION	TOOL	TOOL CLASSIFICATION	MACHINES
1 2	FACING A FACING B	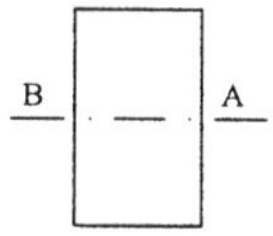 B A	R.H.T.	SANDVIK T-MAX P	C.L.M.
3	EXTERNAL STRAIGHT TURNING FOR DIA ø25mm	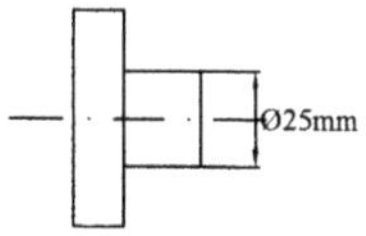 Ø25mm	R.H.T.	SANDVIK T-MAX P	C.L.M.
4	CENTRING	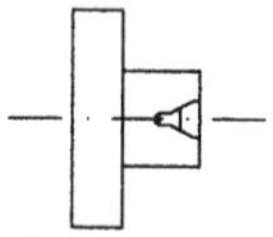	CENTER DRILL	ø 2.5mm DIN333A	C.L.M.
5	DRILLING HOLE ø10mm	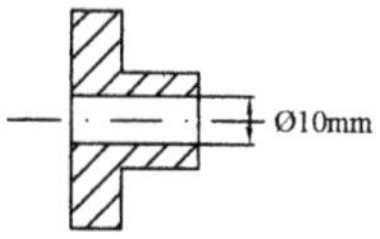 Ø10mm	TWIST DRILL	ø 10mm DIN1897 RN	C.L.M.
6	BORING ø15mm	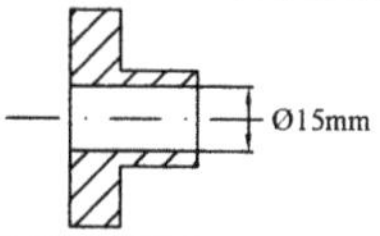 Ø15mm	BORING TOOL	SANDVIK COROTURN 107	C.L.M.
7	DRILLING FOR (6) HOLES ø5mm	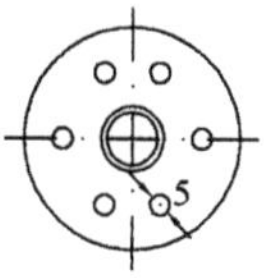 5	TWIST DRILL	ø 5mm DIN1897 RN	DRILLING MACHINE
8	VERTICAL SHAPING FOR SLIDES	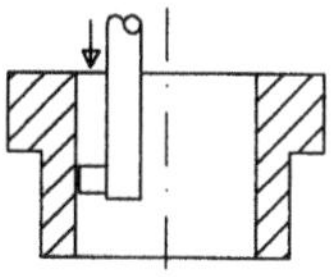	GROOVING TOOL	4mm CUSTOM MADE TOOL FROM LEVICKS HYPEAK 202 HSS BAR	VERTICAL SHAPING MACHINE

C.L.M. CENTER LATHE MACHINE

TABLE 7.30
Procedure Sheet for Motor Base

PROCEDURE SHEET

SUB ASSEMBLY NAME	COLUMN	No. OF PART	1
PART NAME	MOTOR BASE	PART No.	-
DRAWING No.	MHRS.0041	MATERIAL	Al

S.No.	OPERATION NAME	SKETCH OF OPERATION	TOOL	TOOL CLASSIFICATION	MACHINES
1	RAW MATERIAL PLATE 20X110X110mm	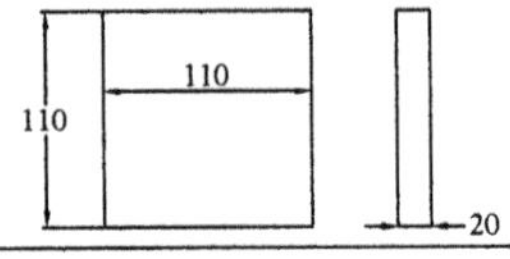 110; 110; 20			
2	ENDMILL ALL SIDES TO GET A SQUARE OF 106mm	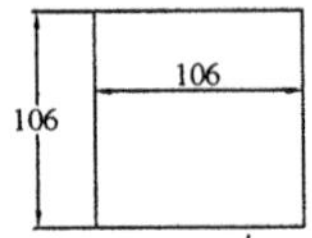 106; 106	ENDMILL	BS122/4	MILLING MACHINE
3	MILLING A	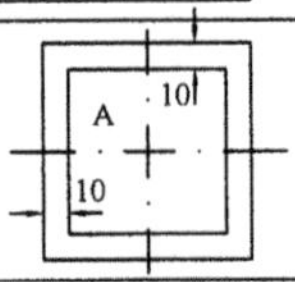 A; 10; 10	ENDMILL	BS122/4	MILLING MACHINE
4	DRILLING (1) HOLE ø15mm AND (4) HOLES ø8mm	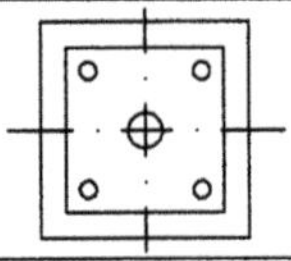	TWIST DRILL	ø 15mm ø 8mm DIN1897 RN	DRILLING MACHINE

TABLE 7.31
Procedure Sheet for Base Motor Column

PROCEDURE SHEET

SUB ASSEMBLY NAME	COLUMN	No. OF PART	4
PART NAME	BASE MOTOR COLUMN	PART No.	–
DRAWING No.	MHRS.0042	MATERIAL	Al

S.No.	OPERATION NAME	SKETCH OF OPERATION	TOOL	TOOL CLASSIFICATION	MACHINES
1	RAW MATERIAL SQUARE ROD 10x10X75mm				
2	FACING A FACING B	B A	R.H.T.	SANDVIK T-MAX P	C.L.M.
3	CENTERING THEN DRILLING		CENTER DRILL& TWIST DRILL	ø 2.5mm DIN333A DIN1897 RN	C.L.M.
4	INTERNAL THREADING		TAP	ISO 529	MANUAL
5	CENTERING DRILLING FOR ANOTHER SIDE		CENTER DRILL& TWIST DRILL	ø 2.5mm DIN333A DIN1897 RN	C.L.M.
6	INTERNAL THREADING		TAP	ISO 529	MANUAL

C.L.M. CENTER LATHE MACHINE

TABLE 7.32
Procedure Sheet for Base Gear Shaft

PROCEDURE SHEET

SUB ASSEMBLY NAME	COLUMN	No. OF PART	1
PART NAME	BASE GEAR SHAFT	PART No.	-
DRAWING No.	MHRS.0037	MATERIAL	Al

S.No.	OPERATION NAME	SKETCH OF OPERATION	TOOL	TOOL CLASSIFICATION	MACHINES
1	FACING A FACING B	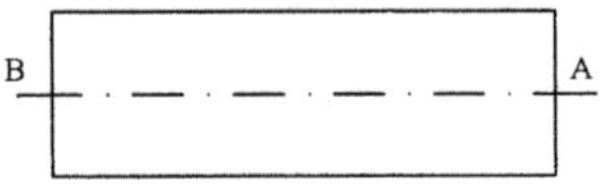 B A	R.H.T.	SANDVIK T-MAX P	C.L.M.
2	EXTERNAL STRAIGHT TURNING FOR DIA Ø10mm	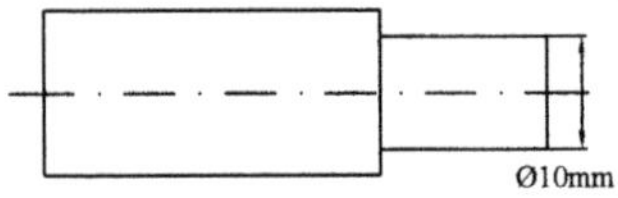 Ø10mm	R.H.T.	SANDVIK T-MAX P	C.L.M.
3	EXTERNAL STRAIGHT TURNING FOR DIA TEETH	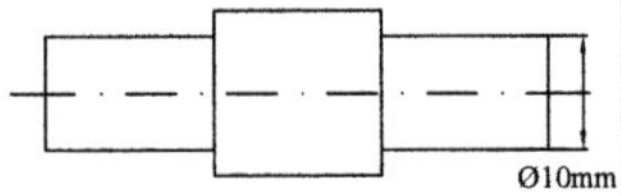 Ø10mm	R.H.T.	SANDVIK T-MAX P	C.L.M.
4	MILLING	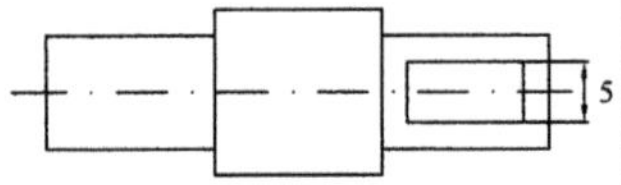 5	ENDMILL	Ø 5mm BS122/4	VERTICAL MILLING MACHINE
5	MILLING	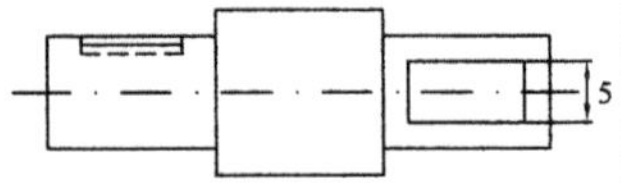 5	ENDMILL	Ø 5mm BS122/4	VERTICAL MILLING MACHINE

C.L.M. CENTER LATHE MACHINE

TABLE 7.33
Procedure Sheet for the Pinion

PROCEDURE SHEET					
SUB ASSEMBLY NAME	COLUMN	No. OF PART	1		
PART NAME	PINION	PART No.	-		
DRAWING No.	MHRS.0035	MATERIAL	Al		
S.No.	**OPERATION NAME**	**SKETCH OF OPERATION**	**TOOL**	**TOOL CLASSIFICATION**	**MACHINES**
1 2	FACING A FACING B	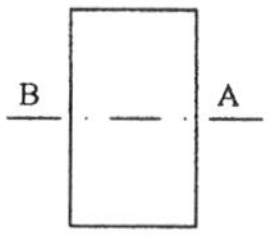 B A	R.H.T.	SANDVIK T-MAX P	C.L.M.
3	CENTRING	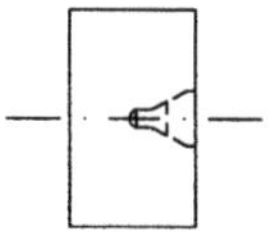	CENTER DRILL	ø 2.5mm DIN333A	C.L.M.
4	DRILLING FOR HOLE ø10mm	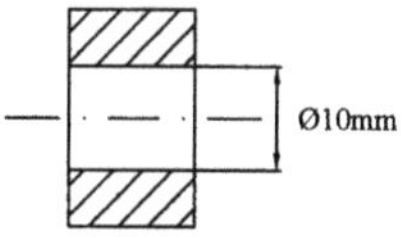 Ø10mm	TWIST DRILL	ø 10mm DIN1897 RN	C.L.M.
5	EXTERNAL STRAIGHT TURNING FOR O.D	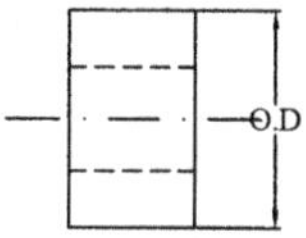 O.D	R.H.T.	SANDVIK T-MAX P	C.L.M.
6	SHAPING FOR SLIDE	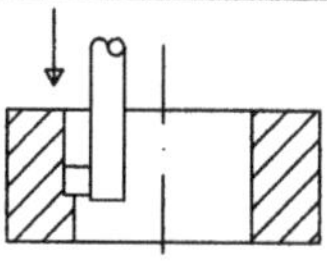	GROOVING TOOL	4mm CUSTOM MADE TOOL FROM LEVICKS HYPEAK 202 HSS BAR	VERTICAL SHAPING MACHINE
7	MILLIING FOR TEETH	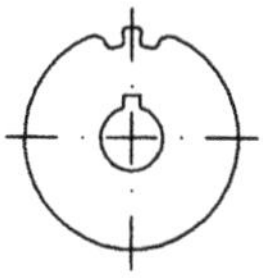	GEAR CUTTER		HORIZON-TAL MILLING MACHINE

C.L.M. CENTER LATHE MACHINE

TABLE 7.34
Procedure Sheet for the Column Base

PROCEDURE SHEET					
SUB ASSEMBLY NAME: COLUMN		No. OF PART: 1			
PART NAME: COLUMN BASE		PART No.: -			
DRAWING No.: MHRS.0036		MATERIAL: Al			

S.No.	OPERATION NAME	SKETCH OF OPERATION	TOOL	TOOL CLASSIFICATION	MACHINES
1	RAW MATERIAL SHEET 510X510X5mm	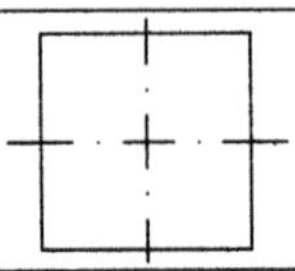			
2	MARKING	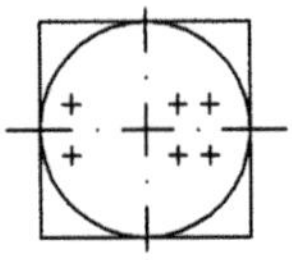	SCRIBER CENTER PUNCH RULER	SUTTON SINGLE POINT SCRIBER (MT007), SUTTON CENTER PUNCH (MT010), SUTTON RULER 12'STAN-LESS STEEL (MT035)	MANUAL
3	PARTING OFF	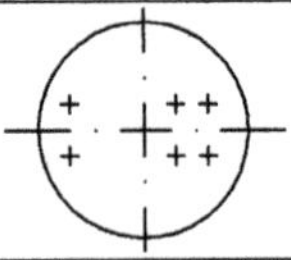	PARTING TOOL	SANDVIK COROCUT1 1	C.L.M.
4	EXTERNAL STRAIGHT TURNING FOR O.D	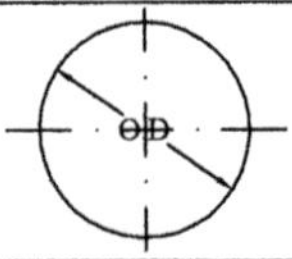 OD	R.H.T.	SANDVIK T-MAX P	C.L.M.
5	GROOVING (A) AND (B)	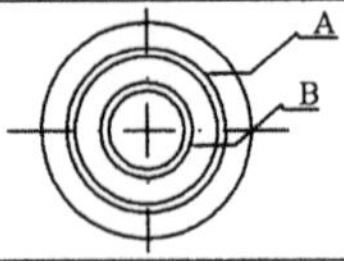 A B	GROOVING TOOL	SANDVIK COROCUT1 1	C.L.M.
6	DRILLING (2) HOLE ø15mm AND (5) HOLES ø10mm	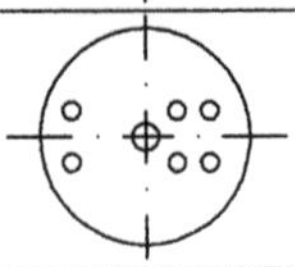	TWIST DRILL	ø 10mm ø 15mm DIN1897 RN	DRILLING MACHINE

C.L.M. CENTER LATHE MACHINE

TABLE 7.35
Procedure Sheet for the Internal Gear

PROCEDURE SHEET

SUB ASSEMBLY NAME	BASE ASSY.	No. OF PART	1
PART NAME	INTERNAL GEAR	PART No.	1
DRAWING No.	MHRS.0049	MATERIAL	Al

S.No.	OPERATION NAME	SKETCH OF OPERATION	TOOL	TOOL CLASSIFICATION	MACHINES
1	RAW MATERIAL PLATE 430X430X20mm				
2	PARTING OFF A RING		PARTING TOOL	SANDVIK COROCUT 1	C.L.M.
3	FACING A FACING B FOR RING	B A	R.H.T.	SANDVIK T-MAX P	C.L.M.
4	EXTERNAL STRAIGHT TURNING FOR OUTER DIA	D	R.H.T.	SANDVIK T-MAX P	C.L.M.
5	BORING FOR INNER DIA	ID	BORING TOOL	SANDVIK COROTURN 107	C.L.M.
6	VERTICAL SHAPING FOR THE GEAR TOOTH		GROOVING TOOL	CUSTOM MADE TOOL FROM LEVICKS HYPEAK 202 HSS BAR	VERTICAL SHAPING MACHINE

C.L.M. CENTER LATHE MACHINE

TABLE 7.36
Procedure Sheet for the Ring

PROCEDURE SHEET

SUB ASSEMBLY NAME BASE ASSY. No. OF PART 1
PART NAME RING PART No. 2
DRAWING No. MHRS.0048 MATERIAL Al

S.No.	OPERATION NAME	SKETCH OF OPERATION	TOOL	TOOL CLASSIFICATION	MACHINES
1	RAW MATERIAL PLATE 430X430X5mm				
2	PARTING OFF A RING		PARTING OF TOOL	SANDVIK COROCUT 1	C.L.M.
3	EXTERNAL STRAIGHT TURNING FOR O.D	OD	R.H.T.	SANDVIK T-MAX P	C.L.M.
4	BORING FOR I.D	I D	BORING TOOL	SANDVIK COROTURN 107	C.L.M.

C.L.M. CENTER LATHE MACHINE

TABLE 7.37
Procedure Sheet for Top Base Plate

PROCEDURE SHEET

SUB ASSEMBLY NAME	BASE ASSY.	No. OF PART	1
PART NAME	TOP BASE PLATE	PART No.	3
DRAWING No.	MHRS.0047	MATERIAL	Al

S.No.	OPERATION NAME	SKETCH OF OPERATION	TOOL	TOOL CLASSIFICATION	MACHINES
1	RAW MATERIAL PLATE 500X500X30mm	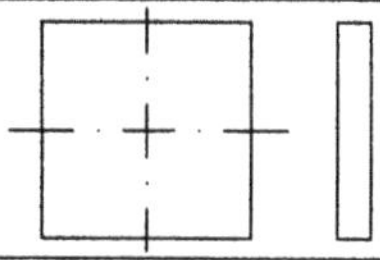			
2	SQUARE THE WORK PIECE	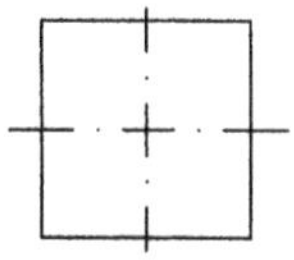	R.H.T.	CUSTOM MADE TOOL FROM MOMAX CLEVELAND 1/2" HSS BAR	HORIZONTAL SHAPING MACHINE
3	FACING A FACING B	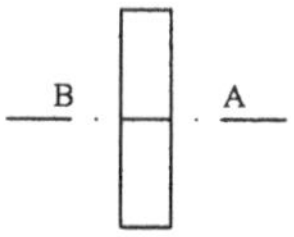 B A	R.H.T.	SANDVIK T-MAX P	C.L.M.
4	CENTRING	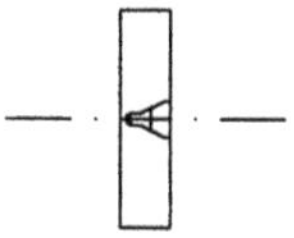	CENTER DRILL	⌀ 2.5mm DIN333A	C.L.M.
5	DRILLING	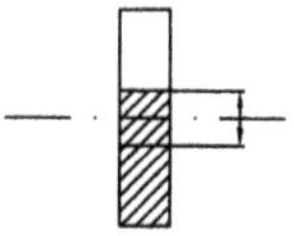	TWIST DRILL	DIN1897 RN	C.L.M.
6	GROOVING FACE	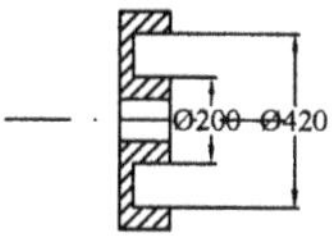 Ø200 Ø420	GROOVING TOOL	CUSTOM MADE TOOL FROM MOMAX CLEVELAND 1/2" HSS BAR	C.L.M.
7	DRILLING (4) HOLES	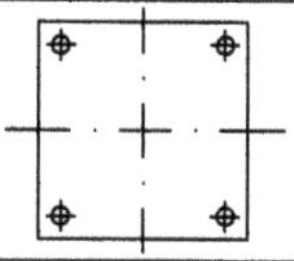	TWIST DRILL	DIN1897 RN	C.L.M.

C.L.M. CENTER LATHE MACHINE

TABLE 7.38
Procedure Sheet for Base Column

PROCEDURE SHEET

SUB ASSEMBLY NAME	BASE ASSY.	No. OF PART	8
PART NAME	BASE COLUMN	PART No.	5
DRAWING No.	MHRS.0046	MATERIAL	Al

S.No.	OPERATION NAME	SKETCH OF OPERATION	TOOL	TOOL CLASSIFICATION	MACHINES
1	RAW MATERIAL SQUARE 20X20X152mm	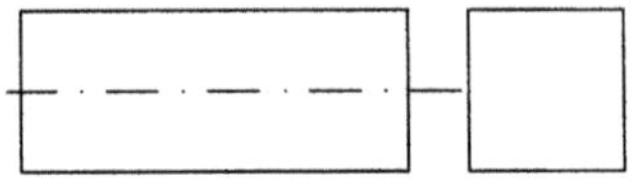			
3	FACING A FACING B FOR L=150mm	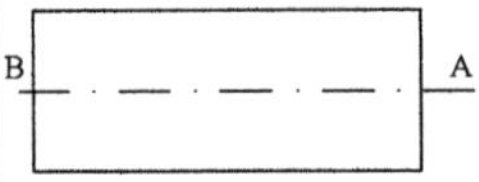 B A	R.H.T.	SANDVIK T-MAX P	C.L.M.
4	CENTRING A	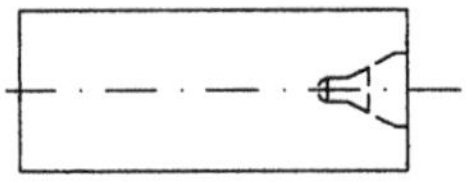	CENTER DRILL	ø 2.5mm DIN333A	C.L.M.
5	DRILLING	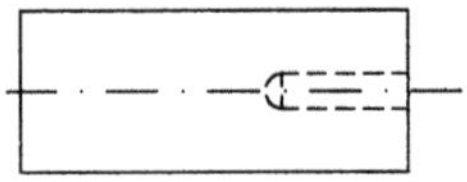	TWIST DRILL	DIN1897 RN	C.L.M.
6	INTERNAL THREADING	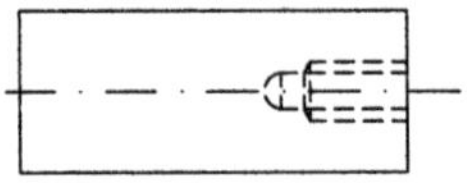	TAP	ISO 529	MANUAL
7	CENTRING B	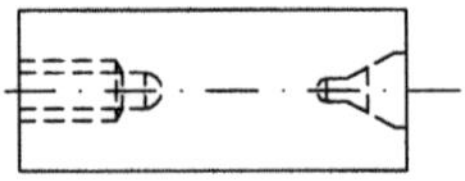	CENTER DRILL	ø 2.5mm DIN333A	C.L.M.
8	DRILLING	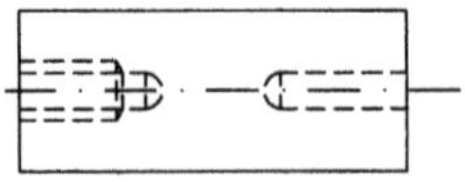	TWIST DRILL	DIN1897 RN	C.L.M.
9	INTERNAL THREADING	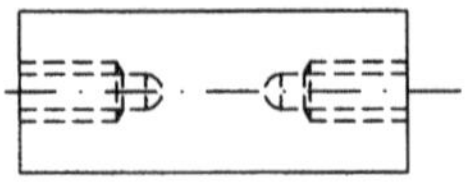	TAP	ISO 529	C.L.M.

C.L.M. CENTER LATHE MACHINE

TABLE 7.39
Procedure Sheet for Bottom Base Plate

PROCEDURE SHEET

SUB ASSEMBLY NAME	BASE ASSY.	No. OF PART	1
PART NAME	BOTTOM BASE PLATE	PART No.	4
DRAWING No.	MHRS.0050	MATERIAL	Al

S.No.	OPERATION NAME	SKETCH OF OPERATION	TOOL	TOOL CLASSIFICATION	MACHINES
1	RAW MATERIAL PLATE 500X500X30mm				
2	SHAPING FOR ALL SIDES		R.H.T.	CUSTOM MADE TOOL FROM MOMAX CLEVELAND 1/2" HSS BAR	HORIZONTAL SHAPING MACHINE
3	FACING A FACING B	B A	R.H.T.	SANDVIK T-MAX P	C.L.M.
4	DRILLING FOR (4) HOLES		TWIST DRILL	DIN1897 RN	DRILING MACHINE

C.L.M. CENTER LATHE MACHINE

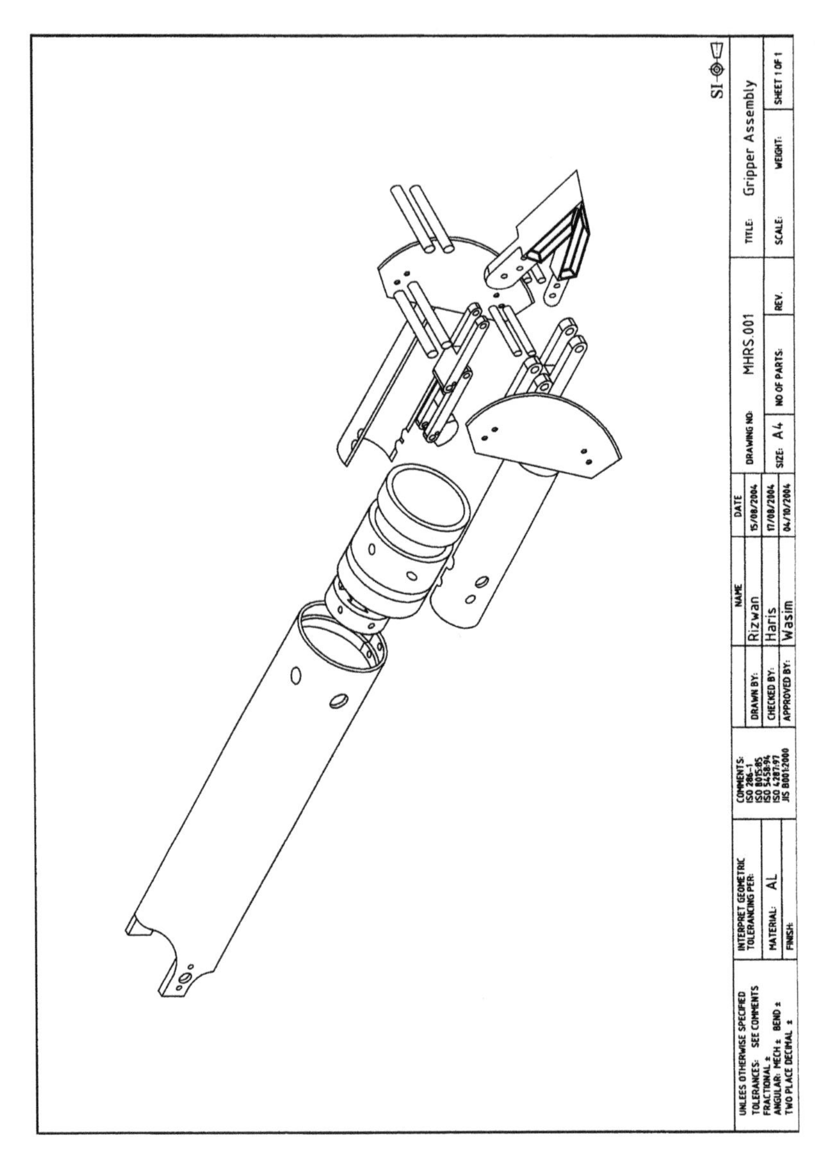

FIGURE 7.41 Gripper assembly.

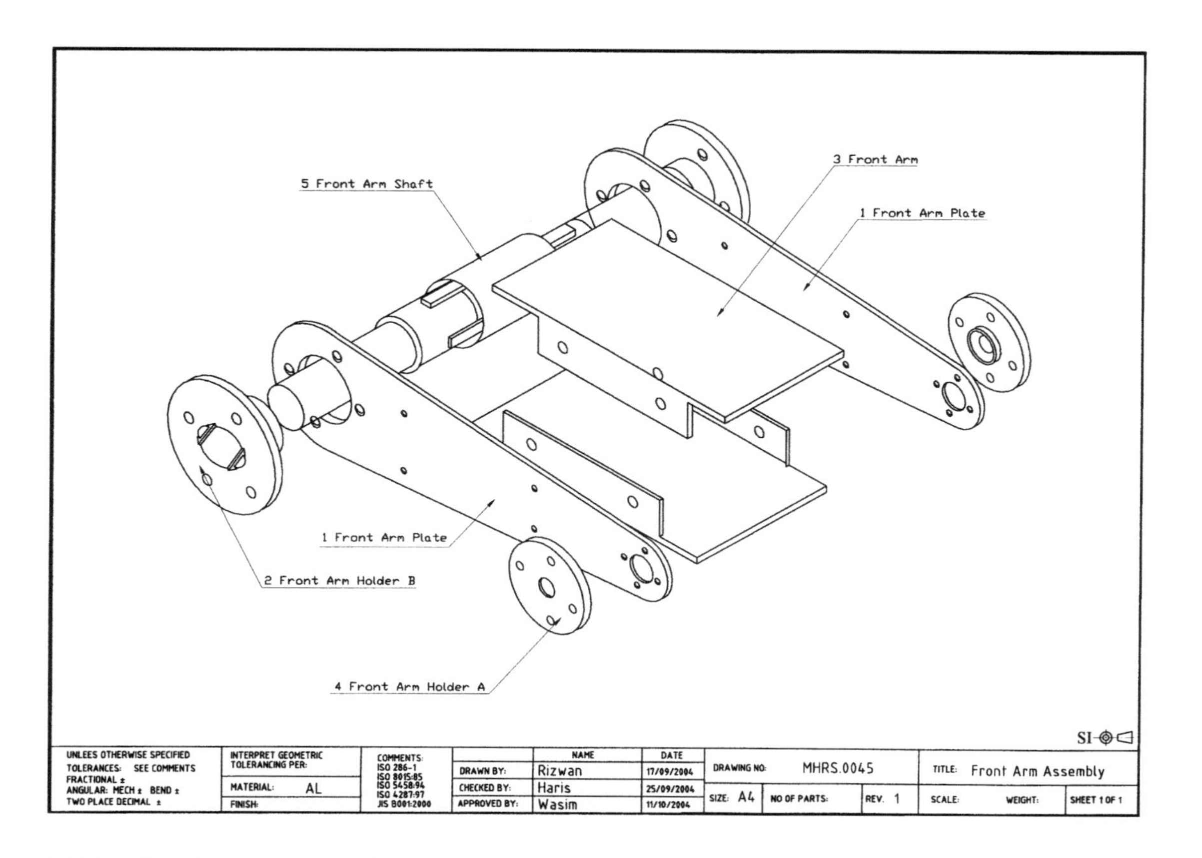

FIGURE 7.42 Front arm assembly.

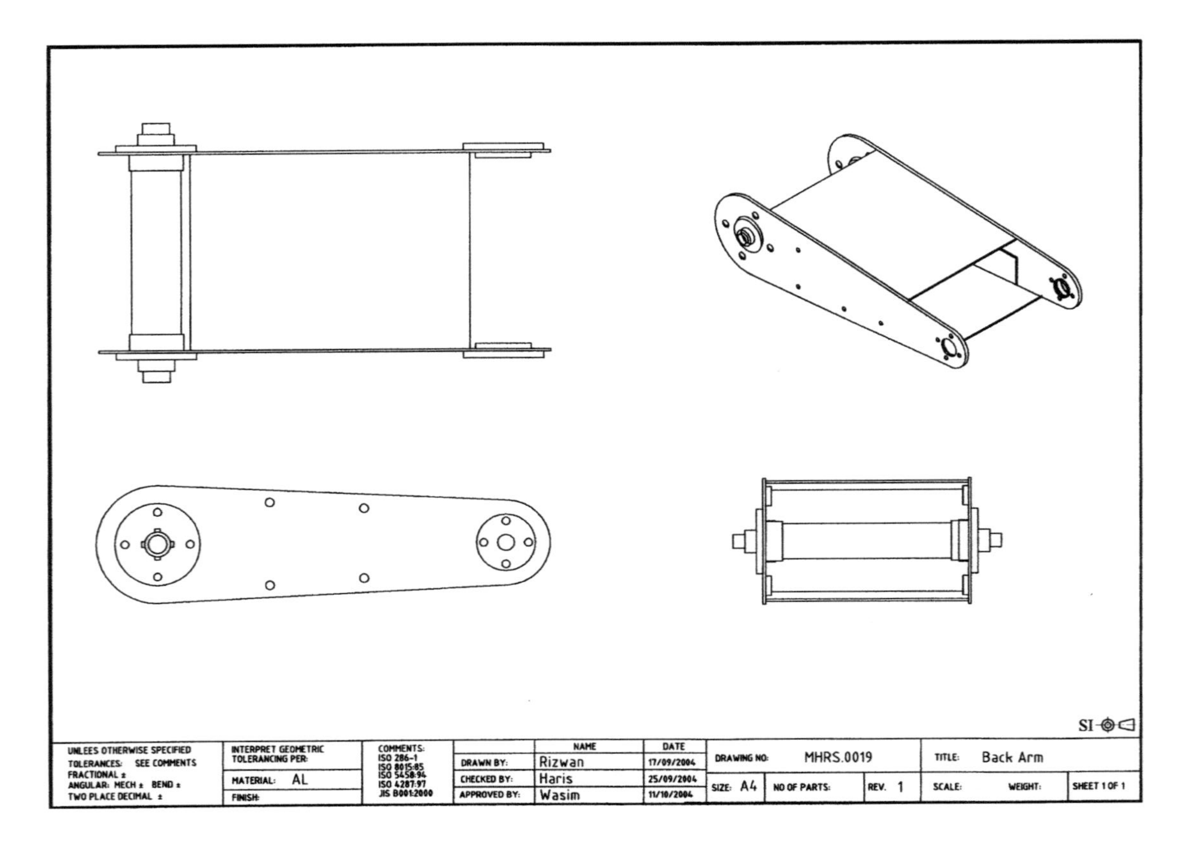

FIGURE 7.43 Back arm assembly.

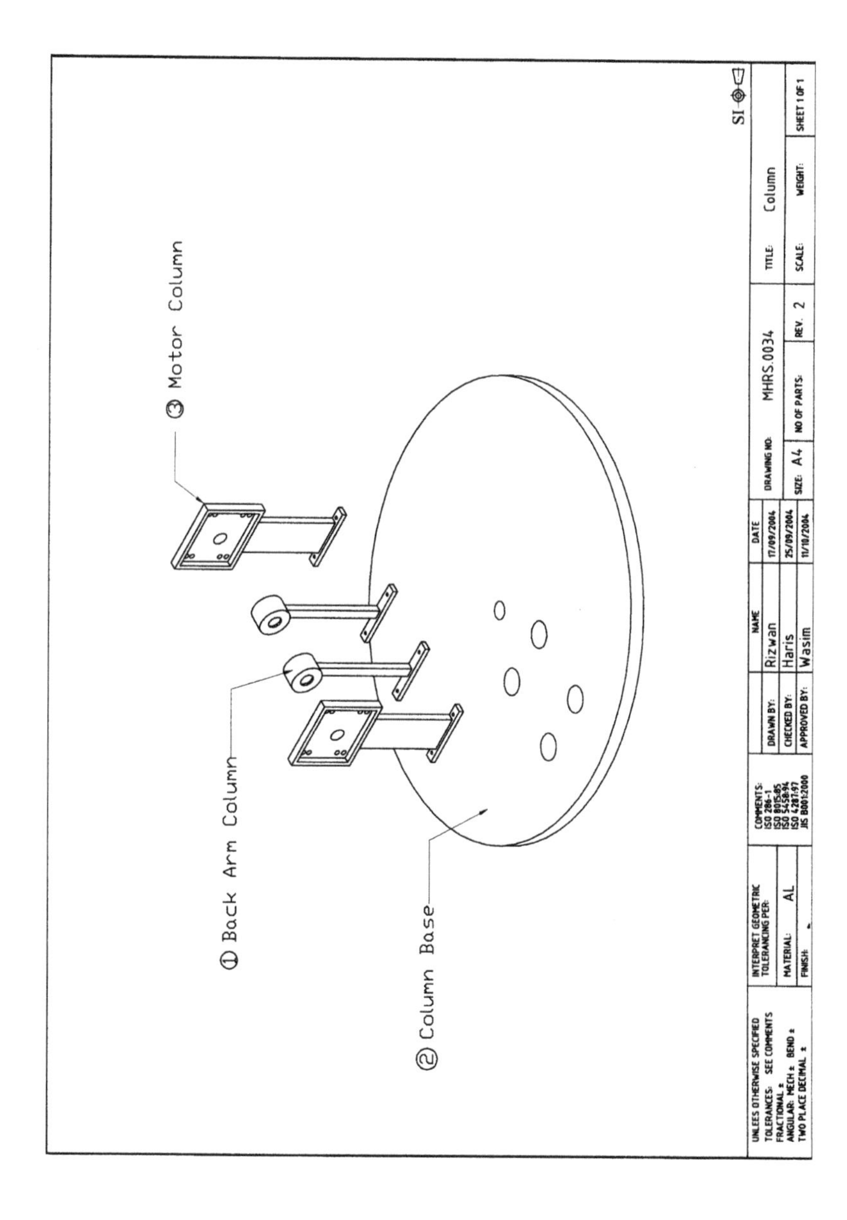

FIGURE 7.44 Column assembly.

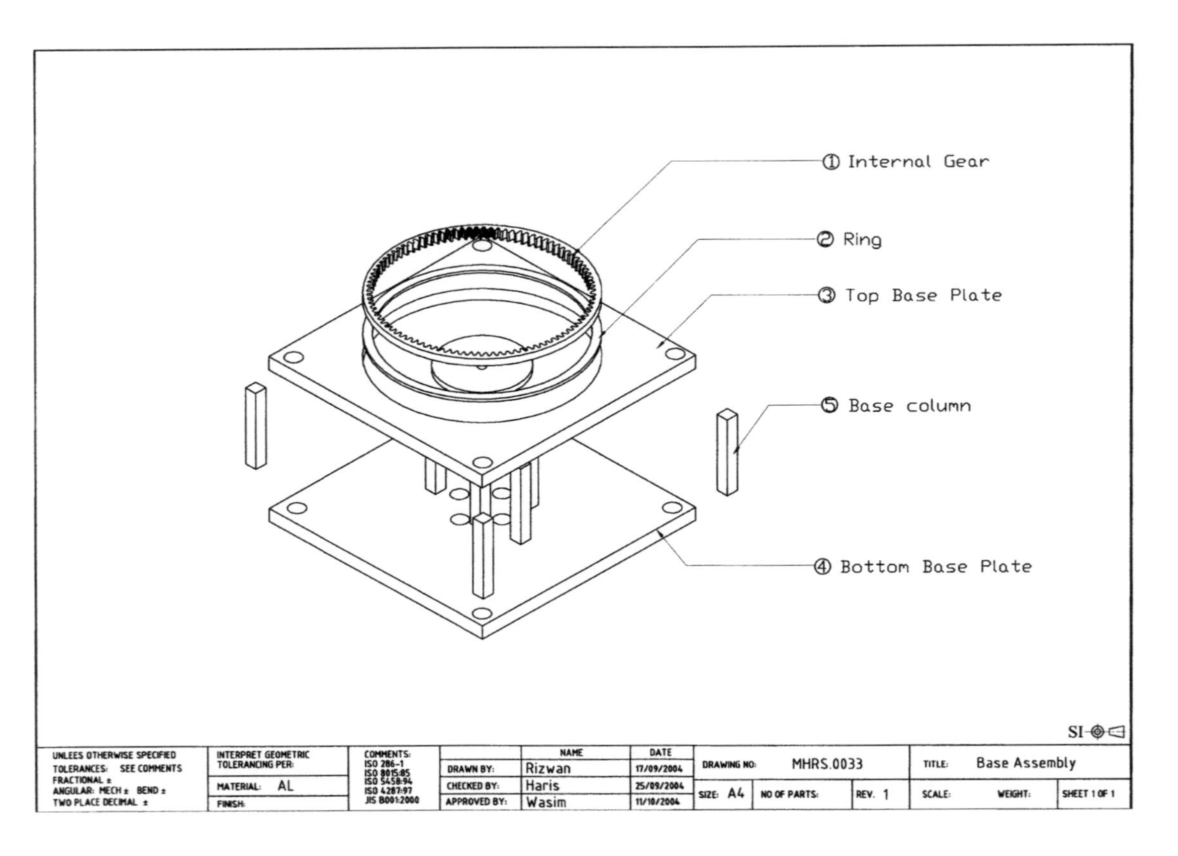

FIGURE 7.45 Base assembly.

TABLE 7.40A
Assembly Procedure for Gripper

ASSY. NAME: GRIPPER ASSEMBLY		PROCEDURE SHEET FOR ASSEMBLY OF GRIPPER	
S. NO.	OPERATION NAME	SKETCH OF OPERATION	
1	ASSEMBLY OF PART NO (7) (PROFILE B) WITH PART NO 6 (PROFILE A) WITH PART NO (5) (CLUTCH A)	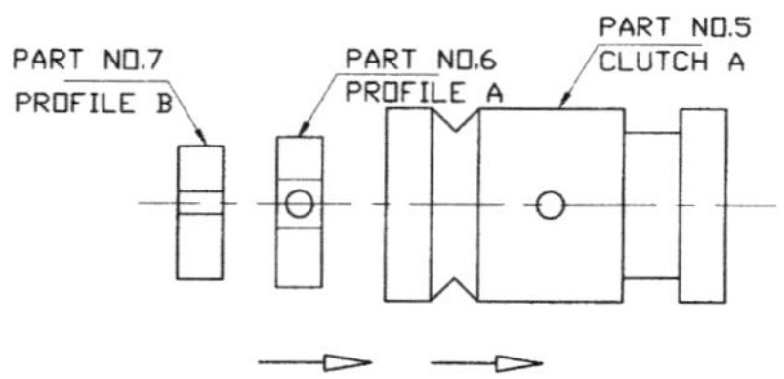 PART NO.7 PROFILE B PART NO.6 PROFILE A PART NO.5 CLUTCH A	
2	ASSEMBLY PART NO.(9) REAR GRIPPER CYLINDER WITH PART NO.(8) CLUTCH B THEN ADDED SUB ASSEMBLY (PART NO.5.6.7)	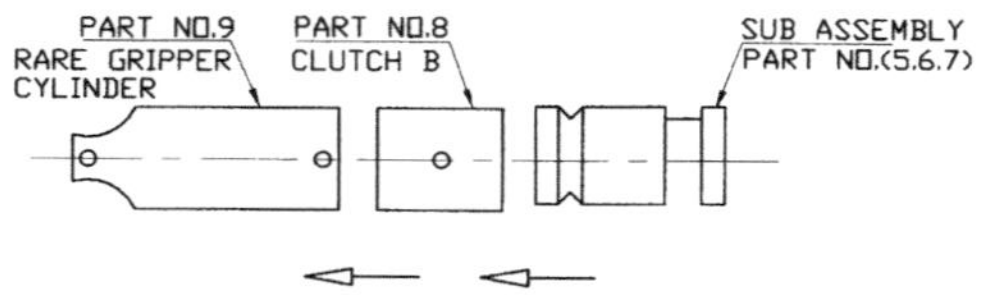 PART NO.9 RARE GRIPPER CYLINDER PART NO.8 CLUTCH B SUB ASSEMBLY PART NO.(5.6.7)	

TABLE 7.40B (Continued)
Assembly Procedure for Gripper

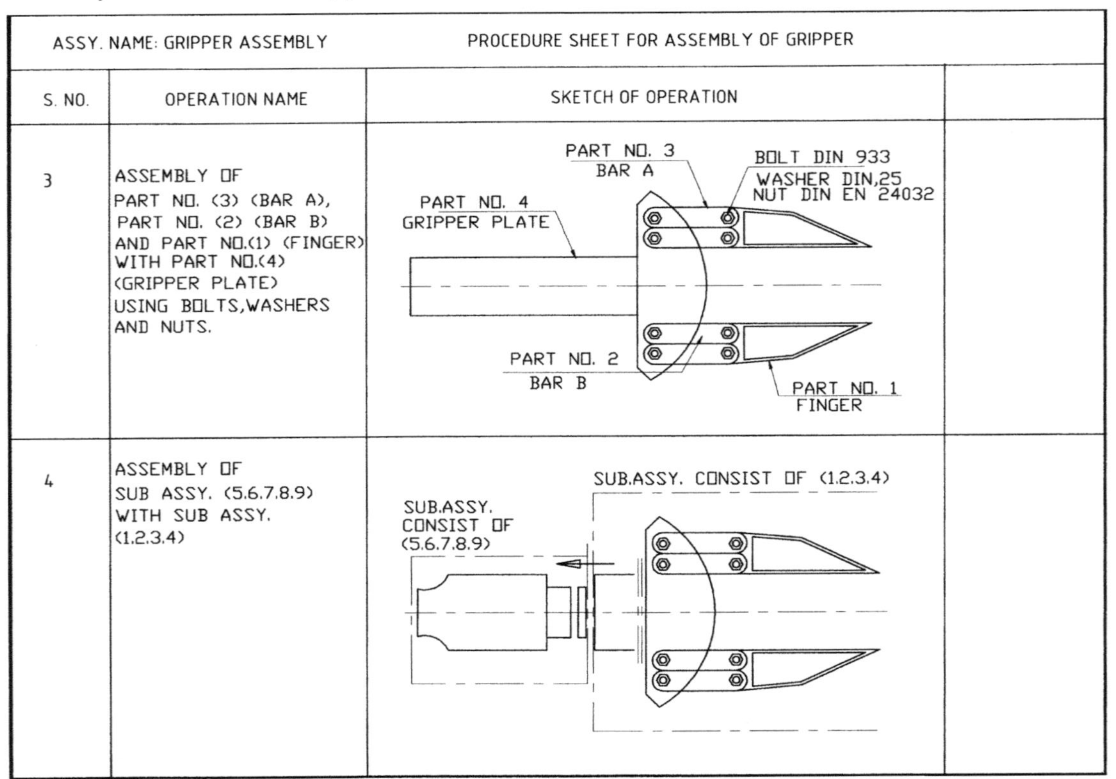

ASSY. NAME: GRIPPER ASSEMBLY		PROCEDURE SHEET FOR ASSEMBLY OF GRIPPER	
S. NO.	OPERATION NAME	SKETCH OF OPERATION	
3	ASSEMBLY OF PART NO. (3) (BAR A), PART NO. (2) (BAR B) AND PART NO.(1) (FINGER) WITH PART NO.(4) (GRIPPER PLATE) USING BOLTS,WASHERS AND NUTS.	PART NO. 3 BAR A; BOLT DIN 933; WASHER DIN,25; NUT DIN EN 24032; PART NO. 4 GRIPPER PLATE; PART NO. 2 BAR B; PART NO. 1 FINGER	
4	ASSEMBLY OF SUB ASSY. (5.6.7.8.9) WITH SUB ASSY. (1.2.3.4)	SUB.ASSY. CONSIST OF (1.2.3.4); SUB.ASSY. CONSIST OF (5.6.7.8.9)	

TABLE 7.41A
Assembly Procedure for Front Arm

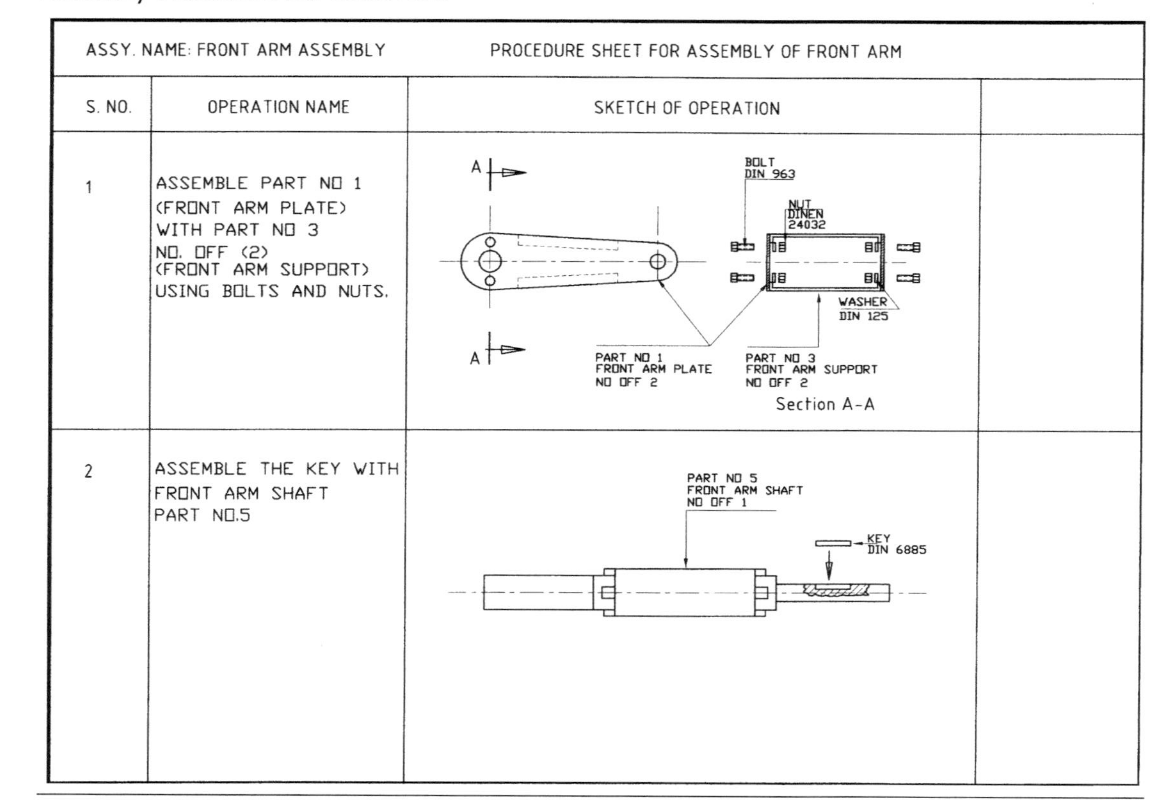

ASSY. NAME: FRONT ARM ASSEMBLY		PROCEDURE SHEET FOR ASSEMBLY OF FRONT ARM	
S. NO.	OPERATION NAME	SKETCH OF OPERATION	
1	ASSEMBLE PART NO 1 (FRONT ARM PLATE) WITH PART NO 3 NO. OFF (2) (FRONT ARM SUPPORT) USING BOLTS AND NUTS.	A A BOLT DIN 963 NUT DINEN 24032 WASHER DIN 125 PART NO 1 FRONT ARM PLATE NO OFF 2 PART NO 3 FRONT ARM SUPPORT NO OFF 2 Section A-A	
2	ASSEMBLE THE KEY WITH FRONT ARM SHAFT PART NO.5	PART NO 5 FRONT ARM SHAFT NO OFF 1 KEY DIN 6885	

TABLE 7.41B (Continued)
Assembly Procedure for Front Arm

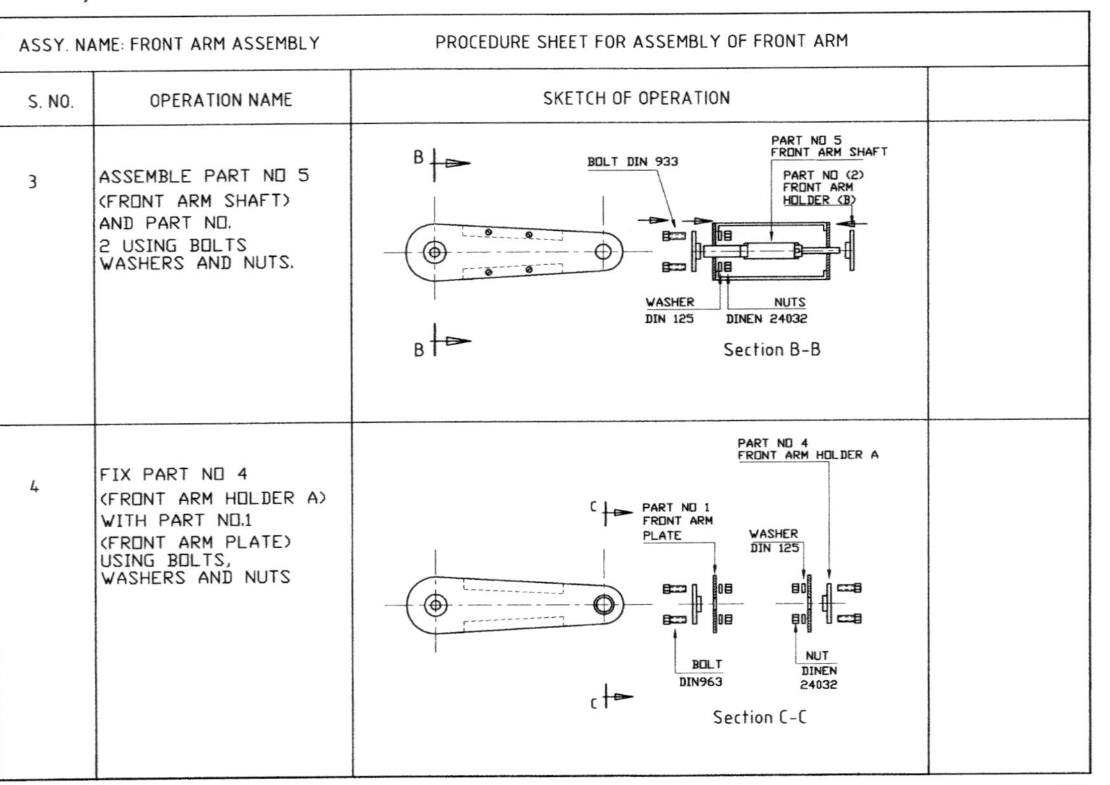

ASSY. NAME: FRONT ARM ASSEMBLY		PROCEDURE SHEET FOR ASSEMBLY OF FRONT ARM	
S. NO.	OPERATION NAME	SKETCH OF OPERATION	
3	ASSEMBLE PART NO 5 (FRONT ARM SHAFT) AND PART NO. 2 USING BOLTS WASHERS AND NUTS.	Section B-B	
4	FIX PART NO 4 (FRONT ARM HOLDER A) WITH PART NO.1 (FRONT ARM PLATE) USING BOLTS, WASHERS AND NUTS	Section C-C	

TABLE 7.42A
Assembly Procedure for Back Arm

ASSY. NAME: BACK ARM ASSEMBLY		PROCEDURE SHEET FOR ASSEMBLY OF BACK ARM	
S. NO.	OPERATION NAME	SKETCH OF OPERATION	
1	ASSEMBLE PART NO 1 (FRONT ARM PLATE) WITH PART NO 3 NO. OFF (2) (FRONT ARM SUPPORT) USING BOLTS AND NUTS.	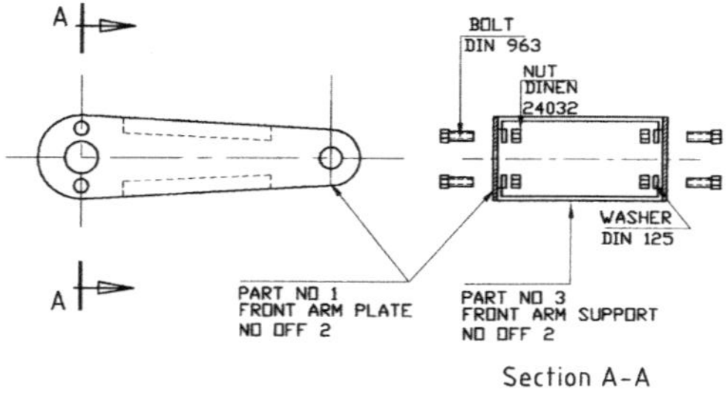 A A BOLT DIN 963 NUT DINEN 24032 WASHER DIN 125 PART NO 1 FRONT ARM PLATE NO OFF 2 PART NO 3 FRONT ARM SUPPORT NO OFF 2 Section A-A	
2	ASSEMBLE THE KEY WITH FRONT ARM SHAFT PART NO.5	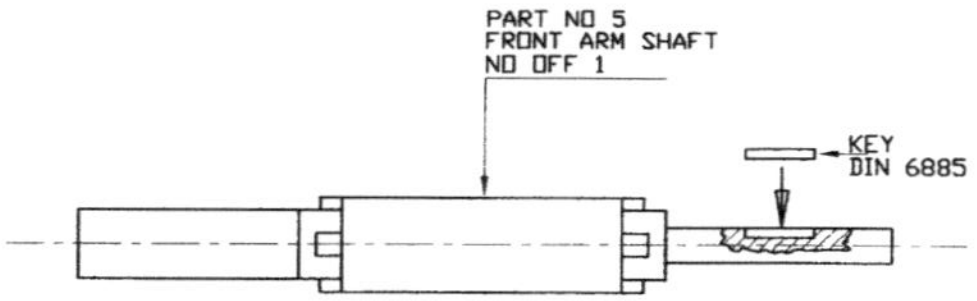 PART NO 5 FRONT ARM SHAFT NO OFF 1 KEY DIN 6885	

TABLE 7.42B (Continued)
Assembly Procedure for Back Arm

ASSY. NAME: BACK ARM ASSEMBLY		PROCEDURE SHEET FOR ASSEMBLY OF BACK ARM	
S. NO.	OPERATION NAME	SKETCH OF OPERATION	
3	ASSEMBLE PART NO 5 (FRONT ARM SHAFT) AND PART NO. 2 USING BOLT WASHER AND NUT.	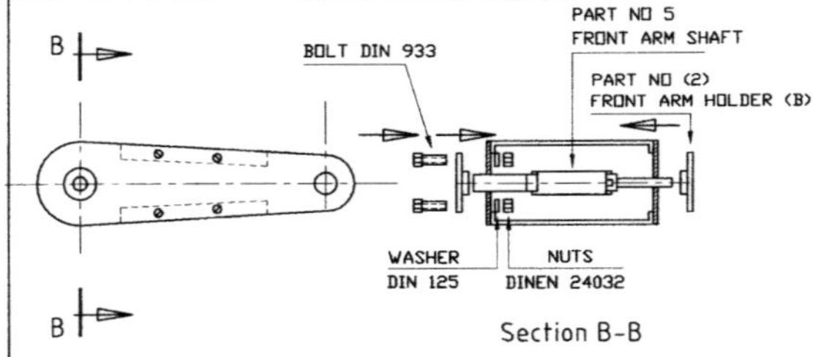 B B BOLT DIN 933 PART NO 5 FRONT ARM SHAFT PART NO (2) FRONT ARM HOLDER (B) WASHER DIN 125 NUTS DINEN 24032 Section B-B	
4	FIX PART NO 4 (FRONT ARM HOLDER A) WITH PART NO.1 (FRONT ARM PLATE) USING BOLTS, WASHERS & NUTS	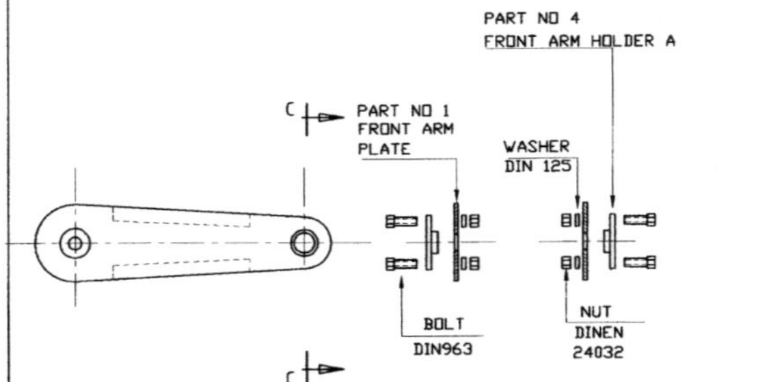 C C PART NO 1 FRONT ARM PLATE PART NO 4 FRONT ARM HOLDER A WASHER DIN 125 BOLT DIN963 NUT DINEN 24032 Section C-C	

TABLE 7.43
Assembly Procedure for Column

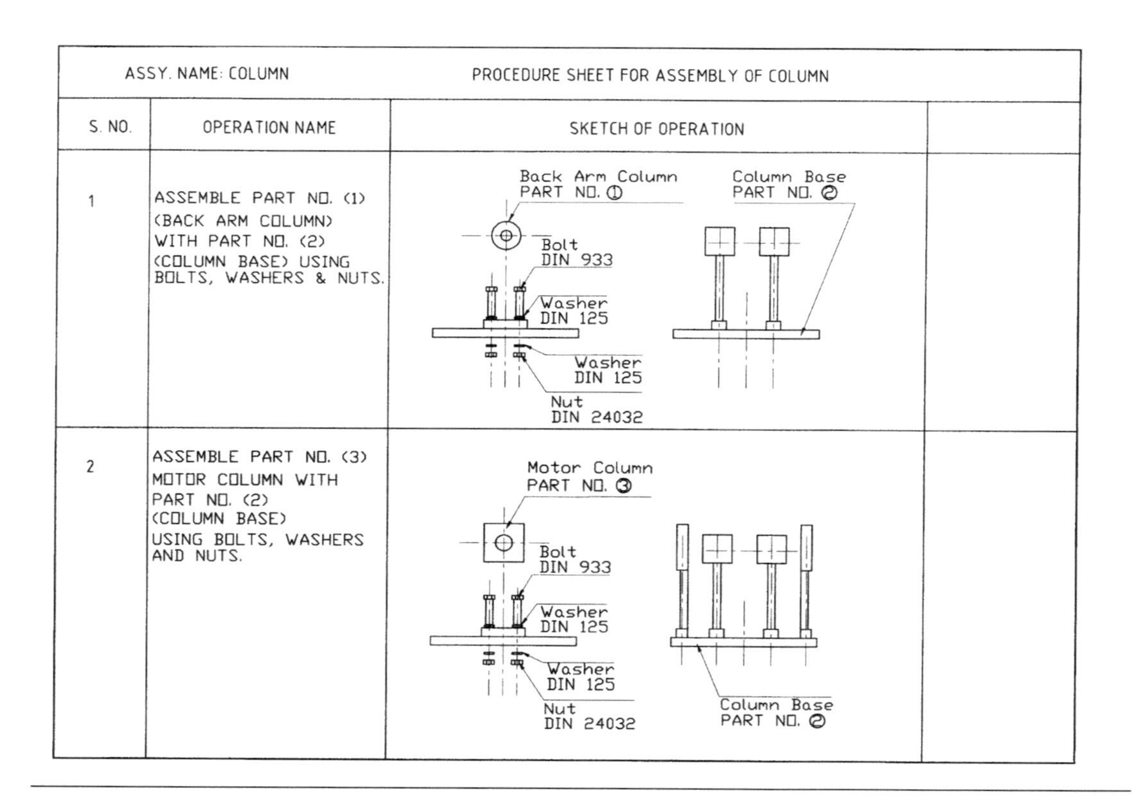

ASSY. NAME: COLUMN		PROCEDURE SHEET FOR ASSEMBLY OF COLUMN	
S. NO.	OPERATION NAME	SKETCH OF OPERATION	
1	ASSEMBLE PART NO. (1) (BACK ARM COLUMN) WITH PART NO. (2) (COLUMN BASE) USING BOLTS, WASHERS & NUTS.	Back Arm Column PART NO. ①; Bolt DIN 933; Washer DIN 125; Washer DIN 125; Nut DIN 24032; Column Base PART NO. ②	
2	ASSEMBLE PART NO. (3) MOTOR COLUMN WITH PART NO. (2) (COLUMN BASE) USING BOLTS, WASHERS AND NUTS.	Motor Column PART NO. ③; Bolt DIN 933; Washer DIN 125; Washer DIN 125; Nut DIN 24032; Column Base PART NO. ②	

TABLE 7.44
Assembly Procedure for Base

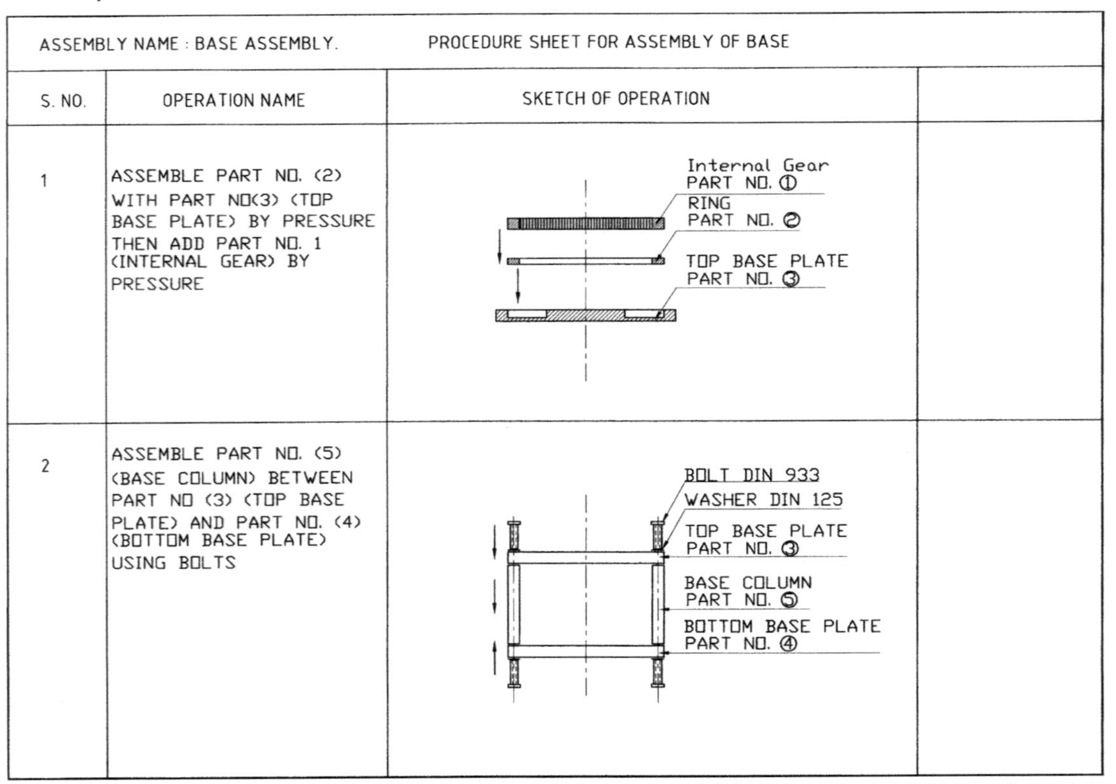

ASSEMBLY NAME : BASE ASSEMBLY.	PROCEDURE SHEET FOR ASSEMBLY OF BASE		
S. NO.	OPERATION NAME	SKETCH OF OPERATION	
1	ASSEMBLE PART NO. (2) WITH PART NO(3) (TOP BASE PLATE) BY PRESSURE THEN ADD PART NO. 1 (INTERNAL GEAR) BY PRESSURE	Internal Gear PART NO. ① RING PART NO. ② TOP BASE PLATE PART NO. ③	
2	ASSEMBLE PART NO. (5) (BASE COLUMN) BETWEEN PART NO (3) (TOP BASE PLATE) AND PART NO. (4) (BOTTOM BASE PLATE) USING BOLTS	BOLT DIN 933 WASHER DIN 125 TOP BASE PLATE PART NO. ③ BASE COLUMN PART NO. ⑤ BOTTOM BASE PLATE PART NO. ④	

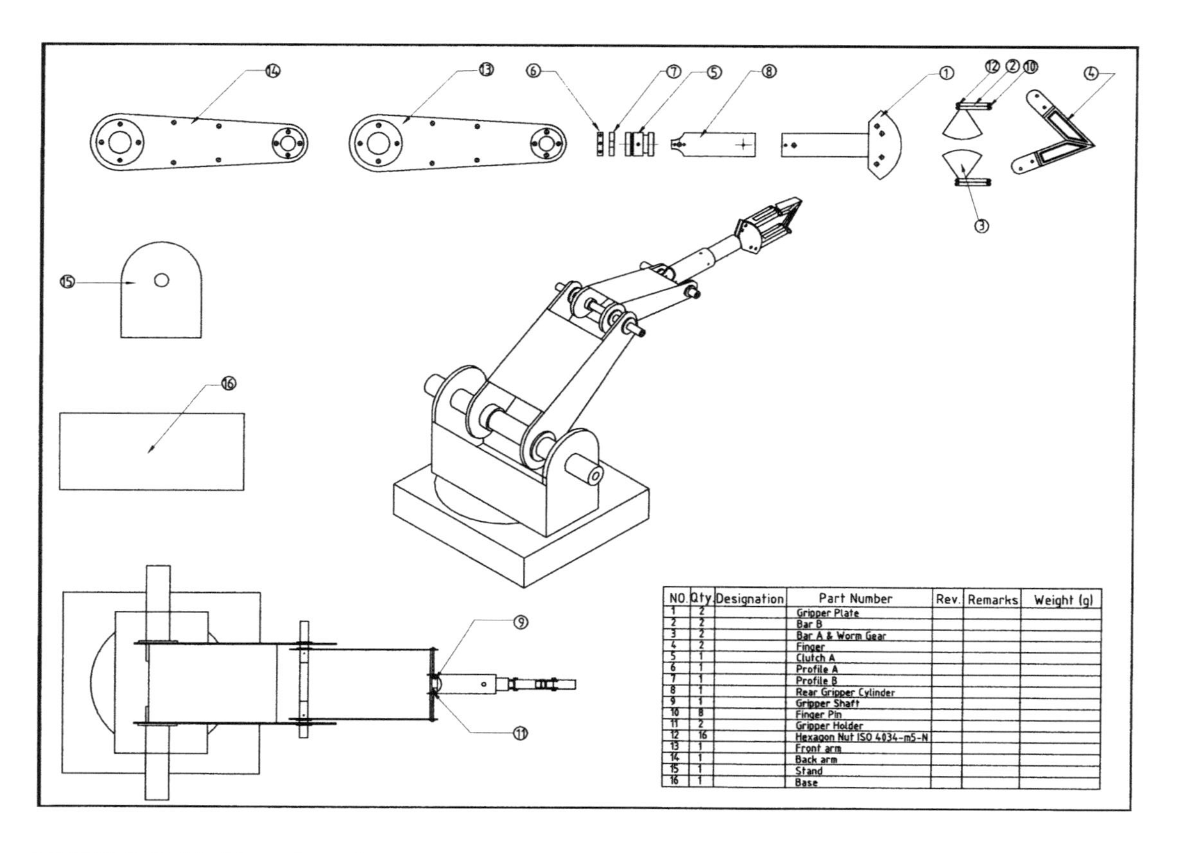

NO.	Qty	Designation	Part Number	Rev.	Remarks	Weight (g)
1	2		Gripper Plate			
2	2		Bar B			
3	2		Bar A & Worm Gear			
4	2		Finger			
5	1		Clutch A			
6	1		Profile A			
7	1		Profile B			
8	1		Rear Gripper Cylinder			
9	1		Gripper Shaft			
10	8		Finger Pin			
11	2		Gripper Holder			
12	16		Hexagon Nut ISO 4034-m5-N			
13	1		Front arm			
14	1		Back arm			
15	1		Stand			
16	1		Base			

FIGURE 7.46 Model revolute robot.

TABLE 7.45A
Assembly Procedure for Robot

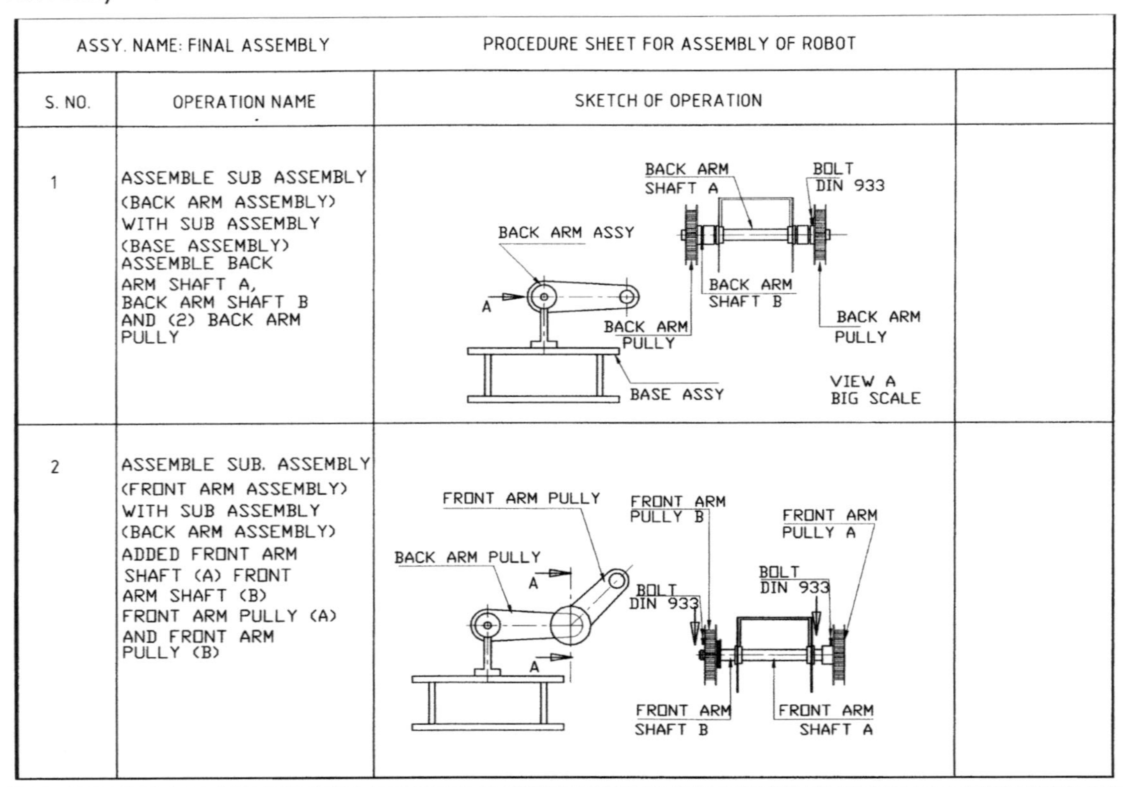

ASSY. NAME: FINAL ASSEMBLY		PROCEDURE SHEET FOR ASSEMBLY OF ROBOT	
S. NO.	OPERATION NAME	SKETCH OF OPERATION	
1	ASSEMBLE SUB ASSEMBLY (BACK ARM ASSEMBLY) WITH SUB ASSEMBLY (BASE ASSEMBLY) ASSEMBLE BACK ARM SHAFT A, BACK ARM SHAFT B AND (2) BACK ARM PULLY		
2	ASSEMBLE SUB. ASSEMBLY (FRONT ARM ASSEMBLY) WITH SUB ASSEMBLY (BACK ARM ASSEMBLY) ADDED FRONT ARM SHAFT (A) FRONT ARM SHAFT (B) FRONT ARM PULLY (A) AND FRONT ARM PULLY (B)		

TABLE 7.45B
Assembly Procedure for Robot

ASSY. NAME: FINAL ASSEMBLY		PROCEDURE SHEET FOR ASSEMBLY OF ROBOT	
S. NO.	OPERATION NAME	SKETCH OF OPERATION	
3	ASSEMBLE SUB. ASSEMBLY (GRIPPER ASSY) WITH SUB. ASSEMBLY (FRONT ARM ASSY) ADDED GRIPPER SHAFT, GRIPPER PULLY AND BELTS.	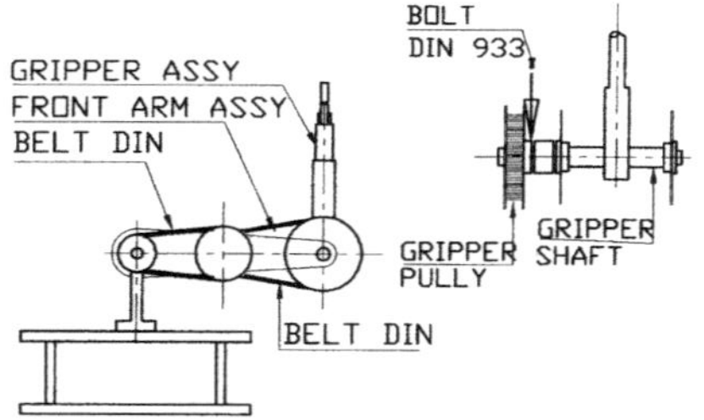 GRIPPER ASSY FRONT ARM ASSY BELT DIN BELT DIN BOLT DIN 933 GRIPPER PULLY GRIPPER SHAFT	

BIBLIOGRAPHY

Adlard, E.J., Use of Flexible Group Technology Codes in Process Planning, Proceedings of the Annual Meeting and Technical Conference, Numerical Control Society, 1982, 154–161.

Alexander, S.M. and Jagannathan, V., Computer aided process planning systems: Current and future directions, *IEEE,* Bombay and Delhi, India, 462–466, 1983.

Blore, D., Computer aided process planning, *CME,* 31(5), 31–34, 1984.

Burgess, J.D., Review of computer aided process planning systems, *SME,* 87, 76, 1984.

Bushong, S., Capacity/process planning—Way to go! *Manufacturing Syst.* 3(11), 16–20, 1985.

Chang, T. and Wysk, R.A. *An Introduction to the Automated Process Planning System* , Prentice Hall, N.J.,1985.

Davies, B.J., Darbyshire, I.L., and Wright, A.J., *Expert Systems in Process Planning*, Butterworth's, London, 1986, 7–14.

Dereli, T., and Filiz, H., A note on the use of STEP for interfacing design to process planning, *CAD,* 34(14), 1075–1085, 2002.

Doran, J.M. and Sechrist, N.W., Computer-Aided Process Planning and Work Measurement, Proceedings of the Annual Meeting and Technical Conference, Numerical Control Society, 1978, 391–409.

Feng, S.C. and Zhang, Y., Conceptual process planning—A definition and functional decomposition, *Am. Soc. Mechanical Engineers MED,* 10, 97–106, 1999.

Halevi G., Weill R.D., and Weill R., *Principles of Process Planning: A Logical Approach,* 1st ed., Kluwer Academic Publishers, *Dordrecht*, 1995.

Tulkoff J., *CAPP: Computer Aided Process Planning,* Manufacturing Update Series, 1st ed., Society of Manufacturing, Dearborn, Mich. 1985.

Jovanoski, D. and Muthsam, H., Workpiece modeling for computer-aided process planning, *Int. J. Advanced Manufacturing Technol.,* 10(6), 404–410, 1995.

Kayacan, M.C. and Celik, S.A., Process planning system for prismatic parts, *Integrated Manufacturing Syst.* 14(2), 75–86, 2003.

Malarkey, L.D., Process planning systems as the CIM integrator, *Am. Soc. Mechanical Engineers PED,* 21, 407–412, 1986.

Mark A.C., *Process Planning,* John Wiley & Sons, New York, 1988.

Ming, X.G., Mak, K.L. and Yan, J.Q., PDES/STEP-based information model for computer-aided process planning, *Robotics Computer-Integrated Manufacturing,* 14(5–6), 347–361, 1998.

Scallan P., *Process Planning: The Design/Manufacture Interface* , Butterworth-Heinemann, Oxford, 2003.

Qiao, L.-H., Zhang, C., Liu, T.-H., Wang, H.-P. B., and Fischer, G.W., PDES/STEP-based product data preparation procedure for computer-aided process planning, *Comput. Ind.* 21(1), 11–22, 1993.

Regli, W.C. and Gaines, D.M., Repository for design, process planning and assembly, *CAD,* 29(12), 895–905, 1997.

Reynolds, B.I., Fink, P.K., and Moehle, C.J., Intelligent Assistance for Variant Process Planning, SME Technical Paper MS93-252, 1993, 1–12.

Sharma, R. and Gao, J.X., Implementation of STEP application protocol 224 in an automated manufacturing planning system, *Proc. Inst. Mechanical Engineers* Part B *J. Eng. Manufacture,* 216(9), 1277–1289, 2002.

Soles, A., CAE—Linking N/C, Process Planning, Computer Aided Design and Standards Programs, Proceedings of the Annual Meeting and Technical Conference, Numerical Control Society, 1983, 167–191.

Sturrock, D.T. and Higley, H.B., Use of simulation for gross planning, scheduling, standards, and tracking, *Comput. Ind. Eng.* 11(1–4), 411–415, 1986.

Tien-Chien C., *Introduction to Automated Process Planning Systems,* Prentice Hall, New York, 1985.

Zandin, K.B., Computer Aided Process Planning and Productivity, SME Technical Paper, 1982.

Zha, X.F. and Ji, P., Assembly process sequence planning for STEP-based mechanical products: An integrated model and system implementation, *Int. J. Industrial Eng. Theory Appl. Pract.* 10(3), 279–288, 2003.

Part 4

Standards for the Use of Discrete Products

8 Review of Standards Related to the Usage of Discrete Product—A User's Perspective

8.1 INTRODUCTION

The user of the discrete product is faced with simple to intricate interactions with the product. The scope of this interaction ranges from man–machine interface (MMI) to human–computer interaction (HCI). MMI and HCI provide ease and safety in operation of products by regulating the input parameters within a specified range. HCI involves a comparatively more automatic interface than does MMI. In the case of MMI, the user interface to the product is more dependent on the operator's presence on the site. The user interface in the case of HCI is comparatively less dependent on humans and, in many cases, can be controlled remotely through either wired or wireless communication networks.

The input and output variables for discrete products are generally identified in Figure 1.4 to Figure 1.6. The output from the product provides the type of benefits shown in Figure 1.6: convenience, pleasure, and better living standards. The scope of the user interface is presented in Figure 8.1.

This chapter deals with standards used for the operation of discrete products. Because of the enormous number of discrete products, listing of all the standards related to usage of every discrete product that are available from various standards-developing organizations is not possible. To show the diversity of available standards, only man–machine interface, human–computer interaction, ergonomics, and output, including emission, from discrete products are selected as domain specifications. The International Organization for Standardization (ISO) is chosen as the target organization. The following search parameters are used:

Man–machine interface
Human–computer interaction
Ergonomics
Emission

The results obtained from the online catalog of the ISO were edited before being listing here.

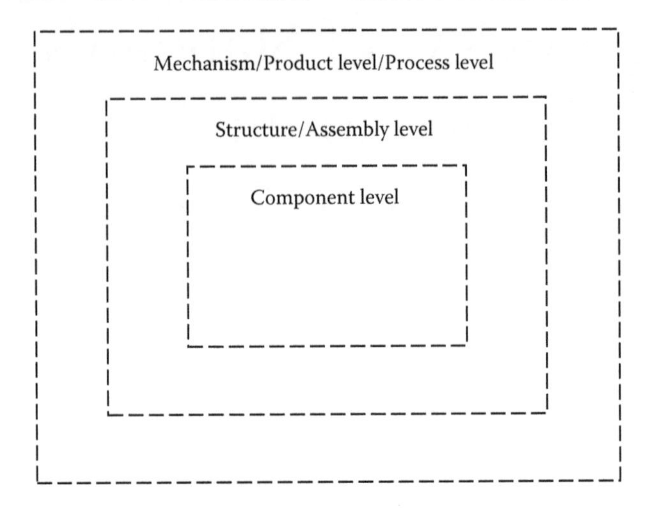

FIGURE 8.1 User interface in automata.

8.2 STANDARDS RELATED TO MAN–MACHINE INTERFACE IN DISCRETE PRODUCTS

To provide ease of operation and safety is a common goal in the design of MMIs for discrete products. MMIs are generally available at manual systems. They are ergonomically designed and adhere to applied standardization. Figure 8.2 details the usage requirements of MMIs for discrete products. Selected MMI standards related to the operation (usage) of discrete products from the ISO's digital catalog are provided in Table 8.1.

8.3 STANDARDS RELATED TO HUMAN–COMPUTER INTERACTION IN DISCRETE PRODUCTS

Modern discrete products extensively utilize microprocessor-based control of features. These characteristics require design of an HCI that allows more automation in the operation of discrete products. A description of the scope of HCI in discrete products is presented in Figure 8.3. Table 8.2 lists standards from ISO's digital catalog that show the diversity in use of standardized HCI in discrete products.

8.4 STANDARDS RELATED TO ERGONOMICS OF DISCRETE PRODUCTS

Ergonomics is a subject that is being given more and more importance in the design of discrete products. It allows use of anthropometrical, physiological, anatomical, and psychological values in the design of discrete products. Because

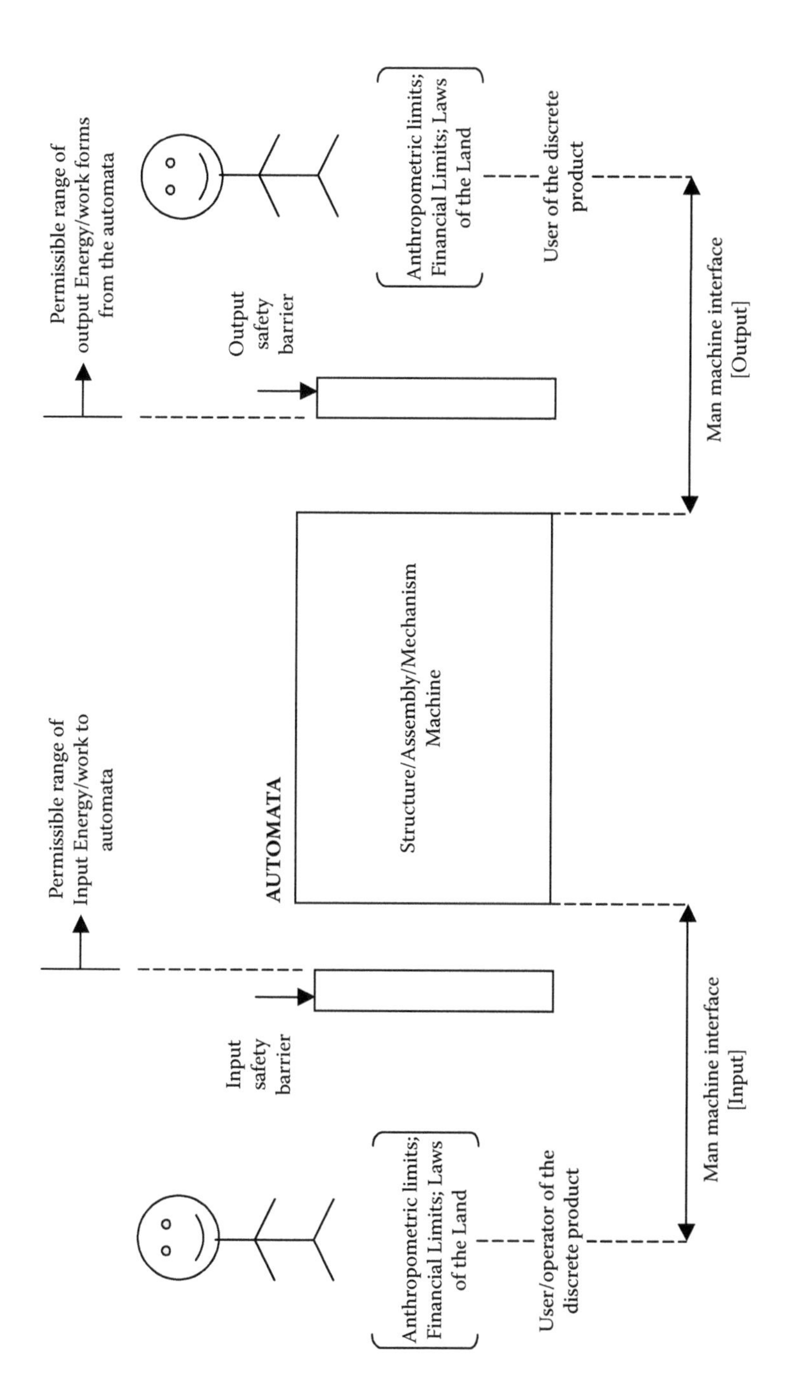

FIGURE 8.2 Man–machine interface for discrete products.

TABLE 8.1
Standards Related to Man–Machine Interface in Discrete Products

ISO 3600:1996
Tractors, Machinery for Agriculture and Forestry, Powered Lawn and Garden Equipment—Operator's Manuals—Content and Presentation
Edition: 3 (monolingual)
Number of pages: 16
Technical committee/subcommittee: TC 23/SC 14
ICS: 65.060.01
Stage: 90.93
Stage date: 2001-07-25

ISO 5805:1997
Mechanical Vibration and Shock—Human Exposure—Vocabulary
Edition: 2 (bilingual)
Number of pages: 21
Technical committee/subcommittee: TC 108/SC 4
ICS: 13.160
Stage: 90.93
Stage date: 2002-11-29

ISO 8655-1:2002
Piston-Operated Volumetric Apparatus—Part 1: Terminology, General Requirements, and User Recommendations
Edition: 1 (monolingual)
Number of pages: 9
Technical committee/subcommittee: TC 48
ICS: 01.040.17; 17.060
Stage: 60.60
Stage date: 2002-10-10

ISO 8835-3:1997
Inhalational Anaesthesia Systems—Part 3: Anaesthetic Gas Scavenging Systems—Transfer and Receiving Systems
Available in English only
Edition: 1 (monolingual)
Number of pages: 16
Technical committee/subcommittee: TC 121/SC 1
ICS: 11.040.10
Stage: 90.92
Stage date: 2003-07-01

TABLE 8.1 (Continued)
Standards Related to Man–Machine Interface in Discrete Products

ISO 9996:1996
Mechanical Vibration and Shock—Disturbance to Human Activity and Performance—Classification
Edition: 1 (monolingual)
Number of pages: 8
Technical committee/subcommittee: TC 108/SC 4
ICS: 13.160
Stage: 90.93
Stage date: 2001-10-16

ISO 10326-2:2001
Mechanical Vibration—Laboratory Method for Evaluating Vehicle Seat Vibration—Part 2: Application to Railway Vehicle
Edition: 1 (monolingual)
Number of pages: 17
Technical committee/subcommittee: TC 108/SC 2
ICS: 45.060.01; 13.160
Stage: 60.60
Stage date: 2001-08-23

ISO 11161:1994
Industrial Automation Systems—Safety of Integrated Manufacturing Systems—Basic Requirements
Available in English only
Edition: 1 (monolingual)
Number of pages: 25
Technical committee/subcommittee: TC 184
ICS: 25.040.01
Stage: 90.92
Stage date: 2001-06-15

ISO 15001:2003
Anesthetic and Respiratory Equipment—Compatibility with Oxygen
Edition: 1 (monolingual)
Number of pages: 39
Technical committee/subcommittee: TC 121/SC 6
ICS: 11.040.10
Stage: 60.60
Stage date: 2003-05-21

(*Continued*)

TABLE 8.1 (Continued)
Standards Related to Man–Machine Interface in Discrete Products

ISO 17776:2000
Petroleum and Natural Gas Industries—Offshore Production Installations—Guidelines on Tools and Techniques for Hazard Identification and Risk Assessment
Available in English only
Edition: 1 (monolingual)
Number of pages: 59
Technical committee/subcommittee: TC 67/SC 6
ICS: 75.180.10
Stage: 60.60
Stage date: 2000-10-19

ISO 15190:2003
Medical Laboratories—Requirements for Safety
Available in English only
Edition: 1 (monolingual)
Number of pages: 39
Technical committee/subcommittee: TC 212
ICS: 11.100
Stage: 60.60
Stage date: 2003-10-27

ISO 11690-1:1996
Acoustics—Recommended Practice for the Design of Low-Noise Workplaces Containing Machinery—Part 1: Noise Control Strategies
Edition: 1 (monolingual)
Number of pages: 23
Technical committee/subcommittee: TC 43/SC 1
ICS: 13.140
Stage: 90.93
Stage date: 2001-12-31

ISO 14152:2001
Neutron Radiation Protection Shielding—Design Principles and Considerations for the Choice of Appropriate Materials
Edition: 1 (monolingual)
Number of pages: 75
Technical committee/subcommittee: TC 85/SC 2
ICS: 13.280
Stage: 60.60
Stage date: 2001-12-20

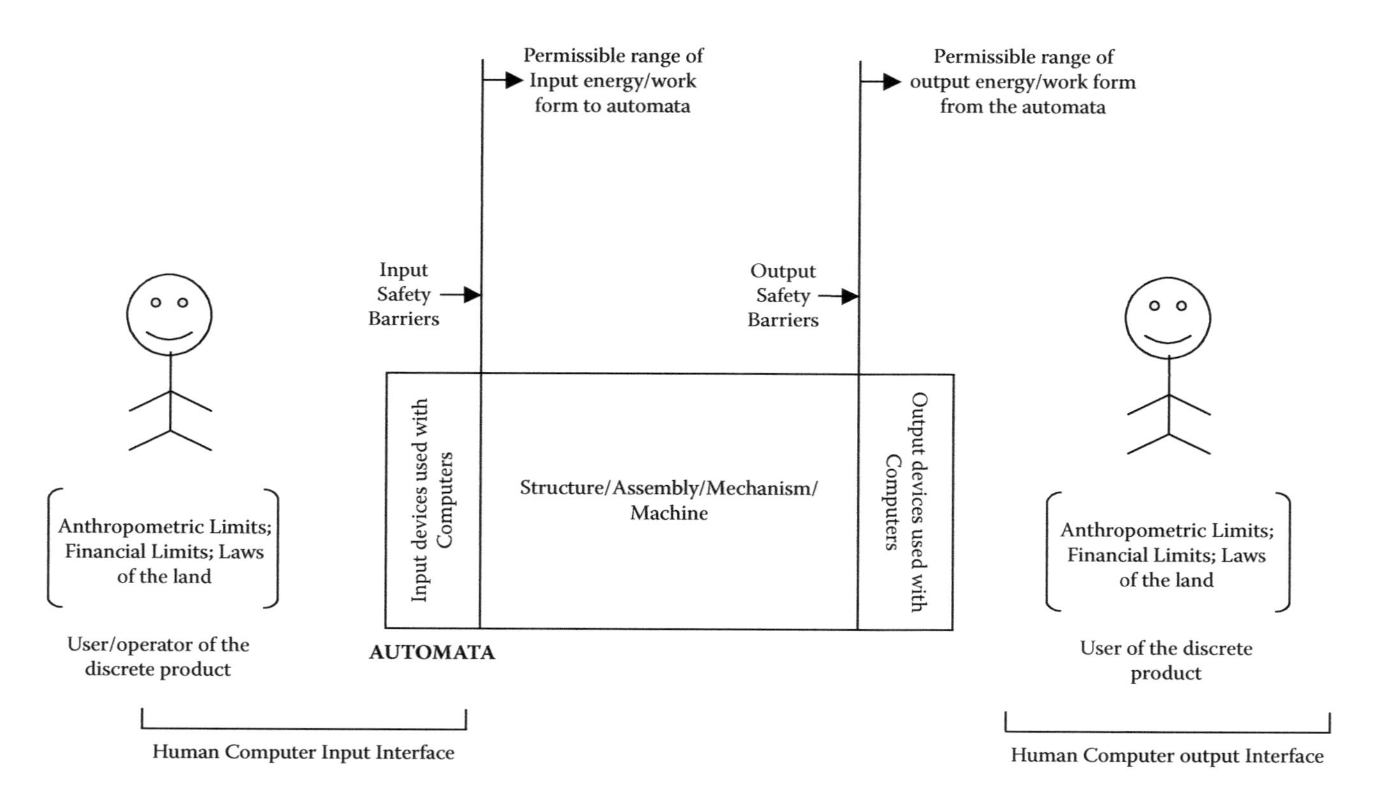

FIGURE 8.3 Human–computer interface for discrete products.

TABLE 8.2
Standards Related to Human–Computer Interaction in Discrete Products

ISO 13407:1999
Human-Centered Design Processes for Interactive Systems
Edition: 1 (monolingual)
Number of pages: 26
Technical committee/subcommittee: TC 159/SC 4
ICS: 13.180
Stage: 90.20
Stage date: 2003-11-25

ISO/TR 18307:2001
Health Informatics—Interoperability and Compatibility in Messaging and Communication Standards—Key Characteristics
Available in English only
Edition: 1 (monolingual)
Number of pages: 93
Technical committee/subcommittee: TC 215
ICS: 35.240.80
Stage: 60.60
Stage date: 2001-12-20

ISO/IEC 9506-5:1999
Industrial Automation Systems—Manufacturing Message Specification—Part 5: Companion Standard for Programmable Controllers
Edition: 1 (bilingual)
Number of pages: 129
Technical committee/subcommittee: IEC/TC 65
ICS: 25.040.40
Stage: 90.20
Stage date: 2003-11-25

ISO 14915-3:2002
Software Ergonomics for Multimedia User Interfaces—Part 3: Media Selection and Combination
Edition: 1 (monolingual)
Number of pages: 46
Technical committee/subcommittee: TC 159/SC 4
ICS: 35.200; 13.180
Stage: 60.60
Stage date: 2002-10-10

ISO/IEC 15411:1999
Information Technology—Segmented Keyboard Layouts
Available in English only
Edition: 1 (monolingual)
Number of pages: 15

TABLE 8.2 (Continued)
Standards Related to Human–Computer Interaction in Discrete Products

Technical committee/subcommittee: JTC 1/SC 35
ICS: 35.180
Stage: 90.20
Stage date: 2004-02-12

ISO/TS 16071:2003
Ergonomics of Human-System Interaction—Guidance on Accessibility for Human-Computer Interfaces
Edition: 1 (monolingual)
Number of pages: 29
Technical committee/subcommittee: TC 159/SC 4
ICS: 13.180
Stage: 90.92
Stage date: 2003-05-12

ISO 14971:2000
Medical Devices—Application of Risk Management to Medical Devices
Edition: 1 (monolingual)
Number of pages: 32
Technical committee/subcommittee: TC 210
ICS: 11.040.01
Stage: 90.92
Stage date: 2002-12-02

ISO 15623:2002
Transport Information and Control Systems—Forward Vehicle Collision Warning Systems—Performance Requirements and Test Procedures
Available in English only
Edition: 1 (monolingual)
Number of pages: 25
Technical committee/subcommittee: TC 204
ICS: 03.220.01; 43.040.99; 35.240.60
Stage: 60.60
Stage date: 2002-10-10

ISO 10218:1992
Manipulating Industrial Robots—Safety
Edition: 1 (monolingual)
Number of pages: 10
Technical committee/ subcommittee: TC 184/SC 2
ICS: 25.040.30
Stage: 90.92
Stage date: 2000-06-13

(Continued)

TABLE 8.2 (Continued)
Standards Related to Human–Computer Interaction in Discrete Products

ISO 10993-1:2003
Biological Evaluation of Medical Devices—Part 1: Evaluation and Testing
Edition: 3 (monolingual)
Number of pages: 14
Technical committee/subcommittee: TC 194
ICS: 11.100
Stage: 60.60
Stage date: 2003-07-28

ISO/TR 14105:2001
Electronic Imaging—Human and Organizational Issues for Successful Electronic Image Management (EIM) Implementation
Available in English only
Edition: 1 (monolingual)
Number of pages: 20
Technical committee/subcommittee: TC 171/SC 2
ICS: 37.080
Stage: 60.60
Stage date: 2001-10-18

of its increased use, many ergonomics features are standardized. The general scope of standards related to ergonomics principles is presented in Figure 8.4. Table 8.3 lists a selection of ergonomics standards that can be used in a diverse range of discrete products.

8.5 STANDARDS RELATED TO OUTPUT FROM THE DISCRETE PRODUCTS

Each discrete product utilizes an input, process, and output cycle. The safe working limit of the input parameters, process parameters, and output parameters is the basic requirement of the design of discrete products. Standards exist that cover the permissible range of parameters related to a variety of discrete products. Figure 8.5 defines the scope of standards related to the output from discrete products. Table 8.4 provides a selection of ISO standards for the category.

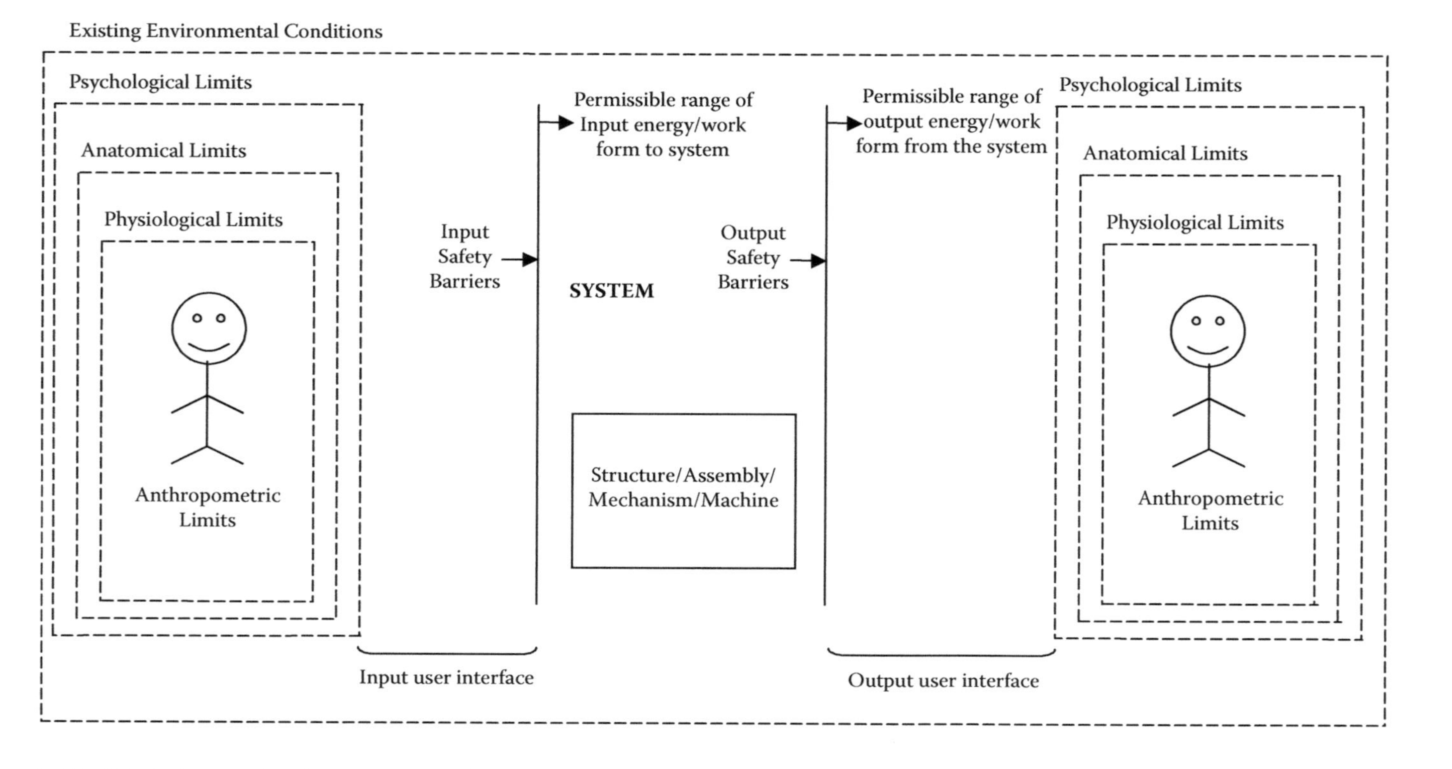

FIGURE 8.4 Ergonomics for a discrete product.

TABLE 8.3
Standards Related to Ergonomics of Discrete Products

ISO 3411:1995
Earth-Moving Machinery—Human Physical Dimensions of Operators and Minimum Operator Space Envelope
Edition: 3 (monolingual)
Number of pages: 10
Technical committee/subcommittee: TC 127/SC 2
ICS: 53.100
Stage: 90.92
Stage date: 2003-04-30

ISO 3958:1996
Passenger Cars—Driver Hand-Control Reach
Edition: 2 (monolingual)
Number of pages: 49
Technical committee/subcommittee: TC 22/SC 13
ICS: 43.040.30
Stage: 90.93
Stage date: 2001-10-12

ISO 4040:2001
Road Vehicles—Location of Hand Controls, Indicators, and Tell-Tales in Motor Vehicles
Available in English only
Edition: 4 (monolingual)
Number of pages: 9
Technical committee/subcommittee: TC 22/SC 13
ICS: 43.040.30
Stage: 60.60
Stage date: 2001-12-13

ISO 5721:1989
Tractors for Agriculture—Operator's Field of Vision
Edition: 2 (monolingual)
Number of pages: 7
Technical committee/subcommittee: TC 23/SC 4
ICS: 65.060.10
Stage: 90.93
Stage date: 2000-12-15

ISO 6385:2004
Ergonomic Principles in the Design of Work Systems
Edition: 2 (monolingual)
Number of pages: 11
Technical committee/subcommittee: TC 159/SC 1
ICS: 13.180
Stage: 60.60
Stage date: 2004-01-13

TABLE 8.3 (Continued)
Standards Related to Ergonomics of Discrete Products

ISO 6682:1986
Earth-Moving Machinery—Zones of Comfort and Reach for Controls
Edition: 2 (monolingual)
Number of pages: 10
Technical committee/subcommittee: TC 127/SC 2
ICS: 53.100
Stage: 90.93
Stage date: 2001-07-27

ISO 7243:1989
Hot Environments—Estimation of the Heat Stress on Working Man, Based on the WBGT-Index (Wet Bulb Globe Temperature)
Edition: 2 (monolingual)
Number of pages: 9
Technical committee/subcommittee: TC 159/SC 5
ICS: 13.180
Stage: 90.93
Stage date: 2003-06-17

ISO 7250:1996
Basic Human Body Measurements for Technological Design
Edition: 1 (monolingual)
Number of pages: 21
Technical committee/subcommittee: TC 159/SC 3
ICS: 13.180
Stage: 90.93
Stage date: 2001-07-09

ISO 7726:1998
Ergonomics of the Thermal Environment—Instruments for Measuring Physical Quantities
Edition: 2 (monolingual)
Number of pages: 51
Technical committee/subcommittee: TC 159/SC 5
ICS: 13.180
Stage: 90.93
Stage date: 2003-05-12

ISO 7731:2003
Ergonomics—Danger Signals for Public and Work Areas—Auditory Danger Signals
Edition: 2 (monolingual)
Number of pages: 17
Technical committee/subcommittee: TC 159/SC 5
ICS: 13.320; 13.180
Stage: 60.60
Stage date: 2003-10-30

(*Continued*)

TABLE 8.3 (Continued)
Standards Related to Ergonomics of Discrete Products

ISO/CIE 8995:2002
Lighting of Indoor Work Places
Available in English only
Edition: 2 (monolingual)
Number of pages: 19
Technical committee/subcommittee: CIE
ICS: 91.160.10; 13.180
Stage: 60.60
Stage date: 2002-06-13

ISO 8996:1990
Ergonomics—Determination of Metabolic Heat Production
Edition: 1 (monolingual)
Number of pages: 17
Technical committee/subcommittee: TC 159/SC 5
ICS: 13.180
Stage: 90.92
Stage date: 2000-05-25

ISO 9241-1:1997
Ergonomic Requirements for Office Work with Visual Display Terminals (VDTs)—Part 1: General Introduction
Edition: 2 (monolingual)
Number of pages: 7
Technical committee/subcommittee: TC 159/SC 4
ICS: 35.180; 13.180
Stage: 90.93
Stage date: 2002-12-11

ISO 9886:2004
Ergonomics—Evaluation of Thermal Strain by Physiological Measurements
Edition: 2 (monolingual)
Number of pages: 21
Technical committee/subcommittee: TC 159/SC 5
ICS: 13.180
Stage: 60.60
Stage date: 2004-02-09

ISO 9921:2003
Ergonomics—Assessment of Speech Communication
Edition: 1 (monolingual)
Number of pages: 28
Technical committee/subcommittee: TC 159/SC 5
ICS: 13.180
Stage: 60.60
Stage date: 2003-10-22

TABLE 8.3 (Continued)
Standards Related to Ergonomics of Discrete Products

ISO 10075:1991
Ergonomic Principles Related to Mental Work-Load—General Terms and Definitions
Edition: 1 (monolingual)
Number of pages: 5
Technical committee/subcommittee: TC 159/SC 1
ICS: 13.180; 01.040.13
Stage: 90.92
Stage date: 1999-06-30

ISO 10551:1995
Ergonomics of the Thermal Environment—Assessment of the Influence of the Thermal Environment Using Subjective Judgment Scales
Edition: 1 (monolingual)
Number of pages: 18
Technical committee/subcommittee: TC 159/SC 5
ICS: 13.180
Stage: 90.93
Stage date: 2000-10-05

ISO 11064-1:2000
Ergonomic Design of Control Centers—Part 1: Principles for the Design of Control Centers
Edition: 1 (monolingual)
Number of pages: 30
Technical committee/subcommittee: TC 159/SC 4
ICS: 13.180
Stage: 60.60
Stage date: 2000-12-21

ISO 11226:2000
Ergonomics—Evaluation of Static Working Postures
Edition: 1 (monolingual)
Number of pages: 19
Technical committee/subcommittee: TC 159/SC 3
ICS: 13.180
Stage: 60.60
Stage date: 2000-12-21

ISO 11228-1:2003
Ergonomics—Manual Handling—Part 1: Lifting and Carrying
Edition: 1 (monolingual)
Number of pages: 23
Technical committee/subcommittee: TC 159/SC 3
ICS: 13.180
Stage: 60.60
Stage date: 2003-05-20

(Continued)

TABLE 8.3 (Continued)
Standards Related to Ergonomics of Discrete Products

ISO 11428:1996
Ergonomics—Visual Danger Signals—General Requirements, Design and Testing
Edition: 1 (monolingual)
Number of pages: 8
Technical committee/subcommittee: TC 159/SC 5
ICS: 13.180; 13.320
Stage: 90.93
Stage date: 2003-05-12

ISO 11429:1996
Ergonomics—System of Auditory and Visual Danger and Information Signals
Edition: 1 (monolingual)
Number of pages: 7
Technical committee/subcommittee: TC 159/SC 5
ICS: 13.180; 13.320
Stage: 90.93
Stage date: 2003-05-12

ISO 13688:1998
Protective Clothing—General Requirements
Edition: 1 (monolingual)
Number of pages: 11
Technical committee/subcommittee: TC 94/SC 13
ICS: 13.340.10
Stage: 90.93
Stage date: 2003-11-12

ISO 14644-4:2001
Clean Rooms and Associated Controlled Environments—Part 4: Design, Construction, and Start-Up
Edition: 1 (monolingual)
Number of pages: 51
Technical committee/subcommittee: TC 209
ICS: 13.040.35
Stage: 60.60
Stage date: 2001-04-12

ISO 14915-1:2002
Software Ergonomics for Multimedia User Interfaces—Part 1: Design Principles and Framework
Edition: 1 (monolingual)
Number of pages: 12
Technical committee/subcommittee: TC 159/SC 4
ICS: 13.180; 35.200
Stage: 60.60
Stage date: 2002-11-18

TABLE 8.3 (Continued)
Standards Related to Ergonomics of Discrete Products

ISO 15534-1:2000

Ergonomic Design for the Safety of Machinery—Part 1: Principles for Determining the Dimensions Required for Openings for Whole-Body Access into Machinery
Available in English only
Edition: 1 (monolingual)
Number of pages: 12
Technical committee/subcommittee: TC 159/SC 3
ICS: 13.110; 13.180
Stage: 60.60
Stage date: 2000-02-17

ISO/PAS 18152:2003

Ergonomics of Human-System Interaction—Specification for the Process Assessment of Human-System Issues
Available in English only
Edition: 1 (monolingual)
Number of pages: 92
Technical committee/ subcommittee: TC 159/SC 4
ICS: 13.180
Stage: 60.60
Stage date: 2003-10-13

ISO/TR 18529:2000

Ergonomics—Ergonomics of Human-System Interaction—Human-Centered Lifecycle Process Descriptions
Available in English only
Edition: 1 (monolingual)
Number of pages: 28
Technical committee/subcommittee: TC 159/SC 4
ICS: 13.180
Stage: 90.20
Stage date: 2003-11-25

ISO/TR 19358:2002

Ergonomics—Construction and Application of Tests for Speech Technology
Available in English only
Edition: 1 (monolingual)
Number of pages: 15
Technical committee/subcommittee: TC 159/SC 5
ICS: 13.180
Stage: 60.60
Stage date: 2002-10-10

(Continued)

TABLE 8.3 (Continued)
Standards Related to Ergonomics of Discrete Products

ISO 7731:2003
Ergonomics—Danger Signals for Public and Work Areas—Auditory Danger Signals
Edition: 2 (monolingual)
Number of pages: 17
Technical committee/subcommittee: TC 159/SC 5
ICS: 13.320; 13.180
Stage: 60.60
Stage date: 2003-10-30

ISO/TR 16982:2002
Ergonomics of Human-System Interaction—Usability Methods Supporting Human-Centered Design
Edition: 1 (monolingual)
Number of pages: 44
Technical committee/subcommittee: TC 159/SC 4
ICS: 13.180
Stage: 60.60
Stage date: 2002-06-13

ISO/PAS 18152:2003
Ergonomics of Human-System Interaction—Specification for the Process Assessment of Human-System Issues
Available in English only
Edition: 1 (monolingual)
Number of pages: 92
Technical committee/subcommittee: TC 159/SC 4
ICS: 13.180
Stage: 60.60
Stage date: 2003-10-13

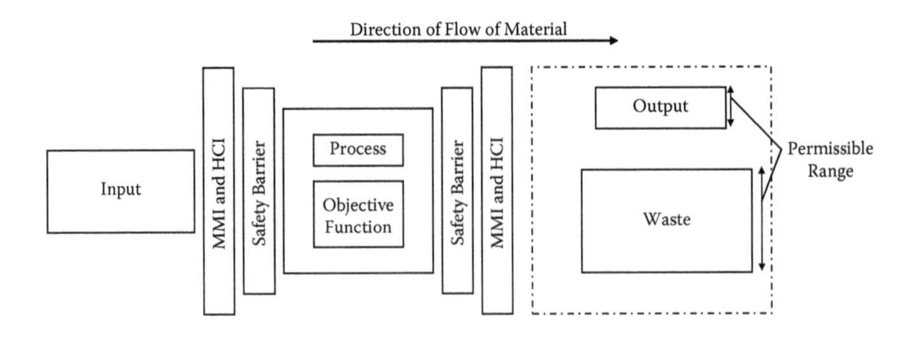

FIGURE 8.5 Scope of standards related to output (emission) from discrete product.

TABLE 8.4
Standards Related to Output from the Discrete Products

ISO 362:1998
Acoustics—Measurement of Noise Emitted By Accelerating Road Vehicles—Engineering Method
Available in English only
Edition: 3 (monolingual)
Number of pages: 12
Technical committee/subcommittee: TC 43/SC 1
ICS: 43.020; 17.140.30
Stage: 90.92
Stage date: 1999-04-22

ISO 789-4:1986
Agricultural Tractors—Test Procedures—Part 4: Measurement of Exhaust Smoke
Edition: 2 (monolingual)
Number of pages: 7
Technical committee/subcommittee: TC 23/SC 2
ICS: 65.060.10; 13.040.50
Stage: 90.93
Stage date: 2001-07-25

ISO 1229:1989
Photography—Expendable Photoflash Lamps—Determination of Light Output
Edition: 2 (monolingual)
Number of pages: 4
Technical committee/subcommittee: TC 42
ICS: 37.040.10
Stage: 90.20
Stage date: 2003-11-25

ISO 1680:1999
Acoustics—Test Code for the Measurement of Airborne Noise Emitted by Rotating Electrical Machines
Edition: 1 (monolingual)
Number of pages: 15
Technical committee/subcommittee: TC 43/SC 1
ICS: 17.140.20; 29.160.01
Stage: 90.20
Stage date: 2003-11-25

ISO 1996-3:1987
Acoustics—Description and Measurement of Environmental Noise—Part 3: Application to Noise Limits
Edition: 1 (monolingual)
Number of pages: 3

(*Conitnued*)

TABLE 8.4 (Continued)
Standards Related to Output from the Discrete Products

Technical committee/subcommittee: TC 43/SC 1
ICS: 13.140
Stage: 90.92
Stage date: 2001-08-31

ISO 2889:1975
General Principles for Sampling Airborne Radioactive Materials
Edition: 1 (monolingual)
Number of pages: 25
Technical committee/subcommittee: TC 85/SC 2
ICS: 13.280
Stage: 90.92
Stage date: 1989-01-15

ISO 3891:1978
Acoustics—Procedure for Describing Aircraft Noise Heard on the Ground
Edition: 1 (monolingual)
Number of pages: 24
Technical committee/subcommittee: TC 43/SC 1
ICS: 49.020; 17.140.30
Stage: 95.92
Stage date: 2002-10-10

ISO 3929:2003
Road Vehicles—Measurement Methods for Exhaust Gas Emissions During Inspection or Maintenance
Edition: 3 (monolingual)
Number of pages: 6
Technical committee/subcommittee: TC 22/SC 5
ICS: 13.040.50; 43.060.20
Stage: 60.60
Stage date: 2003-08-21

ISO 3930:2000
Instruments for Measuring Vehicle Exhaust Emissions
Edition: 3 (monolingual)
Number of pages: 27
Technical committee/subcommittee: TC 22/SC 5
ICS: 43.180; 13.040.50
Stage: 60.60
Stage date: 2000-09-21

ISO 3977-4:2002
Gas Turbines—Procurement—Part 4: Fuels and Environment
Edition: 1 (monolingual)

TABLE 8.4 (Continued)
Standards Related to Output from the Discrete Products

Number of pages: 17
Technical committee/subcommittee: TC 192
ICS: 27.040
Stage: 60.60
Stage date: 2002-06-13

ISO 3999-1:2000
Radiation Protection—Apparatus for Industrial Gamma Radiography—Part 1: Specifications for Performance, Design, and Tests
Edition: 1 (monolingual)
Number of pages: 31
Technical committee/subcommittee: TC 85/SC 2
ICS: 19.100
Stage: 90.92
Stage date: 2004-03-10

ISO 4531-2:1998
Vitreous and Porcelain Enamels—Release of Lead and Cadmium from Enameled Ware in Contact with Food—Part 2: Permissible Limits
Edition: 1 (monolingual)
Number of pages: 3
Technical committee/subcommittee: TC 107
ICS: 25.220.50; 67.250; 97.040.60
Stage: 90.60
Stage date: 2003-07-30

ISO 6879:1995
Air Quality—Performance Characteristics and Related Concepts for Air-Quality Measuring Methods
Available in English only
Edition: 2 (monolingual)
Number of pages: 6
Technical committee/subcommittee: TC 146/SC 4
ICS: 13.040.01
Stage: 90.93
Stage date: 2000-10-19

ISO 7086-2:2000
Glass Hollowware in Contact with Food—Release of Lead and Cadmium—Part 2: Permissible Limits
Edition: 2 (monolingual)
Number of pages: 5
Technical committee/ subcommittee: TC 166
ICS: 81.040.30; 97.040.60; 67.250

(*Continued*)

TABLE 8.4 (Continued)
Standards Related to Output from the Discrete Products

Stage: 60.60
Stage date: 2000-03-02

ISO 7547:2002
Ships and Marine Technology—Air-Conditioning and Ventilation of Accommodation Spaces—Design Conditions and Basis of Calculations
Available in English only
Edition: 2 (monolingual)
Number of pages: 13
Technical committee/subcommittee: TC 8/SC 3
ICS: 47.020.90; 47.020.80
Stage: 60.60
Stage date: 2002-09-26

ISO 7960:1995
Airborne Noise Emitted by Machine Tools—Operating Conditions for Woodworking Machines
Edition: 1 (monolingual)
Number of pages: 115
Technical committee/subcommittee: TC 39/SC 6
ICS: 79.120.10; 17.140.20
Stage: 90.93
Stage date: 2000-12-14

ISO 8579-1:2002
Acceptance Code for Gear Units—Part 1: Test Code for Airborne Sound
Edition: 2 (monolingual)
Number of pages: 53
Technical committee/subcommittee: TC 60
ICS: 17.140.20; 21.200
Stage: 60.60
Stage date: 2002-08-08

ISO 8662-11:1999
Hand-Held Portable Power Tools—Measurement of Vibrations at the Handle—Part 11: Fastener Driving Tools
Edition: 1 (monolingual)
Number of pages: 11
Technical committee/subcommittee: TC 118/SC 3
ICS: 25.140.10; 13.160
Stage: 90.20
Stage date: 2003-11-25

ISO 9207:1995
Manually Portable Chain-Saws with Internal Combustion Engine—Determination of Sound Power Levels—Engineering Method (Grade 2)

TABLE 8.4 (Continued)
Standards Related to Output from the Discrete Products

Edition: 1 (monolingual)
Number of pages: 7
Technical committee/subcommittee: TC 23/SC 17
ICS: 17.140.20; 65.060.80
Stage: 90.93
Stage date: 2000-07-01

ISO 9785:2002
Ships and Marine Technology—Ventilation of Cargo Spaces Where Vehicles with Internal Combustion Engines Are Driven—Calculation of Theoretical Total Airflow Required
Available in English only
Edition: 2 (monolingual)
Number of pages: 10
Technical committee/subcommittee: TC 8/SC 3
ICS: 47.020.90
Stage: 60.60
Stage date: 2002-07-25

ISO 9902-5:2001
Textile Machinery—Noise Test Code—Part 5: Weaving and Knitting Preparatory Machinery
Edition: 1 (monolingual)
Number of pages: 6
Technical committee/subcommittee: TC 72/SC 8
ICS: 59.120.40; 59.120.30; 17.140.20
Stage: 60.60
Stage date: 2001-03-29

ISO 13617:2001
Ships and Marine Technology—Shipboard Incinerators—Requirements
Available in English only
Edition: 2 (monolingual)
Number of pages: 20
Technical committee/subcommittee: TC 8/SC 3
ICS: 47.020.99; 13.030.40
Stage: 60.60
Stage date: 2001-11-15

BIBLIOGRAPHY

Industrial man-machine interfaces, *Electron. Prod. Design*, 15(9), 57–58, 1994.
Aykin, N.M. and Aykin, T., Individual differences in human-computer interaction, *Comput. Industrial Eng.* 20(3), 373–379, 1991.
Bedrosian, S.D., Color and language as man-machine interface parameters, *Int. J. Intelligent Syst.,* 4(1), 45–54, 1989.

Brown, G.R., Control system design, Part 4: The man/machine interface, *InTech,* 36(8), 34–39, 1989.

Butler, K.A., Jacob, R.J.K., and John, B.E., Introduction and overview to Human-Computer Interaction, Conference on Human Factors in Computing Systems, Proceedings, p. 345–346, 1995.

Charwat, H.J., Proper man machine interface for process control, *Process Automation*, 1, 22–29, 1984.

Downton, A.C., Engineering the man-machine interface, *Electron. & Power*, 33(11), 691–694, 1987.

Eason, K.D., Ergonomic perspectives on advances in human-computer interaction, *Ergonomics*, 34(6), 721–741, 1991.

Froehlich, T.J., *Phenomenological Critique of Models of Man-machine Interface, ACM*, New York, 1983, 86.

Halter, R., Man-machine interface design challenges, *Design News*, 41(16), 63–70, 1985.

Kee, D., Jung, E.S., and Chang, S.R., Man-machine interface model for ergonomic design, *Comput Industrial Eng*. 27(1–4), 365–368, 1994.

Lewis, C.H., Research agenda for the nineties in human-computer interaction, *Hum. Comput. Interaction,* 5(2–3), 125–143, 1990.

Matsumura, I., Tatsuno, J., Kokubo, Y., and Kobayashi, H., Man-machine Interface with Verbal Communication, Robot and Human Communication—Proceedings of the IEEE International Workshop, 1997, 398–401.

Nishitani, H., Human-computer interaction in the new process technology, *J. Process Control,* 6(2–3), 111–117, 1996.

Oborski, P., Man-machine interactions in advanced manufacturing systems, *Int. J. Advanced Manufacturing Technol*. 23(3–4), 227–232, 2004.

Obrenovic, Z., and Starcevic, D., Modeling multimodal human-computer interaction, *Computer,* 37(9), 65–72, 2004.

Smith, M.J., Human Factors Safety Considerations in Human/Computer Interaction, American Institute of Chemical Engineers, National Meeting, Seattle, WA, 1985,

Visner, R.J., Console systems: The man-machine interface, *Control Eng.,* 35(Part 2,9), 18–19, 1988.

9 Application of Human Factors to Design and Manufacturing of Discrete Products—A Case Study

9.1 INTRODUCTION

Humans have a better adaptability in the natural world than in a synthetic environment. Living in a synthetic environment versus living in a natural environment creates a contrast that is best handled by centuries-old considerations to subjects such as human factors, biomechanics, engineering psychology, and ergonomics. These overlapping interdisciplinary subjects aim to enhance modern-day living in a world full of discrete products.

Products are designed to be used by human beings. The designers must consider the user's capabilities and capacities if the product or system is to be used appropriately and without significant errors made by the user. For example, while designing an instrument dial, the designer must consider such factors as the correct position of the dial (the operator should not have to make unnecessary head and eye movements to observe it), the size of the markings (the operator with normal vision should be able to read them under available light conditions), the arrangement of controls on the panel (they are easily discernible and should not increase the operator's reaction time).

A person can be viewed as a mechanical structure who can read only so far, lift only so much, sense only so rapidly, and react only so quickly. His or her sensory system has definite limitations. Every human response to a situation involves perception, decision, and response. Difficulties arise when a demand is made for abnormal uses of a person's senses, quick decisions based on inadequate and overwhelming information, excessively rapid responses, and physical capabilities that the person does not have.

A brief description of capabilities and limitations of the user of discrete products is provided. This discussion is followed by descriptions of some human factor–related design criteria.

9.2 CAPABILITIES AND LIMITATIONS OF USERS

A human being, whenever operating in a "man–machine" system, processes information. A signal is presented to him or her and he or she must take the required action. A "man–machine" system could be a child playing with a mechanical toy or a pilot flying the most sophisticated jet plane. Human information-processing capabilities have definite limitations that must be considered in design. The stimulus provided is either visual, auditory, temperature, etc. Each sensory organ responds to its specific energy spectrum and is sensitive to a certain range, band, and magnitude of stimulation.

9.2.1 The Auditory Modality

The hearing mechanism responds to impingement by rapidly oscillating air, solid medium, or liquid medium that is excited by a sound-emitting body. The minimum detectable oscillating force is 0.001 dynes/cm^2; the maximum is 1,000 dynes/cm^2.

9.2.2 The Visual Modality

The visual modality responds to a spectral range of light energy in wavelengths from 400 to 760 μm. The intensity range of sensitivity varies; a completely dark adapted eye has a 50% probability of detecting 10^{-10} fc of power. Acuity, contrast effects, and the ability to differentiate intensity levels become asymptotic around 100 fc.

9.2.3 Other Sensory Limitations

Sensory response of skin is minuscule compared with the magnitude of stimulation possible. Skin responds at around 2 mg of pressure. Odors can be detected at faint molecule levels, but odors rapidly fade because of saturation and fatigue.

9.3 INFORMATION-PROCESSING LIMITATIONS

Sensory information can be of value to the human operator only if it conforms to the output of the nervous system

9.3.1 Expectations

A human becomes used to the ways things are happening. For example, a dial is read clockwise to indicate increase and a control device is pushed up to increase the reading.

9.3.2 Information Processing

A person is likely to be unreliable when recalling something that just happened. A person is severely limited in the number of different operations that he or she can perform in a given time period.

9.3.3 Emotionalism

A person is sensitive to interpersonal conflicts and personal status.

9.3.4 Boredom

Information-processing efficiency is degraded by long cycles and repetitious tasks.

9.3.5 Sensitivity to Stress

Information processing is limited by the amount of stress in an operation. Information-processing capability also falls off when distractions occur because of poor lighting, high noise levels, and irrelevant work requirements.

9.4 LIMITATIONS OF AN EFFECTOR SUBSYSTEM

Human performance as an output is influenced by the size of the body's effector mechanism, its lifting and force-exerting capability, and its reaction time and tracking coordination limitations.

9.4.1 Body Measurement or Anthropometry

A person has body and size characteristics that must be accommodated and that limit the extent of output capacity. Such data are available in various texts on human factors. Such data most often pertains to military populations. However, commercial firms have also compiled anthropometric data on civilian populations.

9.4.2 Lifting and Force Capabilities

A person's lifting strength and force application capabilities are limited by muscle, bone, and tendon structures. Lifting and force capability is a function of torqueing forces that can be applied at the fixed or moveable joints between bones of the human body.

9.4.3 Response-Speed Capability

A person is inherently limited by the maximum possible speed of a human reaction. Minimum reaction time is on the order of 0.2 to 0.3 seconds when the person is anticipating or has a signal expectancy.

9.4.4 Tracking Capabilities

Tracking behavior, or finding and pursuing a signal element, is limited by reaction time. The person is generally limited by a response lag of 0.5 seconds.

The designer must accommodate inherent limitations in human performance and, thus, make the product or system more suitable to humans.

9.5 HUMAN PERFORMANCE CRITERIA

The display and control requirements analysis provides the necessary design information for selecting the best configuration that will serve the person in performing the required task.

Figure 9.1 presents a few conventional type displays. A rule of thumb for good design is to use the simplest configuration to avoid distraction. Auditory display can also be used effectively to reduce the load on vision and provide warning to the operator.

Figure 9.2 presents some of the conventional controls. Other factors that affect human performance are the following:

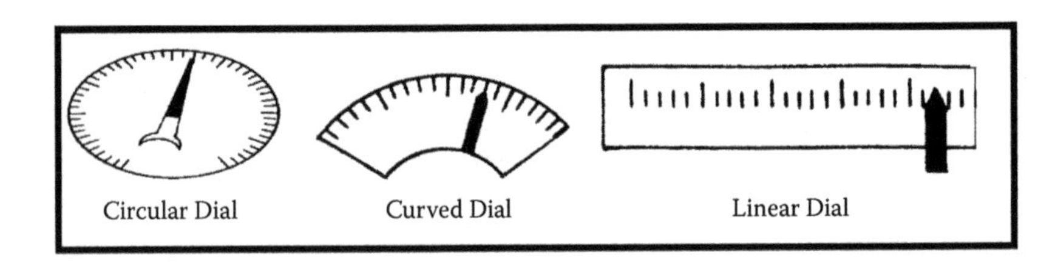

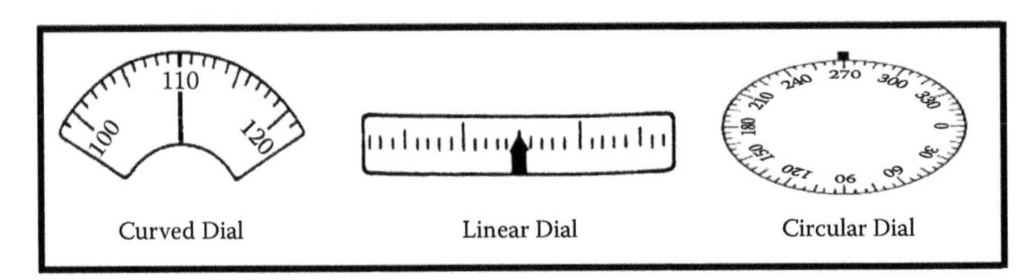

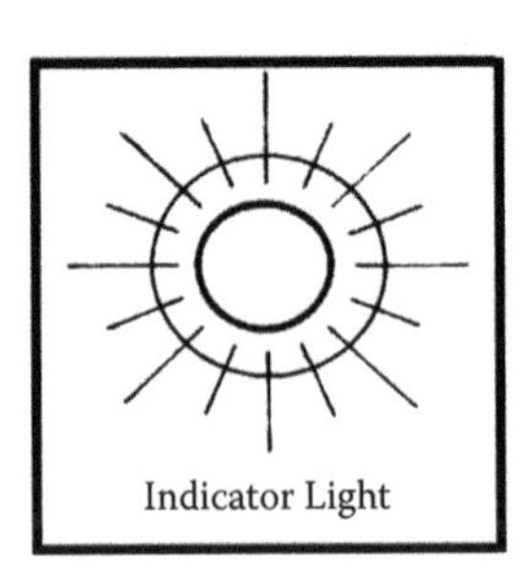

FIGURE 9.1 Common types of displays.

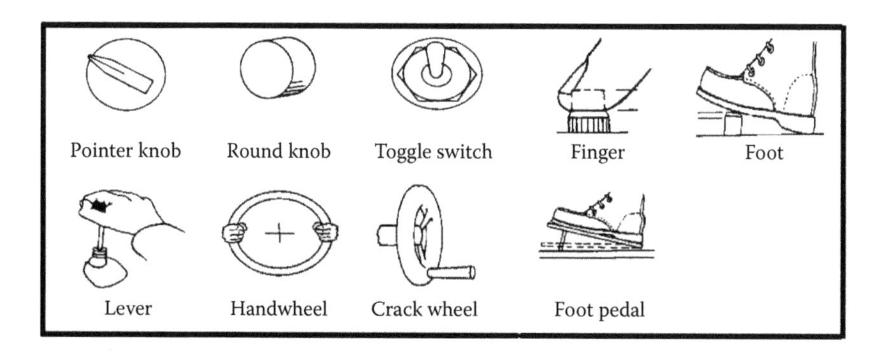

FIGURE 9.2 Common types of controls.

Noise
Vibrations
Illumination
Temperature and limitations
Pollutants in the environment

9.5.1 Workplace Design Criteria

The workplace design must be adjusted to accommodate people of varying anthropometric dimensions. When the largest likely worker for freedom of movement and clearance of obstructions is accommodated, all persons having smaller dimensions will automatically be accommodated.

9.5.2 Human Performance Assessment Criteria

The selection of alternative display and control configurations, as well as different workplace layouts can be based on measurable performance criteria. For further details, please refer to books on human factors.

9.6 THE AVAILABILITY AND APPLICATION OF STANDARD INFORMATION

A large number of design examples that have ergonomics flaws are presented at www.baddesign.com/examples.html. A wide variety of standards is available that apply to ergonomically designed discrete products. Section 8.3 lists a brief selection of these standards. Figure 9.3 to Figure 9.6 detail the adoptability of ergonomic factors in the system approach to design.

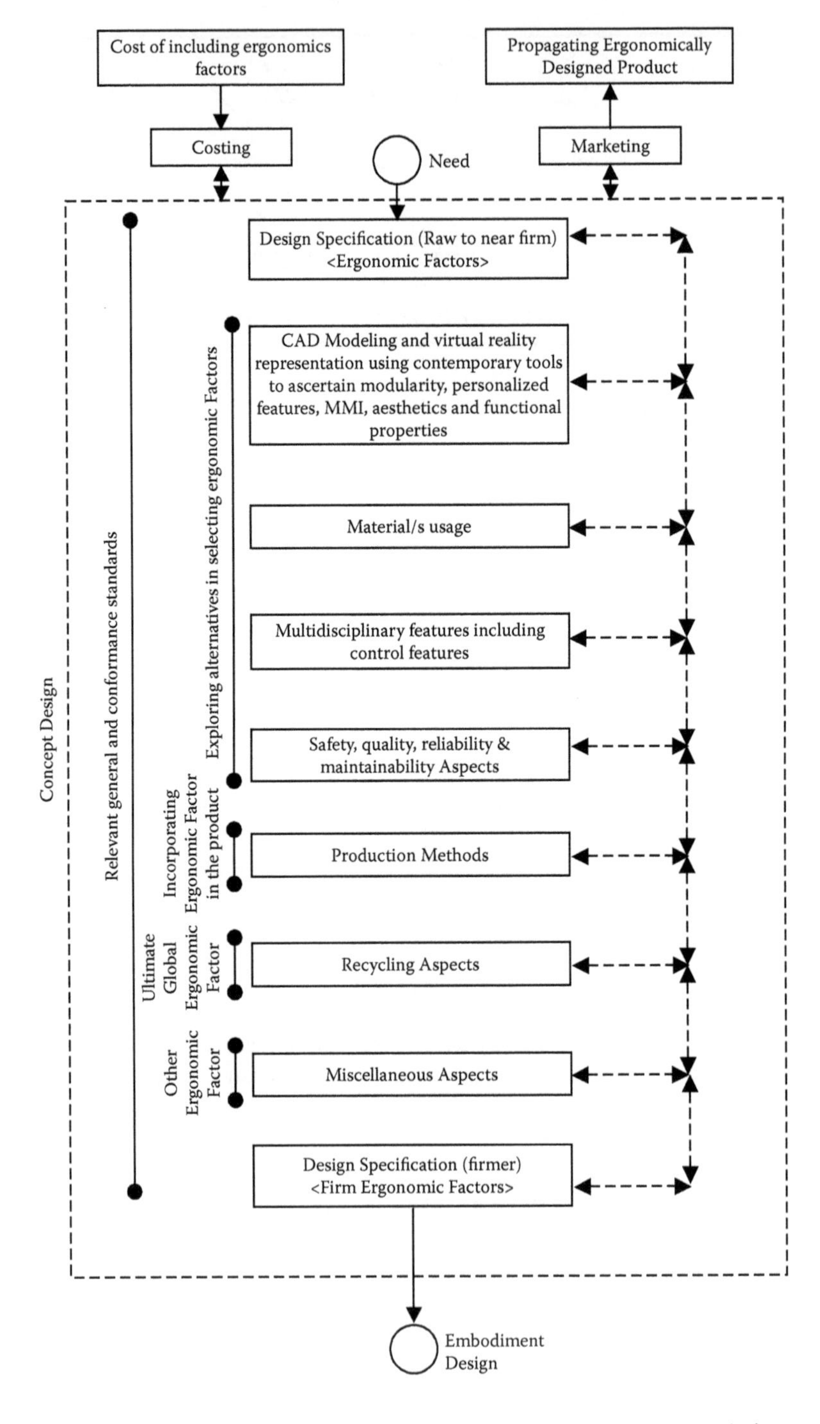

FIGURE 9.3 An ergonomics perspective in the system approach to concept design.

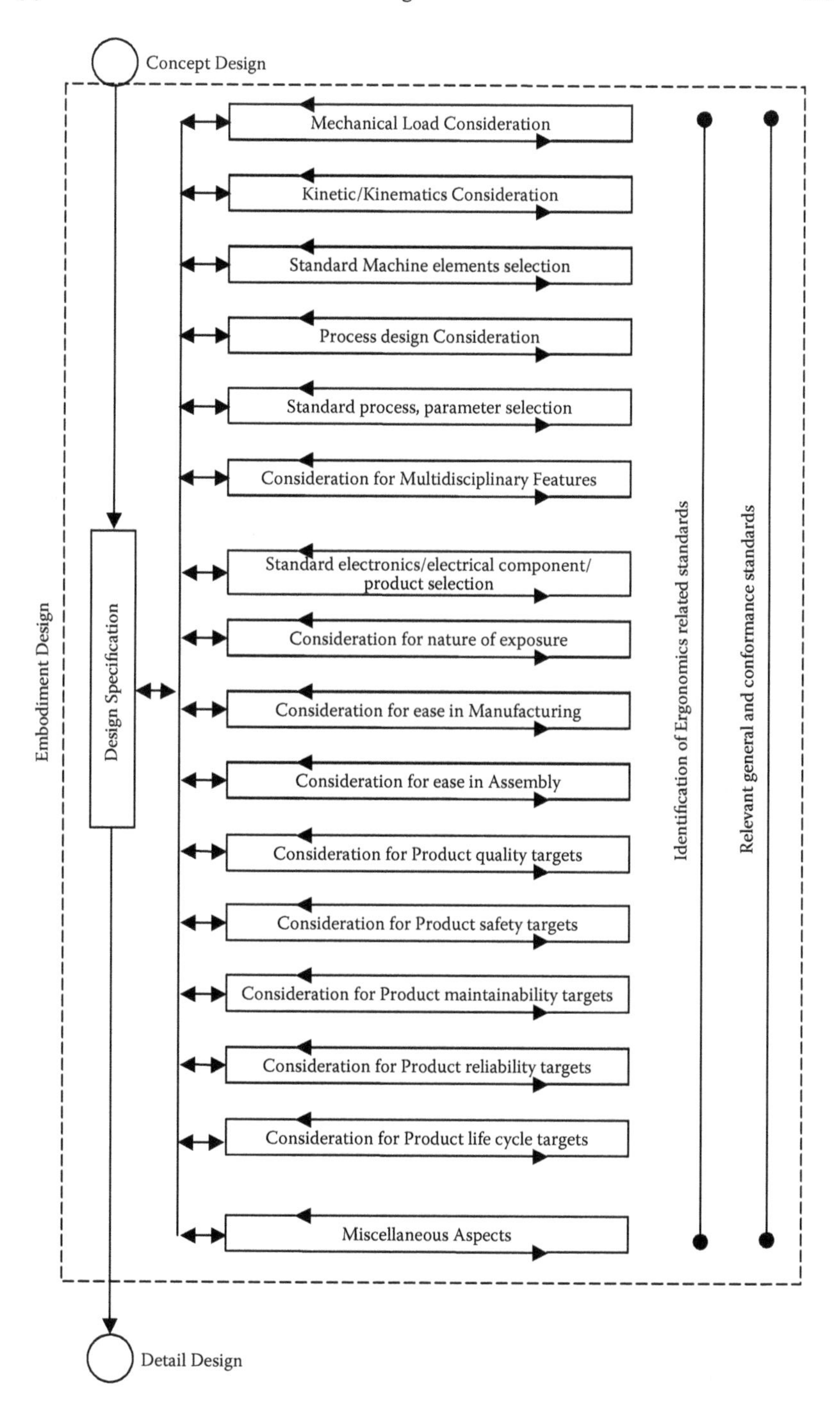

FIGURE 9.4 An ergonomics perspective in the system approach to embodiment design.

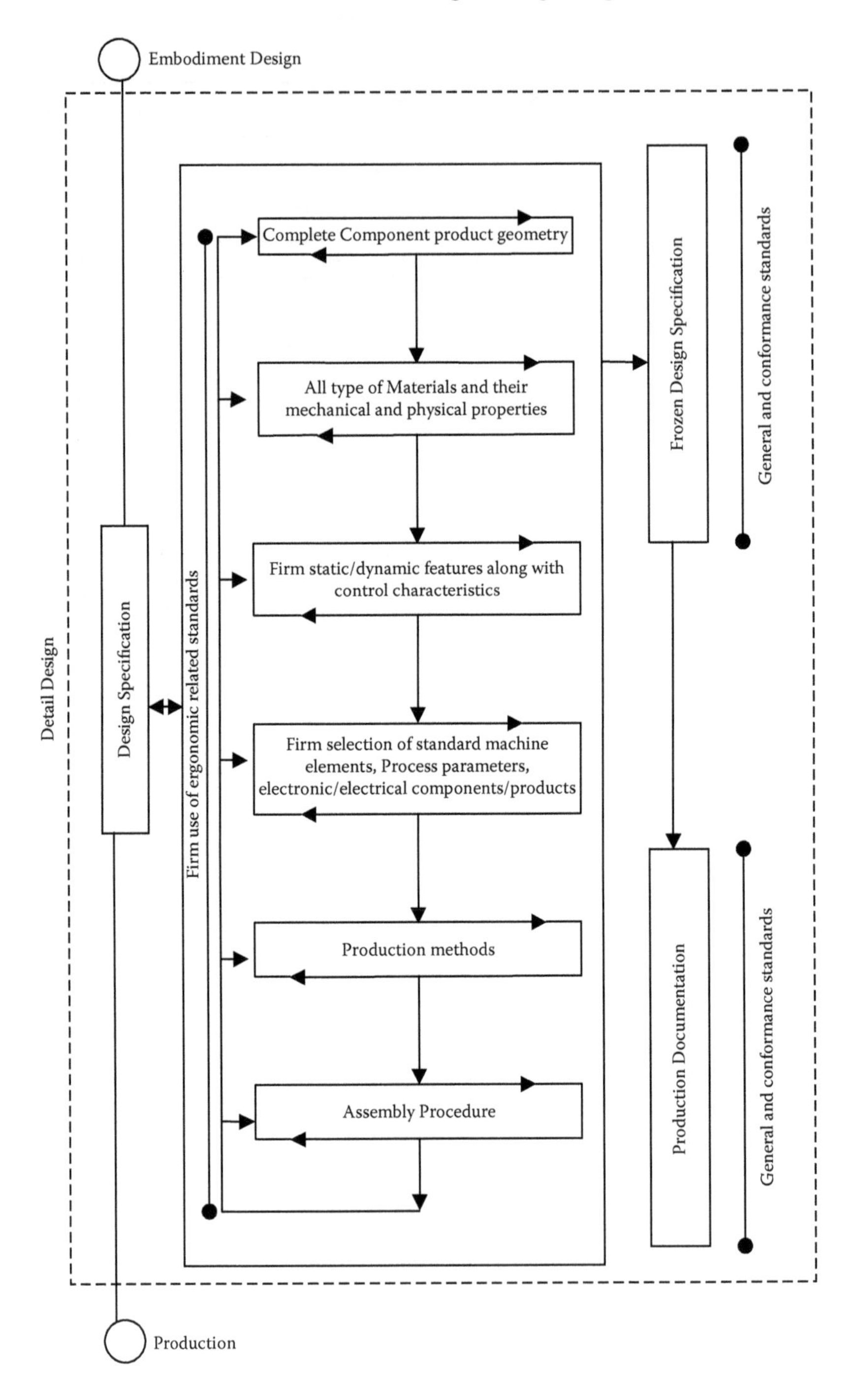

FIGURE 9.5 System approach to detail design.

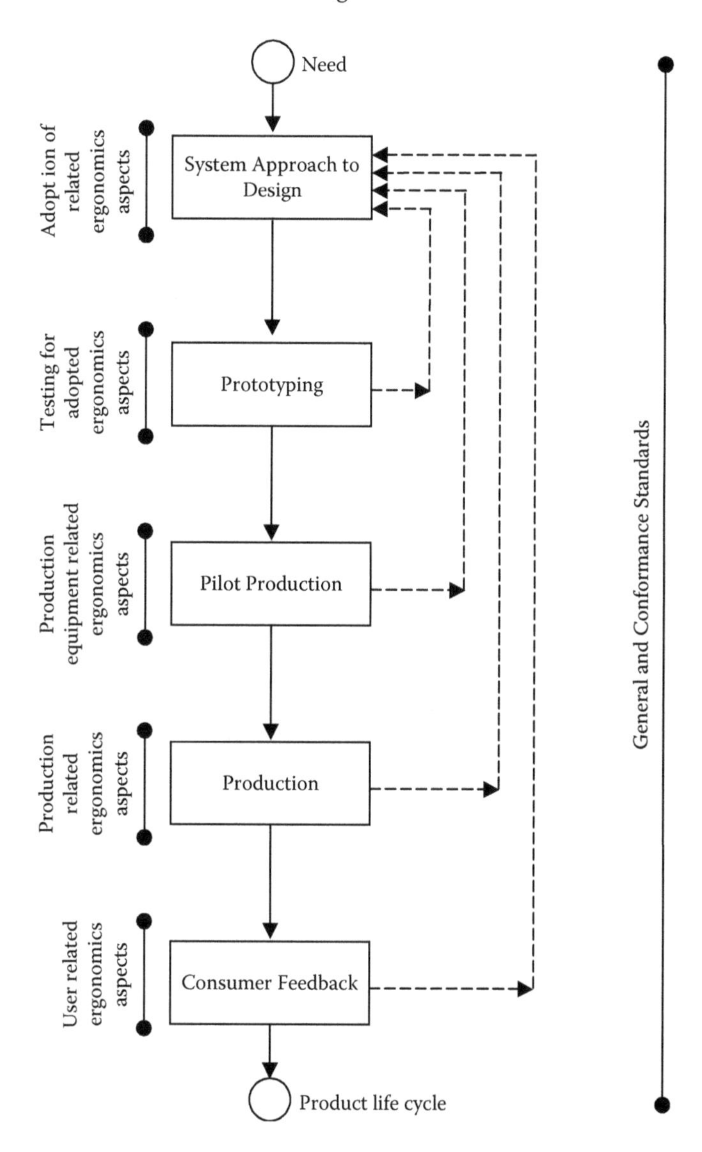

FIGURE 9.6 Universal design cycle.

BIBLIOGRAPHY

Algera, J.A., Reitsma, W.D., Scholtens, S., Vrins, A.A.C., and Wijnen, C.J.D., Ingredients of ergonomic intervention: How to get ergonomics applied, *Ergonomics*, 33(5), 557–578, 1990.

Baybutt, P. and Haight, J.M., Human factors analysis for process safety, *Hydrocarbon Process.,* 82(4), 79–83, 2003.

Beevis, D., Ergonomics—Costs and benefits revisited, *Appl. Ergonomics,* 34(5), 491–496, 2003.

Bias, R.G., Cost-justifying human factors support, *Proc. Hum. Factors Soc.,* 832–833, 1990.

Buck, J.R., Human factors design, *Proc. Hum. Factors Soc.,* 1, 534–535, 1991.

Macleod, I., Cognitive ergonomics, *Industrial Engineer,* 36(3), 26–30, 2004.

Malone, T.B., Human factors and human error, *Proc. Hum. Factors Soc.,* 651–654, 1990.

Rouse, W.B., Cody, W.R., and Boff, K.R., Human factors of system design: Understanding and enhancing the role of human factors engineering, *Int. J. Hum. Factors Manufacturing,* 1(1), 87–104, 1991.

Smith, R.T., Growing an ergonomics culture in manufacturing, *Proc. Institution Mechanical Engineers* (Part B) *J. Eng. Manufacture,* 217(7), 1027–1030, 2003.

Stramler, J.H., Jr., Dictionary for human factors/ergonomics: A significant reference work in human factors, *Proc. Hum. Factors Soc.,* 1, 544–547, 1992.

Strom, G., When was the beginning of ergonomics and human factors? *Ergonomics in Design*, 11(2), 5–6, 2003.

Yates, R.F., Human factors—An overview, *Br. Telecom Technol. J.,* 6(4), 7–16, 1988.

Part 5

Support Standards in Design and Manufacturing of Discrete Products

10 Review of Support Standards for Discrete Products

10.1 AUTOMATION IN DISCRETE PRODUCTS

As mentioned in the earlier chapters, discrete products exist in a large variety and employ features that require well-defined standards. One of the most common features of discrete products is the automation that involves use of a microprocessor in an open-loop or closed-loop control methodology. The design of microprocessor-based control systems for discrete products in general requires methodology to represent data, methodology to code data, and methodology to exchange data; hardware interfacing protocols; data communication protocols; and methodology to design and develop interface-control software. The most common protocol used for representation and exchange of data is the ANSI standard ASCII (American Standard Code for Information Interchange). The sector-specific data exchange protocols, as in computer-integrated manufacturing (CIM), include IGES, STEP, PDES, and VDA-FS. A more advanced XML (Extensible Markup Language) protocol is used for coding of multimedia and hypermedia information. Hardware-interfacing protocols govern the physical interface between the microprocessor and sensors, transducers, and actuators and between the microprocessor and the user interface. Standards exist to define hardware interfacing requirements in general and in application-specific scenarios. With increased reliance on networks, discrete products utilize digital communication network standards, telecommunication standards, and wireless communication standards. Standardized high-level languages and standardized software-development procedures also govern the software interface between the physical world and the microprocessor. Figure 10.1 provides a generalized scope for support standards used in the automation of discrete products.

This chapter includes the following categories of selected support standards from the online catalog of commercial standards supplier Techstreet to demonstrate the availability of related standards in this category:

Information representation, coding, and exchange
Interface hardware
Sensors, transducers, and actuators
Data communication

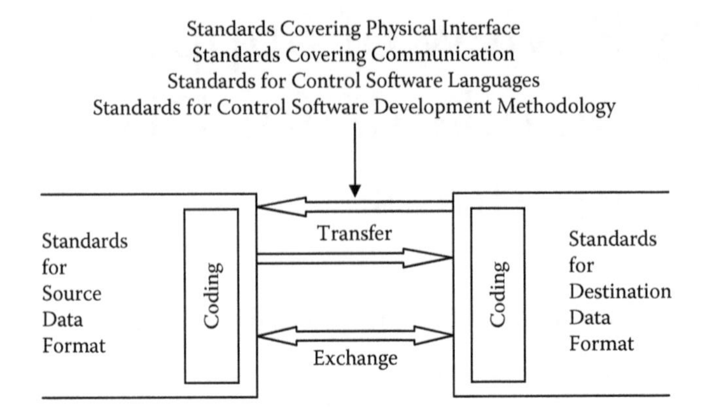

FIGURE 10.1 Domain for standards in automation of discrete products.

Programming languages
Software development

Various key words are used to obtain the edited standards list.

10.1.1 Information Representation, Coding, and Exchange Standards

Several international, national, and professional organizations are active in this high-tech general and sector-specific domain. A short list of these standards is provided in Table 10.1. The online search was conducted on the Techstreet database and results are provided after suitable editing.

10.1.2 Interface Hardware Standards

Interface hardware standards are normally market-oriented protocols from major computer hardware companies and professional organizations involved in development of standards. These standards may be common standards and standards that are built around a chipset (e.g., from Intel or Motorola). International and national standards organizations generally provide the basic standards related to materials, chip design guidelines, printed circuit board development, manufacturing methods, and testing procedures for both digital and analog systems. A brief list of standards related to interface hardware is provided from Techstreet's database in Table 10.2.

10.1.3 Standards Related to Sensors, Transducers, and Actuators

Discrete products that implement feedback control systems make use of a variety of sensors, transducers, and actuators. The specifications of these sensors, transducers, and actuators normally include size, type, range, power rating, and the

TABLE 10.1
Information Representation, Coding, and Exchange Standards

1. **ISO/IEC 13522-8 01-May-2001**
   Information Technology—Coding of Multimedia and Hypermedia Information—Part 8: XML Notation for ISO/IEC 13522-5—International Restrictions
   Product Type: Standard

2. **ETSI TS 101 903 01-Feb-2002**
   XML Advanced Electronic Signatures (XAdES)
   Product Type: Standard

3. **ANSI/IPC 2501 01-Jul-2003**
   Definition for Web-Based Exchange of XML Data
   Product Type: Standard

4. **DIN 16557-5 01-Nov-2002**
   Electronic Data Interchange for Administration, Commerce, and Transport (EDIFACT)—Part 5: Rules for Generation of XML Scheme Files (XSD) on the Basis of EDI(FACT) Implementation Guidelines (ISO/TS 20625:2002)
   Product Type: Standard

5. **ISO/TS 20625 01-May-2002**
   Electronic Data Interchange for Administration, Commerce, and Transport (EDIFACT)—Rules for Generation of XML Scheme Files (XSD) on the Basis of EDI(FACT) Implementation Guidelines—International Restrictions
   Product Type: Standard

6. **API RP 3901 14-Feb-2002**
   RP 3901 PIDX™ XML Transaction Standards Version 1.0
   Product Type: Standard

7. **DIN CWA 13993 01-Dec-2000**
   Recommendations and Guidance on the Use of XML for Electronic Data Interchange
   Product Type: Standard

8. **ETSI TS 132 615 01-Dec-2001**
   Universal Mobile Telecommunications System (UMTS); Telecommunication Management; Configuration Management; 3G Configuration Management: Bulk Configuration Management IRP: XML File Format Definition (3GPP TS 32.615 Version 4.1.0 Release 4)
   Product Type: Standard

9. **ASTM E2210-02 10-May-2002**
   Standard Specification for Guideline Elements Model (GEM)—Document Model for Clinical Practice Guidelines
   Product Type: Standard

TABLE 10.2
Interface Hardware Standards

1. **BS 4727-1: Group 11:1991 29-Mar-1991**
   Glossary of Electrotechnical, Power, Telecommunication, Electronics, Lighting, and Color Terms. Terms Common to Power, Telecommunications, and Electronics. Printed Circuits
   Product Type: Standard

2. **BS 4727-3: Group 16:1992 15-Oct-1992**
   Glossary of Electrotechnical, Power, Telecommunication, Electronics, Lighting, and Color Terms. Terms Particular to Telecommunications and Electronics. Switching and Signaling in Telecommunications
   Product Type: Standard

3. **ASTM Volume 10.04-03 01-Apr-2003**
   ASTM Book of Standards Volume 10.04: Electronics (I)
   Product Type: Standard

4. **ASTM Volume 10.05-03 01-Apr-2003**
   ASTM Book of Standards Volume 10.05: Electronics (II)
   Product Type: Standard

5. **IEC 61954 Amd 1 19-Feb-2003**
   Amendment 1—Power Electronics for Electrical Transmission and Distribution Systems—Testing of Thyristor Valves for Static VAR Compensators—International Restrictions
   Product Type: Standard

6. **IEC 61190-1-1 25-Mar-2002**
   Attachment Materials for Electronic Assembly—Part 1–1: Requirements for Soldering Fluxes for High-Quality Interconnections in Electronics Assembly—International Restrictions
   Product Type: Standard

7. **DIN EN 61190-1-1 01-Jan-2003**
   Attachment Materials for Electronic Assembly—Part 1-1: Requirements for Soldering Fluxes for High-Quality Interconnections in Electronics Assembly (IEC 61190-1-1:2002); German Version EN 61190-1-1:2002
   Product Type: Standard

8. **BS EN 61190-1-1:2002 16-Aug-2002**
   Attachment Materials for Electronic Assembly. Requirements for Soldering Fluxes for High-Quality Interconnections in Electronics Assembly
   Product Type: Standard

installation procedures. These parameters are either company specific or are drawn from general standards provided by the international, national, or professional standards-developing organizations. The standards pertaining to use of materials, manufacturing methods, and calibration procedures in general or to sector-specific applications are also provided by the international, national, and professional standards-developing organizations. A brief list of standards related to sensors, transducers, and actuators is provided in Table 10.3.

10.1.4 Data Communication Standards

General-purpose, sector-specific, and product-specific serial, parallel, and wireless communication methodologies are well in place and are supported by

TABLE 10.3
Standards Related to Sensors, Transducers, and Actuators

1. **SAE J2057/3 01-Aug-2001**
   Class A Multiplexing Sensors
   Product Type: Standard

2. **ISA TP-973019 01-Jan-1997**
   Applying Sensors in Safety Instrumented Systems
   Product Type: Standard

3. **BS EN 50227:1999 01-Oct-1999**
   Control Circuit Devices and Switching Elements Proximity Sensors, D.C. Interface for Proximity Sensors and Switching Amplifiers (NAMUR)
   Product Type: Standard

4. **ASTM E879-93 01-Aug-1993 (withdrawn and replaced by ASTM E879-01)**
   Standard Specification for Thermistor Sensors for Clinical Laboratory Temperature Measurements
   Product Type: Standard

5. **ISA 37.3-1982 (R1995) 01-Apr-1982**
   Specifications and Tests for Strain Gage Pressure Transducers
   Product Type: Standard

6. **NFPA (Fluid) T2.30.4 R1-2002 22-Aug-2002**
   Recommended Practice—Fluid Power Systems—Application Guidelines for Selection of Fluid Power Position Transducers
   Product Type: Standard

7. **ANSI/ASA S2.11-1969 (R2001) 01-Jan-1969 (withdrawn)**
   American National Standard for Selection of Calibrations and Tests for Electrical Transducers Used for Measuring Shock and Vibration
   Product Type: Standard

8. **SAE J2570 01-Sep-2001**
   Performance Specifications for Anthropomorphic Test Device Transducers
   Product Type: Standard

(Continued)

TABLE 10.3 (Continued)
Standards Related to Sensors, Transducers, and Actuators

9. **ASTM E1774-96(2002) 10-Dec-1996**
   Standard Guide for Electromagnetic Acoustic Transducers (EMATs)
   Product Type: Standard

10. **UL 756 01-Jan-1997**
    Coin and Currency Changers and Actuators
    Product Type: Standard

11. **VDE 0199 01-Jan-1994**
    Coding of Indicating Devices and Actuators by Colors Supplementary Means (German only)
    Product Type: Standard

12. **BS 2G 143:1965 14-Apr-1965**
    Specification for A.C. and D.C. Rotary and Linear Actuators for Aircraft
    Product Type: Standard

13. **SAE ARP1281C 01-Sep-2002**
    Actuators: Aircraft Flight Controls, Power Operated, Hydraulic, General Specification For
    Product Type: Standard

14. **SAE ARP4058 01-Aug-1988**
    Actuators: Mechanical, Geared Rotary, General Specification For
    Product Type: Standard

15. **SME MS910331 01-Jun-1991**
    Artificial Muscle: Actuators
    Product Type: Standard

16. **UL 8730-2-14 29-May-1998**
    Automatic Electrical Controls for Household and Similar Use—Part 2: Particular Requirements for Electric Actuators
    Product Type: Standard

17. **SAE SP1418 01-Jan-1999**
    Electronic Engine Controls 1999: Sensors, Actuators, and Development Tools
    Product Type: Standard

18. **IEEE 382-1996 01-May-1996**
    IEEE Standard for Qualification of Actuators for Power-Operated Valve Assemblies with Safety-Related Functions for Nuclear Power Plants
    Product Type: Standard

international, national, and professional organizations. The professional organizations include EIA, IEEE, IEC, and ISO. The following list of standards shows the diversity of these standards and their use. Techstreet's database was used to extract this edited list of standards as given in Table 10.4.

TABLE 10.4
Data Communication Standards

1. **ISO/TR 7477 01-Sep-1985 (withdrawn)**
   Data Communication; Arrangements for DTE to DTE Physical Connection Using V.24 and X.24 Interchange Circuits—International Restrictions
   Product Type: Standard

2. **DIN ISO/IEC 8881 01-Jan-1996**
   Data Communication—Use of the X.25 Packet Level Protocol in Local and Metropolitan Area Networks; Identical with ISO/IEC 8881:1989 (Status of 1991)
   Product Type: Standard

3. **BS ISO/IEC 8327-1:1996 15-Feb-1997**
   Information Technology. Open Systems Interconnection. Connection-Oriented Session Protocol. Protocol Specification
   Product Type: Standard

4. **BS ISO/IEC 11575:1995 15-Dec-1995**
   Information Technology. Telecommunications and Information Exchange Between Systems. Protocol Mappings for the OSI Data Link Service
   Product Type: Standard

5. **BS ISO/IEC 9314-7:1998 15-Dec-1998**
   Information Processing Systems. Fiber Distributed Data Interface (FDDI). Physical Layer Protocol (PHY-2)
   Product Type: Standard

6. **BS ISO 11754:2003 15-May-2003**
   Space Data and Information Transfer Systems. Telemetry Channel Coding
   Product Type: Standard

7. **BS ISO/IEC 10857:1994 15-Jul-1995**
   Information Technology. Microprocessor Systems. Futurebus+. Logical Protocol Specification
   Product Type: Standard

8. **BS ISO 10303-203:1994 15-Jul-1996**
   Industrial Automation Systems and Integration. Product Data Representation and Exchange. Application Protocol. Configuration Controlled Design
   Product Type: Standard

9. **BS ISO/IEC 10861:1994 15-Jul-1995**
   Information Technology. Microprocessor Systems. High-Performance Synchronous 32-Bit Bus: MULTIBUS II
   Product Type: Standard

(Continued)

TABLE 10.4 (Continued)
Data Communication Standards

10. **ISO 16484-5 01-Nov-2003**
Building Automation and Control Systems—Part 5: Data Communication Protocol—International Restrictions
Product Type: Standard

11. **DIN EN ISO 11073-10101 - DRAFT 01-Mar-2004**
Draft Document—Health Informatics—Point-of-Care Medical Device Communication—Part 10101: Nomenclature (ISO/IEEE DIS 11073-10101:2003); English Version prEN ISO 11073-10101:2003
Product Type: Standard

12. **CAN/CSA-ISO/IEC 11518-3-97 (R2001) 19-Feb-2000**
Information Technology—High-Performance Parallel Interface—Part 3: Encapsulation of ISO/IEC 8802-2 (IEEE Std. 802.2) Logical Link Control Protocol Data Units (HIPPI-LE) (adopted ISO/IEC 11518-3:1996)
Product Type: Standard

13. **CAN/CSA-ISO/IEC 10026-1-00**
Information Technology—Open Systems Interconnection—Distributed Transaction Processing—Part 1: OSI TP Model (adopted ISO/IEC 10026-1:1998, Second Edition, 1998-10-15)
Product Type: Standard

14. **CAN/CSA-ISO/IEC 10030-96 (R2001) 19-Feb-2000**
Information Technology—Telecommunications and Information Exchange Between Systems—End System Routing Information Exchange Protocol for Use in Conjunction with ISO/IEC 8878 (adopted ISO/IEC 10030:1995)
Product Type: Standard

15. **CAN/CSA-ISO/IEC ISP 10609-1-99 18-Feb-2000**
Information Technology—International Standardized Profiles TB, TC, TD, and TE—Connection-Mode Transport Service Over Connection-Mode Network Service—Part 1: Subnetwork-Type Independent Requirements for Group TB (adopted ISO/IEC ISP 10609-1:1992, First Edition, 1992-09-15)
Product Type: Standard

16. **ISO/IEC ISP 11183-2 01-Dec-1992**
Information Technology; International Standardized Profiles AOM1n OSI Management; Management Communication; Part 2: CMISE/ROSE for AOM12; Enhanced Management Communications
Product Type: Standard

17. **ISO 16484-5 01-Nov-2003**
Building Automation and Control Systems—Part 5: Data Communication Protocol—International Restrictions
Product Type: Standard

TABLE 10.4 (Continued)
Data Communication Standards

18. **ISO/DIS 15662 01-Mar-2003**
    Intelligent Transport Systems—Wide Area Communication—Protocol Management Information (Draft)—International Restrictions
    Product Type: Standard

19. **ISO 13374-1 01-Mar-2003**
    Condition Monitoring and Diagnostics of Machines—Data Processing, Communication and Presentation—Part 1: General Guidelines—International Restrictions
    Product Type: Standard

10.1.5 Standards Related to Programming Languages

To control the hardware interface between the microprocessor and the process, the user communicates the control parameters to the microprocessor through software developed by use of programming languages. These high-level and low-level languages are standardized in terms of available commands, methodology to use the commands, and debugging procedure. The high-level languages normally operate through an interpreter or compiler. The microprocessors manufacturers develop the low-level languages or assembly language mnemonics, whereas high-level languages are standardized by national, international, and professional organizations. A brief listing of standards related to the programming languages from Techstreet's database is provided in Table 10.5.

10.1.6 Software-Development Standards

Software-development procedures govern the process used to create software in general. Several procedures exist. Two of the most commonly used are computer-aided software engineering tools, such as the Unified Modeling Language procedure, and the Agile Software Development procedure. CMM is the major are two software development verification models that can have conformance to set standards and a verification procedure.

Software development by use of high-level or low-level languages essentially requires use of one of the above procedures. Some of these procedures are standard specific, whereas others work under the guidelines established by different professional organizations. A brief listing of standards related to this area is given in Table 10.6.

10.2 QUALITY ASSURANCE

The achievement of quality or excellence of goods or services is the best way to describe quality assurance. The design, development, operation, maintenance, and disassembly leading to recycling or disposal of systems, processes, and

TABLE 10.5
Standards Related to Programming Languages

1. **ISO/IEC 11404 01-Dec-1996**
   Information Technology—Programming Languages, Their Environments, and System Software Interfaces—Language-Independent Datatypes—International Restrictions

2. **ISO/IEC 13817-1 01-Dec-1996**
   Information Technology—Programming Languages, Their Environments, and System Software Interfaces—Vienna Development Method—Specification Language—Part 1: Base Language—International Restrictions
   Product Type: Standard

3. **BS ISO/IEC 13817-1:1996 15-Apr-1997**
   Information Technology—Programming Languages, Their Environments, and System Software Interfaces—Vienna Development Method—Specification Language—Base Language
   Product Type: Standard

4. **ISO/IEC 13817-1 01-Dec-1996**
   Information Technology—Programming Languages, Their Environments, and System Software Interfaces—Vienna Development Method—Specification Language—Part 1: Base Language—International Restrictions
   Product Type: Standard

5. **BS ISO/IEC TR 10182:1993 15-Aug-1994**
   Information Technology. Programming Languages, Their Environments, and System Software Interfaces. Guidelines for Language Bindings
   Product Type: Standard

6. **BS ISO/IEC TR 14369:1999 15-Mar-2000**
   Information Technology. Programming Languages, Their Environments, and System Software Interfaces. Guidelines for the Preparation of Language-Independent Service Specifications (LISS)
   Product Type: Standard

products require stringent quality-assurance procedures to maintain the competitive edge of an organization. Elaborate quality-assurance standards exist for general use and for sector-specific applications.

10.2.1 Quality-Assurance Standards

This section lists a selection of quality-assurance standards obtained from Techstreet's database as given in Table 10.7.

TABLE 10.6
Software Development Standards

1. **ISO/IEC 12207 01-Aug-1995**
   Information Technology—Software Life Cycle Processes—International Restrictions
   Product Type: Standard

2. **BS ISO/IEC 90003:2004 19-Feb-2004**
   Software Engineering. Guidelines for the Application of ISO 9001:2000 to Computer Software
   Product Type: Standard

3. **BS ISO/IEC 14143-2:2002 13-Dec-2002**
   Information Technology. Software Measurement. Functional Size Measurement. Conformity Evaluation of Software Size Measurement Methods to ISO/IEC 14143-1:1998
   Product Type: Standard

4. **BS ISO/IEC TR 15271:1998 15-Oct-1999**
   Information Technology. Guide for ISO/IEC 12207. (Software Life Cycle Processes)
   Product Type: Standard

5. **BS ISO/IEC 14764:1999 15-Jan-2000**
   Information Technology. Software Maintenance
   Product Type: Standard

6. **BS ISO/IEC TR 14143-4:2002 24-Oct-2002**
   Information Technology. Software Measurement. Functional Size Measurement. Reference Model
   Product Type: Standard

7. **BS ISO/IEC 14598-2:2000 15-Mar-2000**
   Information Technology. Software Production Evaluation. Planning and Management
   Product Type: Standard

8. **BS ISO/IEC 18019:2004 21-Jan-2004**
   Software and System Engineering. Guidelines for the Design and Preparation of User Documentation for Application Software
   Product Type: Standard

9. **BS ISO/IEC 19761:2003 11-Mar-2003**
   Software Engineering. COSMIC-FFP. A Functional Size Measurement Method
   Product Type: Standard

10. **BS ISO/IEC 20926:2003 31-Oct-2003**
    Software Engineering. IFPUG 4.1 Unadjusted Functional Size Measurement Method. Counting Practices Manual
    Product Type: Standard

11. **BS ISO/IEC 20968:2002 19-Dec-2002**
    Software Engineering. Mk II Function Point Analysis. Counting Practices Manual
    Product Type: Standard

(*Continued*)

TABLE 10.6 (Continued)
Software Development Standards

12. **CAN/CSA-ISO/IEC 12207-96 (R2000) 19-Feb-2000**
    Information Technology—Software Life Cycle Processes (adopted ISO/IEC 12207:1995)
    Product Type: Standard

13. **CAN/CSA-ISO/IEC 14143-1-01 26-Mar-2001**
    Information Technology—Software Measurement—Functional Size Measurement—Part 1: Definition of Concepts (adopted ISO/IEC 14143-1:1998, First Edition, 1998-06-15)
    Product Type: Standard

14. **CAN/CSA-ISO/IEC 12119-95 (R2000) 17-Jan-2000**
    Information Technology—Software Packages—Quality Requirements and Testing (adopted ISO/IEC 12119:1994)
    Product Type: Standard

15. **CAN/CSA-ISO/IEC 9126-96 (R2000) 16-Feb-2000**
    Information Technology—Software Product Evaluation—Quality Characteristics and Guidelines for Their Use (adopted ISO/IEC 9126:1991)
    Product Type: Standard

16. **ISO/IEC 14496-5 AMD 1 01-Dec-2002**
    Information Technology—Coding of Audio-Visual Objects—Part 5: Reference Software; Amendment 1: Reference Software for MPEG-4—International Restrictions
    Product Type: Standard

17. **ISO/IEC 14496-5 AMD 2 01-Mar-2003**
    Information Technology—Coding of Audio-Visual Objects—Part 5: Reference Software; Amendment 2: MPEG-4 Reference Software Extensions for XMT and Media Nodes—International Restrictions
    Product Type: Standard

18. **ISO/IEC 15444-5 AMD 1 01-Dec-2003**
    Information Technology—JPEG 2000 Image Coding System—Part 5: Reference Software; Amendment 1: Reference Software for the JP2 File Format—International Restrictions
    Product Type: Standard

19. **ISO/IEC 14764 01-Nov-1999**
    Information Technology—Software Maintenance—International Restrictions
    Product Type: Standard

20. **ISO/IEC 14143-2 01-Nov-2002**
    Information Technology—Software Measurement; Functional Size Measurement—Part 2: Conformity Evaluation of Software Size Measurement Methods to ISO/IEC 14143-1:1998—International Restrictions
    Product Type: Standard

TABLE 10.7
Quality Assurance Standards

1. **ISO 9000-1 01-Jul-1994**
   Quality Management and Quality Assurance Standards—Part 1: Guidelines for Selection and Use—International Restrictions
   Product Type: Standard

2. **DIN ISO 9000-4 01-Jun-1994 (withdrawn)**
   Quality Management and Quality Assurance Standards; Guide to Dependability Program Management (identical to ISO 9000-4:1993 resp. IEC 60300-1:1993); German version EN 60300-1:1993
   Product Type: Standard

3. **ISO/TR 13352 01-Jun-1997**
   Guidelines for Interpretation of ISO 9000 Series for Application Within the Iron Ore Industry—International Restrictions
   Product Type: Standard

4. **ISO 9000 01-Dec-2000**
   Quality Management Systems—Fundamentals and Vocabulary—International Restrictions
   Product Type: Standard

5. **ISO 9000-1 01-Jul-1994**
   Quality Management and Quality Assurance Standards—Part 1: Guidelines for Selection and Use—International Restrictions
   Product Type: Standard

6. **ISO 9000-2 01-Dec-1997**
   Quality Management and Quality Assurance Standards—Part 2: Generic Guidelines for the Application of ISO 9001, ISO 9002, and ISO 9003—International Restrictions
   Product Type: Standard

7. **BS ISO 9000-2:1997 15-Aug-1997**
   Quality Management and Quality Assurance Standards. Generic Guidelines for the Application of ISO 9001, ISO 9002, and ISO 9003
   Product Type: Standard

8. **BS EN ISO 9000-3:1997 15-Feb-1998**
   Quality Management and Quality Assurance Standards. Guidelines for the Application of ISO 9001:1994 to the Development, Supply, Installation, and Maintenance of Computer Software
   Product Type: Standard

9. **BS EN ISO 9000:2000 15-Dec-2000**
   Quality Management Systems. Fundamentals and Vocabulary
   Product Type: Standard

10.3 ENVIRONMENTAL CONSIDERATIONS

The manufacturing and operation of products that result from discrete or continuous manufacturing processes or systems have an environmental impact in terms of disassembly, reuse, recycling, and disposal of these products after the end of their useful lives. This section provides a listing of standards relevant to this area.

10.3.1 STANDARDS RELATED TO DISASSEMBLY, RECYCLING, AND DISPOSAL

In today's world, people are more aware of the need to decrease pollution and make optimum use of all natural resources. Consequently, the contemporary system approach to design for discrete products advocates consideration of disassembly, recycling, and eventual disposal. A wide range of standards in diverse areas is available to the designer of discrete products to these modern parameters into account. Table 10.8 lists a selection of standards from Techstreet's database.

TABLE 10.8
Standards Related to Disassembly, Recycling, and Disposal

1. **SME MS96-132 01-Jun-1996**
   Disassembly Model for Recycling—Personal Computer
   Product Type: Standard

2. **VDI 2343 Sheet 3 01-Feb-2002**
   Recycling of Electrical and Electronic Equipment—Disassembly and Processing
   Product Type: Standard

3. **BS EN 13437:2003 27-Feb-2004**
   Packaging and Material Recycling. Criteria for Recycling Methods. Description of Recycling Processes and Flow Chart
   Product Type: Standard

4. **BS EN 61429:1997 15-Sep-1997**
   Marking of Secondary Cells and Batteries with the International Recycling Symbol ISO 7000-1135
   Product Type: Standard

5. **SAE J1770 01-Oct-1995**
   Automotive Refrigerant Recovery/Recycling Equipment Intended for Use with Both R12 and R134a
   Product Type: Standard

TABLE 10.8 (Continued)
Standards Related to Disassembly, Recycling, and Disposal

6. **ISRI RADRPP**
   Radioactivity in the Scrap Recycling Industry—Recommended Practice and Procedures
   Product Type: Standard

7. **VDI 2074 01-Mar-2000**
   Recycling in the Building Services
   Product Type: Standard

8. **ACI RDCM-94 01-Jan-1994**
   Recycling of Demolished Concrete and Masonry
   Product Type: Standard

9. **VDI 2343 Sheet 3 01-Feb-2002**
   Recycling of Electrical and Electronic Equipment—Disassembly and Processing
   Product Type: Standard

10. **ASTM D5834-95 01-Jan-1995**
    Standard Guide for Source Reduction Reuse, Recycling, and Disposal of Solid and Corrugated Fiberboard (Cardboard)
    Product Type: Standard

11. **ASTM D5833-95e1 01-Jan-1995**
    Standard Guide for Source Reduction Reuse, Recycling, or Disposal of Steel Cans
    Product Type: Standard

12. **SME FC900648 01-Jun-1990**
    Waste Minimization and Resource Recycling in Paint Booth Operations
    Product Type: Standard

13. **BS EN 12258-3:2003 09-Sep-2003**
    Aluminum and Aluminum Alloys. Terms and Definitions. Scrap
    Product Type: Standard

14. **DIN 30706-1 01-May-1991**
    Waste Disposal Technique; Terms for Household Waste Disposal and Waste Disposal Vehicles
    Product Type: Standard

15. **ASTM E1451-98 10-Apr-1998**
    Standard Guide for Disposal of Wastes Containing Silicon Carbide Whiskers and Fibers
    Product Type: Standard

(*Continued*)

TABLE 10.8 (Continued)
Standards Related to Disassembly, Recycling, and Disposal

16. **AWWA REUSE52138 01-Jan-2000**
Beneficial Reuse as a Wastewater Disposal Alternative for the Center Region of Pennsylvania—International Restrictions
Product Type: Conference Proceeding

17. **BS EN 12832:1999 15-Oct-1999**
Characterization of Sludges. Utilization and Disposal of Sludges. Vocabulary
Product Type: Standard

18. **VDI 4413 01-Nov-2003**
Logistic of Waste Disposal in Producing Enterprises
Product Type: Standard

19. **DIN 6647-4 01-Apr-2004**
Means of Packaging—Cylindrical Beverage Containers—Part 4: Disposal Pack with Allowable Operating Pressure Up to 3 Bar, Nominal Volume Up to 60 Liters
Product Type: Standard

20. **CAN/CSA Z7396-2-02 28-Mar-2002**
Medical Gas Pipeline Systems—Part 2: Anesthetic Gas Scavenging Disposal Systems (adopted ISO 7396-2:2000, First Edition, 2000-11-15, with Canadian Deviations)
Product Type: Standard

21. **ISO 7396-2 01-Nov-2000**
Medical Gas Pipeline Systems—Part 2: Anesthetic Gas Scavenging Disposal Systems—International Restrictions
Product Type: Standard

22. **DIN EN 737-2 01-Jan-2000**
Medical Gas Pipeline Systems—Part 2: Anesthetic Gas Scavenging Disposal Systems; Basic Requirements (includes Amendment A1:1999); German Version EN 737-2:1998 + A1:1999
Product Type: Standard

23. **DIN V 1264 01-Apr-2003**
Steps for Underground Man Entry Chambers—Application in Construction Works for Wastewater Disposal
Product Type: Standard

10.3.2 Standards Related to Environmental Impact

To minimize the adverse effects on the environment by the manufacturing and operation of discrete products, several organization work actively to formulate standard practices. Table 10.9 logs a brief collection of these standards from Techstreet's database.

TABLE 10.9
Standards Related to Environmental Impact

1. **ISO 14004 01-Sep-1996**
   Environmental Management Systems—General Guidelines on Principles, Systems, and Supporting Techniques—International Restrictions
   Product Type: Standard

2. **ISO 14001 01-Sep-1996**
   Environmental Management Systems—Specification with Guidance for Use—International Restrictions
   Product Type: Standard

3. **ISO 14010 01-Oct-1996**
   Guidelines for Environmental Auditing—General Principles—International Restrictions
   Product Type: Standard

4. **ISO 14011 01-Sep-1996**
   Guidelines for Environmental Auditing—Audit Procedures—Auditing of Environmental Management Systems—International Restrictions
   Product Type: Standard

5. **ISO 14012 27-Nov-1996**
   Guidelines for Environmental Auditing—Qualification Criteria for Environmental Auditors—International Restrictions
   Product Type: Standard

6. **ISO 14020 01-Aug-1998 (replaced by ISO 14020)**
   Environmental Labels and Declarations—General Principles—International Restrictions
   Product Type: Standard

7. **ISO 14024 01-Apr-1999**
   Environmental Labels and Declarations—Type I Environmental Labeling—Principles and Procedures—International Restrictions
   Product Type: Standard

8. **ISO 14040 01-Jun-1997**
   Environmental Management—Life Cycle Assessment—Principles and Framework—International Restrictions
   Product Type: Standard

9. **ANSI/ISO 14010 01-Oct-1996**
   Guidelines for Environmental Auditing—General Principles
   Product Type: Standard

10. **ISO 14010 01-Oct-1996**
    Guidelines for Environmental Auditing—General Principles—International Restrictions
    Product Type: Standard

(Continued)

TABLE 10.9 (Continued)
Standards Related to Environmental Impact

11. ANSI/ISO 14011 01-Sep-1996
Guidelines for Environmental Auditing—Audit Procedures—Auditing of Environmental Management Systems
Product Type: Standard

12. ANSI/ISO 14012 27-Nov-1996
Guidelines for Environmental Auditing—Qualification Criteria for Environmental Auditors
Product Type: Standard

13. ISO 14012 27-Nov-1996
Guidelines for Environmental Auditing—Qualification Criteria for Environmental Auditors—International Restrictions
Product Type: Standard

10.4 E-BUSINESS RELATED STANDARDS

e-Business is the trend of today. Marketing and supply of discrete products is becoming a common e-Business practice. Varieties of standards are available to promote e-Business in an efficient manner. These standards are related to Web development, data exchange, communication, and secure transactions. A small selection of these standards from Techstreet's database is provided in Table 10.10.

TABLE 10.10
e-Business Related Standards

1. Multilingual Catalogue Strategies for e-Commerce and e-Business
   Document Number: BS CWA 15045:2004
   British Standard
   01-Jul-2004
   180 pages
2. Information Technology—Security Techniques; Evaluation Criteria for IT Security—Part2: Security Functional Requirements (ISO/IEC 15408-2:1999)
   Document Number: DIN ISO/IEC 15408-2
   DIN-adopted ISO/IEC Standard
   01-Mar-2001
   189 pages

TABLE 10.10 (Continued)
e-Business Related Standards

3. Information Technology—International Standardized Profiles—OSI Management — Common Information for Management Functions—Part 2: State Management (adopted ISO/IEC ISP 12059-2:1995)
   Document Number: CAN/CSA-ISO/IEC 12059-2-96 (R2000)
   Canada National Standard/Canadian Standards—ISO/IEC
   18-Jan-2000
   26 pages
4. Information Technology—Open Systems Interconnection—Distributed Transaction Processing—Part 2: OSI TP Service (adopted ISO/IEC 10026-2:1998, Third Edition, 1998-10-15)
   Document Number: CAN/CSA-ISO/IEC 10026-2-00
   Canada National Standard/Canadian Standards—ISO/IEC
   01-Apr-2000
5. Information Technology—Security Techniques—Hash-Functions—Part 2: Hash-Functions Using an N-Bit Block Cipher Algorithm (adopted ISO/IEC 10118-2:1994)
   Document Number: CAN/CSA-ISO/IEC 10118-2-96 (R2000)
   Canada National Standard/Canadian Standards—ISO/IEC
   18-Jan-2000
   7 pages
6. Information Technology—Open Systems Interconnection—Remote Database Access—Part 2: SQL Specialization (adopted ISO/IEC 9579-2:1993)
   Document Number: CAN/CSA-ISO/IEC 9579-2-95 (R2000)
   Canada National Standard/Canadian Standards—ISO/IEC
   19-Jan-2000
   59 pages
7. Information Technology—Telecommunications and Information Exchange Between Systems—Local and Metropolitan Area Networks—Common Specifications—Part 2: LAN/MAN Management (adopted ISO/IEC 15802-2:1995, First Edition, 1995-03-10)
   Document Number: CAN/CSA-ISO/IEC 15802-2-00
   Canada National Standard/Canadian Standards—ISO/IEC
   83 pages
8. Information Technology—Security Techniques—Key Management—Part 3: Mechanisms Using Asymmetric Techniques (adopted ISO/IEC 11770-3:1999, First Edition, 1999-11-01)
   Document Number: CAN/CSA-ISO/IEC 11770-3-01
   Canada National Standard/Canadian Standards—ISO/IEC
   26-Mar-2001
   35 pages

BIBLIOGRAPHY

Key NIST contributions result in deployment of new e-commerce standards supporting electronics manufacturing. *J. Res. Natl. Inst. Stand. Technol.*, 106(2), 489, 2001.

Bradley, D.A., Applying mechatronics, *Manufacturing Eng.* 76(3), 117–120, 1997.

Busturia, J.M., Eichberger, A., Gretzschel, M., Merkt, T., Moser, E., Scholz, C., and Schuller, J., European activities and standardization efforts in mechatronics, *Vehicle Syst. Dynamics*, 33, 244–255, 2000.

Hewit, J.R. and King, T.G., Mechatronics design for product enhancement, *IEEE/ASME Trans. Mechatronics*, 1(2), 111–119, 1996.

Hsu, T.-R., Mechatronics—An overview, *IEEE Trans. Components, Packag., Manufacturing Technol. C Manufacturing*, 20(1), 4–7, 1997.

Ishii, T., Future trends in mechatronics, *JSME Int. J. Ser. 3*, 33(1), 1–6, 1990.

Jamshidi, M. and Salminen, V., Mechatronics, *Comput. Electrical Eng.*, 18(1), 1–108, 1992.

Laura, R.A., David, P., Helfferich, W., and Barone, J., Design considerations for ecosystem restoration, *Int. Water Resources Eng. Conf. Proc.* 1, 259–263, 1995.

Leaver, E.W., Process and manufacturing automation—Past, present, and future, *ISA Trans.*, 26(2), 45–50, 1987.

Linde, E., Dolan, D., and Batchelder, M., Mechatronics for Multidisciplinary Teaming, ASEE Annual Conference Proceedings, 2003 ASEE Annual Conference and Exposition: Staying in Tune with Engineering Education, 2003, 6901–6910.

Maes, M.E., and Steffey, J.R., How to improve industrial data communication, *Control Eng.*, 43(4), 4, 1996.

Mufti, A.A., Morris, M.L., and Spencer, W.B., Data exchange standards for computer-aided engineering and manufacturing, *Int. J. Comput. Appl. Technol.*, 3(2), 70–80, 1990.

Nicolson, E.J., Standardizing I/O for mechatronic systems (SIOMS) using real time UNIX device drivers, *Proc. IEEE Int. Conf. Robotics Automation*, 4, 3489–3494, 1994.

Roberts, G., Intelligent mechatronics, *Comput. Control Eng. J.* 9(6), 257–264, 1998.

Stefanac, D.R. and Klager, J.R., Evolution of process controls toward factory automation, *Industrial Heat.*, 53(7), 30–34, 1986.

11 Digital Interfacing and Control Software Development—A Case Study

11.1 MICROPROCESSOR INTERFACING FOR DISCRETE PRODUCTS

One of the most common multidisciplinary features of today's discrete products is their microprocessor-based controls. These discrete products implement open-loop or closed-loop control methodology for the system. In terms of computer hardware, the microprocessor control is implemented by use of a stand-alone personal computer equipped with standard input/output (I/O) ports and by use of other methods, such as special circuits. Interface control software gives access to sensors, transducers, and actuators and provides user interface to the discrete product (Figure 11.1). The user interface, in either case, is carefully designed according to the operator's requirements for the functioning of discrete product.

To control various features of most discrete products, manufacturers commonly use microprocessor interfacing. Microprocessor interfacing encompasses ordinary control features in normal ambient environments as well as specialized control features in, for example, humid, corrosive, or gravity-affected environments. A list of some of the exposures in which control elements of discrete products must function is given in Figure 4.13.

The most common feature of the microprocessor interface design is the availability of I/O channels to control the functions of a discrete product. These I/O channels may be available in desktop or similar computers that are based on a family of electronic chips or through special-purpose I/O chips. These I/O channels may also be available at a purpose-built microcontroller that can be plugged to the extension boards provided at the motherboards of desktop or similar computers, or they may be part of a specialized motherboard that provides all the required facilities on a single printed circuit board (PCB). The interface is operated through control software written in assembly language or an object-oriented programming language, depending on the nature of the application.

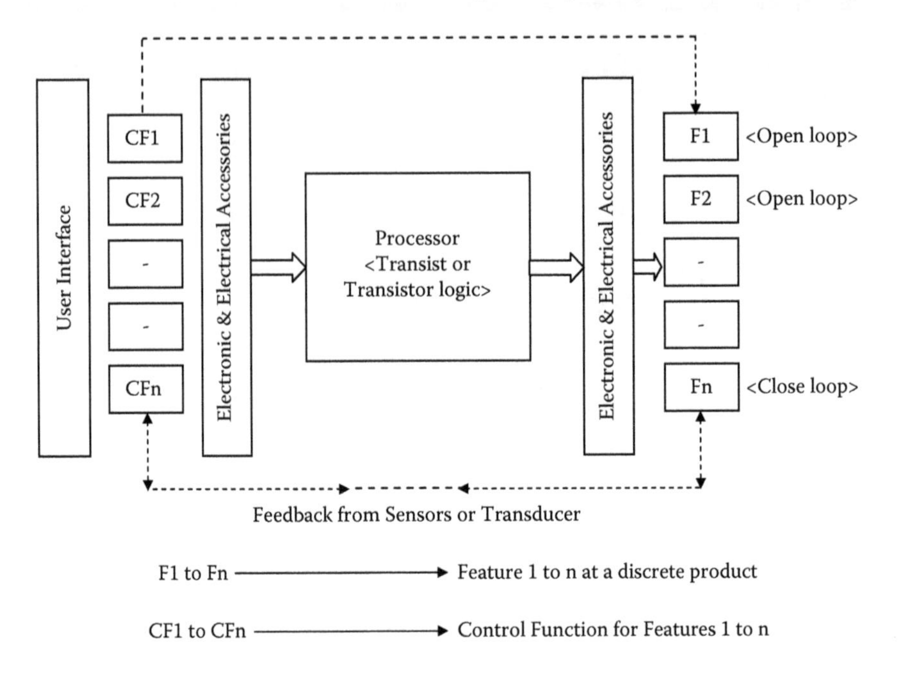

FIGURE 11.1 Control of features of discrete products.

11.2 INTERFACE CIRCUIT DESIGN

Electronic circuit engineering began with Claude Shannon's 1938 thesis. From this point, engineers recognized that a circuit is a rule for a Boolean function. With the advent of highly powerful microprocessors, the explosion of wireless communication, and the development of new generations of integrated sensors and actuators, the way electronic products are conceived, designed, and implemented has undergone a revolution.

To support the electronic-design chain, system designers must establish a design flow. Clean interfaces and unambiguous specifications are essential parts of this design flow. Designers implement algorithms in a selected architecture as software modules or as hardware components. Architecture selection is often an ad hoc process based on experience and extrapolation from existing products. After a certain number of revisions, the function implemented by use of hardware or software is finalized. The hardware implemented must be clearly specified. The software must be cleanly partitioned into application code, communication, design drivers, and basic I/O system (BIOS).

The design methodology includes three main steps: algorithm specification, virtual prototyping, and physical prototyping. It assumes that the designers, given an informal specification of the subsystems, can specify the requirements in a semiformal language such as the Unified Modeling Language (UML).

The overall behavior (functional network) and architecture netlist of the distributed system constitutes the output of this specification phase. This methodology supports the electronic-design chain of the entire discrete-product control system.

Several tools are available to simulate the electronic-circuit design before rapid prototyping to check the system hardware functionality. The hardware design can subsequently be sent to the production line. Innumerable articles, research papers, and books detail electronic-circuit design, from simple applications to the most sophisticated circuits. Section 11.5 provides a detail of the sources of literature on this subject.

11.3 CONTROL SOFTWARE DESIGN

Control software is normally specific to the application (discrete product) in which it will be used. The following steps are required in software design:

1. Declaration of requirement specification
2. Declaration of functional specification
3. High-level design
4. Low-level design
5. Coding for the actual software module
6. Generation and execution of test plan
7. Bug report, which may lead to revision of steps 1 to 6
8. In case of proved functionality, release of the first version of the software, or else repeat of any of steps 1 to 6

Other modern software design processes include rational unified processes and agile software development.

11.3.1 Use of Application-Specific Standard for the Development of Control Software

Standards from various organizations exist that have provisions for operation of several discrete products. These standards include the Electronic Industries Association EIA 274D standard for the control of computerized numerical control (CNC) machinery and the Japanese industrial standard JIS SLIM (Standard Language for Industrial Manipulators) for the control of pick-and-place technology such as robots.

An interpreter is software that is normally based on a standard such as described above. It reads the user's input character by character and generates/receives electronic signals to/from the actuators or sensors to operate a discrete product. It is normally the core element of the control software. Such an interpreter may be embedded in the control software that also provides other functions, such as the graphical user interface (GUI), for the convenience of the user.

The control software development is normally based on one of the software-design methodologies in an appropriately certified environment.

11.4 MICROPROCESSOR INTERFACING FOR A COMPUTERIZED NUMERICAL CONTROL MILLING MACHINE

This case study demonstrates the development of CNC for a milling machine. Both control software and electronic/electrical hardware parts are considered. The general layout of the three-axes milling machine control architecture is given in Figure 11.2. Circuit diagrams provided in Section 11.4.1 details the design of a purpose-built PCB that connects to the parallel port of a personal computer.

The control software design is based on UML, and the UML static structure and use case diagram is given in Figure 11.3. The control software only implements selection of EIA 274D G&M codes to simultaneously control the milling machine and its simulation. The software does not provide access to the complete

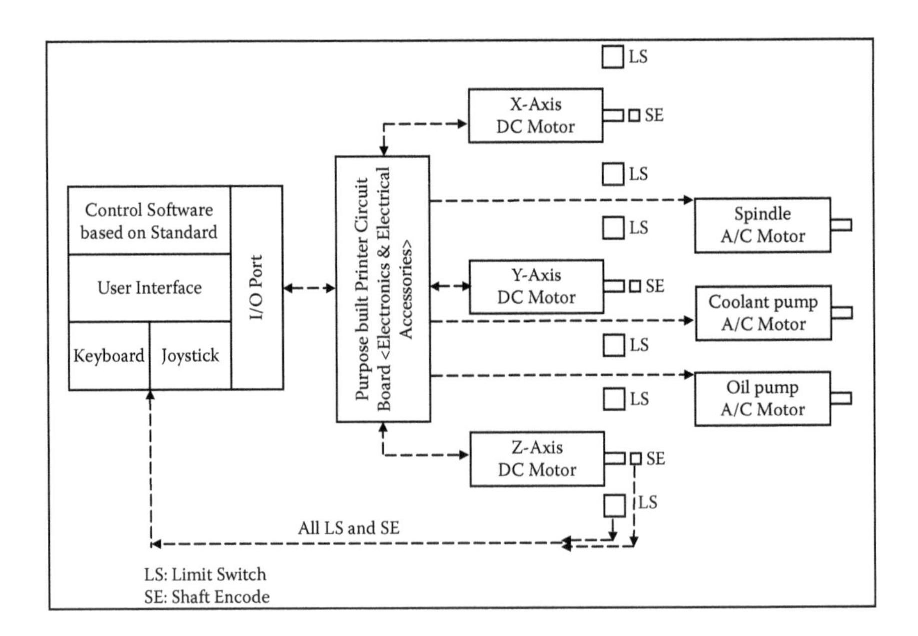

FIGURE 11.2 General layout for milling machine CNC controller.

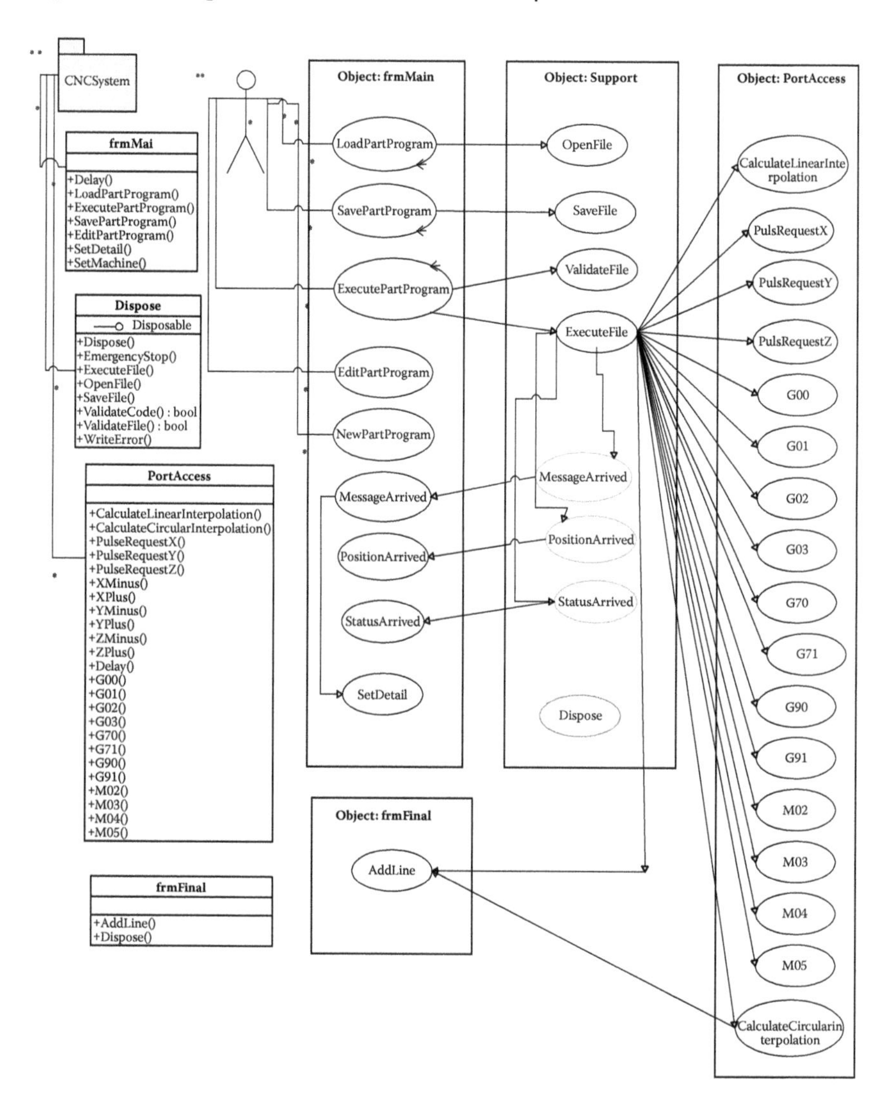

FIGURE 11.3 CNC system UML static structure and case diagram.

hardware listed in Section 11.4.1. The control software uses Visual C#. Net version 1.1. The code for the following modules of the control software is provided in Section 11.4.2.

AssemblyInfo.cs
fromMain.cs
Support.cs
PortAccess.cs
frmfinal.cs

11.4.1 Design of Printed Circuit Boards for Implementation of Computerized Numerical Control at a Milling Machine

The PCBs design for the three-axes milling machine takes into account control of axes motors with limit switches and shaft encoders and spindle and pump controllers. All the circuit diagrams along with bill of materials are provided in Figure 11.4 to Figure 11.16.

11.4.2 Software Modules for a Computerized Numerical Control Milling Machine

Control of the electronic circuit as given in Figure 11.4 to Figure 11.16 is performed through control software based on the EIA 274-D standard. A small group of G&M codes are chosen for the control of CNC milling machines. All the provisions of the electronic circuits are not controlled through control software. The listing of the control software in paper format uses arbitrary addresses for I/O ports. The user must adjust I/O addresses in the software module PortAccess.cs, compile the suit of programs, and then run it.

File Name: AssemblyInfo.cs
Purpose: The code has information about project
Environment: Visual C# .Net Version 1.1

```
using System.Reflection;
using System.Runtime.CompilerServices;

//
// General Information about an assembly is controlled through the
//following set of attributes. Change these attribute values to
//modify the information associated with an assembly.
//
[assembly: AssemblyTitle("CNCSystem")]
[assembly: AssemblyDescription("Programmed by AH")]
[assembly: AssemblyConfiguration("")]
[assembly: AssemblyCompany("HEC")]
[assembly: AssemblyProduct("")]
[assembly: AssemblyCopyright("WAK")]
[assembly: AssemblyTrademark("")]
[assembly: AssemblyCulture("")]

//
// Version information for an assembly consists of the following
four values:
//
//       Major Version
//       Minor Version
//       Build Number
//       Revision
//
```

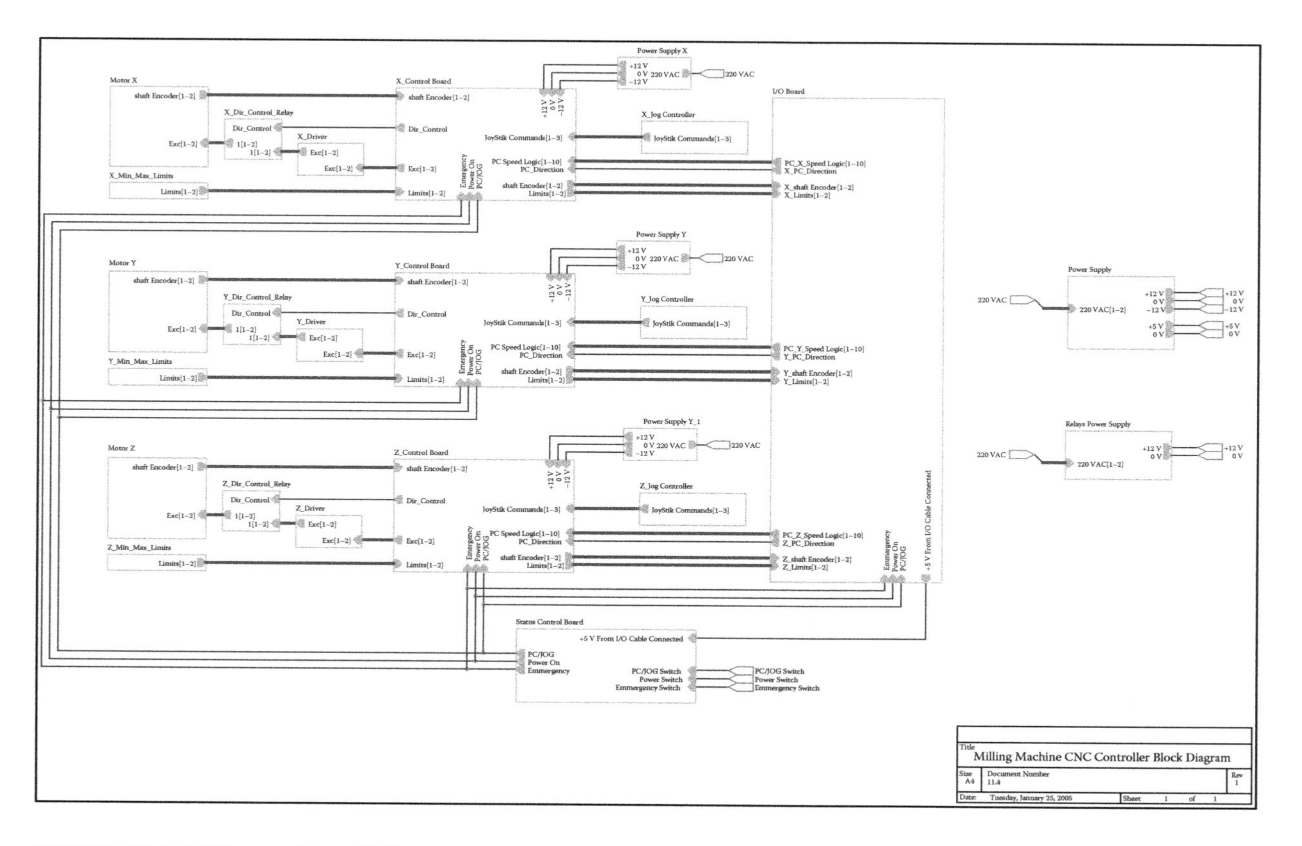

FIGURE 11.4 Milling machine CNC controller block diagram.

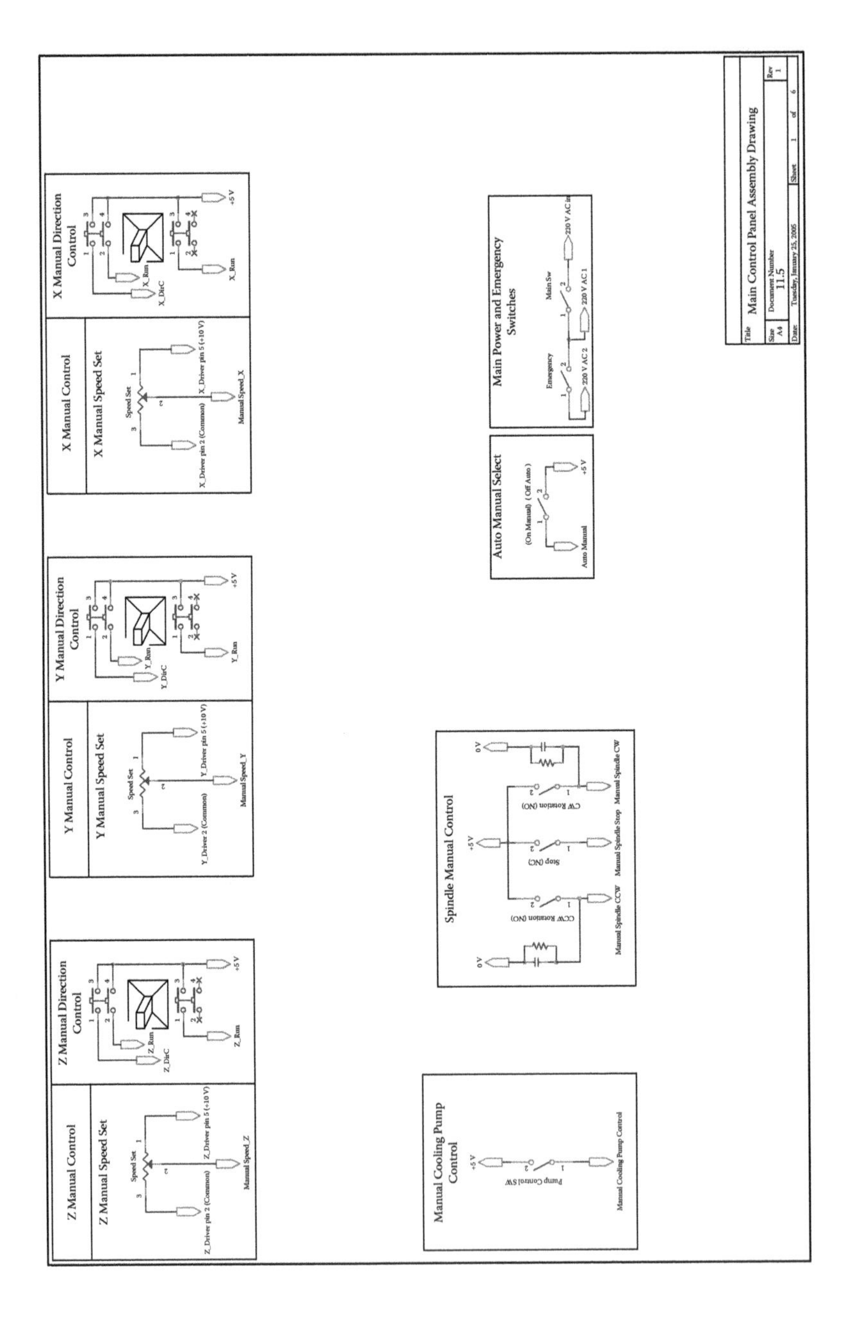

FIGURE 11.5 Main control-panel assembly drawing.

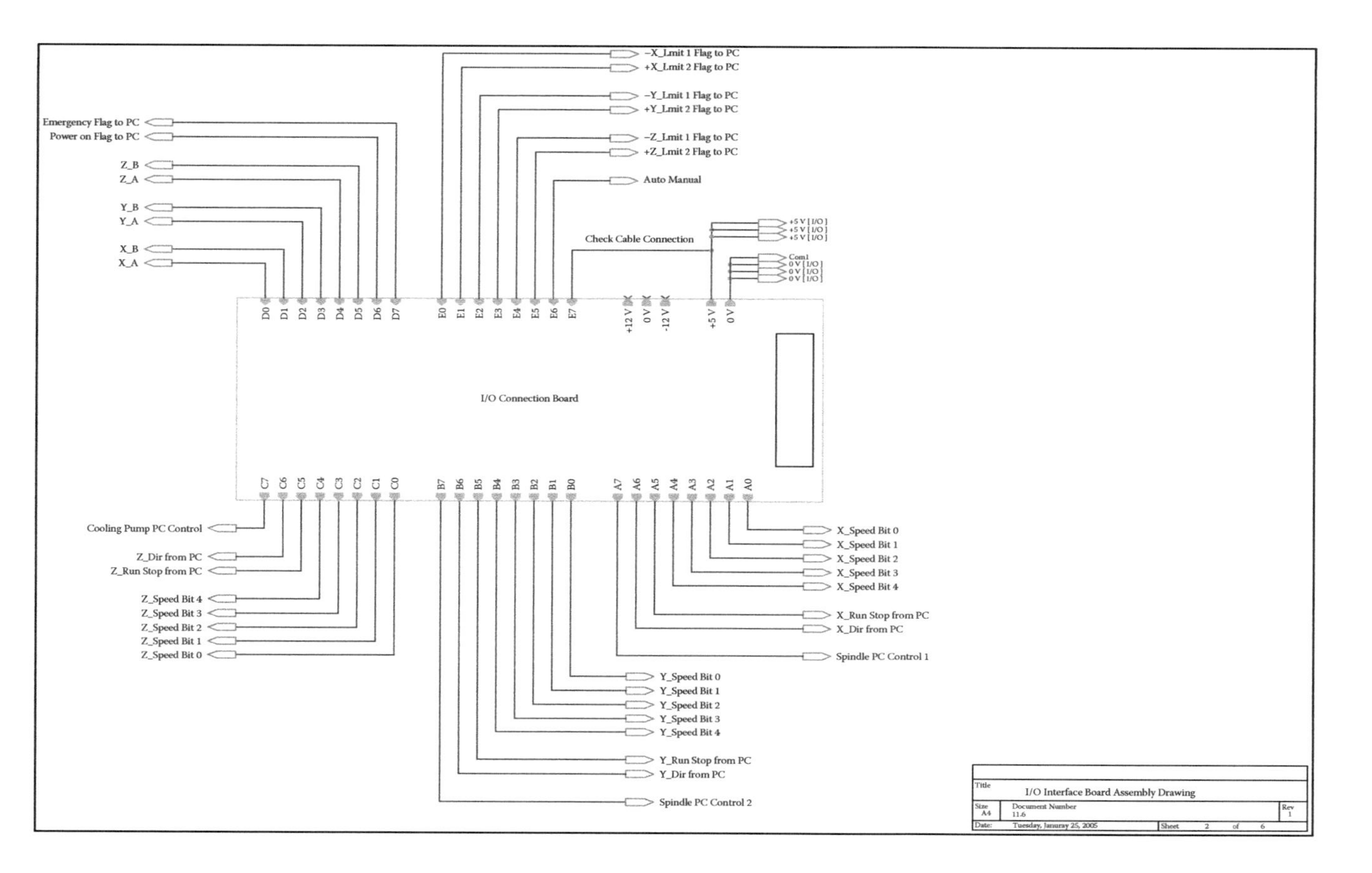

FIGURE 11.6 I/O interface board assembly drawing.

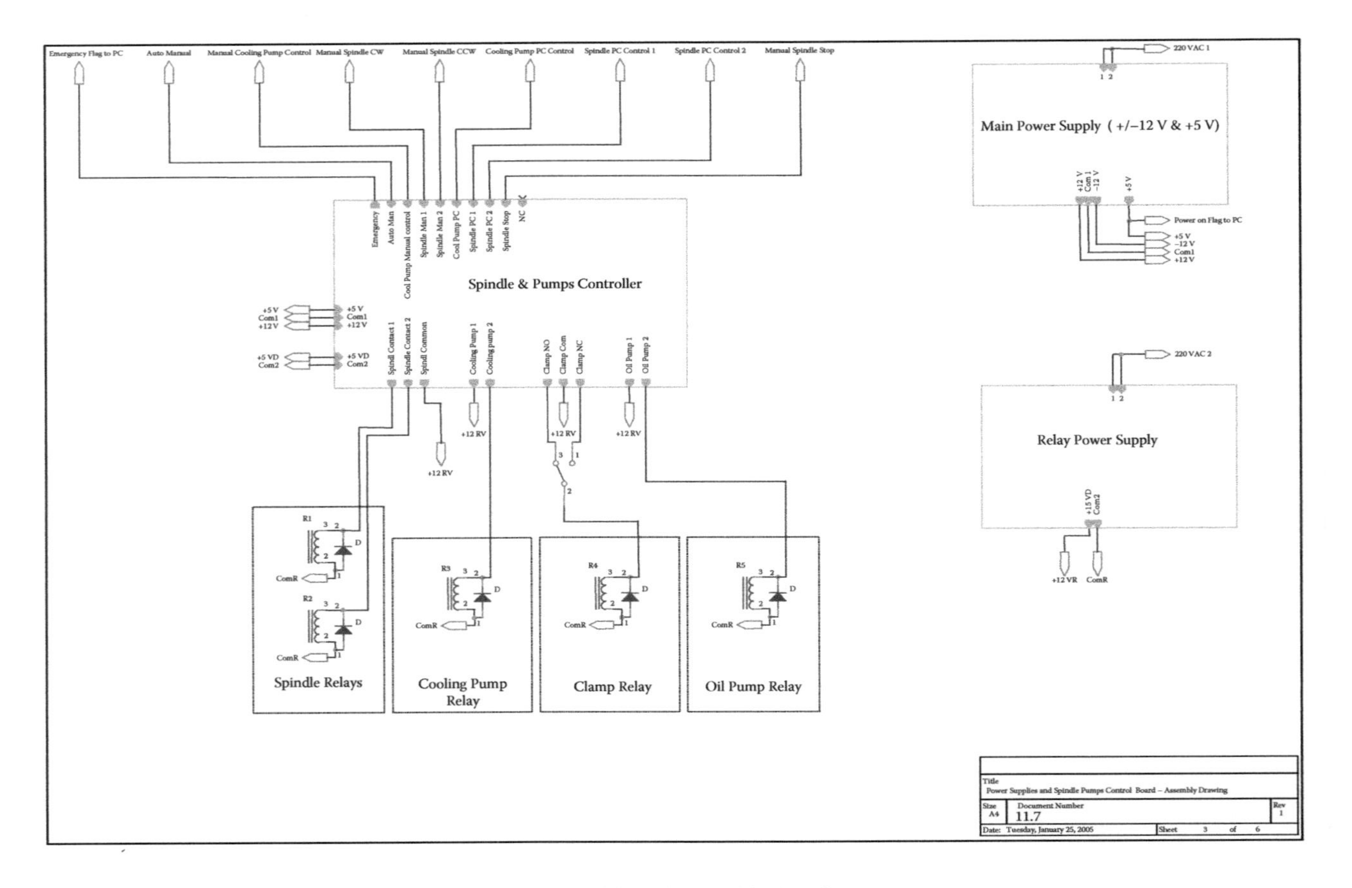

FIGURE 11.7 Power supplies and spindle pumps control-board assembly drawing.

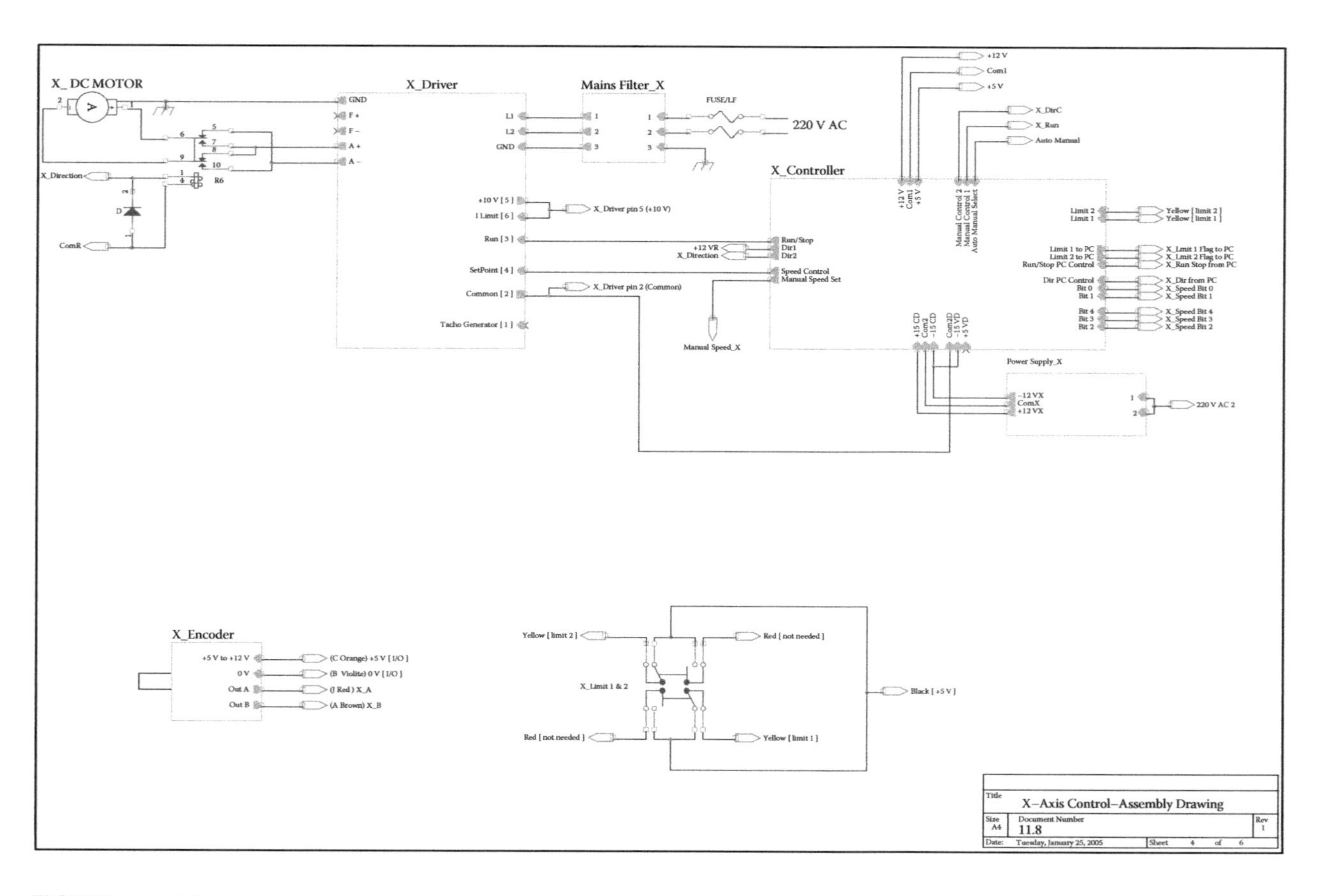

FIGURE 11.8 X-axis control assembly drawing.

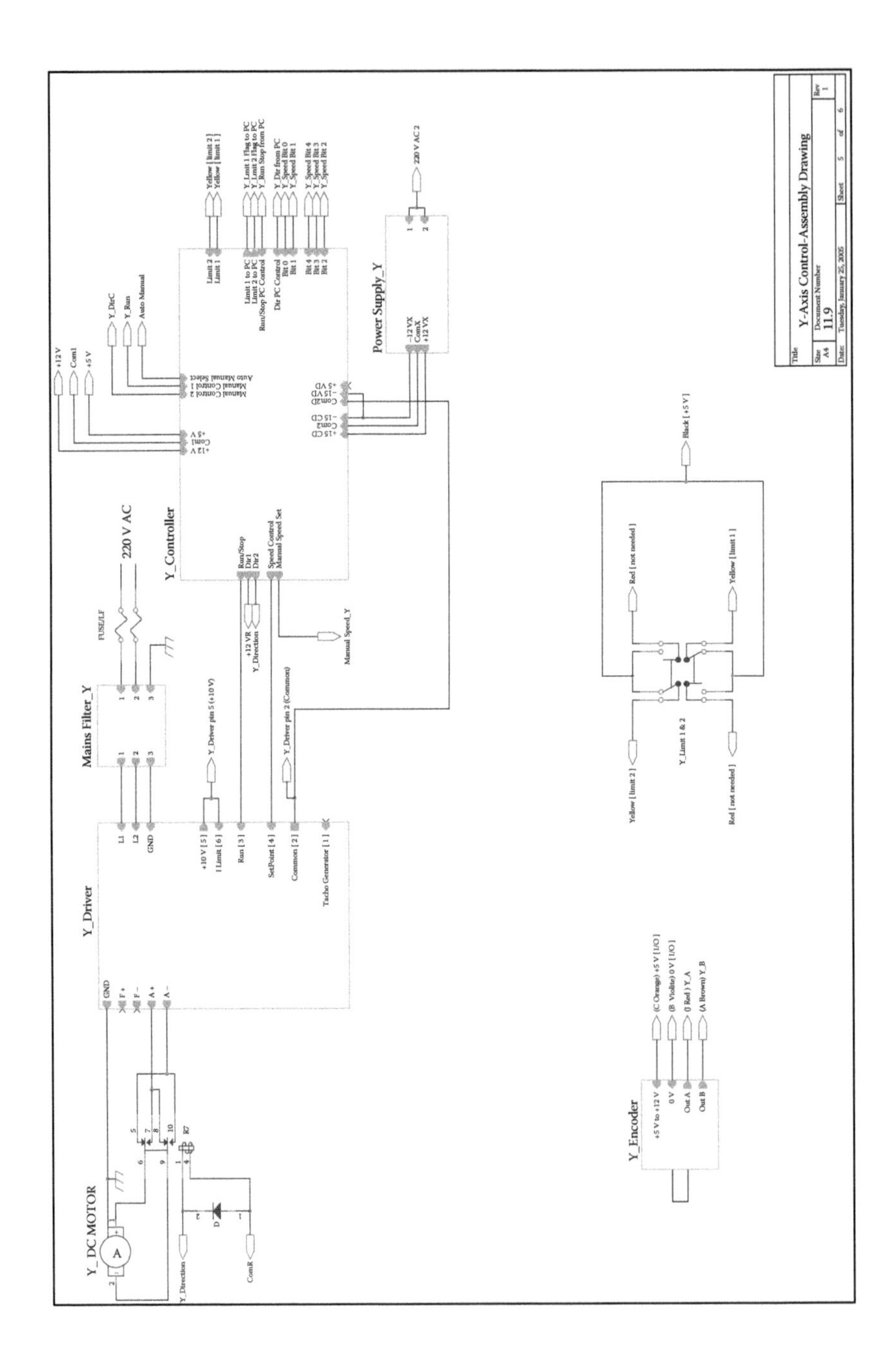

FIGURE 11.9 Y-axis control assembly drawing.

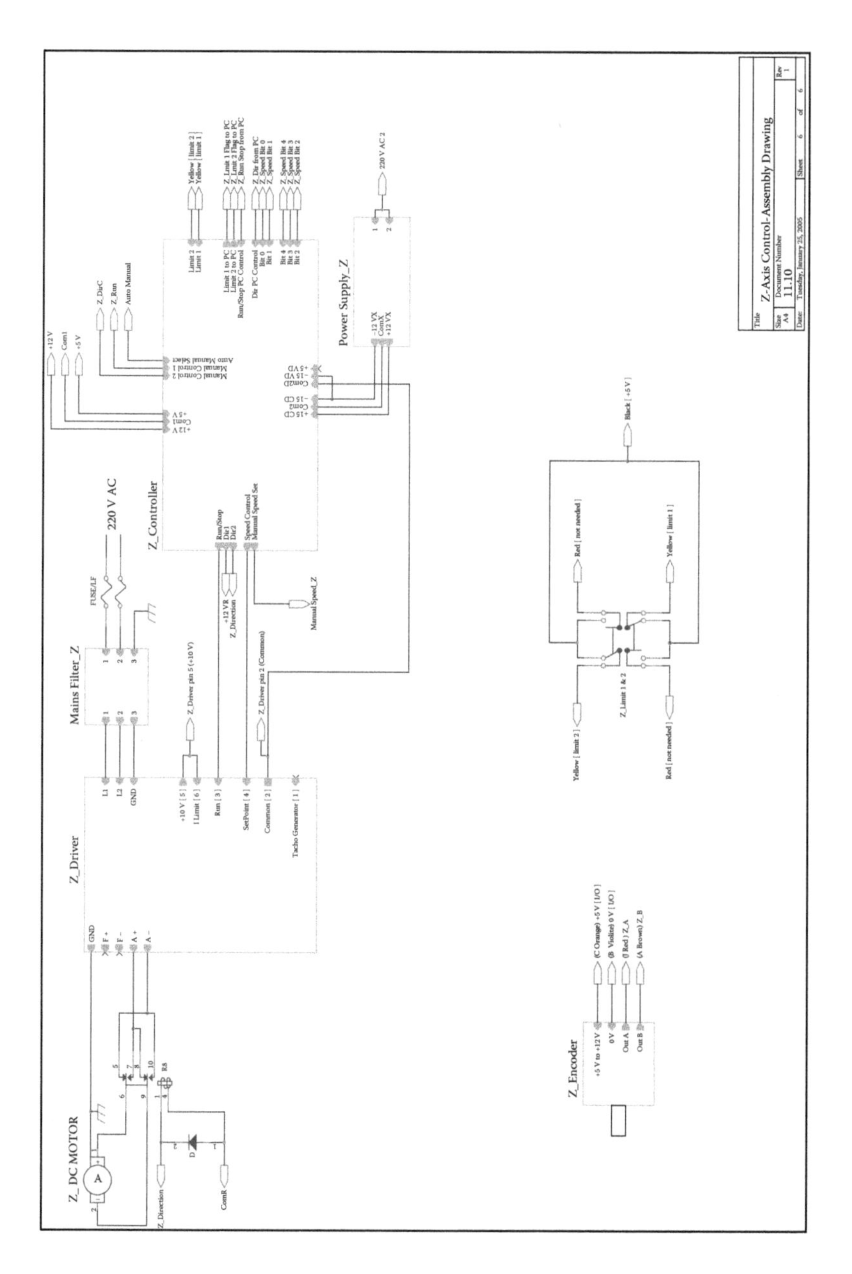

FIGURE 11.10 Z-axis control assembly drawing.

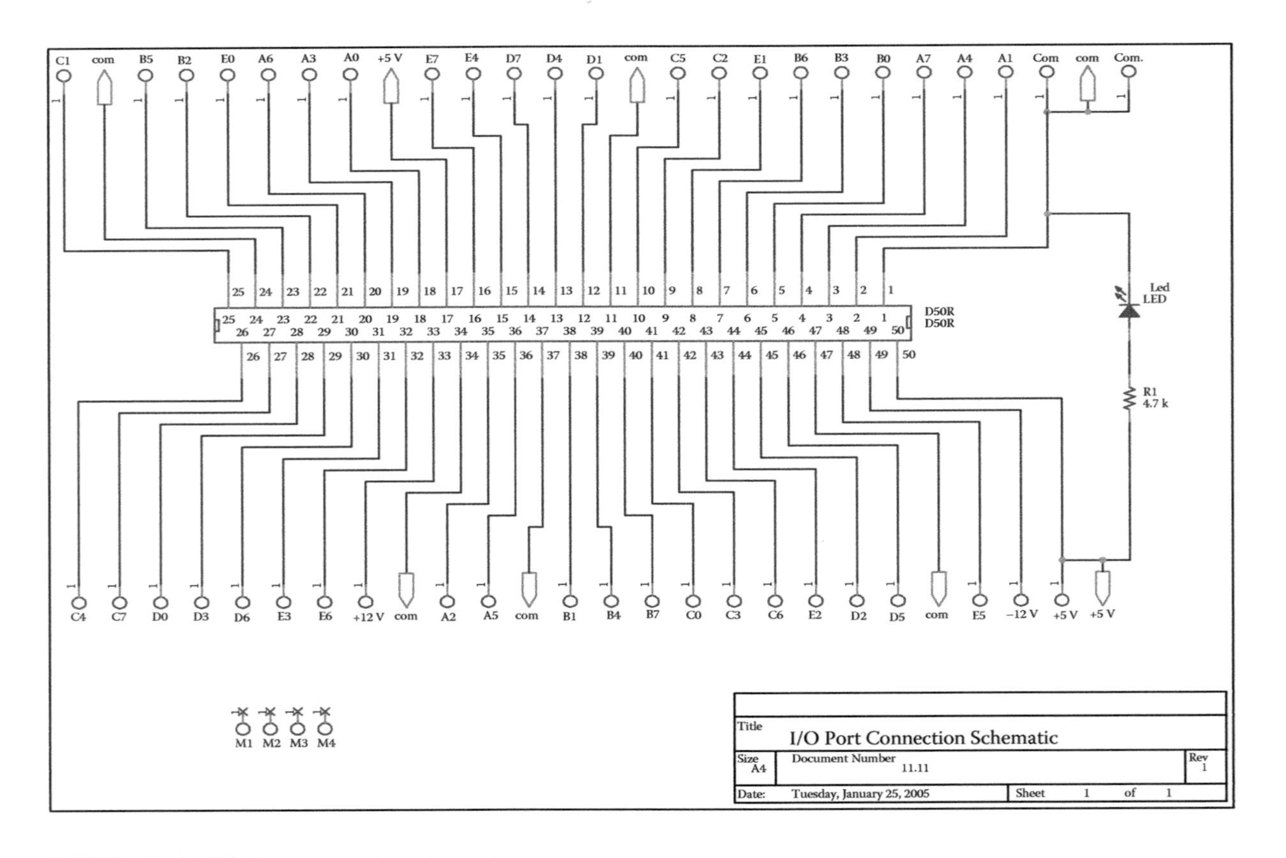

FIGURE 11.11 I/O Port-connection schematic.

```
Bill Of Materials

Item  Quantity    Reference    Part
_______________________________________________

1      49     M1,E1,D1,C1,B1,A1,M2,E2,      T POINT R
              D2,C2,B2,A2,M3,E3,D3,C3,
              B3,A3,M4,E4,D4,C4,B4,A4,
              E5,D5,C5,B5,A5,+5V,E6,D6,
              C6,B6,A6,E7,D7,C7,B7,A7,
              12-V,+12V,E0,D0,Com.,Com
              C0,B0,A0
2      1      D50R  D50R
3      1      Led   LED
4      1      R1    4.7k
```

Title	I/O Port Connection Bill of Materials	
Size A4	Document Number 11.12	Rev 1
Date Tuesday January 25, 2005	Sheet 1 of 1	

FIGURE 11.12 I/O Port-connection bill of materials.

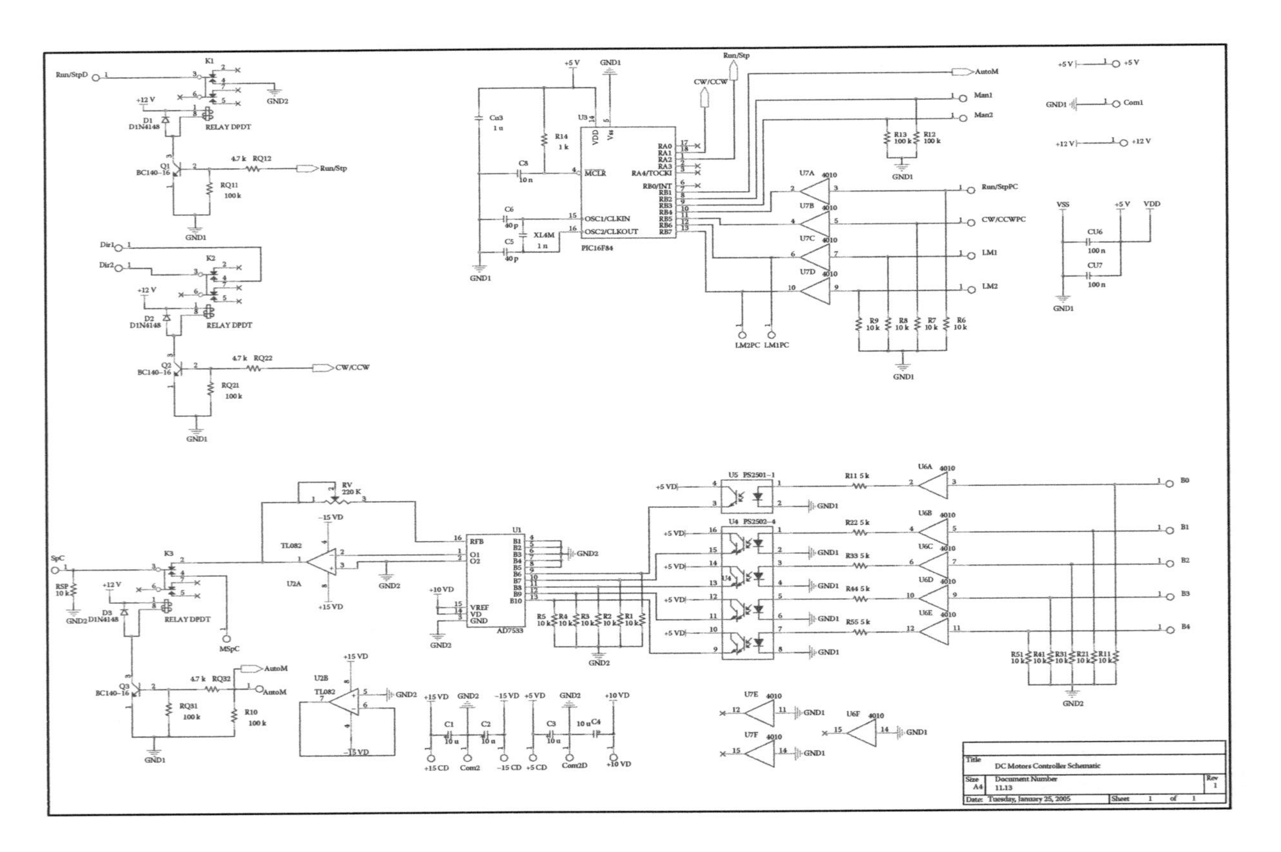

FIGURE 11.13 DC motors controller schematic.

Bill Of Materials

Item	Quantity	Reference	Part
1	28	Man1,LM1PC,LM1,Dir1,Com1, B1,Man2,LM2PC,LM2,Dir2, Com2D,Com2,B2,B3,B4,+5V, 5+CD,+10VD,+12V,-15CD, 15+CD,SpC,Run/StpPC, Run/StpD,MSpC,CW/CCWPC, B0,AutoM	T POINT R
2	2	CU6,CU7	100n
3	1	Cu3	1u
4	4	C1,C2,C3,C4	10u
5	2	C5,C6	40p
6	1	C8	10n
7	3	D1,D2,D3	D1N4148
8	3	K1,K2,K3	RELAY DPDT
9	3	Q1,Q2,Q3	BC140-16
10	6	R10,RQ11,R12,R13,RQ21, RQ31	100k
11	3	RQ12,RQ22,RQ32	4.7k
12	15	R1,R2,R3,R4,R5,R6,R7,R8, R9,R11,R21,R31,R41,R51, RSP	10k
13	1	RV	220K
14	5	R11,R22,R33,R44,R55	5k
15	1	R14	1k
16	1	U1	AD7533
17	1	U2	TL082
18	1	U3	PIC16F84
19	1	U4	PS2502-4
20	1	U5	PS2501-1
21	2	U7,U6	4010
22	1	XL4M	1n

Title DC Motors Controller Bill of Material		
Size A4	Document Number 11.14	Rev 1
Date Tuesday January 25, 2005	Sheet 1 of 1	

FIGURE 11.14 DC motors controller bill of material.

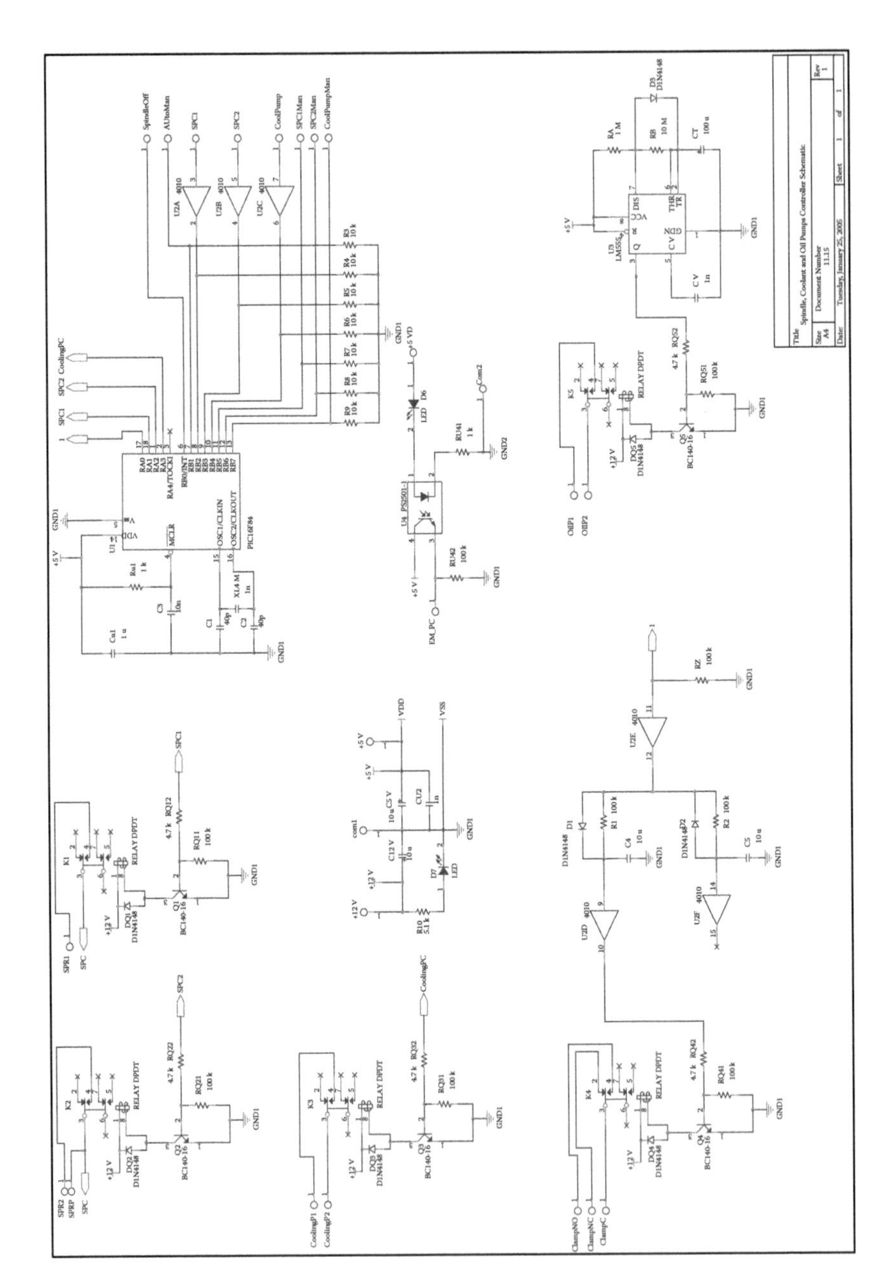

FIGURE 11.15 Spindle, coolant, and oil pumps controller schematic.

```
Bill Of Materials
Item     Quantity        Reference    Part
______________________________________________________

1        24         com1,SPR1,SPC1Man,SPC1,   T POINT R
                    OilP1,CoolingP1,SPR2,
                    SPC2Man,SPC2,OIlP2,
                    CoolingP2,Com2,+5VD,+5V,
                    +12V,SpindleOff,SPRP,
                    EM_PC,CoolPumpMan,
                    CoolPump,ClampNO,ClampNC,
                    ClampC,AUtoMan
2        1          CT     100u
3        3          CU2,XL4M,CV 1n
4        1          Cu1    1u
5        2          C2,C1 40p
6        1          C3     10n
7        4          C4,C5V,C5,C12V     10u
8        8          DQ1,D1,DQ2,D2,DQ3,D3,DQ4,       D1N4148
                    DQ5
9        2          D6,D7 LED
10       5          K1,K2,K3,K4,K5     RELAY DPDT
11       5          Q1,Q2,Q3,Q4,Q5     BC140-16
12       1          RA     1M
13       1          RB     10M
14       9          R1,R2,RQ11,RQ21,RQ31,    100k
                    RQ41,RU42,RQ51,RZ
15       5          RQ12,RQ22,RQ32,RQ42,RQ52 4.7k
16       2          RU41,Ru1     1k
17       7          R3,R4,R5,R6,R7,R8,R9      10k
18       1          R10    5.1k
19       1          U1     PIC16F84
20       1          U2     4010
21       1          U3     LM555
22       1          U4     PS2501-1
```

Title	Spindle and Pumps Controller Bill of Materials	
Size A4	Document Number 11.16	Rev 1
Date Tuesday January 25, 2005	Sheet 1 of 1	

FIGURE 11.16 Spindle and pumps controller bill of materials.

```
// Specify all the values or use default values, for example, the
Revision and Build numbers
// by using the '*' as shown below:

[assembly: AssemblyVersion("1.0.*")]

//
// To sign assembly, specify a key for use. Refer to the
// Microsoft.NET Framework documentation for more information on
assembly signing.
//
// Use the attributes below to control which key is used for signing.
//
// Notes:
//    (*) If no key is specified, the assembly is not signed.
//    (*) Key Name refers to a key that has been installed in the
Crypto Service
//        Provider (CSP) on your machine. Key File refers to a file
that contains
//        a key.
//    (*) If the Key File and the Key Name values are both specified,
//        the following processing occurs:
//        (1) If the Key Name can be found in the CSP, that key is
              used.
//        (2) If the Key Name does not exist and the Key File does
exist,
// the key in the Key File is installed into the CSP and used.
//    (*) To create a Key File, use the sn.exe (Strong Name) utility.
//        When specifying the Key File, the location of the Key
//        File should be relative to the project output directory,
//        which is
//        %Project Directory%\obj\<configuration>. For example, if
Key File is
//        located in the project directory, specify the Assembly
Key File
//        attribute as [assembly: Assembly Key
File("..\\..\\mykey.snk")]
//    (*) Delay Signing is an advanced option. See the Microsoft.
NET Framework
//        documentation for more information.
//
[assembly: AssemblyDelaySign(false)]
[assembly: AssemblyKeyFile("")]
[assembly: AssemblyKeyName("")]
```

File Name: frmMain.cs
Purpose: The code generates main screen of application.
Environment: Visual C# .Net version 1.1

```
using System;
using System.Drawing;
using System.Collections;
using System.ComponentModel;
using System.Windows.Forms;
using System.Drawing.Drawing2D;
using System.Diagnostics;


namespace CNCSystem
{
	/// <summary>
	/// Summary description for frmMain.
	/// </summary>
	#region Enumeration Current Status Type
	/// <summary>
	///This enumeration use to set the
	///current status of x position, y position, and so on.
	/// </summary>
	public enum Current
	{
		X,
		Y,
		Z,
		Spindle,
		Feed,
		Speed,
		Movement,
		Coordinate,
		Unit
	}

	#endregion
	public class frmMain : System.Windows.Forms.Form
	{
		/// <summary>
		/// Required designer variable.
		/// </summary>
		///

		[STAThread]
		static void Main()
		{
			Application.Run(new frmMain());
		}
		#region Variables Declaration
		private System.Windows.Forms.MainMenu mainMenu;
		private System.Windows.Forms.MenuItem menuItem1;
```

```
        private System.Windows.Forms.MenuItem menuItemLoad;
        private System.Windows.Forms.OpenFileDialog
openFileDialog;
        private System.Windows.Forms.MenuItem menuItemExecute;
        private System.Windows.Forms.StatusBar statusBar;
        private System.Windows.Forms.StatusBarPanel Information;
        private System.ComponentModel.IContainer components;
        private System.Windows.Forms.MenuItem menuItem2;
        private System.Windows.Forms.MenuItem menuItem3;
        private System.Windows.Forms.SaveFileDialog
saveFileDialog;
        private string[] sLineNumber = new string[999];
        private System.Windows.Forms.StatusBarPanel PortStatus;
        private System.Windows.Forms.MenuItem menuItem4;
        private System.Windows.Forms.MenuItem menuItem5;
        private System.Windows.Forms.GroupBox groupBox1;
        private System.Windows.Forms.Label label1;
        private System.Windows.Forms.Label label2;
        private System.Windows.Forms.Label label3;
        private System.Windows.Forms.MenuItem menuItem6;
        private System.Windows.Forms.GroupBox groupBox3;
        private System.Windows.Forms.GroupBox groupBox4;
        private System.Windows.Forms.Button txtX;
        private System.Windows.Forms.Button txtZ;
        private System.Windows.Forms.Button txtY;
        private System.Windows.Forms.Button txt4;
        private System.Windows.Forms.Button txt3;
        private System.Windows.Forms.Button txt2;
        private System.Windows.Forms.Button txt1;
        private System.Windows.Forms.Button txt9;
        private System.Windows.Forms.Button txt8;
        private System.Windows.Forms.Button txt7;
        private System.Windows.Forms.Button txt6;
        private System.Windows.Forms.Button txt5;
        private System.Windows.Forms.Button txt0;
        private System.Windows.Forms.Button txtpoint;
        private System.Windows.Forms.Button txtminus;
        private System.Windows.Forms.Button txtplus;
        private System.Windows.Forms.Button txtG1;
        private System.Windows.Forms.Button txtG0;
        private System.Windows.Forms.Button txtG2;
        private System.Windows.Forms.Button txtG3;
        private System.Windows.Forms.Button txtG90;
        private System.Windows.Forms.Button txtG91;
        private System.Windows.Forms.Button txtG70;
        private System.Windows.Forms.Button txtG71;
        private System.Windows.Forms.Button txtM5;
        private System.Windows.Forms.Button txtM3;
        private System.Windows.Forms.Button txtM4;
        private System.Windows.Forms.Button txtT;
        private System.Windows.Forms.Button txtS;
        private System.Windows.Forms.Button txtJ;
        private System.Windows.Forms.Button txtI;
        private System.Windows.Forms.Button txtF;
```

```
        private System.Windows.Forms.TextBox txtMain;
        private System.Windows.Forms.GroupBox groupBox5;
        private System.Windows.Forms.TextBox txtDetail;
        private System.Windows.Forms.Button btnStop;
        private System.Windows.Forms.Label lblUnit;
        private System.Windows.Forms.Label label6;
        private System.Windows.Forms.Label lblCoordinate;
        private System.Windows.Forms.Label label5;
        private System.Windows.Forms.Label lblMovement;
        private System.Windows.Forms.Label label4;
        private System.Windows.Forms.Label lblSpeed;
        private System.Windows.Forms.Label lblFeed;
        private System.Windows.Forms.Label lblSpindle;
        private System.Windows.Forms.Label label7;
        private System.Windows.Forms.Label label8;
        private System.Windows.Forms.Label label9;
        private System.Windows.Forms.Button txtM2;
        private System.Windows.Forms.GroupBox groupBox2;
        private System.Windows.Forms.Button btnReset;
        private System.Windows.Forms.Button btnEdit;
        private System.Windows.Forms.Button btnExit;
        private System.Windows.Forms.Button btnSave;
        private System.Windows.Forms.Button btnNew;
        private System.Windows.Forms.Button btnLoad;
        private System.Windows.Forms.Button btnRun;
        private System.Windows.Forms.Label lblZ;
        private System.Windows.Forms.Label lblY;
        private System.Windows.Forms.Label lblX;
        private System.Windows.Forms.ToolTip toolTip;
        private System.Windows.Forms.PictureBox pictureBox2;
        private Point p = new Point(648,304);/// moving picture
original points in x and y
        private System.Windows.Forms.Timer Timer;
        private const int DevidedBy = 25;
        private System.Windows.Forms.MenuItem menuItem7;
        private System.Windows.Forms.MenuItem menuItem8;
        private System.Windows.Forms.Button btnenter;
        private System.Windows.Forms.Button btnbspace;
        private System.Windows.Forms.Button btnSpace;
        private System.Windows.Forms.PictureBox pictureBox1;
        private System.Windows.Forms.PictureBox pDrill;
        private int TimerCounter;



        #endregion
        public frmMain()
        {
            //
            // Required for Windows Form Designer support
            //
            InitializeComponent();

            //
```

```
			// TODO: Add any constructor code after Initialize
Component call
			//
		}

		/// <summary>
		/// Clean up any resources being used.
		/// </summary>
		protected override void Dispose( bool disposing )
		{
			if( disposing )
			{
				if(components != null)
				{
				components.Dispose();
				}
			}
			base.Dispose( disposing );
		}

		#region Windows Form Designer generated code
		/// <summary>
		/// Required method for Design support. do not modify
		/// the contents of this method with the code editor.
		/// </summary>
		private void InitializeComponent()
		{
			this.components = new
System.ComponentModel.Container();
			System.Resources.ResourceManager resources = new
System.Resources.ResourceManager(typeof(frmMain));
			this.mainMenu = new System.Windows.Forms.MainMenu();
			this.menuItem1 = new System.Windows.Forms.MenuItem();
			this.menuItemLoad = new
System.Windows.Forms.MenuItem();
			this.menuItemExecute = new
System.Windows.Forms.MenuItem();
			this.menuItem2 = new System.Windows.Forms.MenuItem();
			this.menuItem3 = new System.Windows.Forms.MenuItem();
			this.menuItem4 = new System.Windows.Forms.MenuItem();
			this.menuItem5 = new System.Windows.Forms.MenuItem();
			this.menuItem6 = new System.Windows.Forms.MenuItem();
			this.menuItem7 = new System.Windows.Forms.MenuItem();
			this.menuItem8 = new System.Windows.Forms.MenuItem();
			this.openFileDialog = new
System.Windows.Forms.OpenFileDialog();
			this.statusBar = new
System.Windows.Forms.StatusBar();
			this.Information = new
System.Windows.Forms.StatusBarPanel();
			this.PortStatus = new
System.Windows.Forms.StatusBarPanel();
			this.saveFileDialog = new
System.Windows.Forms.SaveFileDialog();
```

```
this.groupBox1 = new System.Windows.Forms.GroupBox();
this.lblUnit = new System.Windows.Forms.Label();
this.label6 = new System.Windows.Forms.Label();
this.lblCoordinate = new System.Windows.Forms.Label();
this.label5 = new System.Windows.Forms.Label();
this.lblMovement = new System.Windows.Forms.Label();
this.label4 = new System.Windows.Forms.Label();
this.lblSpeed = new System.Windows.Forms.Label();
this.lblFeed = new System.Windows.Forms.Label();
this.lblSpindle = new System.Windows.Forms.Label();
this.label7 = new System.Windows.Forms.Label();
this.label8 = new System.Windows.Forms.Label();
this.label9 = new System.Windows.Forms.Label();
this.lblZ = new System.Windows.Forms.Label();
this.lblY = new System.Windows.Forms.Label();
this.lblX = new System.Windows.Forms.Label();
this.label3 = new System.Windows.Forms.Label();
this.label2 = new System.Windows.Forms.Label();
this.label1 = new System.Windows.Forms.Label();
this.groupBox3 = new System.Windows.Forms.GroupBox();
this.groupBox2 = new System.Windows.Forms.GroupBox();
this.btnReset = new System.Windows.Forms.Button();
this.btnEdit = new System.Windows.Forms.Button();
this.btnExit = new System.Windows.Forms.Button();
this.btnSave = new System.Windows.Forms.Button();
this.btnNew = new System.Windows.Forms.Button();
this.btnLoad = new System.Windows.Forms.Button();
this.btnRun = new System.Windows.Forms.Button();
this.btnStop = new System.Windows.Forms.Button();
this.groupBox5 = new System.Windows.Forms.GroupBox();
this.txtDetail = new System.Windows.Forms.TextBox();
this.txtMain = new System.Windows.Forms.TextBox();
this.groupBox4 = new System.Windows.Forms.GroupBox();
this.btnenter = new System.Windows.Forms.Button();
this.btnbspace = new System.Windows.Forms.Button();
this.btnSpace = new System.Windows.Forms.Button();
this.txtS = new System.Windows.Forms.Button();
this.txtT = new System.Windows.Forms.Button();
this.txtI = new System.Windows.Forms.Button();
this.txtF = new System.Windows.Forms.Button();
this.txtJ = new System.Windows.Forms.Button();
this.txtM4 = new System.Windows.Forms.Button();
this.txtM2 = new System.Windows.Forms.Button();
this.txtM3 = new System.Windows.Forms.Button();
this.txtM5 = new System.Windows.Forms.Button();
this.txtG91 = new System.Windows.Forms.Button();
this.txtG0 = new System.Windows.Forms.Button();
this.txtG1 = new System.Windows.Forms.Button();
this.txtG2 = new System.Windows.Forms.Button();
this.txtG3 = new System.Windows.Forms.Button();
this.txtG90 = new System.Windows.Forms.Button();
this.txtG71 = new System.Windows.Forms.Button();
this.txtG70 = new System.Windows.Forms.Button();
this.txtplus = new System.Windows.Forms.Button();
```

```
            this.txtminus = new System.Windows.Forms.Button();
            this.txtpoint = new System.Windows.Forms.Button();
            this.txt5 = new System.Windows.Forms.Button();
            this.txt6 = new System.Windows.Forms.Button();
            this.txt7 = new System.Windows.Forms.Button();
            this.txt8 = new System.Windows.Forms.Button();
            this.txt9 = new System.Windows.Forms.Button();
            this.txt0 = new System.Windows.Forms.Button();
            this.txtY = new System.Windows.Forms.Button();
            this.txtZ = new System.Windows.Forms.Button();
            this.txt1 = new System.Windows.Forms.Button();
            this.txt2 = new System.Windows.Forms.Button();
            this.txt3 = new System.Windows.Forms.Button();
            this.txt4 = new System.Windows.Forms.Button();
            this.txtX = new System.Windows.Forms.Button();
            this.toolTip = new
System.Windows.Forms.ToolTip(this.components);
            this.pictureBox2 = new
System.Windows.Forms.PictureBox();
            this.Timer = new
System.Windows.Forms.Timer(this.components);
            this.pictureBox1 = new
System.Windows.Forms.PictureBox();
            this.pDrill = new System.Windows.Forms.PictureBox();

((System.ComponentModel.ISupportInitialize)(this.Information)).Begi
nInit();

((System.ComponentModel.ISupportInitialize)(this.PortStatus)).Begin
Init();
            this.groupBox1.SuspendLayout();
            this.groupBox3.SuspendLayout();
            this.groupBox2.SuspendLayout();
            this.groupBox5.SuspendLayout();
            this.groupBox4.SuspendLayout();
            this.SuspendLayout();
            //
            // Main Menu
            //
            this.mainMenu.MenuItems.AddRange(new
System.Windows.Forms.MenuItem[] {
                  this.menuItem1,
                  this.menuItem7});
            //
            // menuItem1
            //
            this.menuItem1.Index = 0;
            this.menuItem1.MenuItems.AddRange(new
System.Windows.Forms.MenuItem[] {
                   this.menuItemLoad,
                   this.menuItemExecute,
                   this.menuItem2,
                   this.menuItem3,
```

```
                        this.menuItem4,
                        this.menuItem5,
                        this.menuItem6});
            this.menuItem1.Text = "&Main";
            //
            // menu Item Load
            //
            this.menuItemLoad.Index = 0;
            this.menuItemLoad.Text = "&Load Part Program";
            this.menuItemLoad.Click += new
System.EventHandler(this.menuItemLoad_Click);
            //
            // menu Item Execute
            //
            this.menuItemExecute.Index = 1;
            this.menuItemExecute.Text = "&Execute Part Program";
            this.menuItemExecute.Click += new
System.EventHandler(this.menuItemExecute_Click);
            //
            // menu Item2
            //
            this.menuItem2.Index = 2;
            this.menuItem2.Text = "New Part Program";
            this.menuItem2.Click += new
System.EventHandler(this.menuItem2_Click);
            //
            // menu Item3
            //
            this.menuItem3.Index = 3;
            this.menuItem3.Text = "&Save Part Program";
            this.menuItem3.Click += new
System.EventHandler(this.menuItem3_Click);
            //
            // menu Item4
            //
            this.menuItem4.Index = 4;
            this.menuItem4.Text = "E&dit Part Program";
            this.menuItem4.Click += new
System.EventHandler(this.menuItem4_Click);
            //
            // menu Item5
            //
            this.menuItem5.Index = 5;
            this.menuItem5.Text = "Reset &Machine";
            this.menuItem5.Click += new
System.EventHandler(this.menuItem5_Click);
            //
            // menu Item6
            //
            this.menuItem6.Index = 6;
            this.menuItem6.Text = "E&xit";
            this.menuItem6.Click += new
System.EventHandler(this.menuItem6_Click);
```

```
            //
            // menu Item7
            //
            this.menuItem7.Index = 1;
            this.menuItem7.MenuItems.AddRange(new
System.Windows.Forms.MenuItem[] {
                  this.menuItem8});
            this.menuItem7.Text = "&Help";
            //
            // menu Item8
            //
            this.menuItem8.Index = 0;
            this.menuItem8.Text = "CNCSystem Help";
            this.menuItem8.Click += new
System.EventHandler(this.menuItem8_Click);
            //
            // open File Dialog
            //
            this.openFileDialog.AddExtension = false;
            this.openFileDialog.DefaultExt = "cnc";
            this.openFileDialog.Filter = "*.cnc|*.cnc";
            this.openFileDialog.Title = "Select CNC File.";
            //
            // status Bar
            //
            this.statusBar.Location = new System.Drawing.Point(0,
713);
            this.statusBar.Name = "statusBar";
            this.statusBar.Panels.AddRange(new
System.Windows.Forms.StatusBarPanel[] {
                 this.Information,
                 this.PortStatus});
            this.statusBar.ShowPanels = true;
            this.statusBar.Size = new System.Drawing.Size(1030, 22);
            this.statusBar.TabIndex = 3;
            this.statusBar.Text = "Ready";
            //
            // Information
            //
            this.Information.AutoSize =
System.Windows.Forms.StatusBarPanelAutoSize.Spring;
            this.Information.Text = "Information";
            this.Information.ToolTipText = "Information";
            this.Information.Width = 507;
            //
            // Port Status
            //
            this.PortStatus.AutoSize =
System.Windows.Forms.StatusBarPanelAutoSize.Spring;
            this.PortStatus.Text = "Port Information";
            this.PortStatus.ToolTipText = "Port Informaiton";
            this.PortStatus.Width = 507;
            //
```

```
            // save File Dialog
            //
            this.saveFileDialog.DefaultExt = "cnc";
            this.saveFileDialog.Filter = "*.cnc|*.cnc";
            this.saveFileDialog.Title = "Save Part Programe";
            //
            // group Box1
            //
            this.groupBox1.Anchor =
((System.Windows.Forms.AnchorStyles)(((System.Windows.Forms.Anchor
Styles.Bottom | System.Windows.Forms.AnchorStyles.Left)
                | System.Windows.Forms.AnchorStyles.Right)));
            this.groupBox1.BackColor =
System.Drawing.Color.White;
            this.groupBox1.Controls.Add(this.lblUnit);
            this.groupBox1.Controls.Add(this.label6);
            this.groupBox1.Controls.Add(this.lblCoordinate);
            this.groupBox1.Controls.Add(this.label5);
            this.groupBox1.Controls.Add(this.lblMovement);
            this.groupBox1.Controls.Add(this.label4);
            this.groupBox1.Controls.Add(this.lblSpeed);
            this.groupBox1.Controls.Add(this.lblFeed);
            this.groupBox1.Controls.Add(this.lblSpindle);
            this.groupBox1.Controls.Add(this.label7);
            this.groupBox1.Controls.Add(this.label8);
            this.groupBox1.Controls.Add(this.label9);
            this.groupBox1.Controls.Add(this.lblZ);
            this.groupBox1.Controls.Add(this.lblY);
            this.groupBox1.Controls.Add(this.lblX);
            this.groupBox1.Controls.Add(this.label3);
            this.groupBox1.Controls.Add(this.label2);
            this.groupBox1.Controls.Add(this.label1);
            this.groupBox1.ForeColor =
System.Drawing.SystemColors.HotTrack;
            this.groupBox1.Location = new
System.Drawing.Point(320, 636);
            this.groupBox1.Name = "groupBox1";
            this.groupBox1.Size = new
System.Drawing.Size(704, 72);
            this.groupBox1.TabIndex = 12;
            this.groupBox1.TabStop = false;
            this.groupBox1.Text = "Current Position:";
            //
            // lbl Unit
            //
            this.lblUnit.AutoSize = true;
            this.lblUnit.ForeColor = System.Drawing.Color.Maroon;
            this.lblUnit.Location = new
System.Drawing.Point(496, 48);
            this.lblUnit.Name = "lblUnit";
            this.lblUnit.Size = new System.Drawing.Size(35, 16);
            this.lblUnit.TabIndex = 29;
            this.lblUnit.Text = "Metric";
```

```
            //
            // label6
            //
            this.label6.AutoSize = true;
            this.label6.ForeColor = System.Drawing.Color.Red;
            this.label6.Location = new
System.Drawing.Point(432, 48);
            this.label6.Name = "label6";
            this.label6.Size = new System.Drawing.Size(24, 16);
            this.label6.TabIndex = 28;
            this.label6.Text = "Unit";
            //
            // lbl Coordinate
            //
            this.lblCoordinate.AutoSize = true;
            this.lblCoordinate.ForeColor =
System.Drawing.Color.Maroon;
            this.lblCoordinate.Location = new
System.Drawing.Point(496, 32);
            this.lblCoordinate.Name = "lblCoordinate";
            this.lblCoordinate.Size = new
System.Drawing.Size(48, 16);
            this.lblCoordinate.TabIndex = 27;
            this.lblCoordinate.Text = "Absolute";
            //
            // label5
            //
            this.label5.AutoSize = true;
            this.label5.ForeColor = System.Drawing.Color.Red;
            this.label5.Location = new
System.Drawing.Point(432, 32);
            this.label5.Name = "label5";
            this.label5.Size = new System.Drawing.Size(63, 16);
            this.label5.TabIndex = 26;
            this.label5.Text = "Coordinate:";
            //
            // lbl Movement
            //
            this.lblMovement.AutoSize = true;
            this.lblMovement.ForeColor =
System.Drawing.Color.Maroon;
            this.lblMovement.Location = new
System.Drawing.Point(496, 16);
            this.lblMovement.Name = "lblMovement";
            this.lblMovement.Size = new System.Drawing.Size(34, 16);
            this.lblMovement.TabIndex = 25;
            this.lblMovement.Text = "Rapid";
            //
            // label4
            //
            this.label4.AutoSize = true;
            this.label4.ForeColor = System.Drawing.Color.Red;
            this.label4.Location = new
System.Drawing.Point(432, 16);
```

```
            this.label4.Name = "label4";
            this.label4.Size = new System.Drawing.Size(60, 16);
            this.label4.TabIndex = 24;
            this.label4.Text = "Movement:";
            //
            // lbl Speed
            //
            this.lblSpeed.AutoSize = true;
            this.lblSpeed.ForeColor = System.Drawing.Color.Maroon;
            this.lblSpeed.Location = new
System.Drawing.Point(312, 48);
            this.lblSpeed.Name = "lblSpeed";
            this.lblSpeed.Size = new System.Drawing.Size(10, 16);
            this.lblSpeed.TabIndex = 23;
            this.lblSpeed.Text = "0";
            //
            // lbl Feed
            //
            this.lblFeed.AutoSize = true;
            this.lblFeed.ForeColor = System.Drawing.Color.Maroon;
            this.lblFeed.Location = new
System.Drawing.Point(312, 32);
            this.lblFeed.Name = "lblFeed";
            this.lblFeed.Size = new System.Drawing.Size(20, 16);
            this.lblFeed.TabIndex = 22;
            this.lblFeed.Text = "0.0";
            //
            // lbl Spindle
            //
            this.lblSpindle.AutoSize = true;
            this.lblSpindle.ForeColor =
System.Drawing.Color.Maroon;
            this.lblSpindle.Location = new
System.Drawing.Point(312, 16);
            this.lblSpindle.Name = "lblSpindle";
            this.lblSpindle.Size = new System.Drawing.Size(27, 16);
            this.lblSpindle.TabIndex = 21;
            this.lblSpindle.Text = "OFF";
            //
            // label7
            //
            this.label7.AutoSize = true;
            this.label7.ForeColor = System.Drawing.Color.Red;
            this.label7.Location = new
System.Drawing.Point(232, 16);
            this.label7.Name = "label7";
            this.label7.Size = new System.Drawing.Size(45, 16);
            this.label7.TabIndex = 20;
            this.label7.Text = "Spindle:";
            //
            // label8
            //
            this.label8.AutoSize = true;
            this.label8.ForeColor = System.Drawing.Color.Red;
```

```
                this.label8.Location = new
System.Drawing.Point(232, 48);
                this.label8.Name = "label8";
                this.label8.Size = new System.Drawing.Size(40, 16);
                this.label8.TabIndex = 19;
                this.label8.Text = "Speed:";
                //
                // label9
                //
                this.label9.AutoSize = true;
                this.label9.ForeColor = System.Drawing.Color.Red;
                this.label9.Location = new
System.Drawing.Point(232, 32);
                this.label9.Name = "label9";
                this.label9.Size = new System.Drawing.Size(33, 16);
                this.label9.TabIndex = 18;
                this.label9.Text = "Feed:";
                //
                // lblZ
                //
                this.lblZ.AutoSize = true;
                this.lblZ.ForeColor = System.Drawing.Color.Maroon;
                this.lblZ.Location = new System.Drawing.Point(120, 48);
                this.lblZ.Name = "lblZ";
                this.lblZ.Size = new System.Drawing.Size(20, 16);
                this.lblZ.TabIndex = 5;
                this.lblZ.Text = "0.0";
                //
                // lblY
                //
                this.lblY.AutoSize = true;
                this.lblY.BackColor = System.Drawing.Color.White;
                this.lblY.ForeColor = System.Drawing.Color.Maroon;
                this.lblY.Location = new System.Drawing.Point(120, 32);
                this.lblY.Name = "lblY";
                this.lblY.Size = new System.Drawing.Size(20, 16);
                this.lblY.TabIndex = 4;
                this.lblY.Text = "0.0";
                //
                // lblX
                //
                this.lblX.AutoSize = true;
                this.lblX.ForeColor = System.Drawing.Color.Maroon;
                this.lblX.Location = new System.Drawing.Point(120, 16);
                this.lblX.Name = "lblX";
                this.lblX.Size = new System.Drawing.Size(20, 16);
                this.lblX.TabIndex = 3;
                this.lblX.Text = "0.0";
                //
                // label3
                //
                this.label3.AutoSize = true;
                this.label3.ForeColor = System.Drawing.Color.Red;
                this.label3.Location = new System.Drawing.Point(8, 16);
```

```
            this.label3.Name = "label3";
            this.label3.Size = new System.Drawing.Size(99, 16);
            this.label3.TabIndex = 2;
            this.label3.Text = "Current X Position:";
            //
            // label2
            //
            this.label2.AutoSize = true;
            this.label2.ForeColor = System.Drawing.Color.Red;
            this.label2.Location = new System.Drawing.Point(8, 48);
            this.label2.Name = "label2";
            this.label2.Size = new System.Drawing.Size(98, 16);
            this.label2.TabIndex = 1;
            this.label2.Text = "Current Z Position:";
            //
            // label1
            //
            this.label1.AutoSize = true;
            this.label1.ForeColor = System.Drawing.Color.Red;
            this.label1.Location = new System.Drawing.Point(8, 32);
            this.label1.Name = "label1";
            this.label1.Size = new System.Drawing.Size(96, 16);
            this.label1.TabIndex = 0;
            this.label1.Text = "Current Y Position";
            //
            // group Box3
            //
            this.groupBox3.BackColor =
System.Drawing.Color.DarkGray;
            this.groupBox3.Controls.Add(this.groupBox2);
            this.groupBox3.Controls.Add(this.btnStop);
            this.groupBox3.Controls.Add(this.groupBox5);
            this.groupBox3.Controls.Add(this.txtMain);
            this.groupBox3.Controls.Add(this.groupBox4);
            this.groupBox3.Dock =
System.Windows.Forms.DockStyle.Left;
            this.groupBox3.Location = new
System.Drawing.Point(0, 0);
            this.groupBox3.Name = "groupBox3";
            this.groupBox3.Size = new System.Drawing.Size(312, 713);
            this.groupBox3.TabIndex = 14;
            this.groupBox3.TabStop = false;
            this.groupBox3.Text = "Machine Control";
            //
            // group Box2
            //
            this.groupBox2.Anchor =
System.Windows.Forms.AnchorStyles.Bottom;
            this.groupBox2.Controls.Add(this.btnReset);
            this.groupBox2.Controls.Add(this.btnEdit);
            this.groupBox2.Controls.Add(this.btnExit);
            this.groupBox2.Controls.Add(this.btnSave);
            this.groupBox2.Controls.Add(this.btnNew);
            this.groupBox2.Controls.Add(this.btnLoad);
```

```
                this.groupBox2.Controls.Add(this.btnRun);
                this.groupBox2.Location = new
System.Drawing.Point(8, 568);
                this.groupBox2.Name = "groupBox2";
                this.groupBox2.Size = new System.Drawing.Size(296, 112);
                this.groupBox2.TabIndex = 18;
                this.groupBox2.TabStop = false;
                this.groupBox2.Text = "Control Panel";
                //
                // btn Reset
                //
                this.btnReset.BackColor =
System.Drawing.Color.Gainsboro;
                this.btnReset.FlatStyle =
System.Windows.Forms.FlatStyle.Flat;
                this.btnReset.Location = new
System.Drawing.Point(116, 48);
                this.btnReset.Name = "btnReset";
                this.btnReset.Size = new
System.Drawing.Size(56, 24);
                this.btnReset.TabIndex = 24;
                this.btnReset.Text = "ReSe&t";
                this.btnReset.Click += new
System.EventHandler(this.btnReset_Click);
                //
                // btn Edit
                //
                this.btnEdit.BackColor =
System.Drawing.Color.Gainsboro;
                this.btnEdit.FlatStyle =
System.Windows.Forms.FlatStyle.Flat;
                this.btnEdit.Location = new
System.Drawing.Point(52, 48);
                this.btnEdit.Name = "btnEdit";
                this.btnEdit.Size = new System.Drawing.Size(56, 24);
                this.btnEdit.TabIndex = 23;
                this.btnEdit.Text = "&Edit";
                this.btnEdit.Click += new
System.EventHandler(this.btnEdit_Click);
                //
                // btn Exit
                //
                this.btnExit.BackColor =
System.Drawing.Color.Gainsboro;
                this.btnExit.FlatStyle =
System.Windows.Forms.FlatStyle.Flat;
                this.btnExit.Location = new
System.Drawing.Point(84, 80);
                this.btnExit.Name = "btnExit";
                this.btnExit.Size = new System.Drawing.Size(120, 24);
                this.btnExit.TabIndex = 22;
                this.btnExit.Text = "E&xit";
                this.btnExit.Click += new
System.EventHandler(this.btnExit_Click);
```

```
            //
            // btn Save
            //
            this.btnSave.BackColor =
System.Drawing.Color.Gainsboro;
            this.btnSave.FlatStyle =
System.Windows.Forms.FlatStyle.Flat;
            this.btnSave.Location = new
System.Drawing.Point(180, 48);
            this.btnSave.Name = "btnSave";
            this.btnSave.Size = new System.Drawing.Size(56, 24);
            this.btnSave.TabIndex = 21;
            this.btnSave.Text = "&Save";
            this.btnSave.Click += new
System.EventHandler(this.btnSave_Click);
            //
            // btn New
            //
            this.btnNew.BackColor =
System.Drawing.Color.Gainsboro;
            this.btnNew.FlatStyle =
System.Windows.Forms.FlatStyle.Flat;
            this.btnNew.Location = new
System.Drawing.Point(180, 16);
            this.btnNew.Name = "btnNew";
            this.btnNew.Size = new System.Drawing.Size(56, 24);
            this.btnNew.TabIndex = 20;
            this.btnNew.Text = "&New";
            this.btnNew.Click += new
System.EventHandler(this.btnNew_Click);
            //
            // btn Load
            //
            this.btnLoad.BackColor =
System.Drawing.Color.Gainsboro;
            this.btnLoad.FlatStyle =
System.Windows.Forms.FlatStyle.Flat;
            this.btnLoad.Location = new
System.Drawing.Point(116, 16);
            this.btnLoad.Name = "btnLoad";
            this.btnLoad.Size = new System.Drawing.Size(56, 24);
            this.btnLoad.TabIndex = 19;
            this.btnLoad.Text = "&Load";
            this.btnLoad.Click += new
System.EventHandler(this.btnLoad_Click);
            //
            // btn Run
            //
            this.btnRun.BackColor =
System.Drawing.Color.Gainsboro;
            this.btnRun.FlatStyle =
System.Windows.Forms.FlatStyle.Flat;
            this.btnRun.Location = new
System.Drawing.Point(52, 16);
```

```
            this.btnRun.Name = "btnRun";
            this.btnRun.Size = new System.Drawing.Size(56, 24);
            this.btnRun.TabIndex = 18;
            this.btnRun.Text = "&Run";
            this.btnRun.Click += new
System.EventHandler(this.btnRun_Click);
            //
            // btn Stop
            //
            this.btnStop.BackColor =
System.Drawing.Color.IndianRed;
            this.btnStop.FlatStyle =
System.Windows.Forms.FlatStyle.Flat;
            this.btnStop.Location = new System.Drawing.Point(8, 384);
            this.btnStop.Name = "btnStop";
            this.btnStop.Size = new System.Drawing.Size(112, 24);
            this.btnStop.TabIndex = 12;
            this.btnStop.Text = "Emergency Stop";
            this.btnStop.Click += new
System.EventHandler(this.btnStop_Click);
            //
            // group Box5
            //
            this.groupBox5.Anchor =
System.Windows.Forms.AnchorStyles.Left;
            this.groupBox5.Controls.Add(this.txtDetail);
            this.groupBox5.Location = new
System.Drawing.Point(8, 432);
            this.groupBox5.Name = "groupBox5";
            this.groupBox5.Size = new
System.Drawing.Size(296, 96);
            this.groupBox5.TabIndex = 9;
            this.groupBox5.TabStop = false;
            this.groupBox5.Text = "File Execution Information:";
            //
            // txt Detail
            //
            this.txtDetail.Location = new
System.Drawing.Point(8, 16);
            this.txtDetail.Multiline = true;
            this.txtDetail.Name = "txtDetail";
            this.txtDetail.ScrollBars =
System.Windows.Forms.ScrollBars.Vertical;
            this.txtDetail.Size = new System.Drawing.Size(280, 72);
            this.txtDetail.TabIndex = 0;
            this.txtDetail.Text = "";
            //
            // txt Main
            //
            this.txtMain.BackColor = System.Drawing.Color.Lavender;
            this.txtMain.CharacterCasing =
System.Windows.Forms.CharacterCasing.Upper;
            this.txtMain.Font = new
System.Drawing.Font("Microsoft Sans Serif", 9.75F,
```

```
System.Drawing.FontStyle.Bold, System.Drawing.GraphicsUnit.Point,
((System.Byte)(0)));
            this.txtMain.Location = new System.Drawing.Point(8, 24);
            this.txtMain.MaxLength = 2000;
            this.txtMain.Multiline = true;
            this.txtMain.Name = "txtMain";
            this.txtMain.Size = new System.Drawing.Size(296, 200);
            this.txtMain.TabIndex = 8;
            this.txtMain.Text = "";
            this.txtMain.TextChanged += new
System.EventHandler(this.txtMain_TextChanged);
            //
            // group Box4
            //
            this.groupBox4.Controls.Add(this.btnenter);
            this.groupBox4.Controls.Add(this.btnbspace);
            this.groupBox4.Controls.Add(this.btnSpace);
            this.groupBox4.Controls.Add(this.txtS);
            this.groupBox4.Controls.Add(this.txtT);
            this.groupBox4.Controls.Add(this.txtI);
            this.groupBox4.Controls.Add(this.txtF);
            this.groupBox4.Controls.Add(this.txtJ);
            this.groupBox4.Controls.Add(this.txtM4);
            this.groupBox4.Controls.Add(this.txtM2);
            this.groupBox4.Controls.Add(this.txtM3);
            this.groupBox4.Controls.Add(this.txtM5);
            this.groupBox4.Controls.Add(this.txtG91);
            this.groupBox4.Controls.Add(this.txtG0);
            this.groupBox4.Controls.Add(this.txtG1);
            this.groupBox4.Controls.Add(this.txtG2);
            this.groupBox4.Controls.Add(this.txtG3);
            this.groupBox4.Controls.Add(this.txtG90);
            this.groupBox4.Controls.Add(this.txtG71);
            this.groupBox4.Controls.Add(this.txtG70);
            this.groupBox4.Controls.Add(this.txtplus);
            this.groupBox4.Controls.Add(this.txtminus);
            this.groupBox4.Controls.Add(this.txtpoint);
            this.groupBox4.Controls.Add(this.txt5);
            this.groupBox4.Controls.Add(this.txt6);
            this.groupBox4.Controls.Add(this.txt7);
            this.groupBox4.Controls.Add(this.txt8);
            this.groupBox4.Controls.Add(this.txt9);
            this.groupBox4.Controls.Add(this.txt0);
            this.groupBox4.Controls.Add(this.txtY);
            this.groupBox4.Controls.Add(this.txtZ);
            this.groupBox4.Controls.Add(this.txt1);
            this.groupBox4.Controls.Add(this.txt2);
            this.groupBox4.Controls.Add(this.txt3);
            this.groupBox4.Controls.Add(this.txt4);
            this.groupBox4.Controls.Add(this.txtX);
            this.groupBox4.Location = new
System.Drawing.Point(8, 240);
            this.groupBox4.Name = "groupBox4";
            this.groupBox4.Size = new System.Drawing.Size(296, 120);
```

```
                this.groupBox4.TabIndex = 0;
                this.groupBox4.TabStop = false;
                this.groupBox4.Text = "Keys Contoler:";
                //
                // btn enter
                //
                this.btnenter.BackColor =
System.Drawing.Color.Gainsboro;
                this.btnenter.FlatStyle =
System.Windows.Forms.FlatStyle.Flat;
                this.btnenter.Font = new
System.Drawing.Font("Microsoft Sans Serif", 8.25F,
System.Drawing.FontStyle.Regular,
System.Drawing.GraphicsUnit.Point, ((System.Byte)(0)));
                this.btnenter.Location = new
System.Drawing.Point(160, 88);
                this.btnenter.Name = "btnenter";
                this.btnenter.Size = new System.Drawing.Size(45, 19);
                this.btnenter.TabIndex = 42;
                this.btnenter.Text = "Enter";
                this.btnenter.Click += new
System.EventHandler(this.btnenter_Click);
                //
                // btnb space
                //
                this.btnbspace.BackColor =
System.Drawing.Color.Gainsboro;
                this.btnbspace.FlatStyle =
System.Windows.Forms.FlatStyle.Flat;
                this.btnbspace.Font = new
System.Drawing.Font("Microsoft Sans Serif", 8.25F,
System.Drawing.FontStyle.Regular,
System.Drawing.GraphicsUnit.Point, ((System.Byte)(0)));
                this.btnbspace.Location = new
System.Drawing.Point(112, 88);
                this.btnbspace.Name = "btnbspace";
                this.btnbspace.Size = new System.Drawing.Size(30, 19);
                this.btnbspace.TabIndex = 41;
                this.btnbspace.Text = "<--";
                this.btnbspace.Click += new
System.EventHandler(this.btnbspace_Click);
                //
                // btn Space
                //
                this.btnSpace.BackColor =
System.Drawing.Color.Gainsboro;
                this.btnSpace.FlatStyle =
System.Windows.Forms.FlatStyle.Flat;
                this.btnSpace.Font = new
System.Drawing.Font("Microsoft Sans Serif", 8.25F,
System.Drawing.FontStyle.Regular,
System.Drawing.GraphicsUnit.Point, ((System.Byte)(0)));
                this.btnSpace.Location = new
System.Drawing.Point(80, 88);
```

```
                this.btnSpace.Name = "btnSpace";
                this.btnSpace.Size = new System.Drawing.Size(30, 19);
                this.btnSpace.TabIndex = 40;
                this.btnSpace.Text = "-->";
                this.btnSpace.Click += new
System.EventHandler(this.btnSpace_Click);
                //
                // txtS
                //
                this.txtS.BackColor =
System.Drawing.Color.FromArgb(((System.Byte)(192)),
((System.Byte)(255)), ((System.Byte)(255)));
                this.txtS.FlatStyle =
System.Windows.Forms.FlatStyle.Flat;
                this.txtS.Font = new System.Drawing.Font("Microsoft
Sans Serif", 8.25F, System.Drawing.FontStyle.Regular,
System.Drawing.GraphicsUnit.Point, ((System.Byte)(0)));
                this.txtS.Location = new System.Drawing.Point(128, 64);
                this.txtS.Name = "txtS";
                this.txtS.Size = new System.Drawing.Size(15, 19);
                this.txtS.TabIndex = 35;
                this.txtS.Text = "S";
                this.txtS.Click += new
System.EventHandler(this.txtS_Click);
                //
                // txtT
                //
                this.txtT.BackColor =
System.Drawing.Color.FromArgb(((System.Byte)(192)),
((System.Byte)(255)), ((System.Byte)(255)));
                this.txtT.FlatStyle =
System.Windows.Forms.FlatStyle.Flat;
                this.txtT.Font = new System.Drawing.Font("Microsoft
Sans Serif", 8.25F, System.Drawing.FontStyle.Regular,
System.Drawing.GraphicsUnit.Point, ((System.Byte)(0)));
                this.txtT.Location = new System.Drawing.Point(80, 64);
                this.txtT.Name = "txtT";
                this.txtT.Size = new System.Drawing.Size(15, 19);
                this.txtT.TabIndex = 34;
                this.txtT.Text = "T";
                this.txtT.Click += new
System.EventHandler(this.txtT_Click);
                //
                // txtI
                //
                this.txtI.BackColor =
System.Drawing.Color.FromArgb(((System.Byte)(192)),
((System.Byte)(255)), ((System.Byte)(192)));
                this.txtI.FlatStyle =
System.Windows.Forms.FlatStyle.Flat;
                this.txtI.Font = new System.Drawing.Font("Microsoft
Sans Serif", 8.25F, System.Drawing.FontStyle.Regular,
System.Drawing.GraphicsUnit.Point, ((System.Byte)(0)));
                this.txtI.Location = new System.Drawing.Point(160, 16);
```

```
            this.txtI.Name = "txtI";
            this.txtI.Size = new System.Drawing.Size(20, 19);
            this.txtI.TabIndex = 33;
            this.txtI.Text = "I";
            this.txtI.Click += new
System.EventHandler(this.txtI_Click);
            //
            // txtF
            //
            this.txtF.BackColor =
System.Drawing.Color.FromArgb(((System.Byte)(192)),
((System.Byte)(255)), ((System.Byte)(255)));
            this.txtF.FlatStyle =
System.Windows.Forms.FlatStyle.Flat;
            this.txtF.Font = new System.Drawing.Font("Microsoft
Sans Serif", 8.25F, System.Drawing.FontStyle.Regular,
System.Drawing.GraphicsUnit.Point, ((System.Byte)(0)));
            this.txtF.Location = new System.Drawing.Point(104, 64);
            this.txtF.Name = "txtF";
            this.txtF.Size = new System.Drawing.Size(15, 19);
            this.txtF.TabIndex = 32;
            this.txtF.Text = "F";
            this.txtF.Click += new
System.EventHandler(this.txtF_Click);
            //
            // txtJ
            //
            this.txtJ.BackColor =
System.Drawing.Color.FromArgb(((System.Byte)(192)),
((System.Byte)(255)), ((System.Byte)(192)));
            this.txtJ.FlatStyle =
System.Windows.Forms.FlatStyle.Flat;
            this.txtJ.Font = new System.Drawing.Font("Microsoft
Sans Serif", 8.25F, System.Drawing.FontStyle.Regular,
System.Drawing.GraphicsUnit.Point, ((System.Byte)(0)));
            this.txtJ.Location = new System.Drawing.Point(160, 40);
            this.txtJ.Name = "txtJ";
            this.txtJ.Size = new System.Drawing.Size(20, 19);
            this.txtJ.TabIndex = 31;
            this.txtJ.Text = "J";
            this.txtJ.Click += new
System.EventHandler(this.txtJ_Click);
            //
            // txtM4
            //
            this.txtM4.BackColor =
System.Drawing.Color.FromArgb(((System.Byte)(255)),
((System.Byte)(192)), ((System.Byte)(192)));
            this.txtM4.FlatStyle =
System.Windows.Forms.FlatStyle.Flat;
            this.txtM4.Font = new System.Drawing.Font("Microsoft
Sans Serif", 8.25F, System.Drawing.FontStyle.Regular,
System.Drawing.GraphicsUnit.Point, ((System.Byte)(0)));
            this.txtM4.Location = new System.Drawing.Point(112, 16);
```

```
                this.txtM4.Name = "txtM4";
                this.txtM4.Size = new System.Drawing.Size(30, 19);
                this.txtM4.TabIndex = 29;
                this.txtM4.Text = "M4";
                this.txtM4.Click += new
System.EventHandler(this.txtM4_Click);
                //
                // txtM2
                //
                this.txtM2.BackColor =
System.Drawing.Color.FromArgb(((System.Byte)(255)),
((System.Byte)(192)), ((System.Byte)(192)));
                this.txtM2.FlatStyle =
System.Windows.Forms.FlatStyle.Flat;
                this.txtM2.Font = new System.Drawing.Font("Microsoft
Sans Serif", 8.25F, System.Drawing.FontStyle.Regular,
System.Drawing.GraphicsUnit.Point, ((System.Byte)(0)));
                this.txtM2.Location = new System.Drawing.Point(80, 16);
                this.txtM2.Name = "txtM2";
                this.txtM2.Size = new System.Drawing.Size(30, 19);
                this.txtM2.TabIndex = 37;
                this.txtM2.Text = "M2";
                this.txtM2.Click += new
System.EventHandler(this.txtM2_Click);
                //
                // txtM3
                //
                this.txtM3.BackColor =
System.Drawing.Color.FromArgb(((System.Byte)(255)),
((System.Byte)(192)), ((System.Byte)(192)));
                this.txtM3.FlatStyle =
System.Windows.Forms.FlatStyle.Flat;
                this.txtM3.Font = new System.Drawing.Font("Microsoft
Sans Serif", 8.25F, System.Drawing.FontStyle.Regular,
System.Drawing.GraphicsUnit.Point, ((System.Byte)(0)));
                this.txtM3.Location = new System.Drawing.Point(80, 40);
                this.txtM3.Name = "txtM3";
                this.txtM3.Size = new System.Drawing.Size(30, 19);
                this.txtM3.TabIndex = 27;
                this.txtM3.Text = "M3";
                this.txtM3.Click += new
System.EventHandler(this.txtM3_Click);
                //
                // txtM5
                //
                this.txtM5.BackColor =
System.Drawing.Color.FromArgb(((System.Byte)(255)),
((System.Byte)(192)), ((System.Byte)(192)));
                this.txtM5.FlatStyle =
System.Windows.Forms.FlatStyle.Flat;
                this.txtM5.Font = new System.Drawing.Font("Microsoft
Sans Serif", 8.25F, System.Drawing.FontStyle.Regular,
System.Drawing.GraphicsUnit.Point, ((System.Byte)(0)));
                this.txtM5.Location = new System.Drawing.Point(112, 40);
```

```
                this.txtM5.Name = "txtM5";
                this.txtM5.Size = new System.Drawing.Size(30, 19);
                this.txtM5.TabIndex = 26;
                this.txtM5.Text = "M5";
                this.txtM5.Click += new
System.EventHandler(this.txtM5_Click);
                //
                // txtG91
                //
                this.txtG91.BackColor =
System.Drawing.Color.FromArgb(((System.Byte)(192)),
((System.Byte)(192)), ((System.Byte)(255)));
                this.txtG91.FlatStyle =
System.Windows.Forms.FlatStyle.Flat;
                this.txtG91.Font = new System.Drawing.Font("Microsoft
Sans Serif", 8.25F, System.Drawing.FontStyle.Regular,
System.Drawing.GraphicsUnit.Point, ((System.Byte)(0)));
                this.txtG91.Location = new System.Drawing.Point(40, 40);
                this.txtG91.Name = "txtG91";
                this.txtG91.Size = new System.Drawing.Size(38, 19);
                this.txtG91.TabIndex = 24;
                this.txtG91.Text = "G91";
                this.txtG91.Click += new
System.EventHandler(this.txtG91_Click);
                //
                // txtG0
                //
                this.txtG0.BackColor =
System.Drawing.Color.FromArgb(((System.Byte)(192)),
((System.Byte)(192)), ((System.Byte)(255)));
                this.txtG0.FlatStyle =
System.Windows.Forms.FlatStyle.Flat;
                this.txtG0.Font = new System.Drawing.Font("Microsoft
Sans Serif", 8.25F, System.Drawing.FontStyle.Regular,
System.Drawing.GraphicsUnit.Point, ((System.Byte)(0)));
                this.txtG0.Location = new System.Drawing.Point(8, 16);
                this.txtG0.Name = "txtG0";
                this.txtG0.Size = new System.Drawing.Size(30, 19);
                this.txtG0.TabIndex = 23;
                this.txtG0.Text = "G0";
                this.txtG0.Click += new
System.EventHandler(this.txtG0_Click);
                //
                // txtG1
                //
                this.txtG1.BackColor =
System.Drawing.Color.FromArgb(((System.Byte)(192)),
((System.Byte)(192)), ((System.Byte)(255)));
                this.txtG1.FlatStyle =
System.Windows.Forms.FlatStyle.Flat;
                this.txtG1.Font = new System.Drawing.Font("Microsoft
Sans Serif", 8.25F, System.Drawing.FontStyle.Regular,
System.Drawing.GraphicsUnit.Point, ((System.Byte)(0)));
                this.txtG1.Location = new System.Drawing.Point(8, 40);
```

```
                this.txtG1.Name = "txtG1";
                this.txtG1.Size = new System.Drawing.Size(30, 19);
                this.txtG1.TabIndex = 22;
                this.txtG1.Text = "G1";
                this.txtG1.Click += new
System.EventHandler(this.txtG1_Click);
                //
                // txtG2
                //
                this.txtG2.BackColor =
System.Drawing.Color.FromArgb(((System.Byte)(192)),
((System.Byte)(192)), ((System.Byte)(255)));
                this.txtG2.FlatStyle =
System.Windows.Forms.FlatStyle.Flat;
                this.txtG2.Font = new System.Drawing.Font("Microsoft
Sans Serif", 8.25F, System.Drawing.FontStyle.Regular,
System.Drawing.GraphicsUnit.Point, ((System.Byte)(0)));
                this.txtG2.Location = new System.Drawing.Point(8, 64);
                this.txtG2.Name = "txtG2";
                this.txtG2.Size = new System.Drawing.Size(30, 19);
                this.txtG2.TabIndex = 21;
                this.txtG2.Text = "G2";
                this.txtG2.Click += new
System.EventHandler(this.txtG2_Click);
                //
                // txtG3
                //
                this.txtG3.BackColor =
System.Drawing.Color.FromArgb(((System.Byte)(192)),
((System.Byte)(192)), ((System.Byte)(255)));
                this.txtG3.FlatStyle =
System.Windows.Forms.FlatStyle.Flat;
                this.txtG3.Font = new System.Drawing.Font("Microsoft
Sans Serif", 8.25F, System.Drawing.FontStyle.Regular,
System.Drawing.GraphicsUnit.Point, ((System.Byte)(0)));
                this.txtG3.Location = new System.Drawing.Point(8, 88);
                this.txtG3.Name = "txtG3";
                this.txtG3.Size = new System.Drawing.Size(30, 19);
                this.txtG3.TabIndex = 20;
                this.txtG3.Text = "G3";
                this.txtG3.Click += new
System.EventHandler(this.txtG3_Click);
                //
                // txtG90
                //
                this.txtG90.BackColor =
System.Drawing.Color.FromArgb(((System.Byte)(192)),
((System.Byte)(192)), ((System.Byte)(255)));
                this.txtG90.FlatStyle =
System.Windows.Forms.FlatStyle.Flat;
                this.txtG90.Font = new System.Drawing.Font("Microsoft
Sans Serif", 8.25F, System.Drawing.FontStyle.Regular,
System.Drawing.GraphicsUnit.Point, ((System.Byte)(0)));
                this.txtG90.Location = new System.Drawing.Point(40, 16);
```

```
            this.txtG90.Name = "txtG90";
            this.txtG90.Size = new System.Drawing.Size(38, 19);
            this.txtG90.TabIndex = 19;
            this.txtG90.Text = "G90";
            this.txtG90.Click += new
System.EventHandler(this.txtG90_Click);
            //
            // txtG71
            //
            this.txtG71.BackColor =
System.Drawing.Color.FromArgb(((System.Byte)(192)),
((System.Byte)(192)), ((System.Byte)(255)));
            this.txtG71.FlatStyle =
System.Windows.Forms.FlatStyle.Flat;
            this.txtG71.Font = new System.Drawing.Font("Microsoft
Sans Serif", 8.25F, System.Drawing.FontStyle.Regular,
System.Drawing.GraphicsUnit.Point, ((System.Byte)(0)));
            this.txtG71.Location = new System.Drawing.Point(40, 88);
            this.txtG71.Name = "txtG71";
            this.txtG71.Size = new System.Drawing.Size(38, 19);
            this.txtG71.TabIndex = 18;
            this.txtG71.Text = "G71";
            this.txtG71.Click += new
System.EventHandler(this.txtG71_Click);
            //
            // txtG70
            //
            this.txtG70.BackColor =
System.Drawing.Color.FromArgb(((System.Byte)(192)),
((System.Byte)(192)), ((System.Byte)(255)));
            this.txtG70.FlatStyle =
System.Windows.Forms.FlatStyle.Flat;
            this.txtG70.Font = new System.Drawing.Font("Microsoft
Sans Serif", 8.25F, System.Drawing.FontStyle.Regular,
System.Drawing.GraphicsUnit.Point, ((System.Byte)(0)));
            this.txtG70.Location = new System.Drawing.Point(40, 64);
            this.txtG70.Name = "txtG70";
            this.txtG70.Size = new System.Drawing.Size(38, 19);
            this.txtG70.TabIndex = 17;
            this.txtG70.Text = "G70";
            this.txtG70.Click += new
System.EventHandler(this.txtG70_Click);
            //
            // txtplus
            //
            this.txtplus.BackColor = System.Drawing.Color.Tomato;
            this.txtplus.FlatStyle =
System.Windows.Forms.FlatStyle.Flat;
            this.txtplus.Font = new
System.Drawing.Font("Microsoft Sans Serif", 8.25F,
System.Drawing.FontStyle.Regular,
System.Drawing.GraphicsUnit.Point, ((System.Byte)(0)));
            this.txtplus.Location = new
System.Drawing.Point(184, 64);
```

```
            this.txtplus.Name = "txtplus";
            this.txtplus.Size = new System.Drawing.Size(18, 19);
            this.txtplus.TabIndex = 15;
            this.txtplus.Text = "+";
            this.txtplus.Click += new
System.EventHandler(this.txtplus_Click);
            //
            // txtminus
            //
            this.txtminus.BackColor = System.Drawing.Color.Tomato;
            this.txtminus.FlatStyle =
System.Windows.Forms.FlatStyle.Flat;
            this.txtminus.Font = new
System.Drawing.Font("Microsoft Sans Serif", 8.25F,
System.Drawing.FontStyle.Regular,
System.Drawing.GraphicsUnit.Point, ((System.Byte)(0)));
            this.txtminus.Location = new System.Drawing.Point(208,
88);
            this.txtminus.Name = "txtminus";
            this.txtminus.Size = new System.Drawing.Size(18, 19);
            this.txtminus.TabIndex = 14;
            this.txtminus.Text = "-";
            this.txtminus.Click += new
System.EventHandler(this.txtminus_Click);
            //
            // txtpoint
            //
            this.txtpoint.BackColor =
System.Drawing.Color.LightSkyBlue;
            this.txtpoint.FlatStyle =
System.Windows.Forms.FlatStyle.Flat;
            this.txtpoint.Font = new
System.Drawing.Font("Microsoft Sans Serif", 8.25F,
System.Drawing.FontStyle.Regular,
System.Drawing.GraphicsUnit.Point, ((System.Byte)(0)));
            this.txtpoint.Location = new
System.Drawing.Point(256, 88);
            this.txtpoint.Name = "txtpoint";
            this.txtpoint.Size = new System.Drawing.Size(18, 19);
            this.txtpoint.TabIndex = 13;
            this.txtpoint.Text = ".";
            this.txtpoint.Click += new
System.EventHandler(this.txtpoint_Click);
            //
            // txt5
            //
            this.txt5.BackColor =
System.Drawing.Color.LightSkyBlue;
            this.txt5.FlatStyle =
System.Windows.Forms.FlatStyle.Flat;
            this.txt5.Font = new System.Drawing.Font("Microsoft
Sans Serif", 8.25F, System.Drawing.FontStyle.Regular,
System.Drawing.GraphicsUnit.Point, ((System.Byte)(0)));
            this.txt5.Location = new System.Drawing.Point(232, 40);
```

```
            this.txt5.Name = "txt5";
            this.txt5.Size = new System.Drawing.Size(18, 19);
            this.txt5.TabIndex = 12;
            this.txt5.Text = "5";
            this.txt5.Click += new System.EventHandler
(this.txt5_Click);
            //
            // txt6
            //
            this.txt6.BackColor =
System.Drawing.Color.LightSkyBlue;
            this.txt6.FlatStyle =
System.Windows.Forms.FlatStyle.Flat;
            this.txt6.Font = new System.Drawing.Font("Microsoft
Sans Serif", 8.25F, System.Drawing.FontStyle.Regular,
System.Drawing.GraphicsUnit.Point, ((System.Byte)(0)));
            this.txt6.Location = new System.Drawing.Point(256, 40);
            this.txt6.Name = "txt6";
            this.txt6.Size = new System.Drawing.Size(18, 19);
            this.txt6.TabIndex = 11;
            this.txt6.Text = "6";
            this.txt6.Click += new
System.EventHandler(this.txt6_Click);
            //
            // txt7
            //
            this.txt7.BackColor =
System.Drawing.Color.LightSkyBlue;
            this.txt7.FlatStyle =
System.Windows.Forms.FlatStyle.Flat;
            this.txt7.Font = new System.Drawing.Font("Microsoft
Sans Serif", 8.25F, System.Drawing.FontStyle.Regular,
System.Drawing.GraphicsUnit.Point, ((System.Byte)(0)));
            this.txt7.Location = new System.Drawing.Point(208, 64);
            this.txt7.Name = "txt7";
            this.txt7.Size = new System.Drawing.Size(18, 19);
            this.txt7.TabIndex = 10;
            this.txt7.Text = "7";
            this.txt7.Click += new
System.EventHandler(this.txt7_Click);
            //
            // txt8
            //
            this.txt8.BackColor =
System.Drawing.Color.LightSkyBlue;
            this.txt8.FlatStyle =
System.Windows.Forms.FlatStyle.Flat;
            this.txt8.Font = new System.Drawing.Font("Microsoft
Sans Serif", 8.25F, System.Drawing.FontStyle.Regular,
System.Drawing.GraphicsUnit.Point, ((System.Byte)(0)));
            this.txt8.Location = new System.Drawing.Point(232, 64);
            this.txt8.Name = "txt8";
            this.txt8.Size = new System.Drawing.Size(18, 19);
            this.txt8.TabIndex = 9;
```

```
            this.txt8.Text = "8";
            this.txt8.Click += new
System.EventHandler(this.txt8_Click);
            //
            // txt9
            //
            this.txt9.BackColor =
System.Drawing.Color.LightSkyBlue;
            this.txt9.FlatStyle =
System.Windows.Forms.FlatStyle.Flat;
            this.txt9.Font = new System.Drawing.Font("Microsoft
Sans Serif", 8.25F, System.Drawing.FontStyle.Regular,
System.Drawing.GraphicsUnit.Point, ((System.Byte)(0)));
            this.txt9.Location = new System.Drawing.Point(256, 64);
            this.txt9.Name = "txt9";
            this.txt9.Size = new System.Drawing.Size(18, 19);
            this.txt9.TabIndex = 8;
            this.txt9.Text = "9";
            this.txt9.Click += new
System.EventHandler(this.txt9_Click);
            //
            // txt0
            //
            this.txt0.BackColor =
System.Drawing.Color.LightSkyBlue;
            this.txt0.FlatStyle =
System.Windows.Forms.FlatStyle.Flat;
            this.txt0.Font = new System.Drawing.Font("Microsoft
Sans Serif", 8.25F, System.Drawing.FontStyle.Regular,
System.Drawing.GraphicsUnit.Point, ((System.Byte)(0)));
            this.txt0.Location = new System.Drawing.Point(232, 88);
            this.txt0.Name = "txt0";
            this.txt0.Size = new System.Drawing.Size(18, 19);
            this.txt0.TabIndex = 7;
            this.txt0.Text = "0";
            this.txt0.Click += new
System.EventHandler(this.txt0_Click);
            //
            // txtY
            //
            this.txtY.BackColor =
System.Drawing.Color.PaleGoldenrod;
            this.txtY.FlatStyle =
System.Windows.Forms.FlatStyle.Flat;
            this.txtY.Font = new System.Drawing.Font("Microsoft
Sans Serif", 8.25F, System.Drawing.FontStyle.Regular,
System.Drawing.GraphicsUnit.Point, ((System.Byte)(0)));
            this.txtY.Location = new System.Drawing.Point(184, 40);
            this.txtY.Name = "txtY";
            this.txtY.Size = new System.Drawing.Size(18, 19);
            this.txtY.TabIndex = 6;
            this.txtY.Text = "Y";
            this.txtY.Click += new
System.EventHandler(this.txtY_Click);
```

```
            //
            // txtZ
            //
            this.txtZ.BackColor =
System.Drawing.Color.PaleGoldenrod;
            this.txtZ.FlatStyle =
System.Windows.Forms.FlatStyle.Flat;
            this.txtZ.Font = new System.Drawing.Font("Microsoft
Sans Serif", 8.25F, System.Drawing.FontStyle.Regular,
System.Drawing.GraphicsUnit.Point, ((System.Byte)(0)));
            this.txtZ.Location = new System.Drawing.Point(160, 64);
            this.txtZ.Name = "txtZ";
            this.txtZ.Size = new System.Drawing.Size(18, 19);
            this.txtZ.TabIndex = 5;
            this.txtZ.Text = "Z";
            this.txtZ.Click += new
System.EventHandler(this.txtZ_Click);
            //
            // txt1
            //
            this.txt1.BackColor =
System.Drawing.Color.LightSkyBlue;
            this.txt1.FlatStyle =
System.Windows.Forms.FlatStyle.Flat;
            this.txt1.Font = new System.Drawing.Font("Microsoft
Sans Serif", 8.25F, System.Drawing.FontStyle.Regular,
System.Drawing.GraphicsUnit.Point, ((System.Byte)(0)));
            this.txt1.Location = new System.Drawing.Point(208, 16);
            this.txt1.Name = "txt1";
            this.txt1.Size = new System.Drawing.Size(18, 19);
            this.txt1.TabIndex = 4;
            this.txt1.Text = "1";
            this.txt1.Click += new
System.EventHandler(this.txt1_Click);
            //
            // txt2
            //
            this.txt2.BackColor =
System.Drawing.Color.LightSkyBlue;
            this.txt2.FlatStyle =
System.Windows.Forms.FlatStyle.Flat;
            this.txt2.Font = new System.Drawing.Font("Microsoft
Sans Serif", 8.25F, System.Drawing.FontStyle.Regular,
System.Drawing.GraphicsUnit.Point, ((System.Byte)(0)));
            this.txt2.Location = new System.Drawing.Point(232, 16);
            this.txt2.Name = "txt2";
            this.txt2.Size = new System.Drawing.Size(18, 19);
            this.txt2.TabIndex = 3;
            this.txt2.Text = "2";
            this.txt2.Click += new
System.EventHandler(this.txt2_Click);
```

```
            //
            // txt3
            //
            this.txt3.BackColor =
System.Drawing.Color.LightSkyBlue;
            this.txt3.FlatStyle =
System.Windows.Forms.FlatStyle.Flat;
            this.txt3.Font = new System.Drawing.Font("Microsoft
Sans Serif", 8.25F, System.Drawing.FontStyle.Regular,
System.Drawing.GraphicsUnit.Point, ((System.Byte)(0)));
            this.txt3.Location = new System.Drawing.Point(256, 16);
            this.txt3.Name = "txt3";
            this.txt3.Size = new System.Drawing.Size(18, 19);
            this.txt3.TabIndex = 2;
            this.txt3.Text = "3";
            this.txt3.Click += new
System.EventHandler(this.txt3_Click);
            //
            // txt4
            //
            this.txt4.BackColor =
System.Drawing.Color.LightSkyBlue;
            this.txt4.FlatStyle =
System.Windows.Forms.FlatStyle.Flat;
            this.txt4.Font = new System.Drawing.Font("Microsoft
Sans Serif", 8.25F, System.Drawing.FontStyle.Regular,
System.Drawing.GraphicsUnit.Point, ((System.Byte)(0)));
            this.txt4.Location = new System.Drawing.Point(208, 40);
            this.txt4.Name = "txt4";
            this.txt4.Size = new System.Drawing.Size(18, 19);
            this.txt4.TabIndex = 1;
            this.txt4.Text = "4";
            this.txt4.Click += new
System.EventHandler(this.txt4_Click);
            //
            // txtX
            //
            this.txtX.BackColor =
System.Drawing.Color.PaleGoldenrod;
            this.txtX.FlatStyle =
System.Windows.Forms.FlatStyle.Flat;
            this.txtX.Font = new System.Drawing.Font("Microsoft
Sans Serif", 8.25F, System.Drawing.FontStyle.Regular,
System.Drawing.GraphicsUnit.Point, ((System.Byte)(0)));
            this.txtX.Location = new System.Drawing.Point(184, 16);
            this.txtX.Name = "txtX";
            this.txtX.Size = new System.Drawing.Size(18, 19);
            this.txtX.TabIndex = 0;
            this.txtX.Text = "X";
            this.txtX.Click += new
System.EventHandler(this.txtX_Click);
```

```
            //
            // pictureBox2
            //
            this.pictureBox2.BackColor =
System.Drawing.Color.White;
            this.pictureBox2.Image =
((System.Drawing.Image)(resources.GetObject("pictureBox2.Image")));
            this.pictureBox2.Location = new
System.Drawing.Point(416, 208);
            this.pictureBox2.Name = "pictureBox2";
            this.pictureBox2.Size = new System.Drawing.Size(528,
272);
            this.pictureBox2.TabIndex = 16;
            this.pictureBox2.TabStop = false;
            //
            // Timer
            //
            this.Timer.Tick += new
System.EventHandler(this.Timer_Tick);
            //
            // pictureBox1
            //
            this.pictureBox1.Image =
((System.Drawing.Image)(resources.GetObject("pictureBox1.Image")));
            this.pictureBox1.Location = new
System.Drawing.Point(664, 344);
            this.pictureBox1.Name = "pictureBox1";
            this.pictureBox1.Size = new System.Drawing.Size(84, 24);
            this.pictureBox1.TabIndex = 21;
            this.pictureBox1.TabStop = false;
            //
            // pDrill
            //
            this.pDrill.Image =
((System.Drawing.Image)(resources.GetObject("pDrill.Image")));
            this.pDrill.Location = new System.Drawing.Point(704,
312);
            this.pDrill.Name = "pDrill";
            this.pDrill.Size = new System.Drawing.Size(5, 24);
            this.pDrill.TabIndex = 22;
            this.pDrill.TabStop = false;
            //
            // frmMain
            //
            this.AutoScaleBaseSize = new System.Drawing.Size(5, 13);
            this.BackColor = System.Drawing.Color.White;
            this.ClientSize = new System.Drawing.Size(1030, 735);
            this.Controls.Add(this.pDrill);
            this.Controls.Add(this.pictureBox1);
            this.Controls.Add(this.pictureBox2);
            this.Controls.Add(this.groupBox3);
            this.Controls.Add(this.groupBox1);
            this.Controls.Add(this.statusBar);
```

```
                this.ForeColor =
System.Drawing.SystemColors.ControlText;
                this.FormBorderStyle =
System.Windows.Forms.FormBorderStyle.FixedSingle;
                this.Icon =
((System.Drawing.Icon)(resources.GetObject("$this.Icon")));
                this.Menu = this.mainMenu;
                this.MinimizeBox = false;
                this.Name = "frmMain";
                this.Text = "CNC System";
                this.WindowState =
System.Windows.Forms.FormWindowState.Maximized;
                this.Load += new
System.EventHandler(this.frmMain_Load);

((System.ComponentModel.ISupportInitialize)(this.Information)).
EndInit();

((System.ComponentModel.ISupportInitialize)(this.PortStatus)).
EndInit();
                this.groupBox1.ResumeLayout(false);
                this.groupBox3.ResumeLayout(false);
                this.groupBox2.ResumeLayout(false);
                this.groupBox5.ResumeLayout(false);
                this.groupBox4.ResumeLayout(false);
                this.ResumeLayout(false);

           }
           #endregion

           private void frmMain_Load(object sender, System.EventArgs e)
           {

                //Disable the Main Menu Item 'Execute CNC file'
                #region Enable OR Disable Controls
                menuItemExecute.Enabled = false;
                menuItem3.Enabled = false;
                menuItem4.Enabled = false;
                btnRun.Enabled = false;
                btnSave.Enabled = false;
                #endregion

                /// set tooltip
                #region SetToolTip
                toolTip.SetToolTip(this.btnEdit,"Edit Part Program");
                toolTip.SetToolTip(this.btnExit,"Close the
Application");
                toolTip.SetToolTip(this.btnSave,"Save Part Program");
                toolTip.SetToolTip(this.btnLoad,"Open Part Program");
                toolTip.SetToolTip(this.btnRun,"Execute Part Program");
                toolTip.SetToolTip(this.btnReset,"Reset Machine To
Its Original Position");
                toolTip.SetToolTip(this.txtG0,"Rapid Movement");
                toolTip.SetToolTip(this.txtG1,"Linear Interpolation");
```

```
            toolTip.SetToolTip(this.txtG2,"Circular
Interpolation c/w");
           toolTip.SetToolTip(this.txtG3,"Circular Interpolation
cc/w");
            toolTip.SetToolTip(this.txtG70,"Inch Units");
            toolTip.SetToolTip(this.txtG71,"Metric Units");
            toolTip.SetToolTip(this.txtG90,"Absolute Coordinate");
            toolTip.SetToolTip(this.txtG91,"Incremental
Coordinate");
            toolTip.SetToolTip(this.txtX,"X Direction");
            toolTip.SetToolTip(this.txtY,"Y Direction");
            toolTip.SetToolTip(this.txtZ,"Z Direction");
            toolTip.SetToolTip(this.txtI,"Center of Arc in X
Direction");
            toolTip.SetToolTip(this.txtJ,"Center of Arc in Y
Direction");
            toolTip.SetToolTip(this.txtM4,"Spindle On cc/w");
            toolTip.SetToolTip(this.txtM5,"Spindle Off");
            toolTip.SetToolTip(this.txtM3,"Spindle On c/w");
            toolTip.SetToolTip(this.txtM2,"End Of Part Program");
            toolTip.SetToolTip(this.txtS,"Speed");
            toolTip.SetToolTip(this.txtF,"Feed");
            toolTip.SetToolTip(this.txtT,"Tool");
            toolTip.SetToolTip(this.btnbspace,"Back Space");
            toolTip.SetToolTip(this.btnSpace, "Space Bar");
            toolTip.SetToolTip(this.btnenter,"Enter");
            #endregion
            /// Generate Sequence Number for Part
            /// Program is put in string array
            #region Generate Sequence Number
            for(int iIndex = 1; iIndex < 999; iIndex++)
            {
                if(iIndex.ToString().Length == 1)
                   sLineNumber[iIndex - 1] = "N00" + iIndex;
                if(iIndex.ToString().Length == 2)
                   sLineNumber[iIndex - 1] = "N0" + iIndex;
                if(iIndex.ToString().Length == 3)
                   sLineNumber[iIndex - 1] = "N" + iIndex;

            }
            #endregion
        }

        #region MainMenu
        /// <summary>
        /// MainMenu events hendling
        /// </summary>

        private void menuItemLoad_Click(object sender,
System.EventArgs e)
        {
            LoadPartProgram();
        }
```

```
        private void menuItemExecute_Click(object sender,
System.EventArgs e)
        {
            ExecutePartProgram();
        }

        private void menuItem2_Click(object sender,
System.EventArgs e)
        {
            NewPartProgram();
        }
        private void menuItem3_Click(object sender,
System.EventArgs e)
        {
            SavePartProgram();
        }
        private void menuItem4_Click(object sender,
System.EventArgs e)
        {
            EditPartProgram();
        }
        private void menuItem5_Click(object sender,
System.EventArgs e)
        {
            SetMachine();
        }
        private void menuItem6_Click(object sender,
System.EventArgs e)
        {
            Application.Exit();
        }
        #endregion
        #region Load Part Program File
        /// <summary>
        /// This region contains all the functionality
        /// of loading part program files and displays in txtMain
        /// </summary>
        private void LoadPartProgram()
        {
            ///Check whether user selects valid cnc file and
selects OK button
            if(openFileDialog.ShowDialog() == DialogResult.OK)
            {
                /// Create the Support Class Object
                Support ObjSupport = new Support();
                ///Get the file name
                Support.sFileName = openFileDialog.FileName;
                this.Text = "CNC System" +
openFileDialog.FileName;
                ///Open the file and display in txtMain.
                ObjSupport.OpenFile(txtMain);
                ///set the txtMain ReadOnly
                txtMain.ReadOnly = true;
                ///Enable the MainMenu Item 'Execute CNC file'
```

```
                    menuItemExecute.Enabled = true;
                    menuItem3.Enabled = false;
                    menuItem4.Enabled = true;
                    btnRun.Enabled = true;
                    txtDetail.Clear();
                    SetDetail(openFileDialog.FileName + "Loaded");
                    ///Clean the object
                    ObjSupport.Dispose();
               }
          }
          #endregion
          private void EditPartProgram()
          {
               txtMain.ReadOnly = false;
               txtMain.Focus();
          }
          #region Execute Part Program File
          /// <summary>
          /// This region contains all the fucntionality
          /// of executing Part Program file. First it will check
          /// Part Program and then execute it
          /// </summary>
          private void ExecutePartProgram()
          {
               /// set machine to its original position before
               /// executing part programme
               SetMachine();
               /// Create Support Class Object
               Support ObjSupport = new Support();

               /// Event Handlers
               #region Event Handler
               ObjSupport.MessageArrived +=new
MessageHandler(ObjSupport_MessageArrived);
               ObjSupport.StatusArrived +=new
StatusHandler(ObjSupport_StatusArrived);
               ObjSupport.PositionArrived +=new
PositionHandler(ObjSupport_PositionArrived);
               ObjSupport.SpindleArrived +=new
SpindleHandler(ObjSupport_SpindleArrived);
               #endregion
               ///set the status bar text
               Information.Text = "Please Wait While Part Programe
File is Validating";
               /// validate file
               SetDetail("Validating File");
               if(ObjSupport.ValidateFile())
               {
                    SetDetail("OK");
                    ///Set the status bar text
                    Information.Text = "Part Programe File is
Validated, Now Executing Part Programe";
                    /// execute file
                    SetDetail("Executing File");
```

```
                    ObjSupport.ExecuteFile();

                    ///Set the status bar text
                    Information.Text = "Part Programe Executed
Successfully";
                    SetDetail("End");
                }
                else /// file is not valid
                {
                    Information.Text = "There is an Error in Part
Programe File";
                    SetDetail("File is not Valid");
                }
                /// Clear the memory
                ObjSupport.Dispose();

            }
            #endregion
            #region New Part Program File
            /// <summary>
            /// This region contains all the functionality
            /// of New Part program file.
            /// </summary>
            private void NewPartProgram()
            {
                btnSave.Enabled = false;
                Support.sFileName = "";
                this.Text = "CNC System";
                txtMain.ReadOnly = false;
                menuItem3.Enabled = false;
                Information.Text = "Ready";
                txtMain.Enabled = true;
                txtMain.Focus();
                txtMain.Text = string.Empty;
            }
            #endregion
            #region Save Part Program File
            /// <summary>
            /// This region contains all the functionality
            /// of Saving part program file. First it will check
            /// all commands. If all commands are valid, then it
saves
            /// the file, else informs the nature of error and
            /// will not save the file.
            /// </summary>
            private void SavePartProgram()
            {
                Support ObjSupport = new Support();
                string sAllText = txtMain.Text; /// contains all text
of file
                string sBlock; /// contains one line of file at a time
                string[] sCommand; /// contains command from one line
                bool bIsValid = false; /// use for command is valid
or not
```

```
                string Error = ""; /// hold the error information
and display end of the function.

                /// loop throgh end of the file, one line at a time
                for(int iBlockNumber = 0; iBlockNumber <
txtMain.Lines.Length - 1; iBlockNumber++)
                {
                      /// Get one line from textbox
                      sBlock = txtMain.Lines[iBlockNumber];
                      /// Get separate commands from line separated
by space " "
                      sCommand = sBlock.ToString().Split
(" ".ToCharArray());
                      /// check it sequence number already exists
                      if(sCommand[0].StartsWith("N"))
                      { /// If sequence number exist by pass it and
use only valid commands
                      for(int iCommandNumber = 1; iCommandNumber <
sCommand.Length; iCommandNumber++)
                            {

ObjSupport.ValidateCode(sCommand[iCommandNumber].ToString(),ref
bIsValid);
                            }
                      }
                      else
                      {
                            for(int iCommandNumber = 0;
iCommandNumber < sCommand.Length; iCommandNumber++)
                            {

ObjSupport.ValidateCode(sCommand[iCommandNumber].ToString(),ref
bIsValid);
                            }
                      }
                      /// Check sequence number from beginning of each
line
                      if(!sCommand[0].StartsWith("N"))
                      {
                            /// if not then place sequence number at
beginning of each line
                            if(txtMain.Lines[iBlockNumber].ToString()
!= "")
                            {
                                  sAllText =
sAllText.Replace(txtMain.Lines[iBlockNumber],sLineNumber[iBlockNumber]
+ "" + txtMain.Lines[iBlockNumber]);
                                  txtMain.Text = sAllText;
                            }
                      }
                      /// if any command is invalid, then note the
line number
                      /// where invalid command exist.
```

```
                    if(!bIsValid)
                        Error = Error + " " +
sLineNumber[iBlockNumber];


                }
                /// If all commands are valid
                if(Error.Length == 0)
                {
                    Information.Text = "The file is valid, please
select file name and select save";
                    if(saveFileDialog.ShowDialog() == DialogResult.OK)
                    {
                        ///set the static sFileName of support


                        Support.sFileName =
saveFileDialog.FileName;
                        ///save the file
                        ObjSupport.SaveFile(txtMain);
                        menuItem3.Enabled = false;
                        btnSave.Enabled = false;
                        txtMain.ReadOnly = true;

                        Information.Text = "The file is saved";
                        SetDetail(saveFileDialog.FileName + "
Saved");
                        this.Text = "CNCSystem" +
saveFileDialog.FileName;
                        btnRun.Enabled = true;
                        menuItemExecute.Enabled = true;
                    }

                }
                    /// if any command is invalid
                else
                {   /// set status bar text
                    Information.Text = "The file is not a valid,
and can not be save. Error in lines "+ Error;
                }
                menuItem3.Enabled = false;
                ObjSupport.Dispose();


            }
            #endregion
            #region ReSet Machine
            /// <summary>
            /// This function resets the machine to its
            /// original position
            /// </summary>
            private void SetMachine()
            {
                p.X = 664;
```

```
            p.Y = 344;
            pictureBox1.Location = p;
        }
        #endregion
        private void Delay(int Milliseconds)
        {

            Timer.Interval = Milliseconds;
            Timer.Enabled = true;
            TimerCounter = 0;

            while(TimerCounter < 3)
            {
                Application.DoEvents();
            }
            Timer.Enabled  = false;

        }
        private void txtMain_TextChanged(object sender,
System.EventArgs e)
        {
            menuItem3.Enabled = true;
            btnSave.Enabled = true;
        }

        #region Keys Control
        /// <summary>
        /// This region contains all event handling
        /// generated by part programme commands
        /// </summary>
        private void txtG0_Click(object sender, System.EventArgs e)
        {
            txtMain.Text = txtMain.Text + "G0";
        }

        private void txtG90_Click(object sender, System.EventArgs e)
        {
            txtMain.Text = txtMain.Text + "G90";

        }

        private void txtG91_Click(object sender, System.EventArgs e)
        {
            txtMain.Text = txtMain.Text + "G91";

        }

        private void txtG1_Click(object sender, System.EventArgs e)
        {
            txtMain.Text = txtMain.Text + "G1";

        }
```

```
private void txtG2_Click(object sender, System.EventArgs e)
{
     txtMain.Text = txtMain.Text + "G2";

}

private void txtG3_Click(object sender, System.EventArgs e)
{
     txtMain.Text = txtMain.Text + "G3";

}

private void txtG70_Click(object sender, System.EventArgs e)
{
     txtMain.Text = txtMain.Text + "G70";

}

private void txtG71_Click(object sender, System.EventArgs e)
{
     txtMain.Text = txtMain.Text + "G71";

}


private void txtM4_Click(object sender, System.EventArgs e)
{
     txtMain.Text = txtMain.Text + "M4";

}

private void txtM3_Click(object sender, System.EventArgs e)
{
     txtMain.Text = txtMain.Text + "M3";

}

private void txtM5_Click(object sender, System.EventArgs e)
{
     txtMain.Text = txtMain.Text + "M5";

}

private void txtT_Click(object sender, System.EventArgs e)
{
     txtMain.Text = txtMain.Text + "T";

}

private void txtF_Click(object sender, System.EventArgs e)
{
     txtMain.Text = txtMain.Text + "F";

}
```

```
private void txtS_Click(object sender, System.EventArgs e)
{
     txtMain.Text = txtMain.Text + "S";

}

private void txtI_Click(object sender, System.EventArgs e)
{
     txtMain.Text = txtMain.Text + "I";

}

private void txtJ_Click(object sender, System.EventArgs e)
{
     txtMain.Text = txtMain.Text + "J";

}

private void txtX_Click(object sender, System.EventArgs e)
{
     txtMain.Text = txtMain.Text + "X";

}

private void txtY_Click(object sender, System.EventArgs e)
{
     txtMain.Text = txtMain.Text + "Y";

}

private void txtZ_Click(object sender, System.EventArgs e)
{
     txtMain.Text = txtMain.Text + "Z";

}

private void txtplus_Click(object sender, System.EventArgs e)
{
     txtMain.Text = txtMain.Text + "+";

}

private void txt1_Click(object sender, System.EventArgs e)
{
     txtMain.Text = txtMain.Text + "1";

}

private void txt4_Click(object sender, System.EventArgs e)
{
     txtMain.Text = txtMain.Text + "4";

}
```

```
        private void txt7_Click(object sender, System.EventArgs e)
        {
             txtMain.Text = txtMain.Text + "7";

        }

        private void txt0_Click(object sender, System.EventArgs e)
        {
             txtMain.Text = txtMain.Text + "0";

        }

        private void txt2_Click(object sender, System.EventArgs e)
        {
             txtMain.Text = txtMain.Text + "2";

        }

        private void txt5_Click(object sender, System.EventArgs e)
        {
             txtMain.Text = txtMain.Text + "5";

        }

        private void txt8_Click(object sender, System.EventArgs e)
        {
             txtMain.Text = txtMain.Text + "8";

        }
        private void btnenter_Click(object sender,
System.EventArgs e)
        {
             txtMain.Text  = txtMain.Text + "\r\n";
        }

        private void btnbspace_Click(object sender,
System.EventArgs e)
        {
             txtMain.Text =
txtMain.Text.Remove(txtMain.Text.Length - 1,1);
        }

        private void btnSpace_Click(object sender, System.EventArgs
e)
        {
             txtMain.Text = txtMain.Text + " ";
        }

        private void txtpoint_Click(object sender,
System.EventArgs e)
        {
             txtMain.Text = txtMain.Text + ".";

        }
```

```
private void txt3_Click(object sender, System.EventArgs e)
{
     txtMain.Text = txtMain.Text + "3";

}

private void txt6_Click(object sender, System.EventArgs e)
{
     txtMain.Text = txtMain.Text + "6";

}

private void txt9_Click(object sender, System.EventArgs e)
{
     txtMain.Text = txtMain.Text + "9";

}

private void txtminus_Click(object sender, System.EventArgs e)
{
     txtMain.Text = txtMain.Text + "-";

}

private void txtM2_Click(object sender, System.EventArgs e)
{
     txtMain.Text = txtMain.Text + "M2";

}
#endregion
#region buttons Control
/// <summary>
/// This region contains all event handling
/// generated by button on the main form
/// </summary>
private void btnRun_Click(object sender, System.EventArgs e)
{
     ExecutePartProgram();
}

private void btnLoad_Click(object sender, System.EventArgs e)
{
     LoadPartProgram();
}

private void btnNew_Click(object sender, System.EventArgs e)
{
     NewPartProgram();
}

private void btnEdit_Click(object sender, System.EventArgs e)
{
     EditPartProgram();
}
```

```
        private void btnReset_Click(object sender, System.EventArgs e)
        {
            SetMachine();
        }

        private void btnSave_Click(object sender, System.EventArgs e)
        {
            SavePartProgram();
        }

        private void btnStop_Click(object sender, System.EventArgs e)
        {
            SetDetail("Emergency Stop");
            Support.EmergencyStop();
        }

        private void btnExit_Click(object sender, System.EventArgs e)
        {
            Application.Exit();
        }

        #endregion
        public void SetDetail(string Message)
        {
            txtDetail.AppendText(Message + "\r\n");
        }
        #region MessageArrived Evnet Handler
        /// <summary>
        /// This function is event handler of MessageArrived event of
        /// support class. When support class needs to inform,
        /// then it generates the event. frmMain class handles
        /// this event and takes the message and calls the
        /// SetDetail.
        /// </summary>

        private void ObjSupport_MessageArrived(string message)
        {
            SetDetail(message);
        }
        #endregion
        #region StatusArrived Event Handler
        /// <summary>
        /// this event handler handles the event generated by
        /// support class and sets the current status of main form
        /// </summary>
        private void ObjSupport_StatusArrived(Current current,
string sValue)
        {
            switch (current)
            {
                case Current.X:
                    lblX.Text = sValue;
                    break;
```

```
                case Current.Y:
                    lblY.Text = sValue;
                    break;
                case Current.Z:
                    lblZ.Text = sValue;
                    break;
                case Current.Spindle:
                    lblSpindle.Text = sValue;
                    break;
                case Current.Speed:
                    lblSpeed.Text = sValue;
                    break;
                case Current.Coordinate:
                    lblCoordinate.Text = sValue;
                    break;
                case Current.Feed:
                    lblFeed.Text = sValue;
                    break;
                case Current.Unit:
                    lblUnit.Text = sValue;
                    break;
                case Current.Movement:
                    lblMovement.Text = sValue;
                    break;
            }

        }
        #endregion
        #region PositionArrived Event Handler
        /// <summary>
        /// this event handler handles the event generated by
        /// support class and sets the position of machine
        /// </summary>

        private void ObjSupport_PositionArrived(string Coordinate,
double Value)
        {

            //Application.DoEvents();
            ///Starting Position Of milling machine
            ///X Starting Position = 648;
            ///Y Starting Position = 304;

            int X = p.X;
            int Y = p.Y;
            int XDirection = 0;
            int ZDirection = 0;

            int ivalue = Convert.ToInt16(Value);
            switch(Coordinate)
            {
                case "X+":
                {
                    if(ivalue == 0)
```

```
                        {
                              X = p.X - 664;
                              pictureBox1.Location = p;
                        }
                        else
                        {
                              XDirection = p.X  + ivalue/DevidedBy;
                              for(; X   < XDirection; X++)
                              {
                                     Delay(1); // Delay of
1 Millisecond
                                     p.X = X;
                                     pictureBox1.Location = p;
                              }
                        }

                  }
                        break;
                  case "X-":
                  {
                        if(ivalue == 0)
                        {
                              p.X = 664;
                              pictureBox1.Location = p;
                        }
                        else
                        {
                              XDirection = X - ivalue/DevidedBy;
                              for(; X   > XDirection; X--)
                              {
                                     Delay(1); // Delay of
1 Millisecond
                                     p.X = X;
                                     pictureBox1.Location = p;
                              }
                        }
                  }
                        break;
                  case "Y+":
                  {
                        for(int number = 1; number < 10; number++)
                        {
                              Delay(1); // Delay of 1 Milliseconds
                              pictureBox1.Image =
System.Drawing.Image.FromFile(Application.StartupPath + @"\Icons\" +
number.ToString() + ".jpg");
                  }


                  }
                        break;
```

```
                    case "Y-":
                    {
                          for(int number = 10; number > 1; number--)
                          {
                                Delay(1); // Delay of 1 Milliseconds
                                pictureBox1.Image =
System.Drawing.Image.FromFile(Application.StartupPath + @"\Icons\" +
number.ToString() + ".jpg");
                          }
                    }
                          break;
                    case "Z+":
                    {
                          if(ivalue == 0)
                          {
                                p.Y = 344;
                                pictureBox1.Location = p;
                          }
                          else
                          {
                                ZDirection = p.Y  - ivalue/10;
                                for(; Y  > ZDirection; Y--)
                                {
                                  Delay(1); // Delay of
1 Milliseconds

                                  p.Y = Y;
                                  pictureBox1.Location = p;
                                }
                          }
                    }
                          break;
                    case "Z-":
                    {
                          if(ivalue == 0)
                          {
                                p.Y = 344;
                                pictureBox1.Location = p;

                          }
                          else
                          {
                                ZDirection = p.Y  + ivalue/10;
                                for(; Y  < ZDirection; Y++)
                                {
                                  Delay(1); // Delay of
1 Milliseconds

                                  p.Y = Y;
                                  pictureBox1.Location = p;
                                }

                          }
                    }
```

```
                        break;
                }
            }
            #endregion

            private void Timer_Tick(object sender, System.EventArgs e)
            {
                TimerCounter++;
                if(TimerCounter > 3)
                    Timer.Enabled = false;
            }
            private void menuItem8_Click(object sender,
System.EventArgs e)
            {
                Process myProcess = new Process();
                myProcess.StartInfo.FileName  =
Application.StartupPath + @"\Help\help.chm";
                myProcess.Start();
            }

            private void ObjSupport_SpindleArrived(bool Status)
            {
                if(Status)
                {
                    pDrill.Image =
System.Drawing.Image.FromFile(Application.StartupPath +
@"\Icons\dril.gif");
                }
                else
                {
                    pDrill.Image =
System.Drawing.Image.FromFile(Application.StartupPath +
@"\Icons\dril_still.gif");
                }

            }
        }
}
```

File Name: Support.cs
Purpose: The code supports routines of CNCSystem.
Environment: Visual C# .Net version 1.1

```
using System;
using System.Windows.Forms;
using System.IO;
using System.Collections;
using System.Drawing;
using System.ComponentModel;

namespace CNCSystem
{
	/// <summary>
	/// Summary description for Support.
	/// Support class is inherited from interface ID is possible to
	/// implement Dispose method.
	/// </summary>
	///
	///event of support class
	public delegate void MessageHandler(string message);
	public delegate void StatusHandler(Current current, string
sValue);
	public delegate void PositionHandler(string Coordinate, double
Value);
	public delegate void SpindleHandler(bool Status);

	public class Support : IDisposable
	{
		public event MessageHandler MessageArrived;/// generate
the event of message
		public event StatusHandler StatusArrived;/// generate the
event of status
		public event PositionHandler PositionArrived;///generate
the event of current position
		public event SpindleHandler SpindleArrived;///generate the
event of spindle on or off

		public static string sFileName; /// Hold the File Name
of .cnc file
		private static ArrayList aList = new ArrayList(); ///Hold
the GM Codes extracted from .cnc file
		private static ArrayList GMCodes = new ArrayList();
/// Hold All GM Codes


		private Component component = new Component();// Other
managed resource this class uses.
		private bool disposed = false;// Track whether Dispose
has been called.
```

```
        PortAccess Port = new PortAccess();/// Create a object
Port of Port Access class


        #region EIA 274-D Commands
        /// <summary>
        /// Constructor of Support Class
        /// </summary>
        public Support()
        {
            ///add all valid G M codes
            GMCodes.Add("G0");
            GMCodes.Add("G00");
            GMCodes.Add("G01");
            GMCodes.Add("G1");
            GMCodes.Add("G02");
            GMCodes.Add("G2");
            GMCodes.Add("G03");
            GMCodes.Add("G3");
            GMCodes.Add("G90");
            GMCodes.Add("G91");
            GMCodes.Add("G70");
            GMCodes.Add("G71");
            GMCodes.Add("M02");
            GMCodes.Add("M2");
            GMCodes.Add("M3");
            GMCodes.Add("M03");
            GMCodes.Add("M4");
            GMCodes.Add("M04");
            GMCodes.Add("M5");
            GMCodes.Add("M05");
            GMCodes.Add("%");
            GMCodes.Add("+");
            GMCodes.Add("-");
            GMCodes.Add("F");
            GMCodes.Add("I");
            GMCodes.Add("J");
            GMCodes.Add("N");
            GMCodes.Add("S");
            GMCodes.Add("T");
            GMCodes.Add("X");
            GMCodes.Add("Y");
            GMCodes.Add("Z");
            GMCodes.Add("X+");
            GMCodes.Add("Y+");
            GMCodes.Add("Z+");
            GMCodes.Add("X-");
            GMCodes.Add("Y-");
            GMCodes.Add("Z-");

            GMCodes.Sort();

        }
        #endregion
```

```
        public void Dispose()
        {
             Dispose(true);
             // This object will be cleared up by the Dispose
method.
             // Therefore, call GC.SupressFinalize to
             // take this object off the finalization queue
             // and prevent finalization code for this object
             // from executing second time.
             GC.SuppressFinalize(this);
        }
        private void Dispose(bool disposing)
        {
             // Check to see if Dispose has already been called.
             if(!this.disposed)
             {
                  // If disposing equals true, dispose all managed
                  // and unmanaged resources.
                  if(disposing)
                  {
                       // Dispose managed resources.
                       component.Dispose();
                  }
             }
             disposed = true;
        }



        #region OpenFile
        /// <summary>
        /// This region contains the definition of Open File
        /// function used for open cnc file and displays in
        /// textbox object
        /// </summary>

        public void OpenFile(TextBox textBox)
        {
             ///try to handle the error during open file
             try
             {    /// Open the file
                  using(StreamReader reader =
File.OpenText(sFileName))
                  {
                       ///clear the textbox
                       textBox.Clear();
                       textBox.Text = reader.ReadToEnd();
                  }
             }
                  /// catch the error
             catch(IOException ex)
             {
                  /// write the error information to application
error log
```

```
                Support.WriteError(ex.Message,ex.Source,"Not
Apply","OpenFile");
            }


        }
        #endregion
        #region WriteError
        /// <summary>
        /// This region contain Write Error Function
        /// to use write errors in error log
        /// </summary>


        public static void WriteError(string sErrorMessage,string
sErrorSource,string sErrorNumber,string sFunctionThatGenerateError)
        {

            string slines;
            ///create new file Error Log.text, if already exist
the append the tex.
            using(System.IO.StreamWriter ofile = new
System.IO.StreamWriter(Application.StartupPath +
"\\ErrorLog.txt",true))
            {
                slines = "New Error Generated by CNC System " +
DateTime.Now + "\r\n";
                // Write the string to a file.
                ofile.WriteLine(slines);
                slines = "Error Message:" + sErrorMessage +
"\r\n";
                ofile.WriteLine(slines);
                slines = "Error Source:"+ sErrorSource + "r\n";
                ofile.WriteLine(slines);
                slines = "Error Number:" + sErrorNumber + "r\n";
                ofile.WriteLine(slines);
                slines = "Function That Generate Error:" +
sFunctionThatGenerateError + "\r\n";
                ofile.WriteLine(slines);
                ofile.Close();

            }

        }
        #endregion
        #region ExecuteFile
        /// <summary>
        /// This region contains Execute File Function's Routine
        /// This function takes commands from local array a List
one by one
        /// and executes.
        /// </summary>
```

```
///
frmFinal Final;
public void ExecuteFile()
{


     Final = new frmFinal(); /// Final Product form
     /// lGM Code string variable for G M Code
     string lGMCode;
     /// a List is an Array List object and fills
     /// Validate File function.
     for(int i =0; i < aList.Count; i++)
     {
          Application.DoEvents();
          /// if Emergency stop, then exit from loop
          if(PortAccess.EmergencyStop)
               break;

          #region Machine Preparation Handling Code
          /// <summary>
          /// In this region, take one GM Code at a time
          /// from a List array in local string variable
lGMCode
          /// and check Against all GM Codes
          /// </summary>
          lGMCode = aList[i].ToString();
          if(lGMCode == "G0" || lGMCode == "G00")
          {
          ///call Rapid Movement function
          Port.G00();
          ///Generates the event, this event will
          ///be handled by calling routine. This
          ///event will give the current status or
          /// activity to object of this class.
          MessageArrived("G0 ok");
          StatusArrived(Current.Movement,"Rapid");

          }
          else if(lGMCode == "G01" || lGMCode == "G1")
          {
               ///call Linear Interpolation function
               Port.G01();
               ///Generate the event, this event will be
               ///handled by calling routine
               MessageArrived("G1 OK");
               StatusArrived(Current.Movement,"Linear");
          }
          else if(lGMCode == "G02" || lGMCode == "G2")
          {
               ///Call Circular Interpolation c/w function
               Port.G02();
               ///Generate the event; this event will be
               ///handled by calling routine
               MessageArrived("G2 OK");
```

```
                        StatusArrived(Current.Movement,"Circular
C/W");
                    }
                    else if(lGMCode == "G03" || lGMCode == "G3")
                    {
                        /// call Circular Interpolation cc/w function
                        Port.G03();
                        ///Generate the event, this event will be
                        ///handled by calling routine
                        MessageArrived("G03 OK");
                        StatusArrived(Current.Movement,"Circular
CC/W");
                    }
                    else if(lGMCode == "G90")
                    {
                        /// call Absolute Coordinate function
                        Port.G90();
                        ///Generate the event, this event will be
                        ///handled by calling routine
                        MessageArrived("G90 OK");
                        StatusArrived(Current.Coordinate,
"Absolute");
                    }
                    else if(lGMCode == "G91")
                    {
                        /// call Incremental Coordinate function
                        Port.G91();
                        ///Generate the event, this event will be
                        ///handled by calling routine
                        MessageArrived("G91 OK");
StatusArrived(Current.Coordinate, "Incremental");
                    }
                    else if(lGMCode == "G70")
                    {
                        /// call Inch Units
                        Port.G70();
                        ///Generate the event; this event will be
                        ///handled by calling routine
                        MessageArrived("G70 OK");
                        StatusArrived(Current.Unit,"Inch");
                    }
                    else if(lGMCode == "G71")
                    {
                        /// call Metric Units function
                        Port.G71();
                        ///Generate the event; this event will be
                        ///handled by calling routine
                        MessageArrived("G71 OK");
                        StatusArrived(Current.Unit,"MM");
                    }
```

```
                    else if(lGMCode == "M02" || lGMCode == "M2")
                    {
                         Port.M02();

                    }
                    else if(lGMCode == "M03" || lGMCode == "M3")
                    {
                         Port.M03();
                         ///Generate the event, this event will be
                         ///handled by calling routine
                         MessageArrived("M3 OK");
                         StatusArrived(Current.Spindle,"ON C/W");
                         SpindleArrived(true);
                    }
                    else if(lGMCode == "M04" || lGMCode == "M4")
                    {
                         Port.M04();
                         ///Generate the event, this event will be
                         ///handled by calling routine
                         MessageArrived("M4 OK");
                         StatusArrived(Current.Spindle,"ON CC/W");
                         SpindleArrived(true);
                    }
                    else if(lGMCode == "M05" || lGMCode == "M5")
                    {
                         Port.M05();
                         ///Generate the event, this event will be
                         ///handled by calling routine
                         MessageArrived("M5 OK");
                         StatusArrived(Current.Spindle,"OFF");
                         SpindleArrived(false);
                    }
                    else if(lGMCode.StartsWith("S"))
                    {
                         Port.Speed =
Convert.ToInt16(lGMCode.Substring(1,lGMCode.Length - 1));

     StatusArrived(Current.Speed,Port.Speed.ToString());
                    }
                    else if(lGMCode.StartsWith("F"))
                    {
                         Port.Feed =
Convert.ToInt16(lGMCode.Substring(1,lGMCode.Length - 1));

     StatusArrived(Current.Feed,Port.Feed.ToString());
                    }
                    else if(lGMCode.StartsWith("T"))
                    {
                         Port.Tool =
Convert.ToInt16(lGMCode.Substring(1,lGMCode.Length - 1));

                    }
```

```
                        #endregion
                        #region Rapid Movement Handling Code
                        /// <summary>
                        /// In this region, Handle Rapid Movement
                        /// And check X Y and Z Coordinate Separately
                        /// </summary>
                        if(Port.RapidMovement)
                        {
                            #region X in RapidMovement

                            if(lGMCode.StartsWith("X"))
                            {
                                /// GMCode first letter is X
                                if(Port.AbsoluteCoordinate)
                                {
                                    /// take the value after X
                                    /// If there is X55, then we
                                    /// take only 55 and assign in
                                    /// Port.Xvalue
                                    Port.XValue =
Convert.ToDouble(lGMCode.Substring(1,lGMCode.Length - 1));


                                    /// and now check the
current postion of x with new
                                    /// value of x; this check
will control the direction of
                                    /// stepper motor in + or -
direction. Suppose x current position is 100 and
                                    /// new value of x is 150,
its mean move 50 mm in x plus direction
                                    /// to reach 150 location.
                                    if(Port.CurrentPositionX >
Port.XValue)

                                    {
                                        ///set the pulses of X
stepper motor

                                        ///xcurrent position
is 100 and new position will be

                                        ///150; it needs to
move only 50 mm

                                        ///so minus the
currentposition from new position
    Port.PulsRequestX(Port.CurrentPositionX - Port.XValue);
                                        /// send the pulses to
stepper motor

                                        Port.Xplus();
                                        /// set the location of
workpiece of graphic of

                                        /// milling machine
                                        PositionArrived("X-",
Port.CurrentPositionX - Port.XValue);
                                    }
```

```
                                        else
                                        {
                                            /// same as above but
reverse direction
    Port.PulsRequestX(Port.XValue - Port.CurrentPositionX);
                                            Port.Xminus();
    PositionArrived("X+",Port.XValue - Port.CurrentPositionX);
                                        }
                                        /// when location of x is
reached,
                                        /// then the current
position of x is 150,
                                        /// so set the
currentposition with new position of x
                                        Port.CurrentPositionX =
Port.XValue;
                                        /// set the x Value of
current position
    StatusArrived(Current.X,Port.XValue.ToString());
                            }
                            else
                            {
                                        /// Incremented Coordinate
coding
                            }
                        }
                        #endregion
                        #region Y in RapidMovement
                        if(lGMCode.StartsWith("Y"))
                        {
                            if(Port.AbsoluteCoordinate)
                            {
                                Port.YValue =
Convert.ToDouble(lGMCode.Substring(1,lGMCode.Length - 1));
                                if(Port.CurrentPositionY >
Port.XValue)
                                {
    Port.PulsRequestY(Port.CurrentPositionY - Port.YValue);
                                            Port.Yplus();
                                            PositionArrived("Y-",
Port.CurrentPositionY - Port.YValue);
                                }
                                else
                                {
    Port.PulsRequestY(Port.YValue - Port.CurrentPositionY);
                                            Port.Yminus();

PositionArrived("Y+",Port.YValue - Port.CurrentPositionY);
                                }
                                Port.CurrentPositionY = Port.YValue;

    StatusArrived(Current.Y,Port.YValue.ToString());
                            }
```

```
                        else
                        {
                            /// Incremented Coordinate coding
                        }
                    }
                    #endregion
                    #region Z in RapidMovement
                    if(lGMCode.StartsWith("Z"))
                    {
                        if(Port.AbsoluteCoordinate)
                        {
                            Port.ZValue =
Convert.ToDouble(lGMCode.Substring(1,lGMCode.Length - 1));
                            if(Port.CurrentPositionZ > Port.ZValue)
                            {
    Port.PulsRequestZ(Port.CurrentPositionZ - Port.ZValue);
                                        Port.Zplus();
                                        if(Port.ZValue == 0)
                        PositionArrived("Z-",0);
                                        else
    PositionArrived("Z-",Port.CurrentPositionZ - Port.ZValue);
                            }
                            else
                            {

    Port.PulsRequestZ(Port.ZValue - Port.CurrentPositionZ);
                                        Port.Zminus();
                                        if(Port.ZValue == 0)
    PositionArrived("Z+",0);
                                        else
                        PositionArrived("Z+",Port.ZValue -
Port.CurrentPositionZ);
                            }

                            Port.CurrentPositionZ = Port.ZValue;

    StatusArrived(Current.Z,Port.ZValue.ToString());
                        }
                        else
                        {
                            /// Incremented Coordinate coding
                        }
                    }
                    #endregion
                }/// End of if Port.RapidMovement
                #endregion
            #region LinearInterPolation Handling Code
            /// <summary>
            /// In this region, We Handle Linear InterPolation
            /// and check X and Y coordinate together and Z Coordinate
Separately
            /// </summary>
```

```
                if(Port.LinearInterPolation)
                {
                     if(lGMCode.StartsWith("X"))
                     {/// GMCode first letter is X
                          if(Port.AbsoluteCoordinate)
                          {
                               ///take the X Value
                               Port.XValue =
Convert.ToDouble(lGMCode.Substring(1,lGMCode.Length - 1));
                               /// In Linear Interpolation, combine
                               /// Y coordinate with X coordinate
                               /// Check that Y coordinate is after
X coordinate
                               /// as the rules of EIA274D standard
     if(aList[i+1].ToString().StartsWith("Y"))
                               { /// If Y coordinate after X
coordinate are there,
                                    i++;
                                    /// take the value of Y
                                    Port.YValue =
Convert.ToDouble(aList[i].ToString().Substring(1,aList[i].ToString(
).Length - 1));
                                    Port.NextPositionX = Port.XValue;
                                    Port.NextPositionY = Port.YValue;

     Port.CalculateLinearInterpolation(Port.CurrentPositionX,
Port.CurrentPositionY,Port.NextPositionX,Port.NextPositionY);

     Final.AddLine(Port.CurrentPositionX,Port.CurrentPositionY,
Port.NextPositionX,Port.NextPositionY);

     StatusArrived(Current.X,Port.XValue.ToString());

     StatusArrived(Current.Y,Port.YValue.ToString());
                                    if(Port.NextPositionX !=
Port.CurrentPositionX)
                                    {
     if(Port.NextPositionX > Port.CurrentPositionX)
     PositionArrived("X+",Port.NextPositionX -
Port.CurrentPositionX);
                                         else
     PositionArrived("X-",Port.CurrentPositionX - Port.NextPositionX);
                                    }
                                    if(Port.NextPositionY !=
Port.CurrentPositionY)
                                    {
     if(Port.NextPositionY > Port.CurrentPositionY)
     PositionArrived("Y+",Port.NextPositionY - Port.CurrentPositionY);
                                         else
     PositionArrived("Y-",Port.CurrentPositionY - Port.NextPositionY);
                                    }
                                    Port.CurrentPositionX =
Port.NextPositionX;
```

```
                                    Port.CurrentPositionY =
Port.NextPositionY;
                              }

                         }
                         else
                         {
                              /// Incremental Coordinate coding
                         }
                    }
                    if(lGMCode.StartsWith("Z"))
                    {
                         if(Port.AbsoluteCoordinate)
                         {
                              Port.ZValue =
Convert.ToDouble(lGMCode.Substring(1,lGMCode.Length - 1));
                              Port.NextPositionZ = Port.ZValue;
                              if(Port.NextPositionZ >
Port.CurrentPositionZ)
                              {
     Port.PulsRequestZ(Port.NextPositionZ - Port.CurrentPositionZ);
                                        Port.Zminus();

     StatusArrived(Current.Z,Port.ZValue.ToString());

     PositionArrived("Z+",Port.NextPositionZ - Port.CurrentPositionZ);
                                        Port.CurrentPositionZ =
Port.NextPositionZ;
                              }
                              else
                                        if(Port.NextPositionZ <
Port.CurrentPositionZ)
                              {
     Port.PulsRequestZ(Port.CurrentPositionZ - Port.NextPositionZ);
                                        Port.Zplus();

     StatusArrived(Current.Z,Port.ZValue.ToString());
                                        PositionArrived("Z-",
Port.CurrentPositionZ - Port.NextPositionZ);
                                        Port.CurrentPositionZ =
Port.NextPositionZ;
                              }

                         }
                    }
               }/// end of if Port.LinearInterPolation
          #endregion
          #region Circular Interpolation Handling Code
     if(Port.CircularInterpolationCounterClockWise)
               {
                    if(Port.AbsoluteCoordinate)
                    {
                         if(lGMCode.StartsWith("X") &&
aList[i+1].ToString().StartsWith("Y"))
```

```
                        {
    if(aList[i+2].ToString().StartsWith("I") &&
aList[i+3].ToString().StartsWith("J"))
                            {
                                        Port.XValue =
Convert.ToDouble(lGMCode.Substring(1,lGMCode.Length - 1));
                                        Port.YValue =
Convert.ToDouble(aList[i+1].ToString().Substring(1,aList[i+1].
ToString().Length - 1));
                                        Port.I =
Convert.ToDouble(aList[i+2].ToString().Substring(1,aList[i+2].
ToString().Length - 1));
                                        Port.J =
Convert.ToDouble(aList[i+3].ToString().Substring(1,aList[i+3].
ToString().Length - 1));
                                        i = i + 3;

                                    }
                                    else
                                    {
                                        Port.XValue =
Convert.ToDouble(lGMCode.Substring(1,lGMCode.Length - 1));
                                        Port.YValue =
Convert.ToDouble(aList[i+1].ToString().Substring(1,aList[i+1].
ToString().Length - 1));
                                        i = i + 1;
                                        if(Port.I > 0 || Port.
J > 0)
                                        {
                                            Port.I = Port.I;
                                            Port.J = Port.J;
                                        }
                                        else
                                        {
                                            Port.I = 0;
                                            Port.J = 0;
                                        }
                                    }

                                    Port.NextPositionY =
Port.YValue;
                                    Port.NextPositionX =
Port.XValue;
                                    Port.Radius =
Math.Sqrt(((Port.CurrentPositionX - Port.I) * (Port.CurrentPositionX -
Port.I)) + ((Port.CurrentPositionY - Port.J) *
(Port.CurrentPositionY - Port.J)));

    Port.CalculateCircularInterpolation(Final);

    StatusArrived(Current.X,Port.XValue.ToString());

    StatusArrived(Current.Y,Port.YValue.ToString());
```

```
                                        if(Port.NextPositionX !=
Port.CurrentPositionX)
                                        {
                                             if(Port.NextPositionX >
Port.CurrentPositionX)
     PositionArrived("X+",Port.NextPositionX - Port.CurrentPositionX);
                                             else
     PositionArrived("X-",Port.CurrentPositionX - Port.NextPositionX);
                                        }
                                        if(Port.NextPositionY !=
Port.CurrentPositionY)
                                        {
                                             if(Port.NextPositionY >
Port.CurrentPositionY)
     PositionArrived("Y+",Port.NextPositionY - Port.CurrentPositionY);
                                             else
     PositionArrived("Y-",Port.CurrentPositionY - Port.NextPositionY);
                                        }
                                        Port.CurrentPositionX =
Port.NextPositionX;
                                        Port.CurrentPositionY =
Port.NextPositionY;
                                   }
                              }
                         }/// end of if
Port.CircularInterpolationCounterClockWise
                    #endregion

               }//end of for loop
               Final.Show();

          }
          #endregion

          #region ValidateCode

          /// <summary>
          /// Function Name: ValidateCode
          /// Parameters: string GMCode that will be validated,
Boolian bIsGMCodeValid
          /// Return Type void
          /// Purpose: This function will validate the GMCode
          /// If GMCode is valid the bIsGMCodeValid set to true
else false
          /// </summary>

          public  void ValidateCode(string sGMCode,ref bool
bIsGMCodeValid)
          {
               int iIndex = GMCodes.BinarySearch(sGMCode.Clone());
               if(iIndex < 0)
               {    ///if sGMCode's first letter is X or Y or Z or I or J
                    if(sGMCode.StartsWith("X") ||
sGMCode.StartsWith("Y") || sGMCode.StartsWith("Z") ||
```

```
sGMCode.StartsWith("I") || sGMCode.StartsWith("J") ||
sGMCode.StartsWith("S") || sGMCode.StartsWith("T") ||
sGMCode.StartsWith("F"))
                    {
                          ///and now check for any + or - sign
                          if(string.Compare
(sGMCode.Substring (1,1),"+") == 0 ||
string.Compare(sGMCode.Substring(1,1),"-") == 0)
                          {
                                ///and now check for any integer after
X+,X-,Y+,Y-,Z+,Z-,I+,I-,J+,J-;
                                string sValue =
sGMCode.Substring(2,sGMCode.Length - 2);
                                try
                                {   /// try to get integer if any exist
                                      double dTest = 0;

      if(Object.ReferenceEquals(Convert.ToDouble(sValue).GetType(),
dTest.GetType()))
                                            bIsGMCodeValid = true;
                                }
                                catch(InvalidCastException ex)
                                {    /// handle the error; if no
integer after + or - sign

      WriteError(ex.Message,ex.Source,"Not Apply","ValidateCode");
                                      bIsGMCodeValid = false;
                                }
                                catch(FormatException ex)
                                {    /// handle the error; if  no
integer after + or - sign

      WriteError(ex.Message,ex.Source,"Not Apply","ValidateCode");
                                      bIsGMCodeValid = false;
                                }
                          }
                          else
                          {
                                //and Now check for any integer after
X,Y,Z,I,J-;
                                string sValue =
sGMCode.Substring(1,sGMCode.Length - 1);
                                try
                                {
                                      double dTest = 0;

      if(Object.ReferenceEquals(Convert.ToDouble(sValue).GetType(),
dTest.GetType()))
                                            bIsGMCodeValid = true;
                                }
                                catch(InvalidCastException ex)
                                {   /// handle the error; There was
no integer after x,y,z sign
```

```
    WriteError(ex.Message,ex.Source,"Not Apply","ValidateCode");
                                bIsGMCodeValid = false;
                            }
                            catch(FormatException ex)
                            {    /// handle the error; there was
no integer after x,y,z sign

    WriteError(ex.Message,ex.Source,"Not Apply","ValidateCode");
                                bIsGMCodeValid = false;
                            }
                        }
                    }
                    else /// the first letter is other then x,y,z,i,j
                        bIsGMCodeValid = false;
                }
                else
                    bIsGMCodeValid = true;
            }
            #endregion
            #region ValidateFile

            /// <summary>
            /// Function Name: ValidateFile
            /// Parameters: None
            /// Return Type: bool
            /// Purpose: This Function shall validate the cnc part
            /// program file. It will take file name from static
            /// variable set by calling routine. If cnc part program
            /// file is valid, true will be returned, else false will
be returned
            /// </summary>

            public bool ValidateFile()
            {
                aList.Clear();
                string sFile;
                string sDelim = "\r\n";
                char [] cDelimiter = sDelim.ToCharArray();
                string [] sLines = null;
                string [] sWords = null;

                using(StreamReader reader = File.OpenText(sFileName))
                {
                    sFile = reader.ReadToEnd();
                    sLines = sFile.Split(cDelimiter);

                        for(int iLineNumber = 0; iLineNumber <
sLines.Length; iLineNumber++)
                        {
                            sWords = sLines[iLineNumber].Split
(" ".ToCharArray());
                            if(sWords[0] != "")
                            {
```

```
                                        for(int iWordNumber = 1;
iWordNumber < sWords.Length; iWordNumber++)
                                        {
                                            bool bIsValid = true;
    ValidateCode(sWords[iWordNumber], ref bIsValid);
                                            if(bIsValid)

    aList.Add(sWords[iWordNumber]);
                                            else
                                                return false;
                                        }
                                    }
                                }
                        }
                        return true;
                }
                #endregion
                #region SaveFile

                /// <summary>
                /// Function Name: SaveFile
                /// Parameters: TextBox
                /// Return Type: None
                /// Purpose: This Function will save the Part
                /// Program file. It will take file name from static
                /// variable set by calling routine.
                /// </summary>
                public void SaveFile(TextBox textBox)
                {
                        try
                        {    /// try to create stream object to write new file
                             File.Delete(sFileName);
                             Stream stream = File.OpenWrite(sFileName);
                             using(StreamWriter writer = new
StreamWriter(stream))
                             {
                                   /// write new file
                                   writer.Write(textBox.Text);
                             }
                        }
                        catch(IOException ex)
                        {    /// got the error during write file
                             WriteError(ex.Message,ex.Source,"Not
Apply","SaveFile");
                        }
                }
                #endregion

                public static void EmergencyStop()
                {
                        PortAccess.EmergencyStop = true;
                }
        }
}
```

```
File Name: PortAccess.cs
Purpose: Communication routines of parallel port.
Environment: Visual C# .Net version 1.1
Comments: For CNCSystem to work properly that is to send signal to
IO Port, the file inpout32.dll should be present in system32 folder
of main windows folder.


using System;
using System.Windows.Forms;
using System.Runtime.InteropServices;

namespace CNCSystem
{
     /// <summary>
     /// This class is part of CNCSystem
     /// </summary>
     public class PortAccess
     {
          private double _CurrentPositionX;
          private double _CurrentPositionY;
          private double _CurrentPositionZ;
          private double _Radius;
          private double _I;
          private double _J;
          private double _NextPositionX;
          private double _NextPositionY;
          private double _NextPositionZ;
          private double _xvalue;
          private double _yvalue;
          private double _zvalue;

          private bool _RapidMovement;
          private bool _LinearInterpolation;
          private bool _CircularInterpolationClockWise;
          private bool _CircularInterpolationCounterClockWise;
          private bool bMetricUnit;
          private bool bAbsoluteCoordinate;

          private short _Speed;
          private short _Feed;
          private short _Tool;

          private int xPulses;//not for property
          private int yPulses;//not for property
          private int zPulses;//not for property
          private int _signflg;
          private int _signflgy;
          private int _signflgx;


          private string _chksign;
          private string _chksignx;
          private string _chksigny;
```

```
        private string _xyzs;
        private string _yxzs;

        public static bool EmergencyStop;


        #region inpout32.dll
        /// <summary>
        /// The following functions will send and receive the
data from
        /// parallel port
        /// </summary>
        [DllImport("inpout32.dll", EntryPoint="Out32")]
        private static extern void OutPort(int adress, int value);
        [DllImport("inpout32.dll", EntryPoint="Inp32")]
        private static extern int InPort(int address);
        #endregion

        public PortAccess()
        {
             ///Constructor of PortAccess Class

             _LinearInterpolation = false;
             _CircularInterpolationClockWise = false;
             _CircularInterpolationCounterClockWise = false;
             _CurrentPositionX = 0;
             _CurrentPositionY = 0;
             _CurrentPositionZ = 0;
             _Radius = 0;
             _NextPositionX = 0;
             _NextPositionY = 0;
             _NextPositionZ = 0;
             bMetricUnit = true;
             bAbsoluteCoordinate = true;
             EmergencyStop = false;

             _RapidMovement = true;
             _I = -0;
             _J = -0;
             _TATO = 13;


        }
        #region G00
        /// <summary>
        /// Function Name: G00
        /// Parameters: None
        /// Return Type: Void
        /// Purpose: Rapid Movement
        /// </summary>
        public void G00()
        {
             _RapidMovement = true;
             _LinearInterpolation = false;
```

```
        _CircularInterpolationClockWise = false;
        _CircularInterpolationCounterClockWise = false;

}
#endregion
#region G01
/// <summary>
/// Function Name: G01
/// Parameters: None
/// Return Type: Void
/// Purpose: Linear Interpolation
/// </summary>
public void G01()
{

        _LinearInterpolation = true;
        _CircularInterpolationClockWise = false;
        _CircularInterpolationCounterClockWise = false;
        _RapidMovement = false;

}
#endregion
#region G02
/// <summary>
/// Function Name: G02
/// Parameters: None
/// Return Type: Void
/// Purpose: Circular Interpolation c/w
/// </summary>
public void G02()
{
        _CircularInterpolationClockWise = true;
        _CircularInterpolationCounterClockWise = false;
        _LinearInterpolation = false;
        _RapidMovement = false;

}
#endregion
#region G03
/// <summary>
/// Function Name: G03
/// Parameters: None
/// Return Type: Void
/// Purpose: Circular Interpolation cc/w
/// </summary>
public void G03()
{

        _CircularInterpolationClockWise = false;
        _CircularInterpolationCounterClockWise = true;
        _LinearInterpolation = false;
        _RapidMovement = false;

}
```

```
#endregion
#region G70
/// <summary>
/// Function Name: G70
/// Parameters: None
/// Return Type: Void
/// Purpose: Inch Units
/// </summary>
public void G70()
{
     bMetricUnit = false;

}
#endregion
#region G71
/// <summary>
/// Function Name: G71
/// Parameters: None
/// Return Type: Void
/// Purpose: Metric Units
/// </summary>
public void G71()
{
     bMetricUnit = true;

}
#endregion
#region G90
/// <summary>
/// Function Name: G90
/// Parameters: None
/// Return Type: Void
/// Purpose: Absolute Coordinate
/// </summary>
public void G90()
{
     bAbsoluteCoordinate = true;

}
#endregion
#region G91
/// <summary>
/// Function Name: G91
/// Parameters: None
/// Return Type: Void
/// Purpose: Incremental Coordinate
/// </summary>
public void G91()
{
     bAbsoluteCoordinate = false;

}
#endregion
#region M02
```

```
        /// <summary>
        /// Function Name: M02
        /// Parameters: None
        /// Return Type: Void
        /// Purpose: End Of Program
        /// </summary>
        public void M02()
        {

        }
        #endregion
        #region M03
        /// <summary>
        /// Function Name: M03
        /// Parameters: None
        /// Return Type: Void
        /// Purpose: Spindle On c/w
        /// </summary>
        public void M03()
        {
            /// <summary>
            /// 4 bit spindle control
            /// 0 0 0 0
            /// | | | |__Speed    00 = Low, 10 = Medium
            /// | | |____ Speed   01 = High, 11 = Very High
            /// | |______ On/Off  0 = Off, 1 = On
            /// |________ CCW/CW  0 = CW, 1 = CCW
            /// </summary>
//          OutPort(888,0100);
        }
        #endregion
        #region M04
        /// <summary>
        /// Function Name: M04
        /// Parameters: None
        /// Return Type: Void
        /// Purpose: Spindle On cc/w
        /// </summary>
        public void M04()
        {
            /// <summary>
            /// 4 bit spindle control
            /// 0 0 0 0
            /// | | | |__Speed    00 = Low, 10 = Medium
            /// | | |____ Speed   01 = High, 11 = Very High
            /// | |______ On/Off  0 = Off, 1 = On
            /// |________ CCW/CW  0 = CW, 1 = CCW
            /// </summary>
            ///

//          OutPort(888,1100);

        }
        #endregion
```

```
        #region M05
        /// <summary>
        /// Function Name: M05
        /// Parameters: None
        /// Return Type: Void
        /// Purpose: Spindle Off
        /// </summary>
        public void M05()
        {
             /// <summary>
             /// 4 bit spindle control
             /// 0 0 0 0
             /// | | | |__ Speed    00 = Low, 10 = Medium
             /// | | |____ Speed    01 = High, 11 = Very High
             /// | |______ On/Off   0 = Off, 1 = On
             /// |________ CCW/CW   0 = CW, 1 = CCW
             /// </summary>
//           OutPort(888,0000);

        }
        #endregion
        #region Public Functions
        /// <summary>
        /// Public Functions
        /// </summary>

        #region LinearInterpolation
        /// <summary>
        /// LinearInterpolation
        /// Formula applied
        /// y = y1+ y2-y1/x2-x1 (x-x1)
        /// </summary>

        public void CalculateLinearInterpolation(double X1, double
Y1, double X2, double Y2)
        {
             double DeltaY = Y2 - Y1;
             double DeltaX = X2 - X1;
             double CurrentX = X1;
             double CurrentY = Y1;
             double NextX = X1 + 1;
             double NextY;
             double Temp;
             double XDif;
             double YDif;

             double YStep; /// Steps of Stepper motor in Y direction;

             if(X1 > X2)
                  XDif = X1 - X2;
             else
                  XDif = X2 - X1;
```

```
            if(Y1 > Y2)
                YDif = Y1 - Y2;
            else
                YDif = Y2 - Y1;
            if(XDif > YDif)
                YStep = 1;
            else
        YStep = YDif/XDif;

            // move in x direction untill we reach final x
position
            while(CurrentX != X2)
            {

                // Check whether we should move in x- direction
or x+
                if(CurrentX > X2)
                    CurrentX = CurrentX - 1;
                else
                    CurrentX = CurrentX + 1;
                // formula application
                Temp = CurrentX - X1;
                NextY = Y1 + (DeltaY/DeltaX) * Temp;
                // end of formula application
                NextX = CurrentX;
                if(X1 != NextX)
                {
                    // send the request for generate the pulses
for x stepper for 1 mm
                    this.PulsRequestX(1);
                    if(X1 > X2)
                        this.Xplus(); // for x- move
                    else
                this.Xminus(); // for x+ move
                }
                if(Y1 != NextY)
                {
                    this.PulsRequestY(YStep);
                    if(Y1 > Y2)
                        this.Yplus(); // for y- move
                    else
                        this.Yminus(); // for y+ move
                }
            }

        }
        #endregion
        #region Circular Interpolation
        /// <summary>
        /// Circular Interpolation
        /// Formula Applied
        /// y = Sqrt(Sqr(Radius) - Sqr(x - i)) + j
        /// </summary>
```

```
        public void CalculateCircularInterpolation(frmFinal  Final)
        {
            /// current position of x
            double CurrentX = _CurrentPositionX;
            /// current position of y
            double CurrentY = _CurrentPositionY;
            /// Calculated Next positon of y
            double NextY;
            /// Calculated Next position of x
            double NextX;
            /// temprary variable
            double Temp;
            /// Squire of radius
            double RadiusSqr = _Radius * _Radius;
            double YPosition = CurrentY;

            /// Difference between CurrentX position and final
value of x position
            double XDif;
            /// Difference between CurrentY position and final
value of y position
            double YDif;

            double YStep; /// Step of Stepper motor in y direction;
            if(_CurrentPositionX > _NextPositionX)
                XDif = _CurrentPositionX - _NextPositionX;
            else
                XDif = _NextPositionX - _CurrentPositionX;

            if(_CurrentPositionY > _NextPositionY)
                YDif = _CurrentPositionY -_NextPositionY;
            else
                YDif = _NextPositionY - _CurrentPositionY;

            if(XDif > YDif)
                YStep = 1;
            else
                YStep = YDif/XDif;

            while(CurrentX != _NextPositionX)
            {
                if(CurrentX > _NextPositionX)
                    CurrentX = CurrentX - 1;
                else
                    CurrentX = CurrentX + 1;

                Temp = CurrentX - _I;
                Temp = Temp * Temp;

                // formula application
                if(_NextPositionY < _J || _CurrentPositionY < _J)
            NextY = Math.Sqrt(RadiusSqr - Temp) - _J;
                else
                    NextY = Math.Sqrt(RadiusSqr - Temp) + _J;
```

```
                // end of formula application

                NextX = CurrentX;
                // move in x direction until current position
of x
                // and final position of x are same
                if(_CurrentPositionX != NextX)
                {
                    // send the request of generate pulses of
x stepper
                    PulsRequestX(1);
                    // check whether current position is bigger
or next position of x
                    if(_CurrentPositionX > _NextPositionX)
                        this.Xplus(); // if current position
is bigger then move in x- direction
                    else
                        this.Xminus(); // if next position of
x is bigger, then move in x+ direction
                }
                if(_CurrentPositionY != NextY)
                {
                    this.PulsRequestY(YStep);
                    if(_CurrentPositionY > _NextPositionY)
                    {
                        this.Yplus();
                        YPosition = YPosition - YStep;
                    }
                    else
                    {
                        this.Yminus();
                        YPosition = YPosition + YStep;
                    }

                }
                /// add the line in picture that will show
                /// end product.
                Final.AddLine(NextX,NextY,NextX+1,NextY+1);
            }

        }

        #endregion
        #region PulsCalculate And PulsGenerate
        /// <summary>
        /// PulsRequest()
        /// Formula Used:
        /// 1 Puls = 1.8 degree rotate of stepper motor
        /// 200 Puls = 360 degree (complete rotation)
        /// 1 complete rotation (360) = 1 mm
        /// </summary>
        public void PulsRequestX(double X)
```

```
{
     ///X value is in mm
     ///convert it into Pulses
     xPulses = Convert.ToInt32(X * 200);
}
public void PulsRequestY(double Y)
{
     ///Y value is in mm
     ///convert it into Pulses
     yPulses = Convert.ToInt32(Y * 200);
}
public void PulsRequestZ(double Z)
{
     ///Z value is in mm
     ///convert it into Pulses
     zPulses = Convert.ToInt32(Z * 200);
}

public void Xplus()
{
     //for X- move
     for(int loop = 1; loop <= xPulses; loop++)
     {
          Delay();
               OutPort(888,8);
               OutPort(888,4);
               OutPort(888,2);
               OutPort(888,1);
     }
}
public void Xminus()
{
     //for X+ move
     for(int loop = 1; loop <= xPulses; loop++)
     {
          Delay();
               OutPort(888,1);
               OutPort(888,2);
               OutPort(888,4);
               OutPort(888,8);
     }
}
public void Yplus()
{
     // for y- move
     for(int loop = 1; loop <= yPulses; loop++)
     {
          Delay();
               OutPort(888,128);
               OutPort(888,64);
               OutPort(888,32);
               OutPort(888,16);
     }
}
```

```
public void Yminus()
{
     // for y+ move
     for(int loop = 1; loop <= yPulses; loop++)
     {
          Delay();
               OutPort(888,16);
               OutPort(888,32);
               OutPort(888,64);
               OutPort(888,128);
     }
}

public void Zplus()
{
     //for Z- move
     for(int loop = 1; loop <= zPulses; loop++)
     {
          Delay();
               OutPort(890,128);
               OutPort(890,64);
               OutPort(890,32);
               OutPort(890,16);
     }
}
public void Zminus()
{
     //for Z+ move
     for(int loop = 1; loop <= zPulses; loop++)
     {
          Delay();
               OutPort(890,16);
               OutPort(890,32);
               OutPort(890,64);
               OutPort(890,128);
     }
}
public void Delay()
{
     for(int loop = 1;loop <= 1000; loop++)
     {

     }

}
#endregion
#endregion
#region Public Property
/// <summary>
/// Public Properties
/// </summary>
public string chksign
{
```

```
    get
    {
        return _chksign;
    }
    set
    {
        _chksign = value;
    }
}
public string chksigny
{
    get
    {
        return _chksigny;
    }
    set
    {
        _chksigny = value;
    }
}
public int signflg
{
    get
    {
        return _signflg;
    }
    set
    {
        _signflg = value;
    }
}
public int signflgy
{
    get
    {
        return _signflgy;
    }
    set
    {
        _signflgy = value;
    }
}
public string xyzs
{
    get
    {
        return _xyzs;
    }
    set
    {
        _xyzs = value;
    }
}
```

```
public string yxzs
{
     get
     {
          return _yxzs;
     }
     set
     {
          _yxzs = value;
     }
}

public double XValue
{
     get
     {
          return _xvalue;
     }
     set
     {
          _xvalue = value;
     }
}
public double YValue
{
     get
     {
          return _yvalue;
     }
     set
     {
          _yvalue = value;
     }
}
public double ZValue
{
     get
     {
          return _zvalue;
     }
     set
     {
          _zvalue = value;
     }
}
public double I
{
     get
     {
          return _I;
     }
```

```
          set
          {
               _I = value;
          }
     }
     public double J
     {
          get
          {
               return _J;
          }
          set
          {
               _J = value;
          }
     }
     public double Radius
     {
          get
          {
               return _Radius;
          }
          set
          {
               _Radius = value;
          }
     }

     public bool MetricUnit
     {
          get
          {
               return bMetricUnit;
          }
          set
          {
               bMetricUnit = value;
          }
     }
     public bool RapidMovement
     {
          get
          {
               return _RapidMovement;
          }
          set
          {
               _RapidMovement = value;
          }
     }
```

```
public bool LinearInterPolation
{
    get
    {
        return _LinearInterpolation;
    }
    set
    {
        _LinearInterpolation = value;
    }
}

public bool CircularInterpolationClockWise
{
    get
    {
        return _CircularInterpolationClockWise;
    }
    set
    {
        _CircularInterpolationClockWise = value;
    }
}
public bool CircularInterpolationCounterClockWise
{
    get
    {
        return _CircularInterpolationCounterClockWise;
    }
    set
    {
        CircularInterpolationCounterClockWise = value;
    }
}
public string chksignx
{
    get
    {
        return _chksignx;
    }
    set
    {
        _chksignx = value;
    }
}
public int signflgx
{
    get
    {
        return _signflgx;
    }
```

```
        set
        {
            _signflgx = value;
        }
    }
    public short Speed
    {
        get
        {
            return _Speed;
        }
        set
        {
            _Speed = value;
        }
    }
    public short Feed
    {
        get
        {
            return _Feed;
        }
        set
        {
            _Feed = value;
        }
    }
    public short Tool
    {
        get
        {
            return _Tool;
        }
        set
        {
            _Tool = value;
        }
    }
    public double CurrentPositionX
    {
        get
        {
            return _CurrentPositionX;
        }
        set
        {
            _CurrentPositionX = value;
        }
    }
    public double CurrentPositionY
    {
```

```
            get
            {
                return _CurrentPositionY;
            }
            set
            {
                _CurrentPositionY = value;
            }
        }
        public double CurrentPositionZ
        {
            get
            {
                return _CurrentPositionZ;
            }
            set
            {
                _CurrentPositionZ = value;
            }
        }

        public double NextPositionX
        {
            get
            {
                return _NextPositionX;
            }
            set
            {
                _NextPositionX = value;
            }
        }
        public double NextPositionY
        {
            get
            {
                return _NextPositionY;
            }
            set
            {
                _NextPositionY = value;
            }
        }
        public double NextPositionZ
        {
            get
            {
                return _NextPositionZ;
            }
            set
            {
                _NextPositionZ = value;
            }
        }
```

```
            public bool AbsoluteCoordinate
            {
                get
                {
                    return bAbsoluteCoordinate;
                }
            }
            #endregion
        }
}
```

File Name: frmFinal.cs
Purpose: The code generates end product of application.
Environment: Visual C# .Net version 1.1

```
using System;
using System.Drawing.Drawing2D;
using System.Drawing;
using System.Collections;
using System.ComponentModel;
using System.Windows.Forms;

namespace CNCSystem
{
	/// <summary>
	/// Summary description for frmFinal.
	/// </summary>
	public class frmFinal : System.Windows.Forms.Form
	{
		/// <summary>
		/// Required designer variable.
		/// </summary>
		private System.ComponentModel.Container components = null;
		private System.Windows.Forms.Label label1;
		GraphicsPath myPath;
		GraphicsPath tablePath;
		GraphicsPath borderPath;
		GraphicsPath ShadePath;
		LinearGradientBrush lb;
		LinearGradientBrush gb;
		LinearGradientBrush sb;
		Matrix translateMatrix;
		Matrix ShadeMatrix;


		public frmFinal()
		{
			//
			// Required for Windows Form Designer support
			//
			InitializeComponent();
			translateMatrix = new Matrix();
			ShadeMatrix = new Matrix();


			tablePath = new GraphicsPath();
			borderPath = new GraphicsPath();
			myPath = new GraphicsPath();
			ShadePath = new GraphicsPath();

			Point p1 = new Point(5,0);
			Point p2 = new Point(320,0);
```

```
            lb = new LinearGradientBrush(p1,p2,Color.Gray,
Color.Silver );
            gb = new
LinearGradientBrush(p1,p2,Color.White,Color.Chocolate);
            sb = new LinearGradientBrush(p1,p2,Color.Black,
Color.Black);


            tablePath.AddLine(5,-140,5,0);
            tablePath.AddLine(5,0,300,0);
            tablePath.AddLine(300,0,300,-140);
            tablePath.AddLine(300,-140,5,-140);
            tablePath.CloseFigure();

            borderPath.AddLine(5,0,5,9);
            borderPath.AddLine(5,9,300,9);
            borderPath.AddLine(300,9,300,0);
            borderPath.AddLine(300,0,300,9);
            borderPath.AddLine(300,9,300,-131);
            borderPath.AddLine(300,-131,300,-140);




            //
            // TODO: Add any constructor code after
InitializeComponent call
            //
        }

        /// <summary>
        /// Clean up any resources being used.
        /// </summary>
        protected override void Dispose( bool disposing )
        {
            if( disposing )
            {

                if(components != null)
                {
                    components.Dispose();
                }
            }
            base.Dispose( disposing );
        }

        #region Windows Form Designer generated code
        /// <summary>
        /// Required method for Designer support; do not modify
        /// the contents of this method with the code editor.
        /// </summary>
        private void InitializeComponent()
```

```
        {
              this.label1 = new System.Windows.Forms.Label();
              this.SuspendLayout();
              //
              // label1
              //
              this.label1.AutoSize = true;
              this. label1 .Location = new System.Drawing.Point(0, 168);
              this.label1.Name = "label1";
              this.label1.Size = new System.Drawing.Size(20, 16);
              this.label1.TabIndex = 0;
              this.label1.Text = "0,0";
              //
              // frmFinal
              //
              this.AutoScaleBaseSize = new System.Drawing.Size(5, 13);
              this.BackColor = System.Drawing.Color.White;
              this.ClientSize = new System.Drawing.Size(354, 194);
              this.Controls.Add(this.label1);
              this.FormBorderStyle =
System.Windows.Forms.FormBorderStyle.FixedToolWindow;
              this.MaximizeBox = false;
              this.Name = "frmFinal";
              this.Text = "End Product";
              this.Paint += new
System.Windows.Forms.PaintEventHandler(this.frmFinal_Paint);
              this.ResumeLayout(false);

        }
        #endregion

        private void frmFinal_Paint(object sender,
System.Windows.Forms.PaintEventArgs e)
        {
              translateMatrix.Translate(0,155);
              ShadeMatrix.Translate(5,159);

              ShadePath = (GraphicsPath)myPath.Clone();

              tablePath.Transform(translateMatrix);
              borderPath.Transform(translateMatrix);
              myPath.Transform(translateMatrix);
              ShadePath.Transform(ShadeMatrix);



              lb.Transform = translateMatrix;
              gb.Transform = translateMatrix;
              sb.Transform = ShadeMatrix;

              e.Graphics.FillPath(lb,tablePath);
              e.Graphics.FillPath(lb,borderPath);
```

```
            e.Graphics.FillPath(sb,ShadePath);
            e.Graphics.FillPath(gb,myPath);

            e.Graphics.DrawPath(new Pen(Color.Black,1), tablePath);
            e.Graphics.DrawPath(new  Pen(Color.Black,1),
borderPath);

            e.Graphics.DrawPath(new  Pen(Color.Silver,1),  myPath);
            e.Graphics.DrawPath(new  Pen(Color.White,1),ShadePath);

        }

        public void AddLine(double x1, double y1, double x2, double y2)
        {
            if(y1 > 0)
            {
                y1  =  Convert.ToDouble("-"  +  y1.ToString())/5;
                y2  =  Convert.ToDouble("-"  +  y2.ToString())/5;
            }
            else
            {
                y1  =  y1  /  5;
                y2  =  y2  /  5;
            }
            x1  =  x1  /  5;
            x2  =  x2  /  5;
myPath.AddLine((float)x1,(float)y1,(float)x2,(float)y2);

        }
    }
}
```

11.4.3 Use of Standards in Developing Control Software and Hardware for Computerized Numerical Control Milling Machine

The development of control software and the electronic hardware for a three-axes milling machine uses pertinent standards. In the case of control software, the programming language, the environment in which the programming language operates, the method of programming the interpreter for EIA 274D control software, and the codes used to write the part programs are all created according to established international, professional, or company standards.

In the case of electronic hardware, the electronic components, the control elements, printed circuit board design, and development procedures are all part of pertinent professional and company standards and are supported by the normal design and testing procedures available in contemporary literature.

BIBLIOGRAPHY

Baird, R.W., Planning for Industrial Automation Standards, Proceedings, Annual Conference—Standards Engineers Society, 1984, 10–17.

Coombs, T., *Programming with C#.NET*, 1st ed., OnWord Press Albany, N.Y. 2002.

Cox, A.B., Numerical control programming with personal computers, *J. Eng. Comput. Appl.*, 2(3), 41–45, 1988.

Electronics Industries Association RS 274D: Part: Programming Standards, 1979.

Gol, G., Microprocessor based washing machine controller, *IEE Colloquium (Dig.)*, 81, 2.1–2.3, 1981.

Grossman, D.D., Opportunities for research on numerical control machining, *Communications ACM*, 29(6), 515–522, 1986.

Hall, D.V., *Experiments in Microprocessors and Interfacing*: *Programming and Hardware*, 2nd ed., Glencoe/McGraw-Hill, New York, 1992.

Hall, D.V., *Microprocessors and Interfacing*: *Programming and Hardware*, 2nd ed., Glencoe/McGraw-Hill, New York, 1991.

Ndinechi, M.C., Computer interfacing for data acquisition and systems control, *Modelling Meas. Control A*, 75(1–2), 25–32, 2002.

Puers, R., Sensor, sensor interfacing and front-end data management for stand-alone microsystems, *J. Micromechanics Microengineering*, 9(2), R1–R7, 1999.

Singh, A. and Triebel, W.A., *16-Bit and 32-Bit Microprocessors*: *Architecture, Software, and Interfacing Techniques*, 1st ed., Pearson Education, N.J. 1997.

Smiley, J., Learn to Program with C#; 1st ed., Osborne/McGraw-Hill, New York, 2002.

Smith, S.R. and Steiner, M., Printer interface using a single microprocessor, *IBM Tech. Disclosure Bull.*, 27(9), 5384–5385, 1985.

Triebel, W.A., *The 80386, 80486, and Pentium Microprocessor*: *Hardware, Software, and Interfacing*, 1st ed., Prentice Hall, New York, 1997.

Venjara, Y., Standards for machine tools, *Stand. News*, 30(2), 20–23, 2002.

BIBLIOGRAPHY

Appendix

A1 DIGITAL CATALOGS FOR STANDARD MECHANICAL COMPONENTS

http://www.industrialproductsfinder.com
http://www.kellysearch.com
http://www.europages.com

A2 DIGITAL CATALOGS FOR STANDARD MACHINE ELEMENTS

http://www.1stindustrialdirectory.com/
http://www.brd-klee.dk/index.php?/BrdKlee/english.php?
http://mfg.asiaep.com/my/mypindex.htm
http://www.kellysearch.com
http://www.hpceurope.com/
http://www.globalspec.com/
http://www.industrialproductsfinder.com

A3 DIGITAL CATALOGS FOR STANDARD CONTROL ELEMENTS

http://www.1stindustrialdirectory.com/
http://motion-controls.globalspec.com/ProductFinder/Motion_Controls/Integrated_Motion_Control
http://www.olympic-controls.com/products_and_suppliers_motion_control.htm
http://mfg.asiaep.com/my/mypindex.htm
http://www.kellysearch.com
http://www.globalspec.com/

A4 DIGITAL CATALOGS FOR ENGINEERING MATERIALS

http://www.mmsonline.com/mall/metals/
http://www.1stindustrialdirectory.com/
http://search3.ec21.com/search/ProductCategoryCount.cgi?category=C000010
http://mfg.asiaep.com/my/mypindex.htm

http://www.kellysearch.com
http://www.matweb.com/reference/colist.asp
http://dir.yahoo.com/business_and_economy/business_to_business/industrial_supplies/materials/
http://www.pyrotek-inc.com/
http://www.memsnet.org/links/material/
http://gaskets.globalspec.com/Industrial-Directory/spring_materials

A5 DIGITAL LIBRARIES FOR STANDARDS

http://www.bsonline.bsi-global.com/server/subscriber.jsp
http://www.ieee.org/products/onlinepubs/
http://www.ansi.org/Library/overview.aspx?menuid=11
http://www.iee.org.OnComms/PN/functionalsafety/standards/cfm
http://www.astm.orgDEMO/index/shtml
http://www.store.sae.org/digitallibrary.htm

A6 WEB SITES RELATED TO STANDARDS

http://www.google.com/Top/Science/Reference/Standards/Organizations/National/
http://infomgmt.homestead.com/files/std.htm
http://www2.rad.com/networks/1996/standorg.htm
http://www.iol.ie/~mattewar/CIPPM/psof.html
http://www.ucalgary.ca/library/subjects/ENGG/stanorgs.html
http://www.bis-global.com/Members/PLUS/Organizations.xalter
http://www.kub.it/dir/117693/
http://www.linktionary.com/s/standards.html

Index

E

F

G

H

I

N

O

S

T

U

V

W

X

Y

Z

For Product Safety Concerns and Information please contact our EU representative GPSR@taylorandfrancis.com
Taylor & Francis Verlag GmbH, Kaufingerstraße 24, 80331 München, Germany

www.ingramcontent.com/pod-product-compliance
Ingram Content Group UK Ltd.
Pitfield, Milton Keynes, MK11 3LW, UK
UKHW021625240425
457818UK00018B/727

* 9 7 8 0 3 6 7 3 9 1 5 4 6 *